OUR PRECARIOUS HABITAT

THE WILEY BICENTENNIAL—KNOWLEDGE FOR GENERATIONS

*E*ach generation has its unique needs and aspirations. When Charles Wiley first opened his small printing shop in lower Manhattan in 1807, it was a generation of boundless potential searching for an identity. And we were there, helping to define a new American literary tradition. Over half a century later, in the midst of the Second Industrial Revolution, it was a generation focused on building the future. Once again, we were there, supplying the critical scientific, technical, and engineering knowledge that helped frame the world. Throughout the 20th Century, and into the new millennium, nations began to reach out beyond their own borders and a new international community was born. Wiley was there, expanding its operations around the world to enable a global exchange of ideas, opinions, and know-how.

For 200 years, Wiley has been an integral part of each generation's journey, enabling the flow of information and understanding necessary to meet their needs and fulfill their aspirations. Today, bold new technologies are changing the way we live and learn. Wiley will be there, providing you the must-have knowledge you need to imagine new worlds, new possibilities, and new opportunities.

Generations come and go, but you can always count on Wiley to provide you the knowledge you need, when and where you need it!

WILLIAM J. PESCE
PRESIDENT AND CHIEF EXECUTIVE OFFICER

PETER BOOTH WILEY
CHAIRMAN OF THE BOARD

For the young people of the country and beyond,
who are ill-prepared for the monumental issues
certain to transform their lives

CONTENTS

PREFACE

Our Precarious Habitat (OPH) is not your common garden-variety environmental issues document. *OPH* examines the relationship between the physical environment and human health. As such, it takes an epidemiologic approach, eschewing anecdotes for evidence derived from human studies. Each chapter asks, what do we know about this issue, and brings either supportive or deposing evidence to bear. Additionally, each chapter raises pertinent questions; questions that you can sink your teeth into, requiring thought and substantiation. Each chapter's substantive material helps either bolster or undermine arguments.

Considering that decisions about hot-button local and national issues are too often emotionally based, it is essential to try to replace emotional decisionmaking with evidence-based decisions, which *OPH* is structured to do.

The beauty of this book is the fact that instructors can enter it anywhere and engage students with an endless number of health-related environmental issues that are current topics in the mass media, and that students will readily see as relevant to their lives, and will galvanize classroom discussion. At the end of the day, *OPH* wants to see students develop a dollop of skepticism, becoming knowledgeable citizens who know what questions to ask regarding statements and proposals placed before them.

OPH is filled with up-to-date, selected material from peer-reviewed science journals, which are available in university libraries, online, or via interlibrary loan.

The contentious evolution–creationism issue is considered in depth, and offers opportunities for wide discussion and debate, as do the chapters on nuclear energy, alternative energy sources, dietary supplements, and alternative health practices, with which students are personally involved. Yet another topic offering hours of pregnant discussion are the differing beliefs held by scientists and the public about everday risks that directly impact their lives and their families, friends, and neighbors.

Use *OPH* to engage and energize students on a persoanl level, which will enhance their understanding of cutting-edge local and national issues, while casting off fear of the environment, realizing that ultimately, their health is in their hands.

The critical decisions that must be made for our country as well as the international community require, nay, demand a well informed citizenry.

Perhaps this book will move many in that direction. However, OPH could not have come into being without the advice, assistance and help of many colleagues, friends and family. I've written a book about a clutch of complex issues in a manner at once accessible yet uncompromising in its scientific accuracy. For this, unbiased appraisal was essential. My wife Anita, my severest critic could be counted on, as could my journalist son Scott, and my two erudite daughters, Dana and Andi who would assure that jargon and science-speak would not rear their ugly heads.

And for advice, counsel and cool-headedness in those times of stress, of which there were far too many, I could count on my colleague Lester Levin, Professor of Environmental Science at Temple University. Special thanks must go to Eilene Palyvoda who took my scribblings and turned them into a worthy manuscript. Kudos must also go to that still unidentified copy editor at Wiley who smoothed the rough edges and quite obviously may know the book better than I do. Rosalyn Farkas, Senior Production Editor, who worked on the 3rd edition, was ever present to continue her super guidance assuring the book's high technical quality. Many thanks to Erica Spiritos for typing the index, a demanding job, if ever there was one.

Finally, I must raise my glass to the librarians, a special breed of people who know where to locate and retrieve that "stuff" so essential for assuring the quality of the subjects covered. To the librarians at Princeton University's Government Documents Section, Rutgers University's Library of Science and Medicine, the Princeton Public Library, and West Windsor Library, I offer my thanks and appreciation. Nevertheless, whatever help was rendered, OPH is, as noted, a complex volume, consequently the interpretations, conclusions, errors and omissions, are mine alone.

Princeton, NJ   MELVIN A. BENARDE, Ph.D.
May 2007

INTRODUCTION

For spin doctors, all human activity is fair game. While politics is arguably their obvious turf, the health of the public has been fertile ground for endless spinning. Through the past three decades, we, the people, were finely tuned, made to believe that we are a sickly lot; made to fear the air we breathe, the food we eat, the water we drink; and that environmental pollution has exacted a devilish toll.

The environment, our habitat, as agent of disease, may be the singular belief shared nationwide. That few have questioned this absurd and shameful deception is distressing as it informs us that the country's collective intelligence has submitted to so thorough a "conning", for so long. To continue to accept this fantasy, sound evidence to the contrary notwithstanding, suggests that the notion of a "vicarious death wish" is far more pervasive than Dr. Freud dared imagine.

The widely held perception of a polluted, read "poisonous," environment as the cause of our ailments is patently false, as the documented evidence in this volume shows. Exposure to potentially toxic chemicals is so small that risks to health are all but impossible to detect. Alvin Weinberg, former Director of the Oak Ridge National Laboratory, said it best: "The connection between low-level insults and bodily harm," he wrote, "is probably as difficult to prove as the connection between witches and failed crops."

We must accept an evident fact—we Americans are among the healthiest people on planet Earth. I suspect that an objective observer would have thought our good health would be cause for rejoicing, shouting from rooftops, and dancing in the streets. But there is nary a sound. Not a whisper. Nonetheless, this edifying story is begging for the telling.

Given a population obsessed with every ache and pain, striving for eternal life and a risk-free world, it may be understandable that the fact of our good health is a message few want, or care to hear, and woe betide the messenger. Nevertheless, as we wend our way through the first decade of the third millennium, it is time to "march to a different drummer," and to proclaim, along with T. S. Eliot, that "We shall not cease from exploration, and the end of our exploring will be to arrive where we started, and know the place for the first time."

In the previous edition of *OPH*, I noted that the Norwegian sculptor Gustave Vigeland, created a statue in Oslo's Frogner Park, depicting humankind's struggle with the environment. Vigeland believed humankind had not yet—nor would in the forseeable future—overcome the many simultaneous forces impinging on and affecting its existence. He saw humankind as unable to break out of confinement of encircling forces.

I have a different vision, and would sculpt a statue portraying a mind in chains: a mind whose chains must be rend asunder. Until we discard the unworthy notion that we are being poisoned by a polluted environment, we shall forever be shackled. Mark Twain knew this. "Loyalty to petrified opinion," he said, "never broke a chain or freed a human soul."

If we don't start reconsidering the unfounded beliefs that many of us hold, solutions to our pressing problems will not emerge. *OPH* wants readers to consider the up-to-date, comprehensive evidence marshaled here, and compare it with their long-held beliefs. Perhaps then, readers will relinquish any number of worn-out, threadbare perceptions, and let the sun shine in. A mind unchained permits the adoption of new paradigms.

For too long, great numbers of people have continued to believe the Cassandran environmentalists, whose monotonous forecasts of doom have not only never materialized, but have been totally impugned by undeniable evidence. If any group can be said to be false prophets, it is the environmentalists who have garnered an impresssively unenviable track record. As a wag so aptly put it, "the past 25 years of their hegemony is like a Red Sea refusing to part." And their handmaidens, the media, promoted their fear mongering.

For some three decades we have been subjected to a floodtide of fright. Three Mile Island has been the media's and environmentalist's poster child, preventing the maturation of a much-needed nuclear industry in the United States; an industry with a record of safety unmatched by any other. The triad of environmentalism, media frenzy, and scientific illiteracy has erected towering infernos of hysteria, panic, litigation, and political demagogy from a welter of presumed threats—threats without substance: food additives, high powerlines, food irradiation, hazardous waste, genetic modification of crops, cancer clusters, a nationwide cancer epidemic, nuclear energy, pesticides, cancer-causing chemicals everywhere, population outstripping food production, and more. Issues that do not warrant our worrying, but for which our country has paid a huge price, financially and emotionally, for the poor decisionmaking fostered by fright-filled voices.

But the public must share a measure of reproach for buying into the environmentalist chaff; unable or unwilling to challenge their absurd assertions, passively accepting their dire predictions, when a few well-chosen, pointed questions would have sent the beggers packing. The essayist, H. L. Menken, knew these folks for what they are. "The whole aim of practical politics," he wrote, "is to keep the populace alarmed—and hence clamorous to be led to safety—by menacing it with an endless series of hobgoblins, all of them imaginary."

Knowledge and scientific literacy are the waves of the future. Come with me on a voyage of discovery, and finally get to know our (not so) precarious habitat, "for the first time."

M.A.B.

AMERICA THE HEALTHFUL: ITS VITAL SIGNS

Only one in a thousand persons dies a natural death;
the rest die because of ignorant or aberrant behavior.
—*Maimonides (1135–1204)*

The best-kept secret of the twentieth century may have been the good health of the public. Even the Russians hadn't gotten onto it. There is no pressing need for this secret to continue into the twenty-first century.

In 1979, Surgeon General Julius Richmond titled his report to President Carter, *Healthy People* [1]. From the first sentence of his report we learn that "the American people have never been healthier." What happened to that tad of good news? Little to nothing. The media deemed it unworthy to pass along. And then, with the new millennium, the Institute of Medicine (IOM), an arm of the National Academy of Sciences, weighed in with its report, *The Future of the Public Health in the 21st Century*, making it satisfyingly clear that the health of the American people at the beginning of the twenty-first century would astonish those living in 1900. "By every measure," it proclaimed, "We are healthier, live longer (as we shall see), and enjoy lives that are less likely to be marked by injuries, ill health or premature death" [2]. If that isn't cause for, and reason aplenty, to shout the good news from our rooftops, what is?

Given the fact that few know this, shouting from rooftops is simply too much to expect. But why the continuing silence by the media? Why haven't our vast communications networks—radio, TV, newspapers, and magazines—

Our Precarious Habitat . . . It's In Your Hands, Fourth Edition. By Melvin A. Benarde
Copyright © 2007 John Wiley & Sons, Inc.

made much of this uplifting news? Is it that good news doesn't sell? Or is it something else again? Why would anyone want to keep the people ignorant of the sea change that has so affected their lives?

Given the silence of the media, and the public's consequent lack of that knowledge, it is incumbent upon us to provide this remarkable information. And by no means does our good health mean that we don't become ill, or that health has taken a sojourn. Indeed not, but it does mean that by comparison with our parents and grandparents generations, we have made remarkable strides, and that illness and death are no longer constant comparisons.

So, with kudos expressed, we can examine the data for ourselves and see how far we've come. As we journey, we shall ask the pregnant question, which, if any, of our leading causes of death can be ascribed to the food we eat, the water we drink, the air we breathe, and which to our personal behaviors, and which to our inheritance, our genetic constitution.

We begin our trek with a look at the leading causes of death in 2002. Table 1.1 provides a mine of information. At far left, causes of death are ranked from 1 to 12, followed by the total number of deaths for each condition. The fourth column renders the number of deaths per 100,000 populations. By moving the decimal point left, the number of deaths per 10,000, 1000, and 100 can be

TABLE 1.1. Leading Causes of Death, United States, 2002

Rank	Cause	Number	Rate/10^{5a}	Percent of Total (%)
1	Heart disease	696,947	242.0	28.5
2	Cancer	557,271	193.5	22.8
3	Stroke	162,672	56.4	6.6
4	COPB	124,816	43.3	5.1
5	Accidents			
	Motor vehicle	42,281	14.7	1.7
	Other	101,537	35.2	4.1
6	Diabetes	73,249	25.4	3.0
7	Influenza/pneumonia	65,681	22.8	2.7
8	Alzheimer's disease	58,866	20.4	2.4
9	Nephritis/nephrosis, nephrotic syndrome	40,974	14.2	1.6
10	Septicemia	3,865	11.7	1.4
11	Suicide	3,062	10.6	1.2
12	Liver disease	27,257	9.5	1.1
	Total	1,958,478		
	All causes	2,443,387	845.3	100.0

[a] Per 100,000 people.
Note: Total US population on July 1, 2002 was 287,941,220.
Source: National Center for Health Statistics [9].

obtained. The far right column indicates the percent or proportion of deaths contributed by each of the 12. The total of the 12 will yield only 82%. The remaining 18% is contributed by the literally hundred-plus slings and arrows that have found chinks in our armor, our genomes.

From this array it is immediately evident that heart disease lays undisputed claim to the top slot. Since the last edition of *OPH* (this book, *Our Precarious Habitat*), in which heart disease held first place with 323 deaths per 100,000, and was answerable for 37% of all deaths, there has been a gratifying decline of 33%; a triumph of public health. For a longer view, at midtwentieth century, 1950, the death rate was 440.1, a tad less than double our current rate. The numbers do reveal a remarkable achievement, well worth shouting about from rooftops.

Also evident is the fact that once beyond heart disease, the figures in each column of Table 1.1 drop sharply. Quick addition indicates that fully 58% of all deaths are due to heart disease, cancer, and stroke. The top five are responsible for 69% of all deaths. We would be remiss in taking leave of Table 1.1 without noting that the overall death rate stands at less than 1% per year. The rate of 845.3 per 100,000 can be translated as 0.84 per 100 individuals—0.8/100 or 0.8%. The essential takeaway message is that the preponderance of us live full and long lives—well beyond the biblical "fourscore and ten." Most everyone will live to collect their social security benefits for years after retirement. That's part of our current economic problem. When FDR signed the social security legislation into law in 1935, few of us were expected to attain the magic number—65—and collect. Hopefully our length of days will not break the bank. Nevertheless, length of days is the new and assuring message. But there are a clutch of potentially—perhaps more than potentially—deadly myths about heart disease. Heart disease is a man's disease. More men than women die of it. Myths are hard-dying, and this one has had too long a run. For over 20 years more women have died of heart disease than have men. Women put off getting appropriate medical attention, and all too often physicians and hospital personnel have not taken women's heart complaints seriously. Of course, the notion that cancer and AIDS snuff out more lives than does heart disease is laid to rest in Table 1.1, where it is obvious that heart disease is far deadlier than all forms of cancer. As for AIDS, we'll deal with that misperception shortly. We shall also dig deeply into the notion—yes, "notion" is the fit term—that high-dose antioxidant vitamins can protect the heart, is just that: notional. Worse yet, there is highly suggestive evidence that ingesting megadoses of vitamins can blunt the effectiveness of anticholesterol medications.

The myth that your high blood pressure occurs only when you step into the presence of your cardiologist may be the deadliest of all. Too many of us mistakenly believe that our blood pressure is perfectly normal until the white coat appears. True, there is a "white coat" hypertension concern—I know; I'm one of those who manifest that reaction regularly—but too many of us, including physicians, brush it off as nervousness and anxiety. Not a good idea.

Misinformation must not be allowed to make any of us a premature mortality statistic.

Abroad in the land is the extreme misconception that a large number of individuals with coronary heart disease lack any of the major coronary heart disease (CHD) risk factors.

Two recent studies conclude that 80–90% of individuals with CHD have conventional risk factors. Numerous epidemiologic studies have identified cigarette smoking, diabetes, hyperlipidemia, and hypertension as independent risk factors for CHD. Curiously enough, treatment of these risk factors has reduced the risk of a second cardiac event. These four risk factors have been referred to as "conventional risks." However, although well established in the medical literature, it is often stated that more than 50% of those with coronary disease do not exhibit one or more of these conventional risks. Consequently a team of medical researchers from the Cleveland Clinic and the University of North Carolina Medical School set out to determine the prevalence of the four conventional risk factors among patients with CHD, because of the claim that nontraditional and genetic factors play a significant role in the acquisition of heart disease. From their analysis of 122,458 patients enrolled in 14 international randomized clinical trials, they found that at least one of the four risk factors was present in 84.6% of women and 80.6% of men. They concluded that "clinical medicine and public health policies should place significant emphasis on the 4 conventional risk factors and the life-style behaviors causing them to reduce the epidemic of CHD" [3].

Motivated by what they believed to be the same false idea that less than 50% of CHD patients lacked the four conventional risk factors, a team of investigators from Northwestern University School of Medicine, the University of Minnesota School of Public Health, and Boston University School of Medicine set up a study "to determine the frequency of exposure to major CHD risk factors." This team followed 376,915 men and women ages 18–59 for 21–30 years. Among their enrollees, at least one of the four conventional risk factors occurred in 87–100%. Among 40–59-year-olds with a fatal coronary event, exposure to at least one risk factor ranged within 87–94%. For a nonfatal event there was exposure to at least one risk factor in 92% of men and 87% of women. For them, "Antecedent major CHD risk factor exposures were very common among those who developed CHD." They also made the point that, "These results challenge claims that CHD events commonly occur in persons without exposure to at least one major CHD risk factor" [4].

For anyone concerned with prevention of a coronary event, these two statistically powerful studies must be taken seriously, and smoking, high blood pressure, diabetes, and elevated blood lipid levels can no longer be ignored. They can be ignored only at our peril. We will not forget to question, to wonder, whether CHD is an environmental issue or a personal behavioral issue.

Cancer, for which we have an uncommon dread, deserves our scrupulous attention. However, we shall hold it in abeyance until we consider life's less terrifying afflictions.

CHRONIC NONNEOPLASTIC MEDICAL DISORDERS

Cerebrovascular disease (CVD), chronic obstructive pulmonary disease (COPD), diabetes, influenza/pneumonia, Alzheimer's disease, nephritis and nephrosis, septicemia, and liver disease are chronic medical conditions that have emerged as substantial death threats as infectious illness has waned.

Cerebrovascular Disease

Although settled in the number 3 position for generations, cerebrovascular disease (CVD or stroke), with 6.6% of the overall deaths, while not nearly the taker of lives as is heart disease or cancer, is a leading cause of serious, long-term disability. From the numbers we learn that CVD is the cause of one death in every 14, killing 167,661 in the year 2000; 63% of these were women. Stroke deaths afflict Afro-Americans with great force. For every 100,000 men and women the rates are 87 black men, 78 black women; 59 white men, and 58 white women. Current thinking holds this striking racial difference to be a function of genetic inheritance. Why this is so will be revealed by ongoing genetic investigations. As genetic studies indicate that different races are uniquely prone to different diseases, public health policies that recognize this could improve both medical care and the use of prescription drugs that we now know do not provide equal benefit across races [5–7].

Also known is the why of stroke, which occurs as a consequence of oxygen deprivation, when blood flow to the brain is disrupted. This deprivation is the most common cause of disabling neurologic damage and death. Not unlike water through a sediment-clogged pipe, a blood clot or fatty deposition (plaque) (atheroma) can block blood flow anywhere should inflammation narrow blood vessels to the brain. Drugs and hypotension—low blood pressure—can also reduce blood flow.

Chronic Obstructive Pulmonary Disease

Whereas CVD is an obstructive circulatory condition, COPD, chronic obstructive pulmonary disease, the fourth leading cause of death and accounting for the loss of some 125 thousand lives, is the consequence of respiratory obstruction. It is a tenacious blockage of oxygen, often the result of emphysema, chronic bronchitis, and/or asthma.

Our lungs consist of hundreds of millions of tiny airsacs-alveoli—whose walls are astonishingly thin, and necessarily so, permitting the passage of oxygen into and carbon dioxide out of our lungs. The grapelike clusters of alveoli maintain a rigidity that holds the airways open. When the thinner than thinnest tissue-paper thin walls erode, most often the result of cigarette smoke, the alveoli collapse or become hole-ridden, making breathing inordinately difficult. Death ensues from asphyxiation. In his report on women and smoking, the U.S. Surgeon General [8], states that "mortality rates for COPD have increased among women over the past 20 to 30 years." We are further informed that between 1979 and 1985, the annual age-adjusted rates for COPD among women 55 years and older increased by 73%, from 46.6 per 100,000 to 80.7 per 100,000, and this steep rise continued during 1980–1992. Furthermore, from the CDC's National Vital Statistics Report [9], we learn that for 2002, for all races and all ages, women had a combined COPD death rate of 43.7 per 100,000. This arresting statistic reflects one of the most unfortunate and unnecessary facts of COPD deaths: the increase in smoking by women since World War II. Prevention appears to be entirely in their hands.

Diabetes

In the sixth slot with 3% of total deaths, is diabetes mellitus (literally, "honey sweet" diabetes), a group of diseases in which levels of the sugar glucose, are abnormally high because the pancreas fails to release adequate amounts of insulin, an enzyme that normally metabolizes glucose, maintaining steady levels. Diabetes shows itself in two main forms. Type 1 was until recently called *insulin-dependent diabetes mellitus* (IDDM) or *juvenile-onset diabetes*. Type 1 occurs when the body's immune system destroys pancreatic beta cells, the only cells in the body that make insulin. This form usually occurs in children and young adults; hence juvenile diabetes.

But why does the body harm itself? Why does the immune system destroy beta cells? The essence of the immune system is its ability to distinguish self from nonself. Self is we; our cells, tissues, and organs. Every cell in our tissues contains specific modules that identify it as self. Nonself are foreign objects or conditions that do not belong among us; bacteria, viruses, fungi, and other parasites. Because our bodies provide an ideal environment, with nourishing fluids, parasites are always dying to break in. Our immune system is usually ready and able to detect the presence of foreign substances. Sometimes, not often, the recognition system falters and atttacks tissue carrying the self-marker molecules. When that happens, the body manufactures T cells and antibodies directed against, in this case, the beta cells in the islets of Langer-hans that are clustered in the pancreas. With the beta cells destroyed, insulin is not produced, and glucose levels remain uncontrolled. Those T cells ("T" for thymus, the small gland behind the breast bone—the sternum—where these cells are produced) are types of white blood cells—lymphocytes—that normally play a major role in defending the body against foreign invaders.

Type 2 was previously referred to as *non-insulin dependent diabetes mellitus* (NIDD), or adult-onset diabetes. It is type 2 that accounts for the great major-ity of diabetes cases. Type 2 is associated with older age, obesity, and impaired glucose metabolism. African Americans, Latinos (Hispanics), American Indians, and Pacific Islanders are at unusually high risk [10]. However, the number of overweight and obese children in the United States has doubled in the past two decades, and 15% of American children are overweight or obese. Type 2 diabetes, which is linked to excess weight, is being found in such high numbers among children that the adult-onset type will need a new designation [11]. We will deal with the obesity problem shortly.

We cannot leave diabetes without considering its future and possible demise. Current immunological research is so vigorous that it is possible to almost see beyond the horizon. In this instance the horizon is T cells, once thought to be of two types, killer T cells that track down virus-infected cells, and helper T cells, which work primarily by secreting lymphokines that accomplish a number of defensive functions. Now, however, a newly found class of T cells, regulatory T cells, are seen as the promise for vanquishing a number of autoimmune diseases including diabetes and multiple sclerosis, diseases in which the body turns on itself. Regulatory T cells are seen as the basis of new therapies, and 5 years on is seen as a reasonable horizon for regulatory T cells to come on line, making diabetes a thing of the past [12]. Hope is on the way.

Influenza/Pneumonia

Influenza/pneumonia, currently in the seventh slot, was the sixth leading cause of death in 1989, which means that life is improving for both the very young and the very old who are at greatest risk for this insidious duo. Influenza and pneumonia are inseparably linked as influenza often leads to pneumonia and death. Influenza is the sixth leading cause of death of infants under one year and the fifth leading cause for those over 65 [13].

Person-to-person spread of the numerous highly contagious and infectious influenza viruses, occurs via inhalation of droplets of saliva expelled when coughing and sneezing. Sneezing is the more troublesome. When someone sneeze, the teeth are clenched, and the intense force of the sneeze squeezes saliva between the teeth, creating a cloud of particles, some of which can hang in the air and drift with the currents—to others nearby, who will breathe them in with their next breath. The risks for complications, hospitalizations, and death increase with such underlying conditions as heart disease, diabetes, stroke, HIV/AIDS, and, of course, pneumonia. Most assuredly the number of influenza deaths will increase along with the number of elderly, unless antiviral medications are discovered and developed. But that is unlikely in the near term, given the complexity of viruses and their adaptive ability. In fact, a new study recently found that influenza vaccination of the elderly may be less effective in preventing death than previously assumed. The possibility is raised that "herd" immunity may be more protective. This would require the vaccination of larger numbers of younger, healthier individuals, children included, the immune herd, to prevent transmission of the viruses to the high-risk elderly [14].

Bacteria, specifically *Streptococcus pneumoniae, Hemophilus influenzae,* and *Klebsiella pneumoniae,* are the microbial villains, among adults, while a bevy of viruses are children's troublemakers. Here, too, the beasties reach both the upper and lower respiratory tract via inhalation of saliva-bearing organisms sneezed or coughed into the surrounding air. Here again, the elderly and the very young are at increased risk of complications, as are African Americans, Alaska natives, and HIV/AIDS patients. Death occurs in 14–15% of hospitalized patients [13].

Alzheimer's Disease

Alzheimer's disease is a newcomer to the inventory of the leading causes of death. It was not present in 1989, having only recently taken over eighth place, with 59,000 documented deaths, and is currently responsible for over 2% of all deaths. The term "documented" suggests, strongly suggests, that there are many more deaths due to dementia than the numbers indicate [15,16].

Alois Alzheimer, born in Marbreit, Germany, became a medical researcher at Munich Medical School, where he created a new laboratory for brain research. In 1906, he identified an unusual disease of the cerebral cortex that caused memory loss, disorientation, hallucinations, and untimely death. Alzheimer died at the tender age of 51 from complications of pnuemonia and endocarditis [17].

Currently Alzheimer's disease is fairly well understood. It is a disorder that occurs gradually, beginning with mild memory loss, changes in behavior and personality, and a decline in thinking ability. It progresses to loss of speech, and movement, then total incapacitation and eventually death. The brains of Alzheimer's patients have an abundance of plaques and tangles, two abnormal proteins. *Plaques* are sticky forms of β-amyloid; *tangles* are twisted protein fibers called *tau* (τ). Together, these two protein clumps block the transport of electrical messages between neurons that normally allow us to think, talk, remember, and move. Until a medication or procedure is found that can dissolve or prevent the occurrence of these blocking proteins, Alzheimer's can be expected to accumulate victims.

Nephritis and Nephrosis

Nephritis is an inflammation of the kidneys, and *nephrotic syndrome* is a collection of symptoms induced by a number of diseases. Allergies can do it; drugs can do it, and perhaps a dozen ailments, including HIV/AIDS, diabetes, and cancer can cause severe kidney impairment. *Nephrosis* is a severe loss of serum protein that can lead to a number of immunodeficiency disorders and death. Nephritis is usually the consequence of either a streptococcal infection or an adverse immune reaction. Children and adults are equally at risk, and with some 41,000 deaths annually, these kidney miseries have been propelled into the ninth position. However, for infants less than a year, this triad is the seventh destroyer of life [13]. Indeed, they are vulnerable.

Septicemia

Some pathogens can grow and multiply specifically in blood plasma, and their waste products can be toxic and produce septicemia (*septikos*, from the Greek meaning putrefaction, and *haima*, blood). Entry into body cells requires a specialized ability used by various pathogens to produce lytic chemicals capable of dissolving membranes and permitting passage, or via such passive means as entry through breaks in the skin, ulcers, burns, and wounds. They are there waiting an opportunity. By whatever route entrance occurs, toxins shed by the organisms produce a toxic condition, a toxemia, which can be extremely difficult to manage, especially now that many pathogens have evolved resistance to a range of antibiotics.

Sepsis can become risky as a result of surgery and/or the insertion of intravenous catheters, urinary catheters, and drainage tubes. The likelihood of sepsis increases as the time of the indwelling catheters increases. Of course, injecting drug users introduce bacteria directly into their bloodstreams. Also at increased risk are individuals taking anticancer medications, as well as HIV/AIDS patients whose immune systems have been compromised.

Septic shock, in which blood pressure falls to life-threateningly low levels, can be a frightening side effect, occurring most often in newborns, the elderly, and those whose immune systems are in disarray. If infant care is inappropriate, and HIV/AIDS continues unabated as it appears to be doing, septicemia and septic shock will remain integral components of the list.

Liver Disease

Liver disease is in the twelfth position, accounting for some 27,000 annual deaths. Cirrhosis, the primary liver disease, appears to be the end stage of several common causes of liver injury. Cirrhosis (Greek, meaning orange-colored) results in nonfunctioning destroyed liver tissue that can and often does surround areas of viable healthy tissue. Until the cirrhotic condition is well advanced, many people remain a symptomatic, not knowing that they have it. The most common risk factor is alcohol abuse. Among the 45–65-age group cirrhosis is now the sixth leading cause of death [13].

So, here we have gathered the third, fourth, sixth, seventh, eighth, ninth and tenth leading causes of death, which between them account for 24% of what's killing us. Are these environmental?

ACCIDENTS

Although motor vehicle accidents kill over 40,000 men, women, and children every year, and maim hundreds of thousands, this carnage on our highways is considered one of life's less terrifying trials. In fact, it is barely considered.

More to the point, it is accepted as a price to pay for our mobile way of life. Perhaps the numbers that follow will help us see the error of our ways.

Deaths by motor vehicle are synonymous with the advent of the horseless carriage, and have been an integral part of our lives since that crisp September day in 1899 when New York real estate broker H. H. Bliss stepped from a trolley car at 74th Street and Central Park West in New York City and was struck down by an electric horseless carriage [18]. By 1990, 3,000,000 men, women, and children had surrendered their lives to motor vehicles. If this slaughter continues, as it appears to be doing, the fourth million will arrive within 6 years—by 2012. To help comprehend the flaws in our thinking about motor vehicle accidents, we need only revisit Table 1.1. Although this table, which provides substantive information about us as a nation, aggregating all causes of death, is obviously useful, it tends to obscure seminal details. Yes, motor vehicle deaths are the fifth leading cause of death. But the data in Table 1.1 fail to divulge the fact that in 2002, crashes on our highways were the leading cause of death for every age from 3 to 34 [13]. Table 1.2 provides an alternative view of the leading causes of death. By teasing out the top killers by age groups, shock and awe awaits. A new set of risks emerge, which suggest a different approach to prevention. Lest we take some measure of joy that toddlers age 1–3 are not members of this select assembly, I hasten to add that for the 1–3-year-olds, death by motor vehicle is the second leading cause of death. The 35–44-year-olds can take little comfort in the fact that motor vehicle death is the third leading cause of their demise.

Furthermore, motorcycles appear to be the most dangerous type of motor vehicle on the road today. These bikers "were involved in fatal crashes at a rate of 35 per million miles of travel compared with a rate of 1.7 for passenger cars. The peak rate of death among alcohol-impaired cyclists shifted from those 20 to 24, to the 40 to 44 age group" [13]. Stronger law enforcement for this age group would save lives, but any attempt in that direction, will be fought tooth and nail. As for older drivers, Table 1.2 offers yet another surprising and impressive message; those 65 and older do not, repeat, do not appear at any level of leading causes of death by motor vehicle. Now that's a startling and powerful statistic that certainly *does not* accord with conventional wisdom or perception.

The data array in Table 1.2 offers a mother lode of substantive data. We see that for ages 1–34, motor vehicle crashes must or should be their worst nightmare. And as already noted, for those 65 snd older, motor vehicle deaths are not the problem widely thought to be. Continuing to delve into this mine of data, we see that while not on our prime list of the 12 leading causes of death, homicide (e.g., see Table 1.3 for comparisons of firearm homicide rates in three countries) is the second leading cause of death for those aged 16–24, and suicide is the second leading cause of death for the 25–34 age group.

Malignant neoplasms rise to the primary slot for the 35–64 age group, while heart disease deaths takes the top spot for those 65 and older. But who would have believed that homicide is the fourth leading cause of death for toddlers and youngsters from less than a year old to age 3, along with those those ages 8–15? Suicide emerges as the third leading cause of death for young people ages 8–24. This chart is pregnant with matchless information and worthy of continued mining. For example, the last column on the right discloses the millions of years of life lost by each cause of death, which can be translated as the country's loss of creativity and productivity, losses that cannot be retrieved. In fact, because of the many young lives consumed, motor vehicle traffic crashes rank third in terms of years of life lost—that is, the number of remaining years that someone is expected to live had that person not died—prematurely—behind only cancer and heart disease [19].

Before moving on, it is imperative to note the National Highway Traffic Safety Administration's most recent statistic: 43,443 people killed in roadway *accidents* in 2005. This increase from 2003 and 2004, was attributed to accidents involving motorcycles. In fact, deaths from motorcycles rose by 13% from 2004, and by almost 50% of those who died were not wearing helmets, which are known to reduce the probability of dying in a motorcycle crash by 35–50%. Take note.

It would be remiss not to mention that as national census figures indicate, the 1–34-year-old group comprises some 139 million men, women, and children; not quite 20% of the total population. At 20% they would constitute fully 1 in 5 of all our citizens, representing the heart, the core of young, working America. They are sorely needed. Consequently, for them, prevention would seem a top priority.

Life is not without risks; we know that. But there are risks that need not be taken. In 2002, 18,000 men, women, and children died in crashes involving alcohol [9]. Men, women, and children are included here, rather than people or individuals to make this more personal and perhaps more meaningful. After all, these deaths include parents, relatives, and friends. Nevertheless, those 18,000 deaths do not account for the 500,000 men, women, and children who were injured; many were maimed for life. To edge this a bit more starkly, 18,000 deaths and 500,000 injuries translate into 49 deaths and 1370 injuries each day of the year. Drunk driving deaths have been on the rise since 1999 [20]. Some 25% of the 15–20-year-olds killed on the highways had blood alcohol concentrations (BACs) of 0.08 grams per deciliter (g/dL).

A recent study in the *Journal Psychological Science* informs us that those who drink and drive are at a higher risk of accidents because alcohol distorts depth perception—judging distance from obstacle [21]. It is also troubling that drinking has become more prevalent among teenage girls because of the increased advertisement of alcoholic beverages to teenage groups. Research at Georgetown University revealed a striking increase in such ads in over 100 national magazines. A larger percentage of girls age 12–20 were exposed to alcohol ads than were women ages 21–34. Apparently teenagers are advertisers' primary target [22]. Get 'em young, and you've got them for life.

Again, the numbers continue to tell a grim story. Drivers under 20 were involved in 1.6 million crashes in 2002, with 7772 of them fatal, including 3700

TABLE 1.2. Leading Causes of Death in the United States for 2002 by Age Group[1]

| | Cause and Number of Deaths | | | | | |
RANK	Infants Under 1	Toddlers 1–3	Children 4–7	Young Children 8–15	Youth 16–20	Young Adults 21–24
1	Perinatal period 14,106	Congenital anomalies 474	MV Traffic crashes 495	MV Traffic crashses 1,584	MV Traffic crashes 6,327	MV Traffic crashes 4,446
2	Congenital anomalies 5,623	MV Traffic crashes 410	Malignant neoplasms 449	Malignant neoplasms 842	Homicide 2,422	Homicide 2,650
3	Heart disease 500	Accidental drowning 380	Congenital anomalies 180	Suicide 428	Suicide 1,810	Suicide 2,036
4	Homicide 303	Homicide 366	Accidental drowning 171	Homicide 426	Malignant neoplasms 805	Accidental poisoning 974
5	Septicemia 296	Malignant neoplasms 285	Exposure to smoke/fire 151	Congenital anomalies 345	Accidental poisoning 679	Malignant neoplasm 823
6	Influenza/ Pneumonia 263	Exposure to smoke/fire 163	Homicide 134	Accidental drowning 270	Heart disease 449	Heart disease 518
7	Nephritis/ Nephrosis 173	Heart disease 144	Heart disease 73	Heart disease 258	Accidental drowning 345	Accidental drowning 238
8	MV Traffic crashes 120	Influenza/ Pneumonia 92	Influenza/ Pneumonia 41	Exposure to smoke/fire 170	Congenital anomalies 254	Congenital anomalies 186
9	Stroke 117	MV Nontraffic crashes[4] 69	Septicemia 38	Chronic lower resp. dis. 131	MV Nontraffic crashes[4] 121	Accidental falls 134
10	Malignant neoplasms 74	Septicemia 63	Benign neoplasms 36	MV Nontraffic crashes[4] 115	Acc. dischg. of firearms 113	HIV 130
All[3]	28,034	4,079	2,586	6,760	16,239	15,390

[1] When ranked by specific ages, motor vehicle crashes are the leading causes of death for age 3 through 33.

[2] Number of years calculated based on remaining life expectancy at time of death; percents calculated as a proportion of total years of life lost due to all cause of death.

[3] Not a total of top 10 causes of death.

[4] A Motor Vehicle Nontraffic crash is any vehicle crash that occurs entirely in any place other than a public highway.

	Other Adults					Years of Life
25–34	35–44	45–64	Elderly 65+	All Ages	Lost[2]
MV Traffic crashes 6,933	Malignant neoplasms 16,085	Malignant neoplasms 143,028	Heart disease 576,301	Heart disease 696,947	Malignant neoplasms 23% (8,686,782)
Suicide 5,046	Heart disease 13,688	Heart disease 101,804	Malignant neoplasms 391,001	Malignant neoplasms 557,271	Heart disease 22% (8,140,300)
Homicide 4,489	MV Traffic crashes 6,883	Stroke 15,952	Stroke 143,293	Stroke 162,672	MV Traffic crashes 5% (1,766,854)
Malignant neoplasms 3,872	Suicide 6,851	Diabetes 15,518	Chronic lower resp. dis. 108,313	Chronic lower resp. dis. 124,816	Stroke 5% (1,682,465)
Heart disease 3,165	Accidental poisoning 6,007	Chronic lower resp. dis. 14,755	Influenza/ Pneumonia 58,826	Diabetes 73,249	Chronic lower resp. dis. 4% (1,466,004)
Accidental poisoning 3,116	HIV 5,707	Chronic liver disease 13,313	Alzheimer's 25,289	Influenza/ Pneumonia 65,681	Suicide 3% (1,109,748)
HIV 1,839	Homicide 3,239	Suicide 9,926	Diabetes 54,715	Alzheimer's 58,866	Perinatal period 3% (1,099,767)
Diabetes 642	Chronic liver disease 3,154	MV Traffic crashes 9,412	Nephritis/ Nephrosis 34,316	MV Traffic crashes 44,065	Diabetes 3% (1,050,798)
Stroke 567	Stroke 2,425	HIV 5,821	Septicemia 26,670	Nephritis/ Nephrosis 33,865	Homicide 2% (822,762)
Congenital anomalies 475	Diabetes 2,164	Accidental poisoning 5,780	Hypertension renal dis. 17,345	Septicemia 33,865	Accidental poisoning 2% (675,348)
41,355	91,140	425,727	1,811,720	2,443,387	All cause 100% (37,341,511)

Source: National Center for Health Statistics (NCHS), CDC, Mortality Data 2002.

Note: The cause of death classification is based on the National Center for Statistics and Analysis (NCSA) Revised 68 Cause of Death Listing. This listing differs from the one used by the NCHS for its reports on leading causes of death by separating out unintentional injuries into separate causes of death, i.e., motor vehicle traffic crashes, accidental falls, motor vehicle nontraffic crashes, etc. Accordingly, the rank of some causes of death will differ from those reported by the NCHS. This difference will mostly be observed for minor causes of death in smaller age groupings.

TABLE 1.3. Cross-Cultural Differences in Firearm Homicides, 2000

Country	Population (million)	Firearm Homicide	Rate/10^{5a}
United States	275	10,801	39.2
European Union	376	1,260	3.3
Japan	127	22	0.17

[a] Per 100,000 people.

teenagers. Far too many teenagers have neither the skills nor the experience to be permitted full driving privileges. Unfortunately too many parents believe otherwise. And then there is the seatbelt problem, which means the failure to wear them while driving. Highway crashes took the lives of over 19,000 drivers and passengers who failed to wear seatbelts. About a thousand of these were children under 8 who were not buckled up and properly secured in the rear seat of the car [20]. Clearly there is much room for improvement, and much can be readily and easily done to reduce the carnage, but it is difficult to overcome inertia and the tragically mistaken belief that we are all superb drivers. Dispelling that myth could have the salutary affect of relegating the accident death rate to the seventh or eighth position. Would it help to know that the motor vehicle death rates for Mississippi and Massachusetts in 2003 were 31.2 per 100,000 and 8.0 per 100,000, respectively [13]?

The accident equation contains yet another variable: one that cannot be directly attributed to our driving skills, as a third party plays a significant role for which drivers are unprepared. According to the National Safety Council, collisions with animals have risen dramatically from 520,000 animal-related accidents in 2001 to 820,000 accidents in 2002. These collisions resulted in more than 100 deaths and 13,000 injuries [23] as drivers collide with deer bounding across roads, or as we swerve into the path of oncoming vehicles attempting to avoid the creatures. As more of our wilderness gives way to housing developments and animals are forced to seek shelter where they can, their road crossings, day or night, will continue to be a risky business for us both. From the University of North Carolina's Highway Safety Research Center we learn that deer crashes occur most frequently in October, November, and December, and are more likely to occur during 5:00–7:00 A.M. and between 6:00 P.M. and midnight [23].

The National Safety Council maintains that death and injury on our roads and highways is directly related to impulsiveness, alcohol consumption, and poor judgment. It has long been my contention that "accident" is a misnomer for the fateful events that cause the slaughter on our highways. The word should be "purposefuls," as "accident", by definition, means an unforeseen event, or one without an apparent cause. I would imagine that most of us would agree that such a definition fails to adequately describe the crashes, the accidents that permanently removes so many of us prematurely. I'm also comfortably certain that many of us could compile lists of good and sufficient risk factors that contribute to untimely motor vehicle deaths.

Unfortunately motor vehicle accidents do not account for all "accidental" deaths and injury. Falls are the greatest single risk; falls from stepladders, and staircases, falls were responsible for 18,044 deaths in 2003. Over 11,000 were in the over 65+ age group [13]. But what actually caused the fall? Was it a toy left on the stairs that a person tripped over; was it a faulty stepladder, or faulty placement of the ladder? Again, "accident" is probably the wrong word. In addition to falls, fires, drownings, poisonings, drugs, knives, and guns drive the nonvehicular deaths to well over 100,000 annually. A chilling statistic. At this point it is worth recalling that although accidents are the fifth leading cause of death nationwide, they are actually the leading cause of death for those ages 1–30—something that Table 1.2 makes abundantly clear. Unfortunately, and to their detriment, they are oblivious of this horrific statistic. It would be unimaginable if they were aware; that would suggest total denial. Why hasn't the media made much of this frightful and wasteful cause of death?

SUICIDE

Suicide, the taking of one's own life, while currently the nation's eleventh leading cause of death—with violent death among the 8–24-year-olds, as Table 1.2 informs us—it is the the third leading terminator of life, but slips into second place for the 25–34-year-olds In 2003, 31,484 young lives were snuffed out across the country: 25,203 young men and 6281 young women [13]. Adding these traumatic deaths to the highway deaths, we are looking at some 75,000 deaths annually. Although it is not treated that way, because trauma is not a rubric, a recognized category, trauma is one of the most notorious problems the country must deal with, but doesn't. It is as if it doesn't exist, as it is seen as two separate issues, when it should be seen and dealt with as a single substantive issue. Until it is, expect the numbers to increase. The number of potentially productive years of life lost should give us pause: 1,100,000 years of life lost to suicide. Add another 2 million for motor vehicle deaths. From a purely economic concern, can the country afford the loss of 3 million potentially productive days annually? It's a no-brainer.

According to the American Association of Suicidology (AAS), suicide is, beyond doubt, preventable. Most suicidal individuals, the AAS tells us, want to live but are simply unable to recognize alternatives. Most offer clues, warning signs, but parents, friends, teachers, and physicians fail to read them. With that type of illiteracy, a suicide occurs every 16.7 minutes. Although 31,000 do commit suicide, according to the AAS there are 787,000 attempts: 25 attempts for every death, with an average 3 : 1 female : male ratio for every attempt. I suspect we would agree that this is an intolerable form of illiteracy. But how do we become literate?

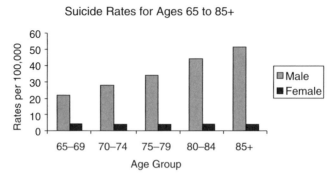

Figure 1.1. Comparison of suicide rates for men and women ages 65–85+. (*Source:* Centers for Disease Control, 2002.)

Surprisingly, winter is not the season of greatest discontent; spring exacts the greatest toll—just the opposite of what I would have imagined. For most of us spring is the season of rebirth and new beginnings, and winter is perceived as cold, damp, and depressing.

At the opposite end of the age spectrum, the elderly, 65 and over, make up 12.4% of the country's total population, yet account for almost 18% of all suicides. According to the AAS, an elderly suicide occurs every 95 minutes. Elderly white men are the highest at-risk group, with some 35 suicides per 100,000. Figure 1.1, shows the rising rates with age and the difference between men and women. Although older adults attempt suicide less often than do others, they have a higher successful completion rate. If all attempts were successful, we would be looking at a 0.75 million suicide deaths a year, which would far exceed the combined death rate due to heart disease and cancer. It boggles the mind. Among the 15–24 age group, which surely must be a terribly fragile age, there is one attempt for every 100–200 attempts. For those over 65, there is one successful suicide for every four attempts. For all age groups, the instruments of most successful attempts are rifles and handguns. Their ready availability speaks for itself. Suffocation and hanging are the second most grizzly choices [24].

The goal of the American Association of Suicidology (AAS) is to understand and prevent suicide. It also serves as a national clearinghouse for information and help, and provides resources and publications on a range of suicide-related topics, including how to read the signs of a potential suicide.

The AAS can be reached at

4201 Connecticut Ave, NW, Suite 408
Washington, DC 20005
202-237-2283
email: infor@suicidology.org
Website: www.suicidology.org

If by nature's way, suicide and homicide need not be violent. When "the elders" decided that 70-year-old Socrates was too disturbing an influence on young Athenian boys, he was handed an extract of *Conium maculatum*, spotted hemlock, in wine, of course, with which to dispatch himself to a more propitious environment [25]. Although hemlock's toxic properties were well established by 399 BCE, it was not until modern times that coniine (2-propylpiperidine) was isolated and found to be the active (dispatching) essence [26]. However, I've never been certain whether Socrates' death was a suicide or a homicide. I tend toward homicide, as there is little evidence that at his age he entertained the idea of ending it. *Conium maculatum* is not alone in bringing about the end of days by either suicide or homicide for any number of imbibers.

Recently, another of nature's wonders made headlines. *Cerbera odollam* appears to be a ready-made suicide tree, the cause of widespread death on the Indian subcontinent and environs. The suicide tree is seen as the agent of death for many more than had been imagined. Cerbera, which grows across India and southeast Asia, has a long history of assisted suicide, but the scientists who recently studied the problem indicate that the authorities have failed, and are failing, to determine how often it is used for murder.

Writing in the journal *Ethnopharmacology*, they inform us that cerbera "belongs to the notoriously poisonous Apocynacea family," and they say that the seeds have long been used as an ordeal poison [27]. Ordeal trials were an ancient test used to determine guilt or innocence of sorcery and other crimes. Belief in the infallibility of the poison to make the correct distinction was so strong that innocent people did not hesitate to take a sip.

Cyberin, cerberoside, and odollin, the deadly glycosides, are contained in the kernel. Between 1989 and 1999, 537 deaths were attributed to odollam poisoning. Among those who knowingly took it, 70–75% were women—which continues to speak to their continued oppression. Because of its ready availability, it is the choice poison for both suicides and homicides. For suicide, the kernels are mashed with gur, a form of sweet molasses. For homicide, the kernels are mashed with food containing capsicum-type peppers to mask the bitter glycosides. Death occurs 3–6 hours after ingestion. Although 50 suicides are recorded annually, the actual numbers are unknown, as are the numbers of homicides.

The authors also tell us that "to the best of our knowledge, no plant in the world is responsible for as many suicides as the odollam tree. Mother nature does work in strange ways. The authors further remark that "this study has made it possible to bring to light an extremely toxic plant that is relatively unknown to western physicians, chemists, analysts and even coronors and forensic toxicologists." Yet another caution for our already full agendas.

HIV/AIDS

During the midtwentieth century, the new pathogens human immunodeficincy viruses HIV-1 and HIV-2, which induce AIDS, autoimmune deficincy syndrome, crossed over to the human population and was first diagnosed in humans in 1981, in Los Angeles. Although these viruses rapidly adapted themselves to human–human transmission, AIDS has yet to be found in nonhuman primates; nevertheless, HIV-1 appears to have evolved from the simian immunodeficiency virus SIV_{cpz}—specifically the chimpanzee, *Pan troglodytes troglodytes* [28]. Furthermore, over the past quarter-century, it has become clear that human genetic differences determine whether susceptibility or resistance to AIDS will have rapid, intermediate, slow, or no progression from initial virus infection.

Although AIDS neither makes it to the top of the charts nor is among the top 15, and hasn't been in years, AIDS requires examination as it is widely perceived as the top one or two leading causes of death in the United States. The perception arises because of the way the media has dealt with this entirely preventable illness.

For reasons yet to be revealed, AIDS has often been presented to the public by the communications media, in the aggregate, as cumulative numbers. Since its initial detection in 1981, each year's total of new cases and deaths has been added to the previous year's total. Thus, from 1981 to 2004, the total number of deaths stood at 529,000+. When such an overwhelming number is presented on TV and radio, or carried in newspapers and magazines, it must shock readers and listeners. But the fact of aggregation is noted nowhere, and it is assumed that these are current cases and deaths, which most assuredly is shocking. Nevertheless, this is a unique bit of calculus as no other illness is aggregated in this way. All other diseases are presented as annual totals. So, as we see in Table 1.1, heart disease took 697,000 lives in 2002 (the last year for complete numbers). Had heart disease deaths been aggregated for the 22 years 1981–2003, as was done for AIDS, we would be looking at 15–25 million heart disease deaths. Simple arithmetic informs us that over those 22 years, for every AIDS death there was approximately 30 deaths from heart disease.

AIDS receives exuberant media coverage, well out of proportion to its actual numbers. Similar accounting divulges 12–25 million cumulative cancer deaths, which would translate to 24 cancer deaths for every AIDS death. The perception that AIDS is a major killer is a media creation, requiring expeditious revamping.

Be that as it may, AIDS takes its greatest toll of 34–54-year-old African American men and women, with the 25–34-year-olds running a close second. Male–male sexual contact is the primary route of viral transmission for gay men, black or white. Women receive the HIV virus by direct injection of drugs, into their bloodstream, and via sexual encounters with infected men, who all too often do not indicate their HIV positivity. Also, although it is well documented that condoms can be an essential preventive, far too many men eschew them as "a sexual killjoy." In March, 2007, WHO officially recommended circumcision of all men as a way to reduce the worldwide heterosexual spread of the AIDS virus. The intention is to vigorously pursue this means of prevention.

Also well documented and established is the fact that the human immuno-deficiency virus (HIV) cannot be picked up by drinking from a water fountain, contact with a toilet seat, or touching an infected person. Neither has saliva been shown to be infectious. HIV can be transmitted by semen and vaginal secretions during intercourse, and is readily transmitted by intravenous injection, by sharing needles used by infected individuals. Accidental needlesticks with a contaminated needle has resulted in infections of health professionals, and infected women can readily transmit the virus to their fetuses during pregnancy.

Early on, as HIV/AIDS spread, scientists discovered that the virus attacks human immune cells. The virus can destroy or disable T cells, which can lay dormant for long periods. As immunity fails, an HIV-infected person becomes prey to life-threatening opportunistic infections and rare cancers. Opportunistic infections, usually bacterial, would be of no consequence to a healthy person, but can be deadly to an individual with a compromised immune system.

HIV tricks the T cell into switching on its copy machine, producing huge numbers of new HIV particles that eventually destroy healthy cells, with the release of vast amounts of virus to continue circulating, infecting, and destroying additional lymphocytes. Over months there is an enormous loss of T lymphocytes (CD4+) cells.

Critical to HIV's lifecycle was protease, a viral enzyme. Researchers expected that by blocking this enzyme virus spread could be prevented. Accordingly, protease inhibitors, saquinivir, ritonavir, indinavir, became available and quickly approved by the Food and Drug Administration (FDA). Unfortunately those protease inhibitors and others that followed did not become the "miracle" cures many had placed their hopes in. HIV produces a variety of versions of itself in a host's cell. Protease inhibitors can kill most, but there are always a resistant few. Not unlike the effect of antibiotics on bacteria, the resistant ones continue the cycle of reproduction, and soon the drug, the inhibitor, is no longer effective. Thus far HIV has eluded all attempts to destroy it [29–31]. So, what remains? Studies have demonstrated that condoms are not a 100% deterrent. It is also evident that safe sexual practices can short-circuit HIV's entrance.

Some 1% of all those infected are "slow progressors," who take years to manifest AIDS. Another 1% are "fast progressors" who develop opportunistic infections in months, when the average time between HIV and AIDS is about 10 years. Should a new and highly drug-resistant viral strain begin to spread via sexual activity, the number of fast progressors could multiply sharply. Consequently, abatement of risky sex is again becoming a priority of public health officials.

At the 2004 International AIDS Conference in Bangkok, Uganda's President Yoweri Musaveni explained that his ABC strategy took Uganda from a 30% infection rate to 6% (A = abstinence—delay having sex if young and unmarried; B = be faithful to your partner—zero grazing; C = use a condom properly and consistently if you're going to move around) [32]. But he also noted that condoms have a failure rate, encouraging promiscuity. Conference attendees were turned off by Musaveni's message. Behavior change, self-discipline, and monogamous relationships were not on their agendas.

Although AIDS involves primarily a disruption of the immune system, it can also traumatize the nervous system. While HIV-1 and HIV-2 do not invade nerve cells directly, they do affect their function, causing mental confusion, behavioral changes, migraines, progressive weakness, loss of sensation in arms and legs, and stroke. Additional complications as a consequence of HIV–drug interactions are spinal cord damage, loss of coordination, difficult and painful swallowing, shingles, depression, loss of vision, destruction of brain tissue, and coma. Thus far no single treatment has been able to alter these neurological complications [33].

As of December 2004, an estimated 944,306 individuals had received a diagnosis of AIDS, and of these 529,113 had died: a steep mortality rate of 56% [34]. Furthermore, "since 1994, the annual number of cases among blacks, members of other racial/ethnic minority populations, and those exposed through heterosexual contact has increased" [34], and the number of children reported with AIDS attributed to perinatal HIV transmission peaked at 945 in 1992 and declined 95% to 48 in 2004, primarily because of the identification of HIV-infected pregnant women and the effectiveness of antiretroviral prophylaxis in reducing mother–child transmission of HIV [34].

Of particular importance, 16–22 million people aged 18–64 are tested each year. By 2002, an estimated 38–44% of all US adults had been tested for HIV [34]. Nevertheless, "at the end of 2003, of the approximately 1.0–1.2 million persons estimated to be living with HIV in the United States, an estimated one quarter (250,000–312,000) persons were unable to benefit from clinical care to reduce morbidity and mortality" [34], and "a number of these persons are likely to have transmitted HIV unknowingly." Because treatment has markedly improved survival rates, since the introduction of highly active antiretroviral therapy (HAART), and because progress in motivating earlier diagnosis has been lacking, the National Centers for Disease Control has issued new HIV testing recommendations.

These recommendations, issued in September 2006 for all individuals age 13–64, seek to level the playing field, as previous requirements for written consent and pretest counseling have now been dropped. The federal health officials now see HIV testing as becoming a routine medical/healthcare procedure. HIV testing would be offered by primary care physicians, as well as emergency rooms, substance abuse centers, prisons, and community health centers. Everyone age 13 should be tested at least once, and some sexually active people should be tested annually. According to CDC Director Dr. Julie L. Gerberding, the new recommendations would detect the 250,000 individuals who do not know that they are infected. This would mean saving lives by earlier diagnosis and treatment before the illness advances and becomes more difficult to treat [34]. According to the New York City Health Commisssioner

Dr. Thomas R. Frieden, "The more people who know their status, the fewer infections you're going to get. They're spreading HIV when they wouldn't if they knew" [35].

Which brings us to the future. At the close of the XVI International AIDS Conference (Toronto, Canada, Aug. 13–18, 2006), WHO Acting Director General Anders Nordstrom told the attendees that "This conference has highlighted the importance of an even strongerr focus on women and young people over the world who bear the greatest burden and need particular attention." He concluded by urging the international participants to consider that "we need to invest more in developing new preventive tools, including microbiocides and of course vaccines," but for him, "the most important area to ensure success in achieving universal access, is a skilled and motivated workforce. No improvement in financing or medical products can make a lasting difference to people's lives until the crisis in the health workforce is solved. WHO's "Treat, Train, Retain" plan directly addresses the need for a healthy, skilled, and motivated workforce" [36]. The battle against HIV/AIDS may have begun in earnest—again.

It has become clear that the battle against HIV/AIDS cannot be won by chemical bullets alone, and surely not for years to come. Political correctness has no place in the AIDS equation. Silence is tantamount to death. Ergo, speaking up about this grievous illness that can be readily prevented is long past due. It is time for the country's communications media to take up the issue and challenge of behavior change. If behavior change is the preferred and productive approach for heart disease and cancer, why not HIV/AIDs?

LONGEVITY AND MORTALITY

Life Expectancy

Yet another set of numbers bring a salutary message that can't but elicit delight and satisfaction.

From the National Office of Health Statistics [37], we learn that a person born in 1950 could, on average, be expected to live for 68.2 years. By 1990, life expectancy had climbed to 75.4 years. It is worth recalling that the biblical injunction of "threescore and ten," 70 years, had been attained and passed in 1969. The U.S. Bureau of the Census recently informed us that life expectancy is at an all-time high—77.8 years. Again we want to recall that when FDR signed the Social Security Act of August 14, 1935, few people were expected to make it to 65 when retirement checks would become available. With life expectancy pressing 80, is it any wonder that the country is seeking new ways to ensure that everyone will not only receive their retirement benefits at age 65 but will continue to do so for as long as they live. In 1935, no one would have imagined that most of us would retire in good health and live another 10–30 years. Currently, 12.3% of our population is 65 and older, and that is

TABLE 1.4. Life Expectancy; Gender and Race, United States, 2002

Life expectancy overall	77.3
Female	79.9
Male	74.5
White female	80.3
Black female	75.6
White male	75.1
Black male	68.8

expected to exceed 20% by 2035—when one in every five individuals will be 65 plus. We are indeed experiencing the graying of America. But we are also experiencing great increases in longevity. Between 1950 and 2002, we have gained 9.1 additional years—a stunning 12%. And since 1900, when life expectancy stood at 47 years, the gain has been a bountiful gift of 30+ years—three additional decades! The gains are not universally equal because of gender and racial differences, as we have seen do make a difference, as Table 1.4 shows. Nevertheless, an unprecedented increase in life expectancy has occurred among all segments of our population [13]. However, a note of caution and concern must be injected here. Recently published data indicate that the 77.3 or 78.2 of obesity-related deaths were not the growing problem that they currently are [38]. Obesity deaths, and their prodigious contribution to heart disease, cancer, stroke, and kidney-related deaths has markedly depressed life expectancy. Dr. S. Jay Olshansky, the study's lead author, remarked that the study's projections were "very conservative, and the negative effect is probably greater than we have shown." Obesity shall not go unmentioned. We shall pick it up shortly.

Although we are seeing more gray than ever, the most portentous statistic may just be the proportion of elderly reporting no disabilities. Close to 80% are disability-free, and many are continuing to work—full and part time [39]. Why not? Their experience and judgment serve us well. Cause of elation? You betcha.

Infant Mortality

The National Center for Health Statistics is chock-a-block with good news these days [13]. Having given us upbeat news about longevity and the oldest among us, they come now with lofty data about the youngest. A backward glance at the numbers for 1900 yields the baleful detail that for every 1000 live births, 100 infants died before their first birthday. By 1950, that abysmal statistic had plunged to 28, and the infant mortality rate (IMR) for 2000 was 6.9. What adjective shall we choose to describe this unimaginable reduction? Is "spectacular" overblown?

TABLE 1.5. Infant Mortality Rates per 1000 Live Births, United States, 2002

Race and Gender	Rate
All races, both sexes	6.9
White	5.7
Black	13.5
American Indian	8.3
White female	5.1
Black female	12.1
White male	6.2
Black male	14.8
American Indian male	9.9
American Indian female	6.7

Whenever the subject of the United States IMR is broached, Sweden, Japan, and Norway are trotted out front and center as the class acts of infant survival. True, 6.9 is well above Sweden's 3.0, but 6.9 may be all the more remarkable given the polyglot nature of our country's population. No country in the world has our diversity. Every race, religion, culture, and economic level is represented, and all manner of health/cultural practices arrive with the immigrants. To compare the United States with homogeneous native Swedes or Japanese is to compare apples with onions. Sweden—with a mite over 8 million people, half that of New York State, and 99% white, Lutheran, and highly literate, living in three major population centers, within hailing distance of one another—is both an invidious and ludicrous exercise. Only a glance at Table 1.4 is needed to realize why such comparisons are odious. No other country has our mix of people. These numbers represent a uniquely American experience. No other country, surely neither Japan nor Sweden, has the contrasts evident in Tables 1.4 and 1.5, which must distort the overall IMR. Let us look deeper. Table 1.6 depicts the IMRs for the 10 highest and 10 lowest states. The disparities stand revealed ever more starkly. Clearly, we see a north/south dichotomy. The fact that the District of Columbia, cheek by jowl at the center of political power, has the nation's highest IMR, as well as one of the highest in the Western world, is at once stunning and depressing. Neither Sweden nor Japan has such an enclave. Is it really possible to compare overall rates with such striking national differences? But that is not all. Teasing out additional details provides as with Table 1.7, and yet additional discomfort, as the disparities simply leap off the page. Even among the southern states, the contrasts are awesome. Income levels below the poverty line, high teenage pregnancy rates (accompanied by late or nonexistent prenatal care), and difficult and premature labor with resulting low-weight infants are good and sufficient reasons for the higher rates. Nevertheless, and all the inequalities notwithstanding, and with the stark differences between white and black, we still have

TABLE 1.6. States with the Highest and Lowest Infant Mortallty Rates per 1000 Live Births, 2002

States with the Lowest Rates	
Massachusetts	5.1
Maine	5.3
New Hampshire	5.3
Washington	5.4
Utah	5.4
California	5.5
Oregon	5.5
Minnesota	5.9
Texas	6.0
Iowa	6.2
States with the Highest Rates	
Georgia	8.3
Arkansas	8.4
Tennessee	8.4
North Carolina	9.0
Louisiana	9.1
South Carolina	9.5
Alabama	9.8
Mississippi	10.3
District of Columbia	13.5
Puerto Rico	10.2

TABLE 1.7. IMR's By Gender for Ten Northern and Southern States

State	White	Black
Massachusetts	4.5	9.9
New Jersey	4.9	13.3
Maryland	5.3	13.9
Virginia	5.6	12.5
Wisconsin	5.8	16.7
Iowa	5.8	17.2
Georgia	5.9	13.4
South Carolina	6.3	15.6
Illinois	6.3	17.1
Michigan	6.4	16.4

achieved a single-digit IMR, which must be seen as a triumph of public health. Media take notice.

Furthermore, the precipitous decline from 29.2 in 1950 to the current 6.9 should suggest that "the environment" is not the "ticking bomb" that the spinners have led so many of us to believe it is.

With life expectancy rising to unprecedented levels, and with infant mortality rates falling and substantially decreasing heart disease and cancer rates, is it reasonable to believe that our ambient environment is toxic to children and other growing things? The media have been making much of very little, and not nearly enough of the public's general good health. Why have they not spread the good news of what must be one of the most successful and beneficial accomplishments of the twentieth century—accomplishments that surely blunt the assumption of an environment harmful to our well-being? Overzealous environmentalists have wrought nothing but fear. It's time to repair the damage and realize that we are a healthy people, who will become healthier still as we reduce trauma, and lessen racial and gender disparities. Given the extensive documented data, fear of the environment is unwarranted.

CANCER

It is now altogether fitting and proper that we attend to cancer, which in the hierarchy of mortality is the uncontested occupant of second place. It has been set apart as the very word strikes fear, and for over the past 30 years the so-called war on cancer, initiated by President Richard Nixon, has not been won, and continues unabated. However, new knowledge of the malignant process is beginning to turn the tide of battle. That horizon is coming into view. But let us first consider cancer and its nature.

At the outset, two portentous questions require consideration. Is there a cancer epidemic abroad in the land, as some would have us believe, and, are cancer numbers, cases, and deaths all soaring?

The Chinese scholar who said a good picture is worth 10,000 words, would be pleased with Figures 1.2 and 1.3, which convey literally gobs of information. In seven distinct trendlines, representing major cancer sites, Figure 1.2 conveys the cancer death rates for men over the 72 years 1930–2001. Of the seven major sites, lung cancer makes the most powerful statement. Not only did it rocket upward between 1940 to a peak in 1990, taking many lives with it, but also clearly evident is its decline since 1990. Antismoking campaigns can take well-deserved credit. The stomach cancer trendline tells another wonderful story. If there is a cancer epidemic across the country, stomach cancer surely hasn't contributed, as it has been dropping steadily for 70 years; by 2000 it had the lowest death rates of the seven trendlines. Colorectal cancer, holding steady for 30 years between 1950 and 1980, has also been declining. After a 5-year blip upward, when new screening tests for prostate cancer appeared, it, too, has declined steadily. Hepatic and pancreatic cancers and leukemia have held steady at 5–10 deaths per 100,000 (people) over the past 70 years.

The scenario is much the same for women. From Figure 1.3, we learn that lung cancer is the leading cause of cancer deaths, and still rising. But stomach, uterine, breast, and colorectal cancers have declined sharply, while ovarian and

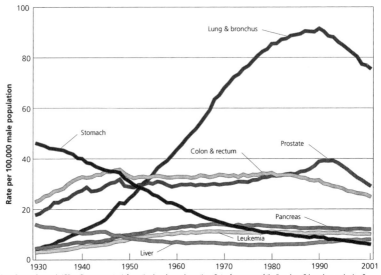

Figure 1.2. Age-adjusted cancer death rates (per 100,000 people, age-adjusted to the 2000 US standard population), males by site, United States, 1930–2001.

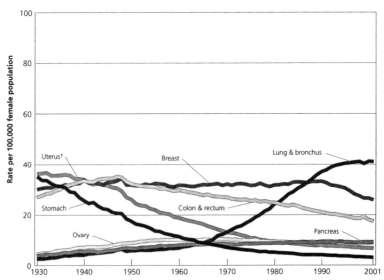

Figure 1.3. Age-adjusted cancer death rates (per 100,000 people, age-adjusted to the 2000 US standard population), females by site, United States, 1930–2001. Uterine cancer death rates are for uterine cervix and uterine corpus combined.

pancreatic cancers have resisted change over 70 years. The answer to the first question seems self-evident. If a cancer epidemic is among us, it is limited to lung cancer in women. We will deal with this shortly. But what is an "epidemic" of any illness or condition? Simply stated, it is a sudden outbreak of an illness above the expected number. Yes, every disease has an expected number of new cases or deaths for each week and month of the year. Should that number be exceeded, it is understood to be of epidemic proportions. Obviously with cancer deaths there have been no sudden increases, and other than lung cancer deaths in women there has been no unusual increase in numbers.

Considering the sweep of time from 1930 to 2001, there appears to be yet another story behind the numbers. Prior to World War II, and well into the 1960s, the United States could be described only as an agriculturally based society. The unprecedented shift to an industrial society, and a giant one at that, was yet to occur. That remarkable shift has occurred over the past 45 years. Yet in these undeniably different environments, most cancer rates have either declined or remained steady. The only soaring cancer rate in sight has been that for lung cancer for both men and women—the result primarily of cigarette smoke.

As for numbers, what we've been experiencing is a statistical artifact—an all-boats-rising phenomenon. Lung cancer is not only the leading cause of cancer deaths; its exceptionally high numbers absolutely skews the rates for all cancer sites combined—an excellent reason for not combining them. This skewing distorts the data and misleads interpretation by falsely implying that cancers of all sites are rising. Can numbers mislead? Indeed, they can. In fact, since 1993, death rates have decreased 1.1% per year—1.5% for men and 0.5% for women—and, perhaps most significantly, from 1950 to 2004, with lung cancer excluded from the total, the combined cancer death rate has dropped by 18%! That's the message the American public should have received, but didn't. That's the message that requires national dissemination—a message that will help dissipate the widespread pall of fear, while bringing a message of hope.

The media totally missed the boat on this. They preferred to trumpet the overall increased rate, rather than explain the distorting effects of lung cancer on the combined rate. Readers, viewers, and listeners are not being served. The media appears to have lost touch with the public. Issues such as this are not of the complexity of the Patriot Act, Social Security reform, or free trade, requiring journalists to have in-depth knowledge of the subject in order to provide the public with comprehensible accounting. By comparison, the facts of life and death, the numbers, are both simple and direct.

Much of the discussion has focused on rates because rates bring unique insights and provide a firm basis for comparing populations, especially populations of diverse sizes. Figure 1.4 shows the estimated number of new cases of cancer for 2004, for each of the 50 states. Glancing east to west, west to east, north to south, or south to north, we see that California with 134,000 new cases is far and away the highest. At the opposite coast is Vermont, with some 3000

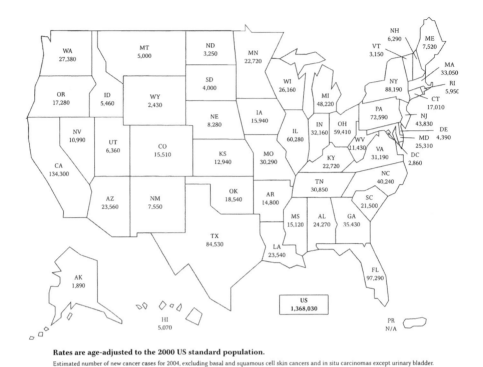

Rates are age-adjusted to the 2000 US standard population.

Estimated number of new cancer cases for 2004, excluding basal and squamous cell skin cancers and in situ carcinomas except urinary bladder.
Note: These estimates are offered as a rough guide and should be interpreted with caution. They are calculated according to the distribution of estimated cancer deaths in 2004 by state. State estimates may not add to US total due to rounding.

Figure 1.4. Cancer deaths by state. (Figure courtesy of the American Cancer Society.)

cases. Should you be looking for a place to drop anchor, Vermont seems a better bet than California. But is it? Table 1.8 compares five states with the highest number of cancer cases with five of the lowest. But now the populations of all states need to be introduced, and the rates per 1000 population calculated. Without rates per thousand, California appears cancer-prone. But Florida, New York, Pennsylvania, and Illinois (see Fig. 1.4) are not that far behind, and suggest avoidance compared to North Dakota, Idaho, and Montana. By considering their populations, and calculating rates per thousand, a much different picture emerges. California, with 134,000 new cases, is in fact the state with the lowest new-case rate, and Vermont, with 45 times fewer new cases, does in fact have a far higher case rate than does California. So, do you still prefer Vermont to California for setting down roots? California, with the nation's largest population, would be expected to have far more cases of anything simply because of its larger numbers. In order to appropriately compare California with 50 times the population of Vermont, calculating rates per 1000 provides a reasonable basis for comparison and interpretation.

Yet another concern about cancer is its predilection for the elderly. Indeed, as Figure 1.5 so clearly represents, cancer death rates soar with advancing age. Although cancer can occur at any age, it is primarily a disease of the elderly.

TABLE 1.8. Estimated Cancer Incidence, United States, 2004

Five Highest States	
California	134,300
Florida	97,290
New York	88,190
Texas	84,530
Illinois	60,280
Five Lowest States	
Alaska	1,890
Wyoming	2,340
Vermont	3,140
South Dakota	4,000
Delaware	4,390

Source: Cancer Facts and Figures, American Cancer Society, 2004.

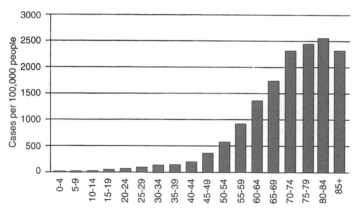

Figure 1.5. Cancer cases by age in the United States. (*Source*: http://seer.cancer. gov.)

As indicated in Figure 1.5, the numbers rise after age 40, and began their steep ascent to the 80s, then decline as the number of available folks over 85 disappear, and cancer along with them. One explanation for the fact that cancer occurs more frequently at the older ages may be that for a tumor to develop, cells must accumulate gene alterations, (mutations), which can occur with each cell division and thus accumulate with advancing age. Before raising the question "Why cancer?" a brief discussion of its nature will buttress our perceptions.

Cancer is a group of diseases. More than 100 types are well documented, each with a distinct character and a different trigger. Ergo, lumping them together gains no understanding, nor does it serve any useful purpose other than gathering numbers. The only commonality among these diseases is that the abnormal cells that they produce have no intention of slowing their runaway division.

Tumors are classified as benign or malignant. Benign tumors are not cancer, and do not spread or metastasize to a new site. They are just lumps. A malignant tumor can and often does enter the bloodstream or lymphatic system to be carried to a site far removed from its original site. Most tumors are named for the organ or cell type in which they began their uncontrolled growth, such as stomach, lung, liver, and breast. Others, such as melanoma, are not as clear. Melanoma is a cancer of melanocytes that produce blue-purple pigments. Melanomas often develop on the skin or in the eyes. Leukemias are cancers of blood cells, and lymphomas are tumors of the lymphatic system.

Around the country, the most common cancers are carcinomas, cancers that develop in the epithelial tissue lining the surfaces of the lung, liver, skin, or breast. Another group of cancers are the sarcomas, which arise in bone, cartilage, fat, connective tissue, and muscle. No tissue or organ has a free pass. Any can become cancerous. And then there is the question "Why?" Why does cancer occur?

We humans have 44 autosomal chromosomes in 22 corresponding pairs. One of each pair is contributed by each parent—which differ in their gene content. In addition to these 22 pairs, normal human cells contain a pair of sex chromosomes. Women carry a pair of X chromosomes, men have an X and a Y, for a total of 23 pairs and 46 chromosomes. A chromosome consists of the body's genetic material, the DNA (deoxyribonucleic acid), along with numbers of other proteins. Within each chromosome, DNA is tightly coiled around these proteins, allowing huge DNA molecules to occupy a tiny space within the cells nucleus. Figure 1.6 shows the tightly coiled DNA strands, which carry the instructions for making proteins. Each chromosome is divided into two segments or "arms"—the short or "p" arm (from the French *petit*, meaning small) and the "q" or long arm. The symbol "q" was chosen simply because it followed "p" in the alphabet and is below the "p" arm The sections are linked at the centromere, the junction where the chromosome attaches during cell division.

Genes are the subunits of DNA. A single chromosome can contain hundreds of protein-encoding genes. Chromosome 16 has 880 genes, including those implicated in breast and prostatic cancers, Crohn's disease, and adult polycystic disease. Chromosome 19, has over 1400 genes, including those that code for cardiovascular disease, insulin-dependent diabetes, and migraines. Cells containing an abnormal number of chromosomes are called *aneuploidic*. It is now evident that cancer cells have either gained or lost entire chromosomes. This loss or gain—this instability, this mutation in chromosome number—can result in cancer. Indeed, the destabilization of a cell's genome is known to initiate cancer. But most cancers are not hereditary, which doesn't end the search for other causes. So, for example, it is also

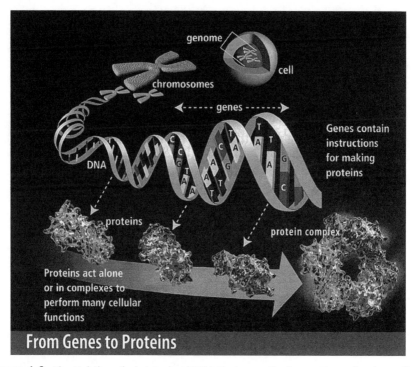

From Genes to Proteins

Figure 1.6. The tightly coiled strands of DNA that carry the instructions allowing cells to make proteins are packaged in chromosomal units. (Figure adapted from *Cancer and the Environment*, National Cancer Institute, publication 03-2039.)

known that alterations in oncogenes, can, as shown in Figure 1.7, signal a cell to divide uncontrollably, rather than repair the DNA or eliminate the injured cell.

One of the cell's main defenses against uncontrolled cell growth is the protein p53. Apparently cancer can occur only when the p53 protein, produced by the p53 gene, is damaged. As p53 may be the key that unlocks the riddle of cancer, we shall consider p53.

According to David Lane [40], director of a cancer research group at the University of Dundee, Scotland, and discoverer of p53 in 1979, p53 may just be "The most important molecule in cancer." He believes, as others now do, that faults in this protein or the processes that it oversees may be the cause of all tumors. Lane also gave the chemical its name: "p" for protein and 53 for its molecular weight of 53,000. It is because of p53's presence and vigilance that cancer is so rare [40]. Who would believe that cancer is rare? In his brief and comely book, *One Renegade Cell*, Robert A. Weinberg, director of MIT's Whitehead Institute, asserts that "One fatal malignancy per hundred million billion cell divisions does not seem so bad at all" [41]. He's not saying that anyone's tumor is okay; rather, he's making the momentous point that with

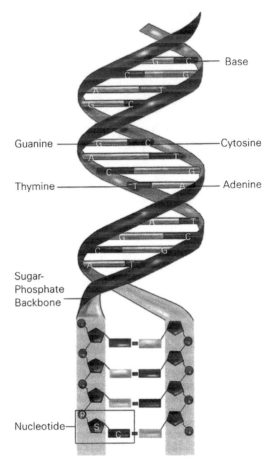

Figure 1.7. DNA—the molecule of life. In double-stranded DNA, the strands are wound about one another in the form of a double helix (spiral) and held together by hydrogen bonds between complementary purine and pyrimidine bases. (Figure adapted from *Genetic Basics*, National Cancer Institute, publication 01-662.)

the body's astronomical number of cells (75–100 trillion) and the ongoing addition of new cells as we live and grow, it is simply remarkable how few cancers actually develop. Given the tremendous number of cells available, one can only gasp and wonder at the incredible fact that we do not get cancer soon after we're born. The stark fact is that youngsters with Li-Fraumeni syndrome, a condition caused by inherited mutations, are prone to develop cancer as young as 2 or 3 years. However, this is an extremely rare condition. It is also known that cancer-associated viruses produce proteins that can shut down p53, leaving cells defenseless.

p53 keeps the process of cell division in check by suppressing cancerous growth. p53 was, and still is, a tumor suppressor gene (TSG). When it was added to cells in culture, those that contained genetic errors made cells cancerous. The normal p53s suppressed cell division. But this protein, which could suppress tumor development, was also the target of cancer-causing viruses and, curiously enough, was found to be mutated in about half of all tumors. It is also odd to find that virologists investigating these unimaginable intracellular events talk of a protein molecule with "godlike properties deciding whether individual cells should live or die." How does this play out? If a cell becomes damaged beyond repair, p53 will force it to self-destruct. Cell suicide or programmed cell death is referred to as *apoptosis* (from the Greek, a "falling off," as leaves from trees) a normal process in which cells perish in a controlled manner. This ability to cause cells to self-destruct is p53's way of protecting us against runaway cell division.

As noted earlier, DNA damage destabilizes genes, promoting mutations. Collections of proteins are constantly traversing genes checking for faulty bases. As shown in Figure 1.6, DNA consists of long, spiral helices—twisted chains—made up of nucleotides. The order of these bases along a strand of DNA is the genome sequence. Each nucleotide contains a single base, one phosphate molecule, and the sugar molecule deoxyribose. The nitrogenous bases in DNA are adenine, thymine, cytosine, and guanine. All instructions in the coded book of life, telling cells what to do, are "written" in an alphabet of just four letters—A, T, C, and G. These bases are strung together in literally billions of ways, which means that billions of coded instructions can be sent to cells. Consider, then, if billions of coded instructions are possible, doesn't this help explain how a single faulty instruction is not only possible but also inevitable? Only a single mutation in the enzyme tyrosinase, an enzyme involved in cat coat color, gives the Siamese cat its dark ears, face, paws, and tail.

So genes do their work by stimulating chemical activity within cells. How? Via proteins, the large complex molecules that require folding into intricate three-dimensional shapes before they can work correctly and provide another possible source of error. (This protein folding ability and requirement will loom large in Chapter 2, during the discussion of several diseases).

These proteins twist and buckle, and only when they settle into their final shape do they become active. Because proteins have many diverse roles, they come in many shapes and sizes. Proteins consist of chains of 20 interlinked amino acids. These chains contain 50–5000 of the 20 amino acids, each with its own amino acid sequence. It is in this sequence that yet additional trouble brews, as an error in just a single amino acid can spell disease. An error, or mutation, can result in an incorrect amino acid at one position in the molecule. So, collections of proteins are searching for faulty bases or breaks in the double helix. If found, they signal p53, which springs into action with an electrifying effect—slamming the brakes on cell division, allowing DNA repair to proceed. As David Lane makes clear, "p53 has earned the title, guardian of the genome." Nevertheless, it can and does malfunction. A variety of triggers can do it. Cigarette smoke and ultraviolet light, among other factors, can damage p53 by twisting the protein out of shape so that it cannot function correctly. Current research seeks to discover ways of blocking the processes that break down p53, or restoring its shape and thereby its function.

An approach taken by a Chinese biotech company was to use gene therapy—adding back normal p53 via injection of viruses primed to reinsert the healthy gene. When combined with radiotherapy, the gene treatment actually eliminated tumors in a number of patients with head and neck tumors, an authentic and epoch-making achievement. Indeed, the creativity of current research is itself mind-boggling. For example, another route of manipulating faulty p53, should its shape be the problem, like humpty-dumpty, it can be brought back together again [42]. Once p53's power source is revealed, there is every reason to believe that cancer will become little more than a chronic illness. The new approaches, based on intimate knowledge of cell mechanisms, will no longer be a one-size-fits-all, shotgun approach, but more akin to a single bullet fired at a specific cellular element. Consequently, I find it quite reasonable to believe that in the fullness of time, 5–7 years down the road, it will have been worked out, incredible as it sounds.

As if this were not sufficiently exciting, recent research at Baylor College of Medicine, in Houston, by Dr. Lawrence A. Donehower and his team, has taken p53 to new heights [43].

In 2002, the Princes of Serendip passed through Houston. As a consequence of a failed experiment, instead of making a protein that Donehower's group wanted, the mice were making tiny fragments of p53. They noticed, too, that the mice were unusually small and were aging prematurely, getting old before their time. As if that weren't startling enough, these mice appeared to be almost cancer-free—highly unusual for mice. As it turned out, the mouse cells contained an unusually high level of p53, which was vigorously suppressing tumors. Dr. Donehower had some 200 mice that were at once innately protected against cancer, but growing old and decrepit well before their time. A reviewer commenting on the Donehower publication in the journal *Nature* said that the condition of the mice "raise[s] the shocking possibility that aging may be a side effect of the natural safeguards that protect us from cancer" [44]. The possibility was suggested that the Baylor mice with extra p53 may be aging prematurely because too many cells are becoming apoptotic and their tissues cannot function properly. These mice do force the issue as to whether human longevity can be increased? In addition to this issue, there is wonderment as to why we can't maintain p53's cancer-fighting potency and also forestall the aging process. A double whammy if ever there was one. So there appears to be a gene that can limit cancer and accelerate aging. Is aging the price to be paid for a cancer-free life?

Can the next development be the outrageous possibility of manipulating p53 to control both cancer and aging? Are we not living in the best of times? In the most exciting time. We need only live long enough to see this all bear fruit. Just down the road, previously inconceivable cancer therapies are being developed. Truly, the tide is running with us. Stay tuned.

TABLE 1.9. Probability (Chance) of Developing Breast Cancer by Specific Ages among US Women

By Age	1 in
15	763,328
20	76,899
30	2,128
45	101
50	53
60	22
70	13
80	9.1
90	7.8

Source: Ries, L. A. G., Eisner, M. P., Kosary, C. L., eds, *SEER Cancer Statistics Review, 1975–2002,* National Cancer Institute, Bethesda, MD, 2005.

Breast cancer in women (men are not immune) is the most frequently diagnosed nonskin cancer. Some 216,000 new cases were estimated to have occurred in 2004. The risks of being diagnosed with breast cancer increases with age, and the risk increases steadily by decade as shown in Table 1.9. Unfortunately the media also got that one wrong. Recent headlines across the country trumpeted the news: "Cancer now the top killer of Americans" and "Cancer passes heart disease as top killer." The implication is that the war on cancer was lost. What the media so glaringly failed to acknowledge, or failed to understand, was that in their most recent annual report (2005), but whose data were limited to those of 2002, the authors extracted deaths by age, which they had never done before [45]. In doing so, they found that although death rates from all cancer sites combined have been falling steadily since 1993 (by 1.1% per year), the rate of death from heart disease, as shown in Figure 1.8, has been declining since the mid-1970s. Nevertheless, in 1999, for those people under age 85, who constitute 85% of the country's population, cancer deaths surpassed heart disease only because heart disease continued its unflagging descent [45]. As for breast cancer (and here the confusion mounts), another severely abused number is the often cited statistic that over a women's lifetime, the risk (the chance, the odds) of her getting breast cancer, on average, is one in eight, or about 13%. Far too many believe that this number is a woman's current risk. No. The risk involved is in fact a woman's lifetime risk, at age 85, and it works this way. If eight women are followed for their entire lives, one of them, on average, is likely to develop breast cancer. Also recall that with a 1 in 8 chance of developing breast cancer, there remain 7 in 8 chances that it will not occur. Again, as we've seen, cancer is a disease of advancing age, and breast cancer is strongly age-related, as Table 1.7 shows. At age 35, as noted in the table, it is 1 in 99, and at age 45 it is 1 in 101, or a 1% chance of developing breast cancer. Perhaps

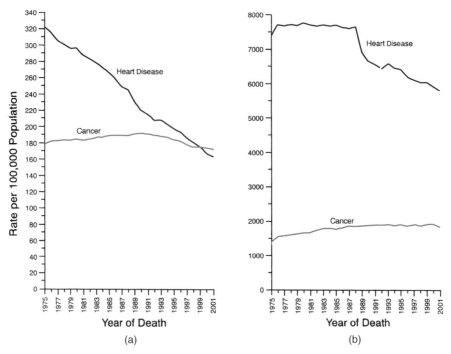

Figure 1.8. Cancer and heart disease death rates (age-adjusted to 2000 US standard population) for individuals younger than (a) and older than (b) age 85. (Figure adapted from American Cancer Society, *CA: A Cancer Journal for Clinicians*.)

more importantly, it is essential to recall that not all women live on to the older ages when breast cancer risk becomes greatest [46].

Much has been made of the fact there are inherited breast cancer suscepti-bility genes—BRCA1 and BRCA2. But these are responsible for no more than 1 in 10 cases of the disease. Yes, 9 out of 10 cases are not inherited. Of addi-tional importance is yet another number: 0.2% the number of women in the United States whose BRCA genes have mutated. These numbers offer a good deal more than cold comfort.

Furthermore, breast cancer activists have consistently flailed their physical environment as the carcinogenic trigger(s) for breast cancer. One of the most politically active areas has been Long Island, New York, where, as in other areas of the country, breast cancer is commonly reported. In 1993, concerned residents got their Congressional representative to push for legislation requiring epidemiologists to investigate a possible environmental carcinogen/ breast cancer link. After a decade of study, the Long Island Breast Cancer Study Project (LIBCSP) began publishing its findings. Among the possible carcinogens under their purview were the polycyclic aromatic hydrocarbons (PAHs). Although the PAHs are potent mammary carcinogens in rodents, their effect on development of human female breast cancer has been equivocal. The LIBCSP wanted to determine whether currently measurable PAH damage to DNA increases breast cancer risk. PAHs are byproducts of the combustion of fossil fuels, cigarette smoke, and grilling of foods and are found in smoked foods. As PAHs can be stored in fatty breast tissue, they were deemed a realistic candidate. The study did not find a relationship between PAH blood levels and exposure to smoked or grilled foods or cigarette smoke, and "no trend in risk was observed" [47]. In addition to PAH, the project studied the relationship between breast cancer and organochlorine pesticide blood levels [48]. Again, no dose–response relationship was uncovered. Nor could they find any support for the hypothesis that organochlorines increase breast cancer risk among the Long Island women.

In another venue, researchers at Maastricht University, in the Netherlands examined the relationship between stressful life events and breast cancer risk [49]. They reported no support for stressful life events and risk of breast cancer.

Although we are most assuredly in an age of breast cancer awareness and breast cancer studies, thus far environmentally related breast cancer carcinogens remain to be discovered. The question at issue is whether heightened awareness and fear are desirable motivators for increasing screening behavior. Clearly the issue is debatable. But overemphasis on breast cancer may well be responsible for inattention to other illnesses. In fact, both heart disease and lung cancer carry greater risks and are greater killers of women than is breast cancer. Shocking though it may be, women worried about breast cancer continue to smoke. According to Dr. Barbara Rimer, Director of the Division of Cancer Control and Population Science at the National Cancer Institute, " We see smokers who are very, very worried about breast cancer, and yet they're continuing to smoke. They have a much better chance of getting and dying of lung cancer than breast cancer, but many women underestimate their chances of getting lung cancer" [50].

Lung cancer is the world's number 1 cancer killer. In the United States, close to 100,000 men and women died of it in 2005. Cigarette smoke is the primary risk. However, another glance at Figures 1.2 and 1.3 shows that men have heeded the antismoking message and their declines in lung cancer deaths are striking whereas women have yet to respond to the messages. Despite the many warnings about the malign affects of smoke, fully 25% continue to do so. Women, especially young women, are the preferred target of cigarette advertisements. And they respond. As many as 20% smoke during their pregnancies. Are they really unaware of the deleterious effects of smoke on the developing fetus? Activists ought to zero in on this curious behavior.

Women and smoking, another cautionary tale that went by the boards, is being given short shrift by the media. However, several of Dr. David Satcher's numbers are devastating. To wit:

- An estimated 27,000+ more women died of lung cancer than breast cancer in 2000.

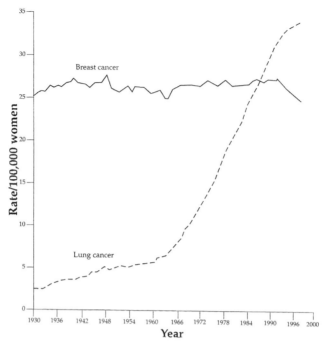

<u>Figure 1.9.</u> Age-adjusted death rates for lung cancer and breast cancer among women, United States, 1930–1997.

- Three million women have died prematurely because of smoking since 1980, and on average, these women died 14 years prematurely.
- For a never-to-be forgotten comparison the US Surgeon General has given us Figure 1.9, for which discussion may even be unnecessary [8].

It has been proposed that there is a higher rate of a specific mutation in the p53 gene in women's lung tumors compared to men. Perhaps. It has also been postulated that women may have a reduced capacity for DNA repair. There is, of course, much yet to be learned. Nevertheless, being female appears to be a factor for extended survival in lung cancer patients. What is not moot is that consequential differences do exist between men and women with lung cancer. Women who have never smoked are more likely to develop lung cancer than are men who have never smoked [51]. The "why" of this and other differences has researchers around the world scurrying for answers. That a number will be found in genes specific to men and in genes specific to women is emerging as a sure bet.

Even though colorectal cancer deaths have been declining over the past 50 years, over 150,000 deaths were expected to occur in 2004. Here again, the primary risk factor is age. Other proposed risks include smoking, alcohol consumption, obesity, and diets high in fat and/or red meats. On the other hand, frequent coffee consumption has been associated with reduced risk of colorectal cancer. Bear that word association in mind. We shall consider this possibility in some depth further along, as it can be easily misinterpreted.

Recently, researchers at Harvard University's School of Public Health probed the relationship between coffee, tea, and caffeine consumption and the incidence (new cases) of colorectal cancer [52]. Using data from the well-established Nurses' Health Study and the Health Professionals' follow-up study (physicians), which together provided 2 million person-years of follow-up and 1438 cases of colorectal cancer. They found that "regular consumption of caffeinated coffee or tea or total caffeine intake was not associated with a reduced incidence of colon and rectal tumors." But they did find that decaffeinated coffee did appear to reduce the incidence of these cancers, but also injected the caveat that this association requires confirmation by other studies. It's a start. Advertising by the tea and coffee producers, especially tea (particularly green tea), would have us believe that these are health-promoting beverages. Would that this were true. We shall see.

Cancer Disparities

In 1991, Dr. Samuel Broder, then Director of the National Cancer Institutes, remarked that "Poverty was a carcinogen" [53]. This suggested an interaction between economic and sociocultural factors that could influence human health. It was his contention that poor and medically underserved communities are at higher risk of developing cancer and have less chance of early diagnosis, treatment, and survival. In 2003 the Institute of Medicine (IOM) published a review describing the disparities that can arise from the interplay between economic and sociocultural factors. For the IOM, poverty was the critical factor affecting health and longevity [54].

As we have seen, African Americans have the highest death rate from all cancer sites combined, as well as cancers of the lung, colon, prostate, female breast, and uterine cervix. For all cancer sites combined, male mortality in 1999 was 13% higher in poorer compared to more affluent counties. Similarly, in the poorer counties there was a 22% higher death rate from prostate cancer.

The prevalence of underlying risk factors for some cancers differs among racial and ethnic groups. The higher rates of stomach cancer among Hispanics (Latinos) and Asian Americans reflects in part the higher rates of *Helicobacter pylori* infections in recent immigrants. Similarly, higher rates of liver cancer are found among Hispanics and Asian Americans, who have a higher prevalence of chronic hepatitis infections [55].

Ethnic differences clearly shows itself among eastern European Jewish families who have an almost exclusive susceptibility to Tay–Sachs disease as well as an inordinately high risk of Gaucher's disease, both of which are the product of mutated genes.

The gap we have seen in black life expectancy compared to that of whites is now believed to be due to the higher rates of heart disease, stroke, kidney disease, and hypertension, the consequence of a genetic predisposition to salt sensitivity—a racial characteristic. Individuals with a higher capacity for salt retention may also retain more water and would tend to be hypertensive (having abnormally high pressure exerted on artery walls), which favors heart disease, stroke, and kidney dysfunction. So, in addition to economic and socio-cultural disparities, racial and ethnic differences are at play in cancer and other morbid conditions [5, 6, 56].

How are these disparities to be dealt with? Can they be dealt with? In principle, equal application of existing knowledge about cancer prevention, early detection, and treatment for all segments of the population should sub-stantially reduce these disparities. However, this will require substantial revi-sions in our healthcare delivery system, which is not known for flexibility. On the other hand, the growing knowledge and acceptance of the idea of racial differences may be a more efficacious stimulus for change, and achieve greater benefits.

OBESITY

"Persons who are naturally fat are apt to die earlier than those who are slender." Hippocrates (circa 460 BCE) was not a man of few words. His many comments have stood the test of time. This quotation is hoary with age, having been written 2500 years ago, and should remind us that fatness is not a new medical concern. What is new is the realization that obesity is a worldwide phenomenon and the consequence of genetic susceptibility, too readily avail-able high-energy foods, and greatly decreased physical activity: a morbid triad [57].

Obesity, unlike AIDS, not only is not on our list of leading causes of death; it is not even in the vicinity of the list. Obesity requires our attention and concern because of its deadly contribution to heart disease, at the top of the charts; to cancer, our second leading cause; to diabetes, the seventh; to hyper-tension, the fifteenth; to sleep-breathing disorders; and osteoarthritis of large and small joints, and we know, as did Hippocrates, that obesity is inversely related to longevity [57].

Obesity can no longer be regarded as a cosmetic problem, but must be seen as a new pandemic that threatens worldwide well-being. What is obesity? For an answer, dictionaries are to no avail as they speak only of excess weight. Obesity goes beyond excess weight, which raises a second question: How fat is too fat? For Peter Paul Rubens (1577–1640), the great Flemish painter, there was no "too fat." Rubens was the master of rotund femininity. As shown in Figure 1.10, the fatter, the healthier, the more beautiful. But that was then. Today, obesity is our number 1 malnutrition problem, and a major contributor to numerous deaths. It has replaced under nutrition and infectious disease as the most significant contribution to poor health [58].

For adults, *overweight* is defined in terms of body mass index (BMI) and calculated as weight in kilograms [2.2 lb (pounds)], divided by the square of

Figure 1.10. *Bacchus*, by Peter Paul Rubens, 1577–1640.

height in meters, is 25 (55 lb over the ideal weight), and obesity entails a BMI of 30, while extreme obesity is BMI 40 or higher. (To calculate your BMI, multiply your weight in pounds by 700, then divide by your height in inches, and repeat that a second time.) Using these numbers, the prevalence of obesity among adults in the United States is understood to be approximately 30.5% of the total population. For children 2–5 years old, it is approximately 10%, and for those 12–19, it is approximately 22% [59]. Paradoxically, these numbers have markedly increased over the past 30 years, during a time of unimaginable preoccupation with diet(s) and weight control. We Americans spent $46 billion on weight loss products and services in 2004. Unfortunately it is now seen that dieting is either ineffective or counterproductive. Those overweight or obese children must not be given short shrift—not taken lightly. The consequences can be enormous. As noted earlier, type 2 diabetes, closely linked to excess weight, is being diagnosed in such high numbers that it can no longer be referred to as "adult-onset diabetes." But that is not the worst of it. In the recent eye-opening report on obesity, Dr. David Ludwig, Director of the Obesity Program at Children's Hospital, Boston, revealed a threat thus far unmentioned. He warned that the current obesity epidemic has had little public impact, "but when these youngsters start developing heart attacks, stroke, kidney failure, amputations, blindness and ultimately death at younger and younger ages, that will have a huge effect on life expectancy." This is not something we want to look forward to. Obesity appears to be the result of multiple causes: genetic, environmental, and psychosocial factors acting synergistically with energy intake and expenditure. Obesity is consistently found in single-gene disorders such as Prader–Willi syndrome (PWS), with its upper body obesity (due to uncontrolled appetite), short stature, mental retardation, hypotonia, and hypogonadism. As for an environmental component, "predictions about possible interactions between genes and the environment are difficult because there may be a delay in an individual's exposure to an 'obesogenic' environment, and/or alteration in life style related to living circumstances and uncertainty about the precise timing of the onset of weight gain" [58]. Not so uncertain is the energy intake/expenditure component. Pima Indians of the American Southwest, with a common genetic heritage to Pimas in Mexico, are an average 50 lb or more heavier than those in Mexico. A similar trend is seen with Nigerians living in the United States, who are obese compared to Nigerians in Africa; the former are also twice as likely to exhibit hypertension [58]. Migrants coming into a new culture pick up the habits of the majority culture and soon reflect their medical problems.

As noted earlier, obesity is a significant public health problem given its substantial contribution to morbidity and mortality, but the health risk could be significantly reduced even with modest weight loss. The peptide hormone leptin, which appears to hold the key to weight loss or gain, is produced by adipose tissue, and a decrease in body fat decreases the amount leptin, which triggers food intake; the reverse is also true. More leptin, less food intake. Clearly, leptin and the brain are in this together. When the system works properly, there is maintenance of weight within a narrow range [60]. This raises yet another question. Why are some of us obese and others not? Although not yet fully crystalized, it appears that obese individuals are leptin-resistant. How to modulate this is a high-priority research activity. Furthermore, clarification of the mechanisms and pathways that control food intake and energy homeostasis are of central and crucial importance, and its neurological and hormonal complexity do not suggest a short timeline. However, the enormous cost to human health attributable to obesity is the engine that will drive basic research, leading ultimately to successful medical treatment, and that includes genetic repair, if need be.

THE ENVIRONMENT? WHAT ENVIRONMENT?

We have traveled far and widely, considering and exploring the greatest threats to our well-being. We have seen, too, that we are healthier than ever before. Nevertheless, the threats are there and will remain, but in ever decreasing numbers if we seriously attend to them. To do so, to take appropriate preventive measures, we need to return to the question posed at the outset: Which, if any, of the adverse conditions that threaten our lives can be attributed to the ambient environment, and what do we mean by "environment"? As we journey into the twenty-first century, is there agreement on the meaning of this word? This is not an idle question, but has profound meaning for our well-being. Consequently it is necessary that it be widely understood.

As has been indicated, the trick of life is to maintain a balance between normal cell division and runaway, uncontrolled growth. For oncogenes to take over, to be turned on, a switch is needed. For the past 30 years, and for far too many people, that switch was the environment.

Misinterpretation and misrepresentation, whether accidental or purposeful, lead directly to misinformation and misunderstanding. This quartet has given "environment" a bad rap, as in "environmental pollution," "tainted environment," "contaminated environment," and "environmental risk." Unrelenting misrepresentation over the past 30 years of "environmental risk factors" has made many of us fear the world. Air, water, food, and soil are seen as polluted and responsible for whatever ails us and as our causes of death.

A 30-year stranglehold on American minds presents some difficulties for extirpating this "cancer" on the body politic. But eliminate it we must, if prevention is to work. Misinterpreting what environmental risks are has a long tradition. As far back as 1964, the World Health Organization (WHO) issued its report declaring that the common fatal cancers occur in large part as a result of lifestyle and are preventable. Here are its words [61]:

> The potential scope of cancer prevention is limited by the proportion of human cancers in which extrinsic factors are responsible. These factors include all environmental carcinogens, whether identified or not, as well as modifying factors that favor neoplasia of apparently intrinis origin (e.g., hormonal imbalance, dietary deficiencies, and metabolic defects). The categories of cancer that are influenced by extrinsic factors including many tumors of the skin and mouth, the respiratory, gastro-intestinal, and urinary tracts, hormone-dependent organs (such as the breast, thyroid, and uterus), haematopoietic and lymphopoietic systems, all of which, collectively, account for more than three-quarters of human cancers. It would seem, therefore, that the majority of human cancer is potentially preventable.

From a cursory reading it is evident that the misinterpretation occurred in the United States, where "extrinsic factors" was deleted and "environmental factors" was substituted. And if that wasn't slippage enough, "environmental factors" was translated once again, becoming "man-made [anthropogenic; synthetic] chemicals," which was never WHO's intent. Extrinsic factors are synonymous with lifestyle, our behavior, or what we choose or don't choose to do. Because many people prefer blaming everyone but themselves, it is understandable that few complained of the transformation of the English language as it moved from Europe to the United States.

At a conference in Canada in 1969, John Higgenson, founding director of the International Agency for Research on Cancer, a WHO affiliate, stated that 60–90% of all cancers were environmentally induced. That remark was to haunt him and the world for decades. He had no inkling that his use of "the environment" would be so bent out of shape. The floodgates opened wide. Soaring cancer rates could hereafter be attributed to a polluted environment.

In 1979, Higgenson was interviewed by an editor of the journal *Science* to further clarify his 60–90% attribution, and to deal with the seemingly intractable fact that so many Americans "believe that cancer-causing agents lurk in everything we eat, drink, and breathe." That such a perception is wrong is evident from Higgenson's responses. He began by noting, "A lot of confusion has arisen in later days because most people have not gone back to the early literature, but have used the word environment purely to mean chemicals." Further along in the interview, he declared that "Environment thus became identified only with industrial chemicals." Then he said, "There's one other thing I should say that has led to the association of the term environment with chemical carcinogens. The ecological movement, I suspect, found the extreme view convenient because of the fear of cancer. If they could possibly make people believe that pollution was going to result in cancer, this would enable them to facilitate the cleanup of water, of the air, or whatever it was"—a remark not calculated to win friends or attract converts. "I think," he continued, "that many people had a gut feeling that pollution ought to cause cancer. They found it hard to accept that general air pollution, smoking factory chimneys, and the like are not the major causes of cancer" [62]. For all the good it did, that interview might well have never occurred. Dynamic denial, on the hand, and the power of the media to shape opinion prevailed, on the other hand, and this false thesis persists. The media and environmentalists are determined to hold their ill-gotten ground, no matter how wrong the association. But the facts will emerge!

In their now classic publication, "The causes of cancer: Quantitative estimates of avoidable risks of cancer in the U.S. today" [63] ("today" being 1981), Doll and Peto placed numbers and percentages on 12 categories of potential risk factors. Their list, shown in Table 1.10, is worth contemplating.

For Doll and Peto, tobacco and diet were so intimately tied to cancer deaths that their estimates of their importance, their contribution to the disease, ranged from 55% to 100%. The uncertainty factor was apparent, but for them this dynamic duo were cancer risks. At the opposite end of the risk spectrum was pollution, to which they assigned a value of less than 1. Recalling that these estimates were made at the beginning of the 1980s, it is reasonable to ask whether they have withstood the test of time.

A research team of Harvard University's School of Public Health took up the challenge, and in 1996 produced its own estimates (Table 1.11). This list has a familiar look.

TABLE 1.10. Proportions of Cancer Death Attributed to Different Risk Factors, 1981

	Percent of All Cancer Deaths	
	Best Estimate	Range of Acceptable Estimate
Tobacco	30	25–40
Alcohol	3	2–4
Diet	35	10–70
Food additives	<1	<0.5–2
Reproductive and sexual behaviors	7	1–13
Occupation	4	2–8
Pollution	2	<1–5
Medicines and medical products	1	0.5–3
Industrial products	>1	<1–2
Infections	10?	1–?

Source: Ref. 63.

TABLE 1.11. Proportions of Cancer Deaths Attributed to Different Risk Factors, 1996

Cancer Risk Factor	Percent Contribution
Tobacco	30
Diet	30
Hardcore[a]	25
Alcohol	3
Microbial (viral, bacterial)	1–2
Pollution	1–2

[a] "Hardcore" are those cancers that would develop even in a world free of external influences simply because of the production of carcinogens within the body, and the occurrence of unrepaired genetic mistakes.
Source: Ref. 64.

The Harvard list echoes Doll and Peto. Tobacco, diet, infectious agents, and sexual behavior are the primary culprits, while pollution, food additives, and ionizing radiation contribute little if anything to cancer risk or death [64]. Any contribution by the ambient environment must be too small to be measured, and thus is of little or no consequence to our health. They also show that the public has overestimated the risk posed by low levels of radiation, obviously encouraged by the constancy of the media and environmentalist mantra. An objective observer could be forgiven her or his lack of comprehension, wondering out loud how it is possible that misunderstanding of "environment" and its risks has become so entrenched.

Yet another Harvard group, this one from the Department of Medicine, and the School of Public Health, has taken up the cudgel. The researchers introduced their recent study on environmental risk factors and female breast cancer with this caveat [65]:

> It is unfortunate that there is confusion as to what constitutes an environmental exposure. Epidemiologists often label as "environment" any risk factor that is not genetic, including diet, body size, exogenous estrogen use, reproductive factors, and medical treatment. Using this definition, most breast cancer is thought to be due to "environment," as only a small proportion is due to inherited mutations in breast cancer susceptibility genes. The general public, however, often interprets this as evidence that much of breast cancer is due to "environmental" pollution. In this review we restrict the definition of environmental exposures to those which a person experinces passively, due to pollution or other charactreristics of the outside world.

Their study, reported in 1998, concerned the possible risk of breast cancer from exposure to ambient environmental chlorinated hydrocarbons (pesticides), ionizing and electromagnetic radiation, and passive cigarette smoke. And their findings? "Based on current evidence, with the exception of ionizing radiation, no environment exposures can be confidently labeled as a cause of breast cancer." The echoes grow louder. But where are the media? Shouldn't women have gotten this information? Shouldn't everyone know this? This predated the Long Island Breast Cancer Project's reports, which obtained similar results, and has received the same silent treatment.

The leading causes of death have been stable for at least 25 years; heart disease, cancer, and stroke have occupied the top three positions, and suicide, homicide, cirrhosis, diabetes, and accidents switch a position or two every so often. Even Inspector Clouseau would look askance on the ambient environment as the source of these conditions. The assault on the environment, which in fact was an assault on us all, was entirely misplaced and unjustified. Neither evidence nor proof supported such a claim, but there was ever-mounting evidence for the lifestyle and behavior paradigm. What will it take to convince and unshackle the American mind?

If better health for all were in fact the nation's goal, the first priority would be modification of our self-destructive behavior. The environment, as commonly understood, does require vigilance, but for reasons other than human health. We have been flailing at windmills that pose minuscule risk and consume our energy, time, and taxes, whereas the major risks, the real killers, languish for lack of individual and institutional concern and support. If we clasped the lifestyle model to our breasts, our country could follow a path to wholesale reductions in illness and death that no manner of medical intervention could ever hope to match. If we are ready to strike out on the veritable road to personal well-being, it is essential to deal with the enemy within.

REFERENCES

1. *Healthy People. Surgeon General's Report on Health Promotion and Disease Prevention*, U.S. Dept. Health and Human Services. No. 79–55011. U.S. Govt. Printing Office, Washington, DC, 1979.

2. *Future of the Public Health in the 21st Century*, Institute of Medicine, National Academy Press, Washington, DC, 1979.

3. Khot, U. M., Khot, M. B., Bajzer, C. T. et al., Prevalence of conventional risk factors in patients with coronary heart disease, *JAMA (Journal of the American Medical Association)* **190**(7):898–904 (2003).

4. Green, P., Knoll, M. D., Stamler, T. et al., Major risk factors as antecedents of fatal and non-fatal coronary heart disease events, *JAMA* **290**(7):891–897 (2003).

5. Collins, F. S., What do we know and don't know about "race," ethnicity, genetics and health at the dawn of the genome era, *Nature Genet. Suppl.* **36**(11):513–515 (2004).

6. Tate, S. K., and Goldstein, D. B., Will tomorrow's medicines work for everyone, *Nature Genet. Suppl.* **36**:S34–S40 (2004).

7. Leroi, A. M., A family tree in every gene, *New York Times* A21 (March 14, 2005).

8. Satcher, D., *Women and Smoking: A Report of the Surgeon General*, U.S. Dept. Health and Human Services, PHS, Rockville, MD, 2001.

9. Anderson, R. N., and Smith, B. L., *Deaths: Leading Causes for 2002*, National Vital Statistics Reports (NVSS) **53**(17), March 17, 2005, CDC, Atlanta, GA.

10. *National Diabetes Fact Sheet*, CDC, National Center for Chronic Disease Prevention and Health Promotion, 2003 (available at `http://www.cdc.gov/diabetes/pubs/general.htm#what`).

11. Warner, M., Guidelines are urged in food ads for children, *New York Times* C7 (March 17, 2005).

12. Wickelgren, I., Policing the immune system, *Science* **306**:596–599 (2004).

13. Hoyert, D. L., Heron, M. P., Murphy, S. L., and Kung, H. C., *Deaths: Final Data for 2003*, National Vital Statistics Reports (NVSS) **54**(13), April 19, 2006, CDC, Atlanta, GA.

14. Simonsen, L., Riechert, T. A., Viboud, C. et al., Impact of influenza vaccine on seasonal mortality in the U.S. elderly population, *Arch. Intern. Med.* **165**:265–272 (2005).

15. Bren, L., Alzheimer's searching for a cure, *FDA Consumer* **37**(4):18–24 (2003).

16. *Alzheimer's Disease*, NCDC, NCH Statistics (`www.cdc.gov/mchs`).

17. *Alois Alzheimer: Who Named It* (`www.whonamedit.com/docter.cfm/177.html`).

18. CBS Worldwide News, *100 Years of Traffic Fatalities* (`www.cbsnews.com/stories/1999/09/13/national/main62061.shtml`); see also Trends in motorcycle fatalities associated with alcohol-impaired driving, United States, 1983–2003, *JAMA* **293**(3):287–288 (2005); also reports from the Centers for Disease Control and Prevention, *Morbidity/Mortality Weekly Report* **53**:1103–1106 (2004).

19. Subramanian, R., *Motor Vehicle Traffic Crashes as a Leading Cause of Death in the United States*, 2002, Traffic Safety Facts, National Highway Transportation Safety Administration, Washington, DC, Jan. 2005.

20. McMillan, A. C., *The State of the Nation in Safety*, National Safety Council speech to the National Press Club, Washington, DC, June 10, 2003.

21. Nowrot, M., Nordenstrom, B., and Olson, A., Disruption of eye movements by ethanol intoxication affects perception of depth from motion parallax, *Psychol. Sci.* **15**:858–865 (2005).

22. Jernigan, D. H., Ostroff, J., Ross, C. et al., Sex differences in adolescent exposure to alcohol advertising in magazines, *Arch. Pediat. Adolesc. Med.* **158**:629–634 (2004).

23. Univ. North Carolina Highway Safety Research Center (`www.hsrc.unc.edu/news_roo/2004_deer_crash.cfm`).

24. American Association of Suicidology, *Fact Sheets*, AAS. Washington, DC, 2004 (`www.suicidology.org`).

25. Scutchfield, F. D., and Genovese, E. N., *Terrible Death of Socrates: Some Medical and Classical Reflections*, The Pharos, 1997, pp. 30–32; see also Stone, I. F., *The Trial of Socrates*, Little Brown, Boston, 1988.

26. Conium maculatum, *The Merck Index*, 10th ed., Merck, Raway, NJ, 1983, item 2473, p. 357.

27. Gaillard, Y., Krishnamoorthy, A., and Bevlot, F., Cerbera odollam: A suicide tree and cause of death in the state of Kerala, India, *J. Ethnopharm.* **95**:123–126 (2004).

28. Heeney, J. L., Dalgleish, A. G., and Weiss, R. A., Origins of HIV and the evolution of resistance to AIDS, *Science* **313**:462–466 (2006).

29. Barry, M., Mulcahy, F., and Black, D. J., Antiretroviral therapy for patients with HIV disease, *Br. J. Clin. Pharmacol.* **115**:221–228 (1998).

30. Wynn, G. H., Zapor, M. J., Smith, B. H. et al., Antiretrovirals. Part I: Overview, history and focus on protease inhibitors, *Psychosomatics* **45**:262–270 (2004).

31. Casiday, R., and Frey, R., *Drug Strategies to Target HIV: Enzyme Kinetics and Enzyme Inhibitors*, Dept. Chemistry, Washington Univ., St. Louis, MO, 2001 (`www.chemistry.wustl.edu/edudev/labTutorials/HIV/drugstrategies.html`) (an excellent overview of key concepts with explanatory graphics that portray the chemical interactions involved with a range of immune cell complexes; highly recommended).

32. AIDS agonistes: Editorial. *The Wall Street Journal* (July 14, 2004).

33. *Neurological Complications of AIDS*, National Institute of Neurological Disorders and Stroke (NINDS), NINDS Neurological Complications of AIDS Information Page (`www.minds.mih.gov/disorders/aids/_aids_pr.htm`).

34. Branson, B. M., Handsfield, H. H., Lanipe, M. H. et al., Revised recommendations for HIV testing of adults, adolescents, and pregnant women in health care settings, *Morbidity/Mortality Weekly Report* **55** (RR-14) (Sept. 22, 2006), CDC, Atlanta, GA.

35. Perez-Pena, R., New federal policy on H.I.V. testing poses unique local challenge, *New York Times* B1, B5 (Oct. 2, 2006).

36. Remarks by Anders Nordstrom, Acting Director General, WHO. XVI International AIDS Conference, Toronto, Aug. 18, 2006 (www.who.int/hiv/toronto2006/en/index.html).

37. Arias, E., *United States Life Tables 2003*, National Vital Statistics Reports (NVSS) **54**(14), April 19, 2006, CDC, Atlanta, GA.

38. Olshansky, S. J., Passaro, D. J., Hershow, R. C. et al., A potential decline in life expectancy in the U.S. in the 21st century, *NEJM (New England Journal of Medicine)* **352**(11):1138–1145 (2005).

39. Spillman, B. C., *Changes in Elderly Disability Rates and the Implications for the Health Care Utilization and Cost*, Office of Disability, Aging and Long-Term Care Policy, U.S. Dept. Health and Human Services, Feb. 3, 2003 (http://aspe.hbs.gov/dal-cp/reports/hcutlcst.htm#chapiii).

40. Lane, D. P., and Crawford, L. V., "T" antigen is bound to a host protein in SV40-transformed cells, *Nature* **278**:261–263 (1979).

41. Weinberg, R. A., *One Renegade Cell*, Basic Books, New York, 1999.

42. Lane, D. P., Dark angel, *New Scientist*, 38–41 (Dec. 18, 2004).

43. Tyner, S. D., Venkatachalam, S., Choi, J. et al., P53. Mutant mice that display early aging associated phenotypis, *Nature* **415**:45–51 (2002).

44. Ferbeyve, G., and Lowe, S. W., The price of tumour supression? *Nature* **415**:26–27 (2002).

45. Twombly, R., Cancer surpasses heart disease as leading cause of death for all but the elderly, *J. Natl. Cancer. Inst.* **97**(5):330–331 (2005).

46. Nelson, J. N., Demystifying statistics. Experts discuss common misunderstandings, *J. Natl. Cancer Inst.* **93**(23):1768–1770 (2001).

47. Gammon, M. D., Santella, R. M., Newgut, A. I. et al., Environmental toxins and breast cancer on Long Island. I. Polycyclic aromatic hydrocarbon DNA adducts, *Cancer Epidemiol. Biomark. Prevent.* **11**:677–685 (2002).

48. Gammon, M. D., Wolfe, M. S., Newgut, A. I. et al., Environmental toxins and breast cancer on Long Island. II. Organochlorine compound levels in blood, *Cancer Epidemiol. Biomark. Prevent.* **11**:686–697 (2002).

49. Duijts, S. F. A., Zeeyars, M. P. A., and Berne, B. V. D., The association between stressful life events and breast cancer risk: A meta analysis, *Int. J. Cancer* **107**:1023–1029 (2003).

50. Gottlieb, N., The age of breast cancer awareness: What is the effect of media coverage, *J. Natl. Cancer Inst.* **93**(20):1520–1522 (2001).

51. Patel, J. D., Bach, P. B., and Kris, M. G., Lung cancer in women: A contemporary epidemic, *JAMA* **291**(14):1763–1768 (2004).

52. Michels, K. B., Willet, W. C., Fuchs, C. S. et al., Coffee, tea and caffeine consumption and incidence of colon and rectal cancer, *J. Natl. Cancer Inst.* **97**(4):282–292 (2005).

53. Broder, S., *Cancer Disparities: Cancer Facts and Figures*, American Cancer Society, Atlanta, GA, 2004, p. 23.

54. Institute of Medicine, *Unequal Treatment: Confronting Racial and Ethnic Disparities in Health Care*, Natl. Academy Press, Washington, DC, 2002.

55. *Cancer Disparities: Cancer Facts and Figures*, American Cancer Society, Atlanta, GA, 2004, pp. 21–26.

56. Dubner, S. J., Toward a unified theory of black America, *New York Times Mag.* 54–59 (March 20, 2005).

57. Sharp, D., The obesity epidemic, *J. Urban Health Bull. NY Acad. Med.* **81**(6):317–318 (2004).

58. Kopelman, P. G., Obesity as a medical problem, *Nature* **404**:635–643 (2000).

59. *Childhood Obesity: Trends and Potential Causes.* The Future of Children (series) Vol. 16, Woodrow Wilson School of Public and International Affairs at Princeton Univ. and the Brookings Institute, Princeton Univ., Princeton, NJ, 2006.

60. Friedman, J. M., Obesity in the new millennium, *Nature* **404**:632–634 (2000).

61. *Preventing Cancer*, WHO Tech. Report Series 276, Geneva, 1964, p. 131.

62. Maugh, T. H. II, Cancer and the environment. Higginson speaks out, *Science* **205**:1363–1366 (1979).

63. Doll, R., and Peto, R., The causes of cancer: Quantitative estimates of avoidable risks of cancer in the U.S., today, *J. Natl. Cancer Inst.* **66**(6):1196–1265 (1981).

64. Trichopoulos, F., Li, F., and Hunter, D. J., What causes cancer? *Sci. Am.* **275**(3):80–87 (1996).

65. Laden, F., and Hunter, D. J., Environmental risk factors and female breast cancer, *Annu. Rev. Publ. Health* **19**:101–123 (1998).

2

NO PROOF IN THE PUDDINGS: ALTERNATIVE HEALTH PRACTICES

"One can't believe impossible things."
"I daresay you haven't had much practice," said the Queen.
"When I was your age, I always did it for half-an-hour a day."
"Why sometimes I've believed as many as six impossible things
before breakfast."
—*Alice in Wonderland, Lewis Carrol*

NEITHER FOOD NOR DRUGS

Dietary supplements, functional foods, probiotics, herbals, botanicals, nutracenticals—is this food we're eating, or are we self-medicating? That is not only the question but also the conundrum. What are we buying, and why are we buying? We must have some idea because we're wiping the products off the shelves faster than they can be replaced. In 2004, dietary supplements were wisked off the shelves to the tune of $19.8 billion; herbals and botanicals, $4.2 billion; and functional foods, $22 billion. These $46 billion's worth is expected to become $50 billion by 2010. Those $46 billion came from some 50% of our citizens—teenagers, seniors, and dieters—both ill and healthy. Are we simply giving our money away, or is the public onto something—something that scientists and physicians have yet to grasp?

Does ginkgo biloba enhance memory, and does St. John's Wort ameliorate depression? Can shark cartilage prevent cancer, and can comfrey cure arthritis? Does garlic reduce serum cholesterol? How would we know? These are not idle questions. They require pursuit and responses. With some 30,000 products available, these are serious questions.

Our first question therefore is: "What is a dietary supplement?" Prior to the Dietary Supplement Health and Education Act (DSHEA) of 1994, of which we will have much to say, the term *dietary supplements* referred to products consisting of one or more of the essential vitamins, minerals, and proteins. With DSHEA, that now includes any, repeat, *any* product intended for consumption—ingestion—as a dietary supplement, which includes botanicals (vegetable), herbs (plant), amino acids, enzymes, concentrates, metabolites, and, of course, vitamins and minerals. It is easy to spot a supplement as the law requires manufacturers to label their products as "dietary supplements." These can be pills, capsules, powders, and liquids, and are sold in supermarkets, grocery stores, discount chain stores, drugstores (pharmacies), hotel resorts, and healthfood stores, as well as on TV and radio, via the Internet, mail order catalog, and direct sales. Think about this range of outlets in terms of FDA and FTC oversight. The Federal Trade Commission is charged with regulating the advertising of dietary supplements, but can either the FDA or the FTC adequately monitor this diversity of outlets?

Dietary supplements (see list of uses in Table 2.1) by law are not drugs, not medicines, not pharmaceuticals. However, and nevertheless, Internet Websites make these claims for dietary supplements:

- "Can improve circulation . . . peripheral vascular insufficiency."
- "Is effective in treatment of mild to moderate depression."
- "Because it has natural antibiotic action, echinacea is an excellent herb for infections of all kinds."
- "This extract of Saw Palmetto is the most widely used herbal preventative and therapeutic agent for benign prostatic hyperplasia."

TABLE 2.1. The Diverse Uses of Supplements

Use	Percentage[a]
Colds	59
Burns	45
Allergies	22
Rashes	18
Insomnia	18
PMS	17
Depression	7
Diarrhea	7
Menopause	4

[a] The fact that these total more than 100% suggests they are being used for more than one condition.

If dietary supplements are medicines—drugs—as these claim simply, they must be preapproved by the FDA prior to marketing, which means that their manufacturers are required to have filed both animal study data along with human clinical trial data supporting efficiency and safety. The food and drug laws place the burden of proof on food and drug manufacturers, as well they should. And it is accepted. But dietary supplements are not being sold in supermarkets, discount chains, and drugstores as drugs or medicines. Are dietary supplements in fact foods? What is a supplement? Here we arrive at our conundrum.

Dietary supplements are not true foods. If they were foods, their manufacturers would also be required to produce the animal and human documentation supporting their safety and efficacy, as drugs do. But they haven't. Because they are not foods. What are they, then? Supplements. According to the dictionaries I've perused, a *supplement* completes or makes an addition. So a dietary supplement is sort of "neither fish nor fowl." Neither a food nor a drug. What are these things that are flying off the shelves? Buyers are grabbing products that I've dubbed the "un"s: untested, unregulated, unstandardized, and of unknown effects: an entirely new concept for the marketplace, and one that makes consumers responsible for knowing what it is they consume— assuming also that what they purchase is safe and efficacious; an erroneous assumption—nevertheless, it is an assumption engendered over long years of purchasing food and drugs, and extended, quite rightly, to all products in the marketplace. Security and reliability had been achieved.

That security was first obtained in 1906, with the passage of the Pure Food and Drug Act, and was further ensconced in 1938 with authorization of the Food, Drug and Cosmetic Act, and reaffirmed yet again in 1960, when Congress decreed that both food and drug manufacturer's provide substantive evidence of efficacy and safety to the FDA's satisfaction prior to marketing—a requirement that has worked well and provided the security that we have come to assume. But trouble was brewing.

In 1989, the amino acid L-tryptophan arrived in the marketplace extolling its talents as an antidote for postmenopausal syndrome, anxiety, and sleep disorders. During that year, 38 people died, while another 1500 incurred connective tissue disorders—underreported estimates—which impelled the FDA to empanel a taskforce to investigate the supplement industry. The panel recommended that over-the-counter sales of amino acids and other supplements be stopped and dosages of vitamins and minerals be sharply limited [1].

With the taskforce's recommendations in hand, the FDA sought Congressional approval for preapproval of all health claims before supplement manufacturers could market their products. They also requested the power to levy fines and penalties. Enter Orrin Hatch. Orrin Hatch, Senator from Utah, is clever and powerful—a wicked and effective combination. In 1992, he forced the passage of legislation barring additional supplement regulations, and orchestrated a highly successful campaign convincing the nation that the government was about to curtail their use of vitamins. Congress was deluged with anti-FDA missives. With an unsuspecting and credulous public behind him, and supplement industry lobbyists and lawyers working with him, they crafted what became Public Law 103-417, the Dietary Supplement Health and Education Act of 1994 (DSHEA), which was to turn established health protection on its head [2].

At the outset, we are informed that "Congress finds that the importance of nutrition and the benefits of dietary supplements to health promotion and disease prevention have been documented increasingly in scientific studies" —a blatant fabrication if ever there was one. But it was Section 403 B(c), Burden of Proof, that left us vulnerable in the marketplace and returned caveat emptor—let the buyer beware—to center stage.

In any proceeding brought under Subsection (a), the new law informs us, "The burden of proof shall be on the United States to establish that an article or other such material is false or misleading" [2]. Short, definitive, absolute. The tables were turned. The supplement manufacturers were off the hook. The government was now responsible for what the supplement fabricators placed on store shelves. Why? Why surrender the public's only source of protection for the benefit of the supplement industry, an industry that had never demonstrated its need in American life? When questioned about this on the Senate floor, Hatch declared that the FDA had "repeatedly attempted to impose unnecessarily stringent standards that would leave many if not most supplement companies with no practical choice but to close their doors" [2]. Clear enough. The public came in last, and would have to fend for itself. The industry was there to make big bucks, which was what really counted. Is it any wonder that Orrin Hatch is referred to as the "Godfather of the supplement industry"? This kindles a cautionary tale. Groucho Marx interviewed Louisiana State Senator Dudley J. LeBlanc about a "miracle" cure-all vitamin (and mineral) tonic called Hadacol that the senator had concocted. When Groucho asked the senator what it was good for, LeBlanc answered with surprising honesty: "It is good for five and a half million for me last year" [3]—enough said.

Indeed, the Act created a new class of products removed from FDA purview and jurisdiction, yet another "un" —the untouchables. Under DSHEA the floodgates opened wide and the products, some 30,000 of them, poured through. And the buyers were there, ready to be the supplement industry's guinea pigs. Why? Simply because none of the manufacturers have done, or care to do, the necessary studies. Why should they? It's not required. And when was the first, or last time anyone asked, "Was this product I've bought even evaluated for safety or effectiveness?" At least safety. Really, there is a great lack of incentive on the part of supplement manufacturers to undertake product studies. They take time. Laboratory animal studies and human clinical trials require professional talent to conduct, and can be quite costly. But most importantly, they can bear ill tidings: un (another "un") favorable results, unwanted, undesirable results—that the product is either unsafe or ineffective, or both. As of 2005, of the literally thousands of products on market shelves across the country, perhaps a half-dozen have been tested for safety. Most disquieting is the fact that since its inception in 1998, the National Center for Complementary and Alternative Medicine, one of NIH's 17 Institutes and Centers, has provided millions of dollars in grants for research studies on many of the most sought after supplements, and has yet to obtain positive beneficial findings for any of them [4]. To be sure, at this point, with supplements sales in the double-digit billions of dollars, it obvious that supplements are a triumph of marketing over science and medicine.

Furthermore—and how this one got away from the FDA is yet another triumph for industry lawyers and lobbyists, over public safety—DSHEA makes it clear that supplement manufacturers are not required to provide reports of adverse events to the FDA. Reporting is voluntary, and the only way the FDA learns of ill effects or deaths is from individual physicians, hospitals, or individuals who take the time to report an untoward event. Currently all negative events are only rough estimates. With the passage of DSHEA the FDA has been running on empty. Until DSHEA is amended, the public is on its own. Caveat emptor should become the country's motto.

Knowing how this anomaly occurred, the answer to our first question may help us with the second: What is it that we are purchasing and ingesting; what are we getting for our money, and how are these supplements being used?

This aggregation is particularly revealing in that these items exhibit a common feature; they are all common medical problems. Evidently supplements are being used as medicines, to treat illness. Users are in fact doing exactly what DSHEA intended: making their own medical decisions, choosing whatever medications, therapies, they prefer in a marketplace unfettered by government regulation, and protection. But do we possess the knowledge to make these decisions? Do we need know only what labels tell us? What in fact do they tell us—or fail to tell us?

Under DSHEA, supplement labels can carry one of three types of claims: nutrient content claims, disease claims, and nutrition support claims, which include structure function claims. Because these diverse claims can and do cause confusion, and offer no assistance in making choices or protection from inaccurate, misleading, or fraudulent claims, we shall parse these in some detail.

Nutrient content claims are the simplest in that they note the level of a nutrient ingredient in a supplement. A supplement containing at least 200 milligrams (mg) of calcium per serving could state "high in calcium." A supplement with at least 12 mg of vitamin C could state "excellent source of vitamin C." The only problem here may be that what the label states in milligrams, may not be in the container. More about this shortly.

Nutrition support claims may describe a link between a nutrient and a deficiency disease that can result if the nutrient is lacking. In this case, the label of such a supplement could state that vitamin C prevents scurvy. But if so stated, the label must also note the prevalence of that deficiency in the United States. Current data suggest one or two cases in the entire country.

Disease claims are intended to connect nutrient or ingredient with a disease- or health-related condition. An example is the water-soluble vitamin folic acid, which is known to decrease the risk of neural tube defects when taken by women during pregnancy. If, and only if, the supplement contains a sufficient amount of folic acid to accomplish this, can it be so stated on the label? Similarly, it is believed that calcium can lower the risk of osteoporosis. If the supplement contains a sufficient amount of calcium, it can make the connection on its label.

Nutrition support claims are not the same as nutrition content claims. Nutrition support claims can describe a link between a nutrient and a deficiency disease that can occur if the nutrient is lacking from the diet.

Ergo, the label of a vitamin C supplement (ascorbic acid) could state that vitamin C prevents scurvey, but, given the rarity of the disease in the United States, the label must also mention the prevalence of that disease in the U.S.

Possibly the most troublesome claim is the *structure–function claim*, which refers to the supplement's effect on the body's structure or function, including its overall effect on a persons well-being. For example, claims such as calcium builds strong bones, antioxidants maintain cell integrity, and fiber maintains bowel regularity, are current examples of structure–function claims.

Lets be clear about this. Structure–function claims describe how consuming the supplement will affect a structure such as the skeletal system, or a function such as the circulatory system, or a person's general well-being, but do not— cannot—claim to reduce the risk, prevention, or treatment of a disease. This is essential. It can be claimed that the supplement "supports the immune system," or "helps support cartilage and joint function." Unfortunately, "supports" is too often translated as "treats," or buyers assume that it treats, which is exactly what the manufacturers intend. Because structure–function claims can be so interpreted, all—repeat, all—structure–function products most carry a disclaimer on their labels. This disclaimer is free of ambiguity. It states that "these statements have not been evaluated by the FDA. This product is not intended to diagnose, treat, cure or prevent any disease." Could anything be clearer? Who would purchase a supplement that could accomplish none of those? Lots of people. Why? Because in the ongoing conflict between the FDA and the industry lawyers and lobbyists, the FDA lost the battle of size of print and placement of the disclaimer. The container on the right in Figure 2.1 clearly shows the required asterisk immediately after the claim, telling the reader to look for a footnote, the disclaimer. But do not expect to see such clarity on labels in supermarkets or other stores. Think of it as a game. *Tip*: Bring a magnifying glass. Finding the disclaimer is a cat-and-mouse affair between the FDA, manufacturers and consumers. But it isn't a game. Not a sport. Not a joke. It may be a comedy of errors that the FDA allowed itself to be part of. On the face of it, the idea of a footnote is nonsense. And the manufacturers took advantage of it by loading their labels with all manner of messages used as camouflage to hide the disclaimer. It would be illuminating and educational to visit your local emporium and dally among its supplement

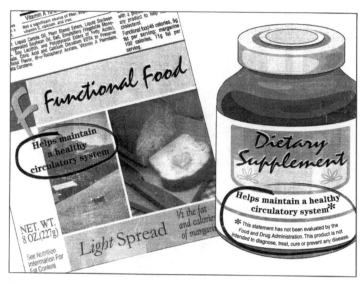

Figure 2.1. Comparison of two products with and without the required FDA disclaimer. (*Source:* U.S. GAO/RCED-00-156, *Dietary Supplements and Functional Foods.*)

aisles. Pick up a selection of containers, boxes, and bottles. Read the labels. Ask yourself, what these messages and statements mean. What in fact is this supplement supposed to accomplish? Check for the disclaimer. I read a label on one product that had a cut-down version. It said, "not intended to diagnose, treat, cure, or prevent any disease." Not a word about the FDA not having evaluated the label statement. That's the last thing they want on their labels.

But that is not the end of it. Manufacturers are tying their products to articles published in scientific journals—no matter how tenuous. Figure 2.2 is a grand example. Juvenon is new to the supplement world. This ad is from the popular and sophisticated magazine *Scientific American*. And here, Juvenon informs us that their product is backed by the research done at UCLA (University of California, Los Angeles), and they cite the article published in one of the most prestigious journals. Fine. But what they don't tell us is that the research was conducted on rats, and doesn't even hint at possible use in humans. But it does sound very scientific, and proceeds from the well-known psychological fact that people overestimate the importance of presented information, and underestimate the importance of information that was not presented [5]. By the way, with third-party information such as this, the FDA requires that the information be "truthful, nonmisleading, and present a balanced view of the available scientific information." The statements in Figure 2.2 are beyond a stretch of the truth, and are decidedly misleading. But who is going to bell the cat? The Internet poses even greater problems. Can the FDA monitor the Internet for dubious references, or protect against double-tongued, direct-to-consumer marketing? So, for example, an Internet pitch

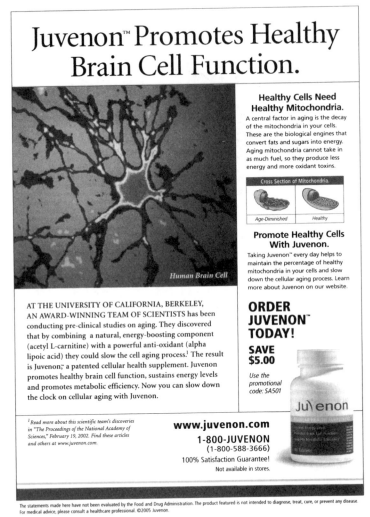

Figure 2.2. Supplement advertisement made to appear scientifically sound. Where is the disclaimer?

referred to a 1931 publication recommending St. John's Wort for pulmonary complaints, and championed the perennial herbal product comfrey, which the FDA had advised manufacturers to remove from markets because of its evident hepatotoxicity [6]. Furthermore, of 443 Internet sites studied, the required disclaimer was omitted on over 50% [7]. No, it's not a game. Caveat emptor (buyer beware) must be taken seriously. Now consider this: The FDA's budget allows only three of its staff to work part time investigating reports of health problems associated with dietary supplements. The FDA, like all government agencies, is a creature of Congress; niceties aside, Congress owns them. That's stating it crudely, but factually. *no agency* can function without a budget, which is Congress' to determine. Congress holds the purse strings. Under DSHEA, the FDA is mandated to protect the health of the public, but not so diligently that manufacturers must toe the line too tightly. Balancing this improbable relationship requires that Congress ensure that the FDA's budget has enough, but not enough to protect the public, from unscrupulous manufacturers. You might want to reread that passage. It's factual, but too hard to believe. But believe it you must.

If that isn't enough to worry about, think fraud. My dictionary defines *fraud* as deliberate deception. Fraudulant products can often be spotted by the type of claims made: a "secret cure," "breakthrough," "miracle cure," "magical," or "a new discovery." Such claims must be seen as signals, tipoffs that the products touting them are palpable frauds. Real cures cannot be kept secret, nor are they miraculous nor magical. Additionally, jargon such as "detoxify," "purify," "energize," and "cleanses the blood" are vague and meaningless terms that should raise red flags, as should claims that a product can cure a range of ills. Not one supplement, or drug for that matter, has all-encompassing capabilities. Moreover, referring back to citations of scientific studies, claims that a product is backed by scientific studies need to pass through a filter of traceable citations, and out-of-date, irrelevant, and/or uncontrolled studies. In addition, claims that the product has only benefits and no side or adverse effects must be taken with the proverbial grain of salt, as any product potent enough to produce physiological effects and responses, will also be potent enough to produce negative side affects.

Finally, a government agency has moved to deal with the egregious marketing of dietary supplements. In March 2006, the Federal Trade Commission slapped the wrist of Garden Life, Inc., and its founder Jules Rubin,with a $225,000 fine for making bogus claims that its products Primal Defense, RM-10, Living Multi, and FYI could cure diseases, and that these products were supported by scientific evidence, which was entirely false. RM-10 was advertised as a concoction of 10 types of mushrooms that could cure cancer. Between 2001 and 2004, Garden Life, Inc., of West Palm Beach, Florida, had sales totaling $58 million. Indeed, $225,000, is no more than a slap on the wrist, and this is the first fine issued against a supplement company in a decade. Whole Foods, GNC, and the Vitamin Shoppe are major sellers of the Garden Lifeline of products.

Wariness must also be extended to products claiming that the government, drug companies, and the medical profession have colluded to keep this beneficial information from you, and that they didn't want you to know about this product; this is patent nonsense simply because it would be impossible for such collusion to occur. Indeed, consumers are faced with a confusing array of claims, some that require rigorous scientific support and others that can be made with less evidence, but with no clear way to distinguish between them. Even when products have made unsupported structure–function claims, the FDA has not taken enforcement action. Consequently consumers can, and do, make inappropriate choices and purchases, and rely on ineffective products to treat their health problems. The line (demarcation) between label requirements and pure advertising has disappeared. So what happens when adverse effects occur, and how does the FDA learn about them?

Do supplements produce adverse effects? Of course, they do. But—and this is a preeminent "but"—firm numbers are exceptionally difficult to come by, and those that are available are estimates, rough estimates at that, due to underreporting. A study published in 2003 noted that of 11 poison control centers around the country, 20% of the total number of centers, and which contribute some 12% of all events, 1466 cases dealt with ingestion of dietary supplements. The researchers interviewed 489 confirmed supplement users and found that 30% had greater than mild adverse reactions, including myocardial infarction, liver failure, bleeding, seizures, and death. Treatment of disease was the reason for supplement use in some 28% of cases. The study's authors were unable to calculate a population-based risk estimate, as they could obtain neither a numerator (events) nor a denominator (numbers using supplements). They were able to estimate that of all the adverse reactions reputed to the 65 poison centers, 17% were contributed by dietary supplements [7]. In another report, the General Accountability Office (GAO) reported that between 1993 and 2003 the FDA received about 3000 reports of adverse affects including 105 deaths. But they also inform us that the FDA's account of adverse affects is at odds with other information. A nationwide consumer survey in 1999 found that 12% of all consumers using a supplement, some 12 million people, said they had experienced an adverse reaction. Obviously few people take the time to report, or don't know where to direct their reports. Also, under DSHEA, the supplement manufacturers are not required to report adverse effects. Is it necessary to repeat that? Ergo, if an adverse event is reported to a company, it is not passed on to the FDA. The entire process is voluntary. Recall that under DSHEA it is up to the FDA to prove a product unsafe. Clearly, the numbers that are available reveal a fraction of the total. How, then, does the FDA identify unsafe supplements? Currently the FDA relies on its Med Watch voluntary reporting system—a system that depends on voluntary reporting of suspected adverse reactions by a person (patient) or healthcare provider who must actually recognize the event as a possible adverse reaction [8]. It is entirely reasonable that most reactions go unrecognized by both patients and professionals as being supplement-related— and even if recognized, less than 10% are reported. When sufficient reports are received, an indication of a possible problem may emerge, but proving causality from case reports is extremely difficult [8]. How anyone can assume that supplements are innocuous is beyond belief. Under DSHEA, it is the marketplace, we, the people, that must prove efficacy and safety, an approach that failed early in the twentieth century and was the reason for establishment of the FDA in the first place. Nevertheless, Med Watch needs support. (It can be reached at 1-800-FDA-1088, or by fax at 1-800-FDA-0178. Their Website can be accessed at www.fda.gov/medwatch.)

A chemical is a chemical, is a chemical; all is chemical. We have yet to inquire as to what in fact are supplements, and how they differ from traditional drugs or medications. *Drugs* are biologically active single-chemical substances formulated to achieve a specific physiological response in tissues and cells. Are supplements drugs?

The roots, stems, leaves, and flowers of hundreds of plants sold as powders, gels, and liquids are in fact mixtures of many chemicals. We all know that when ground and roasted coffee is added to hot water, we get a brew of splendid odor, taste, and color. Close to 1000 well-defined chemicals have been identified as contributing to this fragrant beverage. Tea and cocoa are much the same, loaded with chemicals. That's the nature of all plants—of all living things. They don't just contain a great number and diversity of chemicals; they are nothing but chemicals—some benign, some harmful, some deadly. Because they are natural does not mean that they are safe. Safety and "naturalness" are not necessarily bedfellows. Bear in mind that chemicals obtained from plants were made by plants, for plants, for the many and varied reactions required of the plant to survive and function. Food plants, crops, and herbs naturally account for 99.99% of the pesticides that we ingest. These natural pesticides are manufactured by plants in part to defend themselves against an array of microbes, insects, and animals that crave their succulent tissues. Thus, with the ever-increasing consumption of herbs, regular herb users are among the greatest consumers of dietary pesticides, paralleling the pesticide consumption of vegetarians. It has been estimated that "Americans consume about 1.5 grams of natural pesticides per person per day, which is about 10,000 times more than the synthetic pesticide residues consuned [9].

Any chemical, natural or synthetic, that has an effect on some body function is, ipso facto, a drug. It is simply wrong to think that herbs are an alternative to drugs. They are in fact alternative drugs, a different way of taking a drug—a crude and untested drug, at that.

Taxol (paclitaxel), a chemical and highly effective anticancer drug, was originally obtained from the bark and needles of the Pacific yew tree. Unfortunately, there are not enough yews for all the Taxol cancer patients required because the demand was so great. So Taxol, like so many other drugs originally obtained from plants, was chemically analyzed to produce a synthetic Taxol, indistinguishable from the original, thereby allowing the trees to grow undisturbed. Such synthesis produces chemicals, drugs, whose identity and purity are known, whose quality and potency are assured, and can be made in the quantities needed, time after time. Reproducibility. That's essential.

Herbs, with their naturally occurring mixtures of hundreds of chemicals must have variable potency from season to season, and country to country. With their potpourri of chemicals, can have a spectrum of effects other than, or along with, those of their so-called active principle. There is no denying that drugs they are. Nevertheless, supplement manufacturers scrupulously avoid even a hint that their products be called or thought of as drugs. To do so would place them squarely in the FDA's purview and require full disclosure of efficacy and safety for the purpose intended, prior to marketing. Supplement manufacturers would sooner pet a cobra than put themselves in that position.

With all the St. John's Wort, Saw Palmetto, ginkgo biloba, and echinacea flying off the shelves in the United States, and around the world, is there really enough to go around? Tests show that many supplements do not contain what they say they do. Potency of pills and capsules vary markedly within the same box or bottle.

Ginseng (*Panax ginseng*) supplements are among the top 10 supplements sold worldwide, and are prepared from a variety of plants, primarily *panaxginseng*, the Korean and Chinese ginseng. Another species in the marketplace is the Japanese and Vietnamese *Panax quinquefolius*. Each has different uses, but all contain ginsenosides (or panaxosides) steroidal ring compounds. Siberian ginseng, *Eleutherococcus senticosus*, contains eleutherosides that are chemically distinct from ginsenosides.

Researchers at the University of California, Davis, tested 25 preparations of ginseng for purity and potency. Their pharmacologists found that the commercial products tested were properly labeled as to plant genus, but that there was considerable variability in genocide content among products. They also found that the amounts noted on labels were markedly different from the amounts found in their analyses. From their mass spectrographic procedures, they also found that the Siberian ginseng products varied more than did the Asian. Indeed, as noted above, the high demand for ginseng appears to lead to harvesting of immature plants with lower ginseng concentrations, which will affect the overall quality of the products.

Six of the products tested had almost no ginseng. In addition, concentrations of ginsenosides varied by 15–36 times in capsules and liquids and concentrations of eleutherosides varied by 43–200 times in capsules and liquids. In addition, they noted that previous studies had identified bioactive chemicals such as salicyclic acid, vanicillic acid, peptides, polysaccharides, polyacetylenic alcohols, fatty acids, and vitamins and the minerals copper, cobalt, manganese, and arsenic, in the commercial supplements [10]. Note well that these are all undeclared chemicals, none of which are listed on labels.

In another recent publication, researchers at the Faculty of Medicine, University of Toronto, fed ginseng supplements to a dozen healthy test subjects to determine their affects on glycemic and insulinemic regulation. Their findings confirmed the hypoglycemic effects of a batch of American ginseng but not the hyperglycemic effects of a batch of Asian ginseng. They say that "We do not know whether the batches studied are representative, and we do not know whether reproducing the measured ginsenoside profiles will lead to the same (opposite) results." They concluded by stating that consumers and practitioners alike should be warned of the inherent uncertainty of what commercial batches yield [11]. We do not know what we are actually buying!

An earlier report found that the wild-grown Chinese ginseng is different from the farmed or field-grown variety. The wild type sells for up to $1300 per pound [12]. Potency is the name of the game. Each time, every time.

DRUGS AND HERBAL SUPPLEMENTS

Contaminants, adulterants, detrimental and ineffective supplements. Overview and sampling. Balm from the east.

Ayurvedic Medications

Ayurvedic medicine originated in India some thousands of years ago, and relies heavily on herbal remedies. The *Ayurveda* is a collection of medical information in which disease is believed to be the consequence of an imbalance of the three elements vatta, pitta, and kapha. *Vatta* maintains the body's integrity and the proper functioning of its constituent elements. *Pitta* is the primary constituent having the functions of vision, digestion heat production, hunger, thirst, body softness and suppleness, cheerfulness, and intelligence, while *kapha* organizes the tissues into their macro- and microscopic forms. Imbalances of any one or a combination of these elements causes sickness and the need for specific herbs to restore balance.

Given that lead, mercury, and arsenic intoxication have been associated with the use of ayurvedic supplements, researchers at Harvard and Boston Universities investigated and reported on the prevalence and concentration of heavy metals in ayurvedic products available in Boston. After a systematic search to identify all stores selling these products, they revisited the stores and purchased all ayurvedic products. They found that 14 of 70 products, 1 in 5, contained mercury at levels ranging from 28,000 to 104,000 micorgrams (μg) per gram (g); lead, 5000–37,000 μg/g; and arsenic, 37–8130 μg/g. If taken as label directions recommended, each of the 14 could result in heavy-metal toxicity. The authors of the study say that testing of ayurvedic products for toxic heavy metals should be mandatory [13].

In another account, the National Centers for Disease Control reported 21 cases of lead poisoning in five states among adults who used ayurvedic supplements for rheumatoid arthritis, infertility (men and women), menstrual health, and diabetes. All were seen in hospital emergency rooms with fatigue, vomiting, nausea, and abdominal pain. The ayurvedic remedies they were taking had lead levels of 12,000–78,000 parts per million (ppm). They note, too, that lead levels in many of these supplements ranged from 0.4 to 260,000 parts per million [14].

Actra-Rx-Yilishen

In 1998, the FDA approved Viagra, a pill to treat male impotence. It also warned potential consumers not to use the drug in combination with nitrates to avoid sudden drops in blood pressure. Recently, the FDA warned against purchasing the nonprescription supplements Actra-Rx-Yilishen entering the

United States from China, because they contain Viagra, and as they do not require a prescription, purchasers get no warning. Despite an "all natural" label, Actra-Rx-Yilishen capsules contain prescription-strength sildenafiil citrate, the active ingredient in Viagra. Actra-Rx has been advertised on the Internet as "a natural sexual enhancer." Men at risk include those with diabetes, elevated blood pressure, elevated cholesterol levels, or heart disease [15].

PC-SPES

Over the past several years PC-SPES (PC for prostate cancer; SPES, Latin for hope) had become a potential prospect for treatment of prostate cancer, and sold as a dietary supplement for "prostate health." Labeled as a mixture of seven chinese herbs plus Saw Palmetto, it gained wide popularity among prostate cancer patients. However, by 2001, reports of severe bleeding episodes, abnormal clotting times, and breast tenderness and enlargement, appeared in the literature. In 2002, the California Department of Health posted a warning on its Website that PC-SPES was contaminated with both (1) diethylstilbestrol (DES), a synthetic estrogen, indomethacin, an antiinflammatory compound, and (2) warfarin, a blood thinner. This was verified by radiology researchers at University of California, San Diego, and Palacky University in the Czech Republic. They recommended that "the utility of phytoestrogens and other phytochemicals be investigated for the management of prostate cancer," and "that manufacturer's of such supplement should provide reliable analytical quality insurance" [16]. PC-SPES has since been removed from the marketplace.

St. John's Wort

St. John's Wort (a *wort* is a bryophyte (moss)—*Hypericum perforatum*—is an herbal supplement widely used to treat depression. It is readily available without a prescription, and taken without medical supervision, a splendid example of self-medication. Yet a steady drumbeat of reports indicates that St. John's Wort is involved in dangerous drug interactions. A number of reports have suggested that it has an inductive metabolic effect on the cytochrome P450 (CYP3A4) enzyme. St. John's Wort induces C4P34A expression in both test tubes and human subjects. CYP is scientific notation for cytochrome P or CYP (pronounced "SIP"), a family of some 30 enzymes having an absorbance peak at 450 nanometers (nm); hence the 450 designation. Among the 30 species are such related examples of cytochrome pigments as CYP2s, (2A6, 2C18), CYP3s (3A4,3A7), CYP4s, CYP19, and others, each with its specific amino acid sequence. Lab studies with expressed human recombinant cytochrome P450 proteins will identify which cytochrome P450 form catalyzes the metabolism of a test drug, which products are formed, whether the test drug inhibits the metabolism of other drugs, or whether other drugs reduce or stimulate the metabolism of the test drug. With that as a primer on P450, we can consider a recently published study dealing with the effect of St. John's Wort on drug metabolism. In this investigation of healthy individuals, 6 men and 6 women were given St. John's Wort for 14 days to assess its effect on CYP3A4. CYP3A4 was chosen because it is known to be involved in the metabolism of some 70% of prescription and over-the-counter medications. The researchers found that after 14 days St. John's Wort significantly induced the activity of CYP3A4 as measured by changes in the pharmacokinetics of a test substrate. This indicated that long-term administration of St. John's Wort can result in diminshed clinical effectiveness, or increased dosage requirements for any drug inducing CYP34A, which represent some 50% of all marketed medications [17]. More on pharmacokinetics shortly.

Antagonistic metabolic interactions have left a clear trail to St. John's Wort. Cases of metabolic immunosuppression between St. John's Wort and cyclosporin were recently reported by a team of cardiologists and pharmacologists at the clinic for cardiac surgery at the University Hospital Zurich, Switzerland. Two patients in stable condition after heart transplants a year earlier required treatment at the hospital. In the first case, a 61-year-old man had self-medicated himself with St. John's Wort for a mild depression 3 weeks before his hospitalization. In the second case, a psychiatrist prescribed St. John's Wort to a 63-year-old man for his depression.

The singular untoward finding on admission was depressed cyclosporin levels in both patients. After use of the St. Johns' Wort was discontinued, the cyclosporin levels returned to their normal values. The researchers found that St. John's Wort contained at least 10 different chemicals that could contribute to its pharmacological effects: flavinoids, xanthones, and naphthodiantrons, which are enzyme inducers and can break down cyclosporin, thereby reducing its bioavailability, another instance of CYP3A4 induction [18]. In its wisdom, the body senses the presence of potentially dangerous chemicals and induces the production of CYP3A4 to cleave and dispose of a harmful chemical. However, St. John's Wort can also induce CYP3A4, which then destroys the cyclosporin. Not a happy combination.

Soy Proteins and Isoflavones

The sudden decline in estrogen levels after menopause coincides with decreases in bone mineral density and cognitive function, along with a rise in total cholesterol and low-density lipoprotein. Phytoestrogens, estrogen-like chemicals found in grains, fruits, soybeans, and peas, have been suggested as having a positive effect on the aging process with fewer adverse side effects. A recently published study sought to determine whether soy protein, an isoflavone antioxidant, improves cognitive function, bone mineral density, and plasma lipids in postmenopausal women. This was a randomized, double-blind, controlled clinical trial, the gold standard of epidemologic studies—which we shall explore in depth in Chapter 9—conducted in the Netherlands, and which fed soy protein to 202 healthy postmenopausal women ages 60–75 for 12 months. The researchers concluded that the study "does not support the hypothesis that the use of soy protein supplement containing isoflavones improves cognitive function, bone mineral density or plasma lipids in healthy post menopausal women when started at the age of 60 or later" [19].

Yet another study from the same university hospital in Utrecht, the Netherlands, looked at the relationship between phytoestrogen consumption and breast cancer risk in a Dutch population with a habitually low pytoestrogen intake. This study enrolled 15,555 women aged 49–70. A total of 280 women were newly diagonal with breast cancer during the study. As the authors tell it, "In western populations, a high intake of isoflavones is not significantly related to breast cancer risk" [20]. A third study, done at the University of Perugia's (Italy) Department of Gynecological, Obstetrical and Pediatric Sciences, tested the long-term effects of soy phytoestrogen on histological changes to the endometrium of postmenopausal women. Here too, a double-blind, randomized, placebo-controlled design was employed. After fully 60 months of ingesting 150 mg of isoflavons daily, they found that soy protein produced a precancerous hyperplasia (a thickening) of the endometrial surfaces, and they noted that "These findings call onto question the long-term safety of phytoestrogens with regard to the endometrium" [21].

Genistein and soy are intimately related. Genistein, a phytoestrogen, and naturally present as 4,5,7-trihydroflavone in soy products, is readily metabolized by intestinal bacteria to genistin, which exhibits estrogenic activity. A critical review of published medical studies of genistein/genistin by oncologists at Georgetown University School of Medicine, Washington, DC, cautions that "Although the evidence of the range of genistein's effects is far from conclusive, it is tempting for some in the scientific community to tout genistein as a potential chemopreventive or alternative to hormone-replacement therapy. However, studies indicating a potential cancer-promoting effect should not be taken lightly. Further studies must be before the true scope of genistein's actions can be understood" [22]. That surely sounds like support for the work done in Perugia.

Perioperative Care and Herbal Supplements

Because of the widespread use of herbal supplements (read "medications"), especially among presurgical populations, members of the Department of Anesthesia, Pritzker School of Medicine, University of Chicago, sought to determine whether herbal use could have a negative effect on perioperative patient care. Theirs was an in-depth review of published studies over the period 1996–2000, and their study was limited to eight herbs that potentially pose the greatest impact on patients undergoing surgery. These eight herbs account for over 50% percent of all single-herb preparations among the 2000 sold in the United States. They found that echinacea, a member of the daisy family, must be used with caution in patients with asthma, allergies, and liver dysfunction. Since the pharmacokinetics of echinacea has not been studied, it would be prudent for patients to discontinue its use weeks in advance of surgery. Ephedra (ma huang), used to promote weight loss and treat respiratory conditions, contains a number of alkaloids that can increase blood pressure and heart rate. Because of its pharmacokinetics and cardiovascular risks, it should be discontinued prior to surgery. With the current FDA ban on ephedra, it should no longer pose a problem to surgeons and patients.

Garlic, one of the most widely used and studied plants, is known to inhibit platelet aggregation, and can lower blood pressure; consequently, if postoperative bleeding is a concern, its use should be discontinued at least a week prior to surgery. Ginkgo, from the leaf of the chinese maiden hair tree, *Ginkgo biloba*, acts as an antioxidant and can alter vasoregulation. Because of the risk of bleeding, it should be discontinued several days before surgery. Ginseng can also increase bleeding and should be discontinued at least a week before surgery.

St. John's Wort and Valerian (*Valeriana officinalis*) are both troubling. Because St. John's Wort can interfere with many drugs, it is of especial risk both during and after surgery and should be stopped a full week before surgery. Valerian, used as a sedative, can produce sedation and hypnosis, and can be expected to increase the sedative effects of anesthetics. Prudence dictates that it be discontinued several weeks before surgery. Given this litany, can there still be doubt that herbals and other supplements are no less drugs and chemicals than are drugs sold as drugs. The sobriquet herb, or herbal, should not mask the fact that natural or synthetic, a chemical is a chemical, is a chemical.

As many patients fail to volunteer their use of herbals and other supplements, healthcare professionals must elicit and document its use. The inexcusably limited information on supplement labels about efficacy and safety, becomes a dicey medical issue in patient care, especially when complications arise during and after surgery [23].

And what about vitamins in the treatment and prevention of disease?

Vitamin E

Cardiovascular benefits have consistently been shown for vitamin E in observational, epidemiologic studies. Unfortunately, observational studies are the least dependable of the various epidemiologic investigations. Nevertheless, even the observational studies have shown only small to moderate benefits, and confounding can create interpretational bias. Researchers at Harvard University's School of Public Health reviewed and analyzed seven large-scale randomized clinical trials for effectiveness of vitamin E in the treatment and prevention of heart disease. They found strong support for a lack of both statistically and clinically significant effects of vitamin E on cardiovascular disease. They were concerned that the use of vitamin E could well "contribute to the under use of medications of proven benefit and failure to adopt healthy lifestyles" [24].

Investigators at Wageningen Univeristy, the Netherlands, used a randomized, controlled clinical trial to determine whether daily doses of vitamin E and multivitamin mineral supplementation had a positive effect on acute respiratory tract infections in elderly individuals. They fed 200 mg of vitamin E, and standardized doses of multivitamin minerals, or both, to 652 noninstitutionalized individuals 60 years or older, over a 2-year period. They found that neither daily multivitamin mineral supplementation nor 200 mg of vitamin E had a favorable effect on the incidence or severity of respiratory tract infections. Rather, they were surprised to find such adverse reactions to vitamin E as "longer total illness duration, more symptoms, and a higher frequency of fever and restriction of activity." They conclude on a cautionary note. "If," they state, "our results are confirmed and vitamin E exacerbates respiratory tract infections, elderly people, especially those who are already well-nourished, should be cautious about taking vitamin E supplements" [25].

In a recent study conducted at Minnesota's Mayo Clinic, vitamin E was found no more effective than a sugar-based placebo for delaying the onset of Alzheimer's disease. The investigators studied 769 patients with mild cognitive impairment ("mild" means serious forgetfulness—forgetting things you really want to remember, and "cognitive" refers to awareness, reasoning). In this investigation patients were randomly assigned to one of three groups; vitamin E; placebo, and Aricept, a new drug believed to delay the onset of Alzheimer's. All were observed for three years. At the end of the first year, 38 patients on the placebo, and 33 given vitamin E had progressed to Alzheimer's, compared to 16 taking Aricept [26]. In an accompanying editorial comment, Dr. Deborah Blacker, Professor of Psychiatry at Harvard University Medical School, remarked that the study provides clear-cut evidence that vitamin E is ineffective in the prevention of Alzheimer's [27].

Considering that sales of vitamin E topped $700 million in 2003, yet another recently published study requires exposition. A group of medical researchers at Johns Hopkin's School of Medicine reported on an analysis of 19 studies involving 136,000 patients. They found that those getting a daily dose of at least 400 international units (IUs) of vitamin E had higher overall mortality rates than did those in the control group taking a placebo—a sugar pill. There was no increased risk of death at ≤150 IU. Dr. Miller, the lead author of the study noted that people easily get enough vitamin E, an antioxidant, from vegetable oils, nuts, and green vegetables, and maintains that we need no more than 10 IU per day. This study of 136,000 participants included North Americans, Europeans, and Chinese [28].

Again, a recently published study in the *Journal of the National Cancer Institute*, found that when cancer patients took high doses of vitamin E, their risk of a new cancer increased. The researchers at Laval University, Quebec,

Canada concluded that "our results suggest that caution should be advised regarding the use of high-dose alpha-tocopheral supplements for cancer prevention" [29]. Their 540 participants, all with head and neck cancer, took 400 IU per day. On the first day of radiation therapy, they began a 3-year course of vitamin E or a placebo. Compared with those receiving the placebo, they found those given the supplements were about twice as likely (20% vs. 10%) to develop a new cancer. In the period after the supplements were discontinued, those who had received vitamin E did a little better in avoiding cancer than those on a placebo. But after 8 years, the rate of new cancers was the same in both groups. The authors say that people who want to reduce their cancer risk might be better off improving their diets than taking supplements.

But that's not the end of it. A slate of vitamin E trials that have been in progress for years have now been completed and published. One of the most recent is the HOPE and HOPE-TOO Trial. The *heart outcomes prevention evaluation* (HOPE) was a long term, randomized, placebo-controlled international study begun in December 1993 and completed in April 1999. HOPE-TOO was conducted during an extended period from 1999 to May 2003. The HOPE trial included 247 centers around the United States and world, and enrolled 9514 patients. HOPE-TOO involved 174 centers with 3994 participants. The object of the two trials was to determine the effects of long-term vitamin E supplementation on heart disease and cancer. Each participant in the test group received 400 IU daily, and there were over 100 participating physicians from the United States, England, Brazil, Canada, Austria, Belgium, and others countries.

Because oxidative injury has been suggested in both atherosclerotic heart disease and cancer, antioxidant vitamins have been and are being evaluated in their prevention. alpha-Tocopherol acetate, the predominant and most active form of vitamin E in humans, is the major lipid antioxidant.

The preeminent outcome of the HOPE trials, including the initial and the extension, was the lack of benefit of vitamin E in preventing cancer or major cardiovascular events over an extended period. But let the authors speak for themselves. "After a median of 7 years of follow-up for the entire study population, and a median of 7.2 years for patients at centers continuing in the trial extension, we observed no overall effect of vitamin E on the incidence of fatal and non-fatal cancers. . . . There were no reductions in incident prostate, colorectal, oral/pharyngeal, and gastrointestinal cancers" [30].

Furthermore, vitamin E had no significant effect on myocardial infarction, stroke, cardiovascular death, unstable angina, revascularization, and total mortality. Perhaps the most troubling outcome was the unexpected increase in heart failure rates in the patients taking vitamin E. As other long-term trials of vitamin E did not identify heart failure as a result of long-term ingestion, the authors suggest that the older trials be reviewed for incidents of heart failure. They conclude with the cautionary admonition that vitamin E supplements should not be used by patients with vascular disease or diabetes mellitus.

So, what hope for vitamin E? Over the past 15 years a plethora of studies have supported the hypothesis that antioxidants protect against heart disease, by limiting low-density lipoprotein oxidation in the arterial wall. As for cancer, biological mechanisms have been identified in the cancer process that may be prevented by antioxidants. On the other hand, another group of studies using vitamins E, C, beta-carotene, and combinations of these have failed to show benefits.

"While there is solid evidence linking oxidative processes to human disease (as well as normal biological function) the details of these processes and of proposed therapeutic or preventive interventions appear to need considerable rethinking" [31]. Physicians have been trying to wean their patients off vitamin E since 2000 without success. Perhaps these latest well conducted and controlled studies will have a motivating effect. Why do I doubt it?

Pharmacokinetics

A morsel of food, any food, starts the process. Our teeth grind food into small particles, and the tongue compacts them into lumps, a *bolus*, while saliva containing the digestive enzyme amylase metabolizes any starch in the food. The smaller the chewed (masticated) particles, the faster chemical digestion occurs. Our tongue moves the bolus toward the back of the mouth to the throat, where throat muscles constrict, forcing the bolus into the esophagus. Muscle contraction (peristalsis) carries the bolus down into the stomach, where strong hydrochloric acid digests the bolus and stomach muscles break it into still smaller particles. The protein-splitting enzyme pepsin continues the digestion with the aid of lipase, a fat-splitting enzyme. After several hours in the churning acidic environment, the partially digested food particles, the *chyme*, passes into the duodenum, the small intestine nearest the stomach. As the digesting food passes along, it is hit by additional enzymes secreted by the gallbladder and pancreas. The protein-splitting enzymes yield the amino acids, which will be absorbed through the small intestine and on into the bloodstream and to the liver, the body's largest gland where food particles can be stored and/or converted sending nutrients on to tissue cells. Dietary supplements take the same route, and are given the same chemical treatments, another reason why few supplements function as advertised. Most (not all, but most) are digested, disrupted, in the stomach's acid, and pass into the small intestine and liver as innocuous particles, to be eliminated in urine and feces, without further ado.

At this juncture, we shall see why drugs make it to tissues, where they are needed, and why dietary supplements do not. Think hypodermics. Hypodermic syringes are another way of delivering medications, chemicals (remember, all is chemical—that steak, that carrot, that granola bar) into the body, but avoiding the rigors and hazards of the digestive tract. Drugs, medications, are not immune to the panoply of enzymes and acids that patrol and oversee the digestive process.

Not much gets past these sentinels. A drug's metabolism via the digestive tract, or poor absorbtion through the small intestine results in little to no availability by tissues and cells. To bypass this formidable and vigilant system, many drugs must be injected—into skin, muscle, or bloodstream. Some 30% of all drugs must be injected if they are to do what they were intended to do. All new potential drugs must undergo pharmacokinetic testing; that is, before they can be made available to health professionals, they are tested to determine how soon, and in what condition, they reach their target organ; how rapidly they are metabolized and excreted; their blood levels over time providing information about levels in the bloodstream after 1, 2, 5, or 7 hours, and so on, allowing the determination of how much is required to be injected to achieve the levels needed to produce a desired effect. Without this information, as noted in the perioperative discussion, surgeons and anesthetists know nothing of the physiological dynamics of supplements. How could they? That's why they recommend discontinuance of supplements prior to surgery, protecting the patients and the operating staff. For most supplements, pharmacokinetics remain foreign—and when they are ingested, orally of course, always orally, their fate is unknown; no one can predict what happens to pills, powders, capsules, gels, or teas. Nevertheless, it is a good bet that they meet their waterloo before they ever see the duodenum. If a supplement is touted as an anticancer agent or antidepressant, or "anti"anything, and if it is to be taken orally, it is probably worthless. If this bit of incidental biochemical intelligence could be dispersed about the country, lots of green (dollar bills), otherwise spent on supplements, could be shelled out for chocolates, cheese, and chicklets.

Furthermore, if a supplement is a carbohydrate or a protein (both large molecules), and if it were to be injected, it would call forth a formidable immune response similar to the rejection that occurs with organ transplants as the body, in its wisdom, does not welcome foreign proteins gracefully. Large molecules have an awful time trying to pass through the walls of the small intestine. If too large, carbohydrates, for example, could be subjected to bacterial fermentation in the large intestine, which would totally demolish any unique activity it brought into the body. If not attacked by bacteria, it would be excreted, never making it to the circulating blood or tissue cells. If this further piece of incidental microbial intelligence could make it around the country, fewer phone calls would needed to be made to poison control centers and the FDA hotline.

Over the past half-century, no carbohydrate or protein supplement has ever made it as an anticancer agent, for the simple and wholly understandable reason that carbohydrates have only two ways to go: conversion in the digestive tract or elimination. Save your money. Proteins will, of course, be split, cleaved, into their constituent amino acids. There is no other way. Why can't the FDA make this clear to the public, or biology teachers to their students? The goings-on in the digestive system is old news. Is the country in denial?

So, the pharmacokinetics of all drugs is known; must be prior to use. That's the law. For herbal remedies and supplements, none of this is known, and worse, not required to be known. So, no surprise that most are not efficacious. But not to be surprised if a nasty reaction occurs. Bear in mind that with any single supplement we are dealing with dozens of crude chemicals, and what they do, no one knows. Is that any way to treat a body?

Echinacea

People swear by it. For many, echinacea is the only thing for the common cold, less severe and of shorter duration. Black-eyed susan is one of the most common homegrown flowers of summer on the United States east coast. Everyone seems to have the coneflower, or purple coneflower, as it is often called. Apparently native American Indians used it for centuries. Extracts of *Echinacea purpura* are contained in the most popular supplements.

Although people swear by it, the question remains, whether it really works. In a randomized, double-blind, placebo-controlled clinical trial—and that is as good as it gets—physicians in the Department of Internal Medicine of the Marshfield Clinic, Marshfield, Wisconsin, sought to determine the efficacy of a standard preparation of *E. purpura* in reducing symptom severity and duration of the common cold—one of the most prevalent acute illnesses in the United States Apparently adults contact 2–4 colds a year, and children as many as 10. If echinacea can short-stop colds, in economic terms alone, it would be worth billions. Literally.

At Marshfield, they enrolled 128 individuals who received either 100 mg of *E. purpura* (freeze-dried pressed juice from the aeriol portion of the plant) or a lactose placebo 3 times daily until cold symptoms were relieved or until the end of 14 days, whichever came first. At the conclusion of the study they found that echinacea was no better than the lactose placebo in reducing symptoms or severity. They described echinacea as "an ineffective therapy for the common cold," but noted that their sample of 125 individuals may have been too small to detect benefits [32].

More recently, another randomized trial, this one done at the University of Washington's Child Health Institute, included 400 children with colds. But the results were similar. Echinacea and the placebo were equally poor in reducing severity and duration of symptoms. The authors also noted that 1 in every 22 children developed a rash during the trial [33].

Yet a third study, this one using the randomized, double-blind design, enrolled 289 volunteers age 18–65, and free of acute illness, who were recruited from four military posts and an industrial plant in Munich, Germany. One group was given *E. purpura*; a second received *E. angustifolia*, and the third was given a placebo. Each group took their solutions twice daily, 5 days a week for 12 weeks during the winter. The outcome measure was time to first upper respiratory infection. The study's authors reported that the study "does not support the effectiveness of Echinacea in the prevention of upper respiratory infections" [34]. But they also noted that the participants in the two echinacea groups believed that they had benefited from the echinacea.

In July 2005, physicians at the University of Virginia School of Medicine reported on their evaluation of *E. angustifolia* in experimental rhinovirus infections. In this study, 437 healthy student volunteers susceptible to rhinovirus type 39 participated in a double-blind trial. After receiving the virus, each was isolated in a hotel room for the duration of the study, and the effect of echinacea on rhinovirus-induced inflammation was determined regularly. The physicians reported that the results "indicate that extracts of E. angustifolia root do not have clinically significant affects on infection with a rhinovirus or on the clinical illness that results from it" [35]. They further maintain that "The burden of proof should lie with those who advocate this treatment." A reasonable requirement.

Perhaps of greater concern is the fact that Airborne, a new cold remedy, is advertised as having been "created by a schoolteacher." Victoria Knight-McDowell, an elementary-school teacher in Sprechels, California, and her husband decided to market her "natural formula of 17 ingredients." After Kevin Costner declared his confidence in Airborne, the sales skyrocketed. By 2004, annual sales hit $90 million. There is, of course, no independent medical evidence of its value, and it does carry the asterisk and FDA disclaimer, but these have little efffect on decisions to buy and use [36].

So, given the many negative studies, we are left with the question as to why people continue to use these nostrums (panaceas). "Colds" are for the most part self-limiting. It's either 7 days or 1 week, treatment or no treatment. People must believe that they work; it's based on trust; or distrust, of mainstream drug producers.

Lorenzo's Oil

The movie *Lorenzo's Oil* portrays Lorenzo Odone, a young boy with a rare, inherited (defective gene) disease, adrenomyeloneuropathy, which destroys the sheath covering nerve fibers. The disease, a buildup of long-chain fatty acids in the bloodstream, affects 1 or 2 of every 100,000 people in the United States. Lorenzo's parents, Augusto and Micela Odone, concocted a cure, a mixture of oleic and erucic (fatty) acids that they believed would prevent the disease in healthy boys and delay further decline in those who had it. The Odones publicized Lorenzo's Oil as a cure, and in the film, with Susan Sarandon and Nick Nolte, it was further hyped as a successful treatment.

A team of physicians headed by Dr. Patrick Aubourg of St. Vincent de Paul Hospital in Paris, France, tested Lorenzo's Oil on 24 people—14 men, 5 women, and 5 boys—in varying stages of the disease. The 24 ate a low-fat diet and received daily doses of glycerol trioleate and glycerol trierucate for 2 years and were followed for another 2 years. After an initial decline in fatty acid blood levels, none of the 14 men improved. Nine showed severe deterioration.

One of the asymptomatic boys developed myelopathy (spinal degeneration); changes did not occur in the five heterozygous women. The study's authors concluded that they had seen "no evidence of clinically relevant benefit from dietary treatment with oleic and erucic acids in patients with adrenomyeloneu-ropathy" [37]. Augusto and Michela Odone continued to believe otherwise.

Ephedra

Although this section is limited to an overview and sampling, ephedra, or ephedrine alkaloids, recently banned by the FDA, requires inclusion.

The shrub *Ephedra sinica*, native to China and India, and known as *ma huang*, is the major source of the natural alkaloids *l*-ephedrine (1-phenyl-2-methyl amino-1-propanol) and *d*-pseudoephedrine. The plant contains as much as 1.5% of ephedrine. But it can readily be produced synthetically. Ephedrine is a bronchodilator and vasoconstrictor, similar to epinephrine.

Ephedra, the most widely used ingredient in dietary supplements for weight loss, is a powerful stimulant that can affect the nervous and cardiovascular systems. Because of concerns about the risks of ephedra, athletic associations, states, and medical organizations have tried to curtail its use in supplements.

A recent study conducted by a team from the Department of Family Medicine, University of Southern California, and the U.S. Veterans Affairs Healthcare System of Los Angeles had as its objective the assessment of ephedra's efficacy and safety. This study was a metaanalysis in which 52 controlled trials and 65 case reports were included. Their analysis speaks for itself. "We found sufficient evidence to conclude that ephedrine—and—ephedra-containing dietary supplements have modest short-term benefits with respect to weight loss and have harms in terms of a 2–3 fold increase in psychiatric symptoms, autonomic symptoms, upper gastrointestinal symptoms, and heart palpita-tions." They went on to remark that "We did not find sufficient evidence to support the use of Ephedra for enhancing athletic performance" [38].

Because of concerns about the safety of supplements containing ephedra, the House of Representatives, Sub-Committee on Oversight and Investiga-tions, Committee on Energy and Commerce, requested the General Account-ability Office (GAO) to present to the subcommittee an update on ephedra including records of health-related calls from consumers to metabolite 356, a weight loss product first marketed in 1995 by Metabolite International, a major manufacturer of ephedra supplements. Adverse event reports can often signal potential health problems that deserve investigation, but on their own, are insufficient to establish a cause–effect relationship.

The GAO informed the subcommittee that the FDA had received more reports of adverse events for dietary supplements containing ephedra than for any other supplement ingredient. Poison control centers and Metabolite Inter-national have received thousands of reports of adverse events associated with ephedra-containing supplements. From February 1993 to July 2003, the FDA received 2277 adverse reports, 15 times more than it received for St. John's Wort. The FDA estimates that it receives reports of less than 1% of adverse events associated with supplements.

The GAO also reported that "in our review of Metabolite 356's call records, we identified 14,684 that contained reports of at least one adverse event among users of Metabolite 356. Within those call records we found 92 reports of heart attacks, strokes, seizures and deaths—among those who used the product within the recommended guidelines" [39].

On February 9, 2004 the FDA issued a final rule prohibiting the sale of supplements containing ephedra alkaloids because "such supplements present an unreasonable risk of illness and injury." Although the rule would take 60 days to become effective, the FDA urged consumers to stop using ephedra immediately. This was after Steve Bechler, a 23-year-old pitcher for the Baltimore Orioles, collapsed and died during a workout at Fort Lauderdale Stadium on February 16th, 2003. It took 6 years for the FDA to gather plausible data to make the decision to ban the product. But that is not the end of the story. On April 14, 2005, a federal judge in Utah struck down the ban. The ruling calls into question whether the FDA can enforce its ban on ephedra anywhere in the United States, stating that the Agency had failed to prove that ephedra at low doses was dangerous, and that under DSHEA, it lacked authority to ban the substance without such proof. The judge called for the FDA to devise new rules for ephedra. Furthermore, under the 1994 law, the burden of proof, you may recall, falls on the FDA, to show lack of safety and efficacy. It will be illuminating to see how both the Senate and House respond to this, and if DSHEA gets a much-needed overhaul. Don't hold your breath.

Perhaps the most mind-boggling of all, is the notion of DSHEA's advocates that DSHEA provides the needed protection for the companies that cannot afford to undertake the expense involved in funding the clinical trials needed to prove their products are safe and effective. Yes, you read that correctly. We must be their test subjects, their guinea pigs, mice, and rats. From our adverse reactions and deaths, the manufacturers will learn about their products. Yet over the past decade with the many reports of illness and death, no product has been removed from the shelves. Arrogance. Indifference. Presumption. All appropriate. The supplement industry people know they have the power of Senators Orrin Hatch and Tom Harkin, champions of the dietary supplement industry. Politics rules!

The Chinese Solution

American fascination with, and for Chinese herbs and medications, is difficult to decipher. The question as to what Chinese herbs have done for the Chinese is not a fanciful one. After all, neither the general health of the Chinese people nor their longevity approaches ours, and their infant mortality rates are more than 5 times ours. Perhaps it is that Chinese herbals have a history of thousands of years of safe and effective use. It is not unusual to read that Chinese medicinal texts hark back 5000 or more years. That does catch the eye, and requires consideration. However, and in fact, few handwritten texts have survived the vicissitudes of time. The Yellow Emperor's classic volume of *Internal Medicine* is an exception and is believed to date around 400 BCE. But a compendium of herbal supplements, it isn't. Nor are there any statistics of adverse events, life expectancy, or mortality, but the Yellow Emperor does provide a clue as to the health of the people circa 400 BCE. "I have heard," he wrote, "that in early times the people lived to be over 100 years old. But these days people only reach half that age and must curtail their activities. Does the world change from generation to generation or does man become negligent of the laws of nature? Today people do not know how to find contentment within they are not skilled in the control of their spirits. For these reasons they reach only half their 100 years and then they degenerate." Degenerate? With all the herbs? Someone has done a remarkable job of marketing China as a paragon of health requiring translation to the United States.

One Chinese herb needs to be kept at arm's length. Aristolochic acid (AA), the plant extract of various *Aristolochia* species (clematis, fangchi, and manshuriensis), is a mixture of structurally related nitrophenanthrene carboxylic acids. For years AA was used in snakebite treatments, and as antiinflammatory agent for arthritis, gout and rheumatism. However, in the 1980s it was shown to be a strong carcinogen in rats [39]. Although products containing AA were withdrawn from markets in Germany, it continues to be found and used elsewhere despite FDA warnings. In the 1990s AA was responsible for cases of Chinese herb nephropathy, a unique type of rapidly progressive renal fibrosis due to prolonged ingestion of AA in slimming (weight loss) supplements. Aristocholic acid is one of the few supplements whose pharmacokinetic activity is now well known, along with its genotoxicity and carcinogenic mechanism. The metabolism of AA has been studied in different species including humans, and has shown that the products of nitroreduction, the aristolactams, are found in both urine and feces. It is now clear that the enzyme prostaglandin H synthase is an important activator of AA to toxic and carcinogenic metabolites, that induces bladder cancer in men. With the widespread use of this dangerous supplement, it will not be surprising to see increases in this malignancy. Aristocholic acid is one supplement to steer clear of.

Two herbs used to treat rheumatism and arthritis, the root of *Aconitum kusnezoffei*, and chuanwa, the root of *Aconitum carmichaeli* have caused poisoning and death, as they contain toxic alkaloids [40]. Similarly, the flowers of *Datura*, used to treat asthma, chronic bronchitis, stomachaches, and toothaches, induce anticholinergic poisoning. Cigarettes containing *Datura* for asthma induce tachycardia (rapid heartbeat), dry mouth, dilated pupils, and an array of other symptoms. *Datura* and *Rhododendri* are known to contain the alkaloids hyoscyamine, atropine, and scopolamine, which have analgesic, sedative, and anticholinergic properties. *Stephania* and *Magnolia*, used in weight loss diets, have also produced nephritis. Licorice extracts contain glycyrrhizic and other acids that can cause sodium and water retention, resulting in diarrhea. Chinese blackballs, used in self-medication for joint pain and sleeplessness, often contain diazepam, indomethacin, and phenylbutazone, as well as the heavy metals lead and cadmium and other non-prescription-type drugs. But you'd never know it. None of them ever appear on labels, because none have been analyzed for their actual contents, and there is no requirement that they be listed. Users are playing Russian roulette.

About 25–50% of Chinese infants are given chuenlin by their mothers to clear them of the "toxic products of pregnancy." One of chuenlin's alkaloids, berberine, can readily displace bilirubin from its serum-binding protein, causing a rise in free bilirubin, which can cause brain damage, along with icterus-induced jaundice. Placing total faith in Chinese herbs or proprietary medicine needs rethinking. Complications are increasing with increasing use among both Asians and Westerners. In a recent review of the use of Chinese herbal medicines in Hong Kong researchers of the Department of Clinical Pharmacology at the Chinese University of Hong Kong inform us that of the 150 most-often used herbs, 10 are out-and-out toxic [41].

Supplement manufacturers have no qualms in stating that their products are medically beneficial and will bring about a cure without using "drugs," and they hope that few read the small print or pay attention to the asterisks on the label. Ah, yes, the label. In my recent tour of a new, huge Chinese market in our area, I chanced upon the herbal "medicines." Shelves and shelves of them, directly from mainland China and Taiwan. Asking a Chinese-speaking friend to check a sampling of boxes and bottles for the FDA disclaimer, I found none seemed to have it. These are being sold in markets where anyone can purchase them. It will be some time before the FDA stumbles onto them.

Probiotics—Was Methuselah a Bulgarian?

It's a Microbial World

We need them, we use them. We can't possibly live without them, but we've grown up believing that microbes, germs, the wee beasties, are bad, bad, bad, and being bad need expunging and eradication from every nook and cranny that we humans could conceivably contact. In fact, life without germs (from the Latin *german*, sprout or seed) is inconceivable, and, as we shall see, would be dull and boring. First, the big picture.

All living things are organic, meaning that they—we—are composed of compounds of carbon; plants, trees, and grasses, including marine flora, use carbon dioxide to assemble carbohydrates via photosynthesis. The green things are a major source of food for grazing animals, fish, and us humans. The end product of all animal and human metabolism is carbon dioxide, which is returned to the atmosphere, and when all mammals die, the carbon in their bodies is returned to the soil. Soil microbes are the primary decomposers of dead organic matter in all human and animal waste, dead plants, and other things organic. This circular event goes by the name "carbon cycle" and is microbe-driven.

For protein to be made, amino acids are needed. To make amino acids, nitrogen is needed. The protein of dead organic matter—plants, animals, and humans, decomposed by microbes—deposits nitrogen in the air and soil. Additional nitrogen reaches the soil in the form of urea from urine, and the microbial decomposition of urine releases ammonia, which is then metabolized into nitrogen. Soil bacteria convert nitrogen to nitrates and nitrites, which plants convert to amino acids, which become protein when eaten by animals and humans. A substantial amount of nitrogen rises into the air, but neither humans nor plants can use it. Almost 80% of the air we breathe is nitrogen, but we can't use it. But microbes in the soil can, and do. They convert or "fix" it to ammonia, and then back to nitrates, which plants use to make amino acids, which will become protein again. This circular event is the nitrogen cycle and is also microbe-driven.

At least three amino acids contain sulfur (methionine, cystine, and cysteine) and are essential components of protein. A third cycle, the sulfur cycle, works to release sulfur from decaying protein amino acids, which yet another group of bacteria can attend to, while others convert sulfur to sulfates that plants can use to make the sulfur-containing amino acids. Around and around it goes, forever cycling, ever since animals, plants, and the microbial world formed their unseen, intimate bond millions of years ago. Clearly, a world without microbes is unimaginable—worse, unworkable. Without decomposition and recycling, there is no life, no growth, and no renewal. Microbes and humans are forever linked, and the great majority of microbes work for us, full time. How?

Consider that milk and milk products have been used as important foods since the walls of Jericho came tumbling down. Fermented milks, bacterially fermented buttermilk, acidophilus, yogurt, leben, kumis, hefir, as well as butter owe their unique flavors and texture to the wee beasties of the unseen world. Bacterial fermentation is the oldest form of food preservation. And where would butter be without bread, which owes its leavening of dough to yeast cells, another microbial species. Without yeast we'd all be eating matzoh, unleavened bread. Rye and pumpernickel are sourdough breads, as is buttermilk bread, the products of combinations of microbes. All our crackers are microbially cultured, and those crackers go exceedingly well with Camembert, Limburger, Roquefort, Brie, Gouda, Edam, and cheddar cheeses and some 1500 others, none of which would exist were it not for molds (fungi) and bacteria. And what of those Spanish, Greek, and Italian olives, black and green, all microbially fermented? Can anyone imagine a world bereft of bacterially fermented pickles (kosher dills!), sauerkraut (I'm salivating worse than Pavlov's dogs), fermented cauliflower, and other veggies? Our banquet table would not be complete without an array of sausages; kielbasa, lebanon, genoa, cervelet, thuringer, and salami owe their singular tastes to our many bacterial friends. And what about mushrooms—fungi all.

To help wash all this down, wine, beer, vodka, sherry, champagne, and cider, the hard kind, could never be without abundant help from a variety of bacterial flora. Don't forget the Mexican pulque (maguey), the fermented wine of the agave. Could our day begin or end without coffee, tea, and cocoa? Fermentation of coffee cherries occurs during drying, as it does for cocoa beans and tea leaves. Vinegar, ginger, and tabasco all come to us via microbial flourishes, and could bakers and ice cream makers live without vanilla, from the fermented and dried pods of several species of orchids of the genus *Vanilla*? Not possible. But our table, already groaning under the delicious weight of all the goodies microbes gave us, needs the final touch of soy sauce, miso, tempeh, and MSG (monosodium glutamate), fermented soybeans all. Life without microbes; never.

So, microbes are all around us—in soil, in our foods, in the air. All natural, many necessary. But what about our bodies? True, we have microbes on our skin. But what of our innards? Of course we ingest them with our foods, but do they benefit us—work for us? And more to the point, can we increase our natural flora accruing additional health benefits?

Let us consider our innards. As noted earlier, in our stomach's highly acidic environment (pH 2–3), most microbes, germs, washing down from our mouths are rapidly destroyed. Consequently our stomachs contain small numbers of bacterial cells. The small intestine, the duodenum, contains few organisms given its proximity to the stomach's acid, the action of bile and pancreatic secretions. The midsection, the jejunum, contains a mixed flora of gram-positive rods and cocci. Lactobacteria begin to appear in numbers. In the distal ileum, the mixed flora begin to resemble the complex flora of the colon, the large intestine. Although still not fully accounted for, some 800 species of bacteria constitute the natural flora of the healthy human gastrointestinal tract. The actual numbers are astronomical: 100,000,000,000,000—trillions (10^{14}). The colon has the body's largest microbial population. At birth our intestinal tracts are bacteria-free. If breastfed, organisms of the genus *Bifidobacterium* colonize the gut. If bottle-fed, with cow's milk, lactobacilli are dominant. With consumption of solid food, anaerobic bacteria become important.

The colon contains a complex and curiously balanced microbiota that appear to prevent infection and are nutritionally beneficial. Abrupt changes in diet, antibiotic therapy, diarrhea, and stress can upset the balance, reducing the efficient use of food, as well as increasing susceptibility to infection by pathogenic bacteria, which can gain attachments to the intestinal walls. Indeed, the human gastrointestinal tract is an active microbial ecosystem—but still not as well understood as might be.

In 1905, French scientists at the Pasteur Institute in Paris discovered lactobacilli, and in 1907, Ilya (Elie) Metchnikoff, the director of the Pasteur Institute, published his epoch-making book, *The Prolongation of Life*, in which he suggested that eating fermented milk containing active cultures of lactobacilli would engender the same long and healthy lives as lived by the Bulgarians. Who was this man, and why has his flawed theory survived to this day?

Ilya Ilyich Metchnikoff, a Russian biologist, was born in the Ukraine in 1845. He studied at Russian and German universities and joined the Pasteur Institute (while Pasteur was there) as a researcher when he became intrigued with the immune system. He found that certain white blood cells could engulf and destroy harmful microbes and coined the term *phagocytosis*, and called the engulfing cells *phagocytes*. In 1908, he and Paul Ehrlich shared the Nobel Prize in Physiology and Medicine. Metchnikoff also founded the medical discipline of *gerontology*—the study of old age. Although he didn't coin the term *probiotics*, or *proboscis*, he can be considered the grandfather of our current concern with *probiosis*, a term not found in most dictionaries, but that can be defined as the ingestion of active, live lactobacilli (or *Bifidobacteria*) to promote health. Yahourth (the origin of yogurt), a soured or fermented milk, was a staple of the Bulgarian diet. Soured or curdled milk is the consequence of metabolism of milk proteins by lactic acid producing bacteria—primarily of the lactobacillus species, of which there are a clutch. For Metchnikoff it was *Lactobacillus bulgaricus*. He believed, with little or no direct evidence, that old age and death were hastened by toxins produced by intestinal putrefaction in the colon, and that putrefaction and toxicity could be subdued by the ingestion of yogurt-containing lactic acid producing bacteria.

The notion—and notion it is—that intestinal toxins need to be removed if we are to thrive and obtain length of days, is a medical myth. Intestinal putrefaction (fermentation) and toxins have remained a basic marketing tool of the probiotic movement, even though Metchnikoff wrote that, "In the meantime, those who wish to preserve their intelligence as long as possible and to make their cycle of life as complete and as normal as is possible under present conditions must depend on general sobriety and habits conforming to the rules of rational hygiene" [42]. Few appear to have read this, preferring to down copious pints of yogurt. Yes, Metchnikoff is responsible for today's yogurt craze.

Just what is being purveyed by the probiotic supplementers? In my local Whole Foods supermarket there were, at last count, 13 different products:

MAXI: *Babydophilus*
Jarro dophilus EPS
Jarro dophilus—six strains
Probiotic acidophilus
Primadophilus
Daily dophilus
Flora more—five strains
Micro flora that normally occur
ABC dophilus powder
Advanced acidophilus plus

Multiacidophilus powder

Sensitive colon support

Sensitive cold and flu

6 LAB strains

6 *Bifidobacteria* strains

As I recall, most carried the required FDA disclaimer, but several required a private eye to find it.

In 1982, Tufts University researchers discovered that the lactobacilli commonly found in yogurt, *L. acidophilus* and *L. bulgaricus*, did not in fact colonize the intestine (attach to the walls) and consequently could not have a beneficial effort. However, they did find some success with a strain they referred to as *lactobacillus GG* (after Drs. Sherwood Gorbach and Barry Goldin). Lactobacillus GG was also known as *L. rhamnosus GG*, which is extremely difficult to find in supermarket products, nor is it often the test organism in published studies.

In addition to intestinal rebalancing, probiotics are often touted for their immunity enhancement activity, especially for the elderly, whose immune systems are ebbing. A study carried out at Massey University, New Zealand, administered *Bifidobacterium lactis* HN019 to healthy volunteers 63–84 years old, for 3 weeks before and 3 weeks after a washout period. At the end of 3 weeks, several types of leucocytes had increased in number. But there was no evidence that this rise in white cell count had affected health in any way. Moreover, the study had too few subjects to be meaningful [43].

When it comes to fermented milks, Scandinavians, especially the Finns, as with eastern Europeans and Middle-easterners, are especially fond of yogurt-type milks, and believe in the idea of probiosis. Consequently many studies emanate from Finland. So, for example, physicians in the Department of Pediatrics, University of Turku, Helsinki, ran a randomized, double-blind, placebo-controlled study to determine whether probiotic microbes could reduce allergic reactions in infants. In this trial, pregnant mothers took two capsules of lactobacillus GG daily for 2–4 weeks before delivery. This was then followed by 6 months of feeding lactobacilli to their infants. After 2 years, infants who received the lactobacilli had a rate of chronic allergic eczema half that of infants who didn't. The authors concluded that "gut microflora might be a hitherto unexplored source of immunomodulators and probiosis for prevention of atopic (allergic) disease" [44]. The fact that infants receiving the lactobacilli had half the eczema is a remarkable difference that must be replicated by others, in far larger trials. If it is confirmed, proboscis would represent a major therapeutic advance.

In another study using lactobacillus GG, a team of physicians from the Hospital for Children and Adolescents, Helsinki University, questioned whether a probiotic milk would reduce the incidence and severity of respiratory and gastrointestinal infections after long-term consumption by children in day-care centers. Here, too, a randomized trial was undertaken and included 18 centers with 571 healthy children ages 1–6. In this study, 282 children received the lactobacilli, while the control group drank only milk. At the end of 7 months they found the effects of the lactobacilli to be "modest" —slightly reducing infections [45]. An accompanying editorial suggests the need for larger and more carefully controlled studies, given the fact that lactobacilli GG are widely consumed in Finland, where some 15% of the children in the control group had lactobacillus GG among their gut flora.

Dr. Sherwood Gorbach has stated that "evidence for positive health benefits of Lactobacilli applies to only a few strains used for commercial applications. To qualify as a probiotic, a bacterial strain must be capable of colonizing the intestinal tract if health benefits are to accrue. This requirement disqualifies many of the strains currently used in commercial fermented milks" [46].

Functional Foods—or Food as Medicine

Is our current knowledge of health being used to modify food? My dictionary defines *functional* as the intended purpose of a thing in a specific role, as the function of food. Are all foods inherently functional? More than likely, if we mean that foods provide nutrients and energy. But is this what the new functional foods are about? If so, they needn't bother as there are more than enough foods of every type and description to satisfy the most discriminating palate. Functional foods are meant to do more, to go beyond the traditional— and just what is that?

The Institute of Food Technologists, a professional organization of academic and industrial food scientists, recently convened an expert panel to advance the science of functional foods. The panel produced a report, "The promise of functional foods: Opportunities and challenges" [47]. The panel's report was also encapsulated in an easily understood chart (Figure 2.3.) At the top of the chart are the various options, from food to drugs, and moving down

Institute of Food Technologists Expert Panel View of Functional Foods*

Delivery Options:	Foods-Enhanced Foods-Supplements-Medical Foods-Drugs
Purpose of Therapy	Reduction of risk--------------------------------> Treatment of Disease
Health Professional	Involvement: Low-----------------------------> High
Individual Participation:	High---> Low
Treatment Costs	Low--> High

Figure 2.3. This chart suggests that functional foods have a place in healthcare, providing a low-cost approach to disease prevention. [*Source:* Adapted, with permission, from *Food Technol.* 58(12):37 (2004).]

on the left, is the "why" of therapy; whose involved, and to what degree, and the cost of treatment in a semiquantitative form, but meant to show the benefit of enhanced or functional foods, which is fine. However, this clearly and obviously deals with medical care. This is not just eating. But it does suggest that an "apple a day will keep the doctor away." Are we back to the future? Would Hippocrates (460 BCE) recognize functional foods? After all, he did say, "Let food be thy medicine." But herbs and supplements are not foods. If they were, they could not be sold or advertised as supplements as functional foods are. They would fall directly under the purview of the FDA, which functional foods do not, nor do they wish to.

As the IFT's expert panel sees it, functional foods have a place in the healthcare continuum that provides low-cost approaches to disease prevention and gives consumers a significant role in maintaining good health. This is a splendid concept, and would have much to recommend it. However, thus far, it remains to be demonstrated which dietary supplements and which functional foods prevent any disease. Our knowledge of dietary supplements cannot now support such a claim. How this eluded the expert panel is curious, indeed, and why functional foods with added supplements, would be expected to perform any better, is unclear. Yet the entire concept is medically oriented. Furthermore, the dietary supplements most frequently associated with adverse reactions, ginseng, St. John's Wort, echinacea and ginkgo, are the ones appearing in functional foods and beverages.

Like dietary supplements, functional foods, also known as "nutraceuticals" and "novel foods," are permitted to carry structure–function claims. So, for example producers of echinacea, added to a conventional food, soup, or tea, can state that this product "supports the body's defense system," but only if there is evidence showing that the claim is truthful and not misleading, but is not—and I repeat, is not—required to carry the disclaimer. Refer back to Figure 2.2, to see the difference. This difference has caused widespread confusion, as well it should, as there appears to be no justification for the different safety standards. In fact, there is no scientific reason supporting different standards for foods and supplements. Both should meet the same standards of safety and efficacy. Both are being ingested, and both have systemic effects.

Here again the FDA has been blindsided as the Food and Drug Act allows companies to market a product if it is determined that ingredients in it meet the "generally recognized as safe" (GRAS) standard, so these companies are not required to notify the FDA before bringing their product to market. The FDA may become aware of the product postmarketing if an individual complains, or other companies view the new product as unfair competition. The FDA must then notify the company that their product containing St. John's Wort, ginkgo, ginseng, or echinacea have not been generally recognized as safe in the specific food product. If the FDA wanted to remove such products, it would have to initiate an enforcement action, an involved, lengthy, and costly procedure that they are loath to do, and that is not lost on manufacturers. Additionally, the U.S. Department of Agriculture (USDA) regulates product label claims for functional foods containing 2% or more of cooked or 3% or more raw meat or poultry: potpies, soups, and prepared meals. With these overlapping jurisdictions and laws governing the safety of functional foods and dietary supplements as well as the implementation of the laws, products of questionable safety and efficacy can and do reach our markets.

Curiously enough, broccoli, carrots, and tomatoes are subsumed under the rubric of functional foods because of their content (natural) of sulforaphane, β-carotene, and lycopene, respectively. The General Mills breakfast cereal Cheerios is referred to as a "functional food" because of its soluble fiber (also naturally present in the grain) which is stated on the container to reduce serum cholesterol levels. Broccoli and sulforaphane have been touted as providing health benefits to humans when it has been shown only to reduce the incidence of mammary tumors in female rats. The John's Hopkin's researchers who conducted the rodent study remarked that "little is known of the metabolism of the glucosinolates in humans" [48]. But that does not discourage the functional food folks. For them, animal studies are equal to human studies. As for β-carotene, it has been shown in human trials to be more harmful than beneficial—and lycopene, as a prostate cancer preventive, appears to be fading rapidly.

Quality control was yet another challenge for functional foods. Drugs, of course, are strictly regulated by the FDA with industry compliance. Few, if any, regulations—government or industrial—monitor functional foods. Ergo, currently consumers are swirling in a wild-West-like environment, where anything goes. Misleading health claims appear to be the rule rather than the exception. How long this will continue is problematic, as intimations of change are hinted at in newspaper editorials and letters to the editor, and by some legislators. But the supplement and novel food folks will not allow their billion-dollar baby to slip away without a struggle. The next few years could be uncommonly educational.

Direct-to-Consumer Marketing: Circumventing the Supplement Marketplace

With its lax standards and poor government oversight notwithstanding, direct-to-consumer marketing seeks yet additional advantage and unfettered freedom. By going directly to potential consumers, supplement manufacturers need only extol their product's virtues. Side issues such as disclaimers can be ignored—until someone blows the whistle, which is not very often. So, for example, Dr. Cherry's Pathway to Healing, a line of 19 condition-specific supplements, marketed through a weekly television program, is the brainchild of NAI (Natural Alternatives International, Inc.), a San Marcos, California, company, listed on the Nasdaq exchange. NAI, Inc. also markets (Deepak) Chopra Center essentials directly to potential customers. You'll never find their products in local supermarkets or discount chains.

For Dr. Phil (Dr. Phillip C. McGraw), a talk show host whose syndicated program, *Dr. Phil*, is second only to Ophrah Winfrey's, where he got his start and garnered a huge TV following with his tough-love brand of psychological advice, his popularity as an advice-giver was not sufficient. Tying up with CSA Nutracenticals of Irving, Texas, he lent his name and image to Shape Up!, a line of supplements, vitamin packets, power bars, and meal replacement drinks. Yet Dr. Phil is not an expert on nutrition, which makes little difference. It's his celebrity that is used to market and sell the products. His pitch is that use of his supplements will ensure weight loss. However, professional nutritionists maintain that there is no sound scientific evidence that supplements help anyone lose weight. Shape up! Containers carry the message that "These products contain scientifically researched levels of ingredients that can help you change your behavior to take control of your weight." Dr. Phil knows he has a popular following, and he's milking it to the hilt.

Of course it's the celebrity who sells the product, as Larry King has done for Coral Advantage, extolling the virtues of coral reef calcium—which both the FDA and FTC have moved against.

Celebrities are well paid for their personas—but this is nothing new. Manufacturers of almost everything have used celebrities—ballplayers, movie stars, TV personalities to help sell their products, and the celebrities, who could care less about the product's safety or efficacy, laugh all the way to the bank.

Although Figure 2.4 is a wonderful example of this, Regis philbin doesn't look all that happy and when I showed this ad to a friend, who said, I can see Regis, but I can't see the disclaimer. Can you?

The Younger Generation

Two recent surveys of university and high school students provide useful information about supplement use, and should be taken note of.

For example, among a technical vocational high school population, which the researchers indicate is not representative of the American high school population, although the sports offered is similar, they found that supplement use is common—74% of the students used either traditional or nontraditional supplements; vitamins and minerals were traditional, while ephedra and echinacea were nontraditional. Vitamin C was the most popular choice for colds, as students believed that it reduced their incidence and severity. Ginseng and golden seal were the most frequently used herbals, and 3% reported using ephedra even though banned. Both boys and girls in sports were more likely to use supplements, and they took them for general good health—36% of the teens used supplements to increase energy. They obtained their information about supplements from family and physicians, but were not generally aware of potential ill effects of use and overuse [49].

A survey of students at Rutgers University was conducted by health professionals at the University Health Center, Newark, NJ. With five campuses and

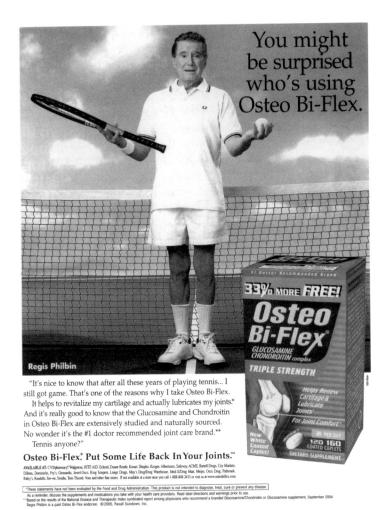

Figure 2.4. Celebrities can sell almost anything. Did you find the disclaimer?

50,000 students, they sent a mass mailing to all University Professors requesting voluntary distribution of the survey to their students. From their 3.5% response (1754 students), they found that more women than men used herbals. Ethnicity did not appear to be a factor. The students using herbals believed them to be safer than prescription drugs, but interestingly enough, were uncertain that herbals were more effective than prescription drugs.

The top five herbals used were, in order of preference, echinacea, ginseng, St. John's Wort, chamomile (a tea), and ginkgo biloba. For the most part their knowledge of supplements came from friends, family, and self. Over 50% of the students who defined themselves as healthy used supplements. The survey authors inform us that because university students are in a period experimentation, and seeking information, and they may be pregnant, and/or may use prescribed, over-the-counter, or recreational drugs, it is necessary that health center personnel be vigilant about their patients' use of supplements and supplement toxicity, and be able to educate their patients about these issues [50]. I can only agree, but it is unfortunate that their survey had so meager a response, as it prevented generalizable statements about supplement use by the huge student body. The small sample may will be unrepresentative. Or it may just be the tip of the iceberg.

Nutrigenomics—an Epigenetic Revolution?

Is it what you eat, or possibly, you are what your grandmother ate?

Classical nutrition treats everyone as genetically similar. Molecular and genomic research assumes that the environment does not influence genetic expression. Nutrigenomics combines these concepts, maintaining that environment can and does influence genetic expression, but does not alter DNA sequences. Mutations are not involved. Shades of Lamark!

Jean Baptiste Lamark (1744–1829), a French biologist, suggested that useful features acquired during an individual's life could pass to the next generation. August Weismann exploded that theory by cutting off the tails of mice and showing that offspring of such mice had intact tails. Nevertheless, scientists have long wondered about such a possibility. It has long been suspected that what pregnant women eat can affect their babies' susceptibility to disease, but no one knew why. A now classic study of the Dutch famine during the war years 1944–1945, when the German Army prevented food from moving into Holland, disclosed that children and grandchildren of Dutch women who were pregnant and on near-starvation diets during the first trimester of their pregnancies produced offspring who during adulthood manifested a wide spectrum of diseases: schizophrenia, diabetes, obesity, heart disease, and breast cancer, as well as low-birth-weight infants, and smaller than normal grandchildren [51]. This implies that a pregnant mother's diet can affect her children, and her children's children, her grandchildren, with a range of illnesses. But Lamark stood in the way of accepting environmentally induced inheritance, and Darwinian evolution required a mutation to effect such a change. Interpretation was impeded.

In 2003, researchers at Duke University unraveled the knot, supporting the Lamarkian concept of acquired inheritance, but in a way that he could never have imagined. *Epigenetics* is the word for today, and is defined as the study of our personal environment, diet, stress, and maternal nutrition, which can alter gene function expression—without altering DNA sequences. So, for their experiment, the Duke investigators chose agouti (*Dasyprocta*) mice that had a transposon (a trigger) abutting the gene that codes for coat color. The transposon induces the gene to overproduce a protein that turns the normally dilute brown mice, yellow or yellow-brown. But the turncoat protein also blocks a feeding control center in the brain, which stimulates the mice to eat excessively and develop diabetes and cancer. This was revolutionary. Diet dramatically altered a heritable phenotypic change, but by changing DNA methylation, not by changing any DNA sequence.

Epigeneticsts are searching for biochemical processes other than mutations, which can modify gene functioning. One mechanism, methylation, appears to be able to speed up or slow down gene expression, depending on the number of methyl groups in the vicinity, or what part of the gene it affects.

During methylation, the methyl groups (CH_3) attach to genes at specific sites and induce changes in gene expression. Methyl groups apparently play an essential role in controlling genes involved in pre- and postnatal development. Methylation is nature's way of permitting environmental factors (diet) to affect gene expression. Even short or transient exposure to methyl groups during fetal development can alter a person's life. But until observed, it will not be known whether the alteration is good or bad. Predictive ability does not yet exist. The question that begs to be asked is, where do methyl groups come from? One answer is food, which places a heavy burden on appropriate diets during pregnancy.

The Duke scientists kept their mice on a diet loaded with methyl groups in the form of vitamin B_{12}, folic acid, choline, and betaine (from sugarbeets, which has three methyl groups, and B_{12} is loaded with methyl groups) before they became pregnant, and throughout the weaning period. According to Drs. Jirtle and Waterland, methyl groups silenced the transposon, which, in turn, affected it's abutting the coat color gene. The mice were born with the normal brown color. Although they had an inherited tendency to obesity, diabetes, and cancer, it was silenced, dampened, by their mother's diet. Also, although the researchers could not determine which supplement or which combination was most effective, they did note that it didn't take a great deal of methylation as they fed the female parents only 3 times as much of the vitamins and betaine as found in normal diets [52,53]. The control group of mice on unsupplemented diets had offspring with yellow coats and high susceptibility to diabetes, cancer, and obesity. This elegant study demonstrates that in mice, appropriate supplementation for pregnant mothers reduces risk of offspring with health problems. "But these findings also suggest that dietary supplementation, long presumed to be purely beneficial, may have unintended deleterious influences on the establishment of epigenetic gene regulation" [54]. So, too, this study "supports the conjecture that population-based supplementation with folic acid, intended to reduce the incidence of neural tube defects, may have unintended (negative?) influences on the establishment of epigenetic gene-regulatory mechanisms during human embryonic development" [55].

Epidemiologic studies over the past decades support the idea that diet at critical stages of fetal development can lead to untoward health problems in the second and third generations. The question currently being asked is whether our modern diets, with their emphasis on fats and sugars, exert epigenetic pressure on our children and grandchildren—positive and/or negative. Although answers are not yet available, one thing is certain—epigenetics is back, and here to stay.

ALTERNATIVE MEDICAL PRACTICES

The Promise and the Peril

If there is one universal desire, it is to be well. No one wants to be sick, especially with a chronic disease for which current traditional, mainstream medicine may not have preventive or curative medication. But people need—must have—hope, and if traditional practitioners can't provide it, people will go elsewhere, for as Shakespeare tells us, "hope springs eternal," no matter how far-fetched or illogical the preferred therapy.

Various surveys indicate that 30–50% of our adult population have used some type of unconventional or alternative medical treatment. According to the National Center for Complementary and Alternative Medicine (NCCAM), complementary and alternative medicine (CAM), are a group of diverse medical and healthcare systems, practices and products that are not presently considered part of conventional medicine.

Before visiting the world of unconventional treatment and practices, let us briefly look at the NCCAM. This center is 1 of 27 institutes and centers of the National Institutes of Health (NIH), which is itself one of eight agencies in the Public Health Service of the Department of Health and Human Services. NCCAM was established in 1998, 7 years after the NIH created the Office of Alternative Medicine under duress of the U.S. Senate, to pull together the patchwork of therapies into a coherent research program. Over the past 14 years it has proved a Sisyphean task that, as we shall see, remains a triumph of politics over science. Nevertheless, the NCCAM supports clinical and basic science research by awarding grants nationally and internationally. It also sponsors conferences and educational programs, and is a clearinghouse for information. Anyone with a question can call or check its Website: www.nccam.mih.gov.

Five major areas that the NCCAM explores and funds include biologically based practices, mind–body therapy, energy therapy, manipulative and body-based practices, and whole medical systems.

Biologically based practices includes the entire realm of dietary supplements and functional foods. The center informs us that after virtually thousands of studies, no single one has proved effective. Clear enough.

Mind–Body Therapy

Mind–body practices traffics with the assumed interactions among the brain, body, and behavior, and the ways in which emotional, mental, social, spiritual, and behavioral factors may (can?) affect health. As the NCCAM makes evident, "It regards as fundamental an approach that respects and enhances each person's capacity for self-knowledge and self-care, and emphasizes techniques grounded in this approach."

With this underpinning, its practices believed to affect health include relaxation, hypnosis, meditation, biofeedback, yoga, reiki, visual imagery, tai chi, qigong, group support, spirituality, and prayer. Curiously enough, mind–body therapeutics views illness as an opportunity for personal growth and transformation, and its practitioners as facilitators and guides in the process. Research issues raised here questions whether emotions influence susceptibility to and severity of infectious disease, as well as rapidity of wound healing; questions that remain to be answered.

Energy Therapy

Therapies involving putative energy fields are based on the belief that we humans are permeated by a subtle form of energy—a vital energy or life force (élan vital), which goes by different names in different cultures: gi, in traditional chinese medicine; ki, in the Japanese kampo system; doshas, in ayurvedic medicine; and elsewhere as prana, etheric energy, fohat, orgone (hypothesized by Wilhelm Reich and supposedly emanating from all organic material and that supposedly can be captured in a boothlike device, the orgone box—used to restore psychological well-being); odic force; mana; and homeopathic resonance.

The NCCAM assigns energy to two major categories: veritable and putative. *Veritable energy* includes sound, electromagnetic energy, and light, each with specific measureable wavelengths and frequencies used to treat medical conditions. Included here are radiation therapy, cardiac pacemakers, magnetic resonance imagery (MRI), laser keratoplasty, and ultraviolet light.

Putative energy, or biofields, which the Center informs us have defied measurement to this day by reproducible methods, are said to flow throughout the human body, but it has not been unequivocally measured by any conventional instruments. Nevertheless, practitioners claim that they can work with this ethereal energy, see it, and use it to affect body changes and health. Examples of practices involving putative energy fields include qigong—a Chinese practice—and healing touch, in which therapists supposedly identify imbalances and correct a person's energy by passing their hands over the individual. Prayer is used specifically for health purposes, such as intercessory prayer, in which a person intercedes through prayer on behalf of another—usually someone they don't know. According to the NCCAM, "These approaches are among the most controversial of CAM practices because neither the external energy fields nor the their therapeutic effects have been demonstrated convincingly by any biophysical means."

The fourth area of involvement and funding consists of the manipulative and body-based practices, including the following:

- *Alexander technique*—patient education/guidance in ways to improve posture and movement, and to use muscles efficiently

- *Bowen technique*—gentle massage of muscles and tendons over acupuncture and reflex points
- *Chiropractic manipulation*—adjustments of the joints of the spine, as well as other joints and muscles
- *Craniosascral therapy*—form of massage using gentle pressure on the plates of the patient's skull
- *Feldenkrais method*—group classes and hands-on lessons designed to improve the coordination of the whole person in comfortable, effective, and intelligent movement
- *Massage therapy*—assortment of techniques involving manipulation of the soft tissues of the body through pressure and movement
- *Osteopathic manipulation*—manipulation of the joints combined with physical therapy and instruction in proper posture
- *Reflexology*—method of foot (and sometimes hand) massage in which pressure is applied to "reflex" zones mapped out on the feet (or hands)
- *Rolfing*—deep-tissue massage (also called *structural integration*)
- *Trager bodywork*—slight rocking and shaking of the patient's trunk and limbs in a rhythmic fashion
- *Tui na*—application of pressure with the fingers and thumb, and manipulation of specific points on the body (acupoints)

The NCCAM reports that of the 43 clinical trials of manipulative techniques, there has been only minimal evidence of short-term relief. Nor has any trial shown effectiveness with asthma, hypertension or dysmenorrhea. As for massage, many studies have shown positive benefits, but these were methodologically flawed studies. Nevertheless, patient satisfaction has been high.

The centers' fifth area of concern deals with whole systems, such as traditional Chinese, homeopathy, naturopathy, and ayurvedic medicine, none of which have ever been vigorously tested for efficacy.

Although over the past decade, the NCCAM has funded studies in these five areas to the tune of hundreds of millions of dollars, but there is little to show for it. Negative studies get little media attention, and positive benefits have yet to emerge. For example, four studies of *reiki*, a form of spiritual self-healing, have been funded and are works in progress. At the University of Washington, a group is studying reiki for the treatment of muscle pain–fibromyalgia. At the University of Michigan, painful neuropathy and cardiovascular risk factors are being investigated. The Cleveland Clinic has a grant to investigate reiki and prostate cancer healing. Temple University in Philadelphia has been funded to study reiki's benefits for advanced AIDS. All four have been works in progress for up to 3 years, with little to show for the effort and cost. Results of studies in many of NCCAM's five funding areas have never been published in reputable medical and scientific journals. However, and nevertheless, many of these areas have been investigated by scientists nationally and internationally. We will deal with a sampling of their published studies.

Therapeutic Touch

A controlled, double-blind test of a therapeutic touch practitioner's claims to be able to feel a "signal" or "fluid" in a nearby person was recently conducted in France [54]. The Observatoire Zetetique, a nonprofit organization of French skeptics, supervised the test. The protocol required that the practitioner try to determine the presence of a person behind a folding screen covered by an opaque cloth. The practitioner would have 100 attempts with the same person—50 with her present, and 50 with her absent; 65 successful determinations would prove the practitioner's ability. Two tries were invalidated, so 98 valid attempts were undertaken, and 64 became the successful number. Of the 98 attempts, the practitioner was successful in 55 and failed in 43. The experiment did not provide a statistically significant result. He failed to demonstrate his claimed ability.

Therapeutic touch is a widely used nursing practice claiming to heal or improve medical problems by manipulation of an energy field above a patient's body. Emily Rosa of Lovelande, Colorado designed and conducted a test of therapeutic touch.

Ms. Rosa's experiment required that she and 21 therapeutic touch practitioners, with up to 27 years of experience, be separated by an opaque screen. Each of 21 were tested to determine whether they could correctly identify which of their hands was closest to Emily's left or right hand. By flipping a coin, she determined which hand to extend. Fourteen practitioners were tested 10 times each, and seven, 20 times each. In 280 tests, they did no better than chance, correctly identifying her hand 123 times (44% of the time), close to what would be expected by chance alone. This experiment was devised by Emily, who was 9 years old at the time, and was working on a project for her school's science fair. Her study was published in the *Journal of the American Medical Association* [55], and showed that therapeutic touch claims were in fact groundless, and that professional use was unjustified.

Intercessory Prayer

Shortly after September 11, 2001, when the United States was staggered by the destruction of the World Trade Center, and prayer was everywhere, *The New York Times* reported that Columbia University researchers had demonstrated that infertile women who were prayed for by Christian prayer groups (who did not know the infertile women) had a doubling of their pregnancy rate, compared to women who were not prayed for—a 100% increase. The infertile women who were prayed for had no idea that groups were praying for them to become pregnant by in vitro fertilization [56]. The study was published in the *Journal of Reproductive Medicine* [57] and was considered nothing short of miraculous. Columbia University was so impressed that, they issued a news release indicating that the study was designed to virtually eliminate bias. This spectacular news raced around the world. How could it not? The implications were beyond belief, and the people doing the praying were thousands of miles away from the infertile women! Simply by praying, fertility not only occurred, but was doubled. No medical procedure could come close. For 2 years after the publication this study was used to buttress the claims of faith healing, until it all crumbled as a fraud—a fraud perpetrated by the authors, and blindly accepted by a medical journal. One of the authors, Daniel Worth, is now in a federal penitentiary because of a long history of fraud and conspiracy. By their abject silence, both Columbia University and the *Journal of Reproductive Medicine* fostered the notion of the power of prayer. How Daniel Worth became a member of the Department of Obstetrics and Gynecology has still not been explained—and because this fraud by an elite university has been hushed up, this "study" continues to be used as a model of the power of prayer.

Iridology

A widely held belief assumes that many diseases can be diagnosed by looking at the iris of the eye. For some, iridology is considered a critical diagnostic supplement to a medical history and conventional physical examination. Consequently, a physician at the University of Limburg, Maastrict, The Netherlands, sought to validate iridology by having iridologists diagnose gallbladder disease [58]. The object was to determine whether skilled iridologists could distinguish between people with and without gallbladder disease, and whether there was any consistency among them.

A series of 39 patients from the Academic Hospital who were scheduled for gallbladder removal, were the study participants. Those with jaundice were excluded. The presence of gallstones and inflammation was confirmed after surgery by both chemical and histologic exams. A similar-size control group matched for age and sex but without gallbladder disease were assembled. None had jaundiced eyes or silent gallstones. Color slides of actual eye size were made of the right iris of all 78 subjects. Studying the slides with a stereo-magnifier obtained a three-dimensional image, a common practice among iridologists. Five experienced iridologists were involved. Each was given the slides along with the information of each person's age, sex and the fact that some had gallbladder disease. Each was asked to score the probability—from 0% to 100%—of gallbladder disease in each of the participants. The results were disappointing. For 15 subjects, 7 with an inflamed gallbladder with gallstones, 4 iridologists considered the iris image to be negative for gallbladder disease. For another 20 subjects, 10 healthy volunteers, 4 iridologists found the subjects to be positive for gallbladder disease. Gallbladder disease was chosen for this test as among iridologists it is considered one of the easiest conditions to diagnose. Conclusion: Iridology is not a valid test for diagnosing gallbladder disease. Whether it is for any other condition remains to be demonstrated.

Homeopathy

Homeopathy is a method of treating disease devised by a German physician Samuel Hahnemann (1755–1843), in which small amounts of a drug that in healthy people produces symptoms similar to those of the disease being treated. Actually, two principles underlie homeopathy: the law of similars and the law of infinitesimals. The *law of similars* maintains that plant, animal, or mineral substances, when given in large quantities to healthy individuals, produce symptoms of disease, but that much smaller doses of the same substances can (theoretically) relieve the same symptoms. The *law of infinitesimals* holds that substances, drugs, treat disease most effectively when they are diluted in water or alcohol to literally undetectable levels. The understanding is that the drugs leave "imprints" in the aqueous or alcoholic solutions and that these "imprints" are sufficient to deal with the disease.

The law of similars is often stated as "like cures like." Modern traditional medicine, allopathy, or allopathic medicine, uses substances to reverse symptoms, not produce them. Many traditional physicians and scientists find it difficult to accept the law of infinitesimals as the solutions are diluted to extinction, and that the idea of imprinting on water or alcohol molecules is illogical and unreasonable. A recent controversy in Italy is instructive.

In July 2000, a popular prime-time radio program presented a brief but highly critical account of homeopathy. Homeopathists were furious and took the program's host to court—filing two lawsuits, one civil, one criminal. This was the first time in Italian legal history that a criminal suit was brought against a remedy. After 3 years, the case was settled. In May 2004, the court handed down its 59-page, highly detailed verdict—homeopathy, it stated, had no scientific validity. This was also the first time a court made a judgment on this topic. Here are the court's words [59]:

> Science, in fact, is not a matter of mere categories of opinion. In the scientific field, either something is, or it is not. Either a treatment works, or it doesn't. And if it works, it is necessary to demonstrate the fact with clear scientific findings backed up by a solid statistical base. Although the international scientific community has always requested such scientific evidence from Homeopathic medicine, it has never received attestation of its validity. It is completely devoid of any such foundation, remaining substantially an emotional medicine.

Supplemental Oxygen

Although oxygen therapy does not fall within the purview of NCCAM, it is worth a moment's consideration as it dramatizes the fraud and nonsense that permeates the marketplace, and attests to the fact that caveat emptor not only remains alive and well, but is essential.

Of course, oxygen does support life and is required full time. But is more better? Does anyone need more; anyone not tapping on death's door? Rose Creek Health Products, Inc., of Kettle Falls, Washington, thinks so. According to Rose Creek, their oxygenated water, with vitamin O, which stabilizes oxygen molecules (whatever that means), would prevent and cure pulmonary ailments, headaches, infections, flu, colds, and cancer. They also claimed that vitamin O aids digestion, relaxes the nervous system, boosts energy, promotes sound sleep, and hones memory and concentration [60]. Was there a market for vitamin O? They were selling upward of 50,000 bottles of water a month when the FTC and FDA had them cease and desist—pulled their plug, telling Rose Creek that they had to stop the unsubstantiated claims, and that as advertised, their vitamin O was a drug under the Food and Drug Act. It does appear that people will buy anything, so long as the label says it will cure something. Not a good omen.

The World Health Organization recently warned that the growing use of alternative medicine poses global health risks, and called on governments around the world to tighten oversight of the natural medicine industry. Reports of adverse reactions and injuries have more than doubled in 3 years, as consumers turn to herbal remedies, such as ginkgo biloba, acupuncture, and thermal therapy. Natural doesn't mean safe, said Xiaorui Zhang, Coordinator of Traditional Medicine at the World Health Organization [61].

At the request of the NIH and the Agency for Health Care Research and Quality, the Institute of Medicine produced their new report, *Complementary and Alternative Medicine*, which assesses what is known about Americans' reliance on those therapies. The report states that healthcare should strive to be both comprehensive and evidence-based, and calls for conventional medical treatments and complementary and alternative treatments to be held to the same standards for demonstrating clinical effectiveness [62]. It surely is about time that a major governmental agency spoke out about the free run CAM had obtained, abetting its skyrocketing popularity—the latest example of the "emperor's new clothes."

A statement by editors of the *Journal of the American Medical Association* has special currency today. They said that, "There is no alternative medicine there is only scientifically proven, evidence-based medicine, supported by solid data or unproven medicine, for which scientific evidence is lacking" [63]. It leads one to wonder why the NCCAM has such difficulty getting its funded studies completed and published, while others around the country and world, do so regularly and frequently. Although NCCAM needs an overhaul, nothing will change as long as senators such as Tom Harkin have the power of the purse. Senator Harkin forced the NIH to establish the NCCAM in order to give CAM standing, status. Holding the budget of the NIH hostage, the senator could easily make the NIH administrators bend to his will. That's an old story along the banks of the Potomac. Alexander Hamilton warned us in Federalist Paper 79 (in which he expressed concern about legislative tampering with judges' salaries) that "A power over a man's subsistence amounts to a power over his will." Current legislators appear to have been up to that type of tampering.

I believe that CAM would fade away if traditional medicine could recapture an old dimension. As I noted earlier, no one wants to be ill. But when sickness occurs, as it will, people want and need two things: the best available medical care and old-fashioned comforting, and hope. Caring, as was the rule when the doctor carrying his little black bag, parked out front and came calling. There wasn't much power in that black bag, but there was the power of his touch—handholding; to comfort and assure. It's that comfort, assurance, and hope that's missing in today's medicial care. But it is not too late to recapture it. Medicine today is in the hands of huge insurance companies that dictate to physicians the number of patients they will see in a day, every day; how much time they will have with each patient, and what their remuneration will be. Profit drives today's medicine. Most physicians are allotted less than 15 minutes per patient. Comforting has no place in this model. Speed is essential. To retake comfort, patients and physicians most rebel against the hegemony of the insurance companies, who really have no business in healthcare. Until physicians and patients pull together, to root out the insurance interlopers, CAM will thrive. Unfortunately.

REFERENCES

1. Meneimer, S., Scorn' with Orrin: How the gentleman from Utah made it easier for kids to buy speed, steroids, and Spanish fly, *Wash. Monthly* 27–35 (Sept. 2001).

2. *Dietary Supplement Health and Education Act of 1994*, Public Law 103–417 (S. 784), Oct. 25, 1994; *U.S. Code and Administrative News*, No. 3.

3. Schrmer, M., Hermits and cranks, *Sci. Am.* **286**(3):36–37 (2002).

4. Atwood, K. C., IV, The ongoing problem with the National Center for Complementary and Alternative Medicine, *Skept. Inq.* pp. 23–29 (Sept./Oct. 2003).

5. Liu, J., Killilea, D. W., and Ames, B. N., Age-associated mitochondrial oxidative decay, *Proc. Natl. Acad. Sci. USA* **99**(4):1876–1881 (2002).

6. Morris, C. A., and Avorn, J., Internet marketing of herbal products, *JAMA* **290**(11):1505–1509 (2003).

7. Palmer, M. E., Haller, C., McKinney, P. E. et al., Adverse events associated with dietary supplements: An observational study, *Lancet* **361**:101–106 (2003).

8. Lewis, J. D., and Strom, B. L., Editorial: Balancing safety of dietary supplements with the free market, *Ann. Intern. Med.* **136**(8):616–618 (2002).

9. Ames, B. N., Profet, M., and Gold, L. S., Dietary pesticides (99.99% all natural), *Proc. Natl. Acad. Sci.* **87**:7777–7781 (1990).

10. Harkey, M. R., Henderson, G. L., Gershwin, M. E. et al., Variability in commercial ginseng products: An analysis of 25 preparations, *Am. J. Clin. Nutr.* **73**:1101–1106 (2001).

11. Sievenpiper, J. L., Arnason, J. T. et al., Decreasing, null and increasing effects of eight popular types of ginseng on acute postprandial glycemic indices in healthy humans, *J. Am. College Nutr.* **23**(3):248–258 (2004).

12. Waggoner, J., Ginseng therapy for farmlands, *New York Times* (Dec. 1, 1999).

13. Saper, R. B., Kales, S. N., Paquin, J. et al., Heavy metal content of ayurvedic herbal medicine products, *JAMA* **292**(23):2868–2873 (2004).

14. Araujo, J., Beelen, A. P., Lewis, L. D. et al., Lead poisoning associated with Ayurvedic medications: Five states, 2000–2003, *Morbidity/Mortality Weekly Report* **53**(26): 582–584 (July 9, 2004).

15. F.D.A. warns of supplements containing Viagra ingredient, *New York Times* (national) (Nov. 3, 2004).

16. Sovak, M., Seligson, A. L., Konas, M. et al., Herbal composition PC-SPES for management of prostate cancer: Identification of active principles, *J. Natl. Cancer Inst.* **94**(17):1275–1281 (2002).

17. Markowitz, J. S., Donovan, J. L., DeVane, D. L. et al., Effect of St. John's Wort on drug metabolism by induction of cytochrome P450 3A4 enzyme, *JAMA* **290**(11):1500–1504 (2003).

18. Ruschitzka, F., Meier, P. J., Turina, M. et al., Acute heart transplant rejection due to Saint John's Wort, *Lancet* **355**:548–549 (2000).

19. Kveijkamp-Kaspers, S., Kok, L., Grobbee, D. E. et al., Effect of soy protein containing isoflavones on cognitive function, bone mineral density, and plasma lipids in postmenopausal women, *JAMA* **292**(1):65–74 (2004).

20. Keinan-Boker, L., Van Der Schouw, Y. T., Grobbee, D. E. et al., Dietary phytoestrogens and breast cancer risk, *Am. J. Clin. Nutr.* **79**:282–288 (2004).

21. Unfer, V., Casini, M. L., Constabile, L. et al., Endometrial effects of long-term treatment with phytoestrogens, *Fertil. and Steril.* **82**(1):145–148 (2004).

22. Bouker, K. B., and Halakivi-Clark, L., Genistein: Does it prevent or promote breast cancer? *Environ. Health Perspect.* **108**:701–708 (2000).

23. Ang-Lee, M. K., Moss, J., Yuan, C. S., Herbal medicines and perioperative care, *JAMA* **286**(2):208–216 (2001).

24. Eidelman, R. S., Hollar, D., Hebert, P. R. et al., Randomized trials of vitamin E in the treatment and prevention of cardiovascular disease, *Arch. Intern. Med.* **164**:1552–1556 (2004).

25. Graat, J. M., Shouten, E. G., and Kok, F. J., Effect of daily vitamin E and multivitamin – mineral supplementation on acute respiratory tract infections in elderly persons, *JAMA* **288**(6):715–721 (2002).

26. Peterson, R. C., Thomas, R. G., Grundman, M. et al., Vitamin E and Donepezil for the treatment of mild cognitive impairments, *NEJM* **352**(23):2379–2388 (2005).

27. Blacker, D., Mild cognitive impairment—no benefit from vitamin E, little from Donepezil, *NEJM* **352**(23):2439–2441 (2005).

28. Miller, E. R., III; Barriuso, R., Dalal, D. et al., Meta analysis: High dosage vitamin E supplementation may increase all-cause mortality, *Ann. Intern. Med.* **142**:37–46 (2005).

29. Bairafi, I., Meyer, F., Gelinas, M. et al., A randomized trial of antioxidant vitamins to prevent second primary cancers in head and neck cancer patients, *J. Natl. Cancer Inst.* **97**(7):481–488 (2005).

30. Lonn, E., Bosch, J., Yusuf, S. et al., Effects of long-term vitamin E supplementation on cardiovascular events and cancer, *JAMA* **293**(11):1338–1347 (2005).

31. Brown, B. G., and Crowley, J., Is there any hope for vitamin E? *JAMA* **293**(11):1387–1390 (2005).

32. Yale, S. H., and Liu, K., Echinacea purpura therapy for the treatment of the common cold, *Arch. Intern. Med.* **164**:1237–1241 (2004).

33. Taylor, J. A., Weber, W., Standish, L. et al., Efficacy and safety of echinacea in treating upper respiratory tract infections in children, *JAMA* **290**(21):2824–2830 (2003).

34. Melchart, D., Walther, E., Linde, K. et al., Echinacea root extracts for the prevention of upper respiratory tract infections, *Arch. Fam. Med.* **7**:541–545 (1998); see also Gunning, K., *J. Fam. Pract.* **48**(2):93 (1999).

35. Turner, R. B., Bauer, R., Woelkart, K. et al., An evaluation of Echinacea angustifolia in experimental rhinovirus infections, *NEJM* **353**(4):341–348 (2005).

36. Aubourg, P., Adamsbaum, C. et al., A two-year trial of oleic and erucic acids ("Lorenzo's Oil") as treatment for adrenomyeloneuropathy, *NEJM* **329**(11):745–752 (1993).

37. Shekelle, P. G., Hardy, M. L., Morton, S. C. et al., Efficacy and safety of ephedra and ephedrine for weight loss and athletic performance, *JAMA* **289**(12):1537–1545 (2003).

38. *Dietary Supplements: Review of Health-Related Call Records for Users of Metabolite 356*, GAO-03-494, General Accounting Office, Washington, DC, March 2003.

39. Arlt, V. M., Stiborova, M., and Schmeiser, H. H., Aristolochic acid as a probable human cancer hazard in herbal remedies: A review, *Mutagenesis* **17**(4):265–277 (2002).

40. Gertner, E., Complications resulting from the use of Chinese herbal medications containing undeclared prescription drugs, *Arthritis Rheum.* **38**(3):614–617 (1995).

41. Chan, T. K. Y., and Chan, J. C., Chinese herbal medicines revisited: A Hong Kong perspective, *Lancet* **342**:1532–1534 (1993).

42. Metchnikoff, I. I., *The Prolongation of Life: Optimistic Studies* (Engl. transl. P. Chalmers Mitchell), Classics in Longevity and Aging (series), International Longevity Center, New York, 2004.

43. Gill, H. S., Rutherford, K. J., Cross, M. L. et al., Enhancement of immunity in the elderly by dietary supplementation with the probiotic Bifidobacterium lactis HN019, *Amer. J. Clin. Nutr.* **74**:833–839 (2001).

44. Kalliomaki, M., Salminen, S., Arvilommi, H. et al., Probiotics in primary prevention of atopic disease, *Lancet* **357**:1076–1079 (2001).

45. Hatakka, K., Savilahti, E., Ponka, A. et al., Effect of long-term consumption of probiotic milk on infections in children attending day care centers, *Br. Med. J.* **322**:1327–1329 (2001).

46. Gorbach, S. L., Probiotics and gastrointestinal health, *Am. J. Gastroenterol.* **95**(1)(Suppl.):S2–S4 (2000).

47. Clydesdale, F., The promise of functional foods: Opportunities and challenges, *Food Technol* **58**(12):35–39 (2004).

48. Fahey, J. W., Zhang, Y., and Talaly, P., Broccoli sprouts: An exceptionally rich source of inducers of enzymes that protect against chemical carcinogens, *Proc. Natl. Acad. Sci. USA* **94**:10367–10372 (1997).

49. Herbold, N. H., Vazquez, I. M., Goodman, E. et al., Vitamin, mineral, herbal and other supplement use by adolescents, *Top. Clin. Nutr.* **19**(4):266–272 (2004).

50. Ambrose, E. T., and Samuels, S., Perception and use of herbals among students and their practitioners in a university setting, *J. Am. Acad. Nurse Pract.* **16**(4):166–173 (2004).

51. Pray, L. A., Epigenetics: Genome, meet your environment, *Scientist* 14–20 (July 5, 2004).

52. Waterland, R. A., and Jirtle, R. L., Transposable elements: Targets for early nutritional effects on epigenetic gene regulation, *Molec. Cell. Biol.* **23**(15):5293–5300 (2003).

53. Stover, P. J., and Garza, C., Bringing individuality to public health recommendations, *J. Nutr.* **132**:2476S–2480S (2002).

54. Frazier, K., French group tests man's claim to sense a person's presence, *Skept. Inq.* 6–7 (Nov./Dec. 2004).

55. Rosa, L., Rosa, E., Sarner, L., and Barrett, S., A close look at therapeutic touch, *JAMA* **279**(13):1005–1010 (1998).

56. Flamm, B., The Columbia University "miracle" study: Flawed and fraud, *Skept. Inq.* 25–31 (Sept./Oct. 2004).

57. Cha, K. Y., Wirth, D. P., and Lobo, R. A., Does prayer influence the success of in-vitro fertilization-embryo transfer? *J. Reprod. Med.* **46**:781–787 (2001).

58. Knipschild, P., Looking for gall bladder disease in a patient's iris, *Br. Med. J.* **297**:1578–1581 (1988).

59. Polidoro, M., Lady homeopathy strikes back ... but science wins out, *Skept. Inq.* 16–17 (Sept./Oct. 2004).

60. Hall, H. A., Oxygen is good—even when its not there, *Skept. Inq.* 48–55 (Jan./Feb. 2004).

61. Spencer, J., Alternative medicine usage holds risks, WHO reports, *Wall Street Journal* D4 (June 24, 2004).

62. Marwick, C., Complementary, alternative therapies should face rigorous testing, IOM concludes, *J. Natl. Cancer Inst.* **97**(4):255–256 (2005).

63. Fontanarosa, P. B., and Lundberg, G. D., Alternative medicine meets science, *JAMA* **280**(18):1618–1619 (1998).

3

MICROBES AMONG US

Great fleas have little fleas upon their backs to bite'em and little fleas
have lesser fleas and so an ad infinitum.
—*Augustus DeGeorge*

THE GOOD, THE BAD, AND THE UGLY

That bad news sells, is a universal truth. Yet, as we have just seen, life without
microbes is neither possible nor pleasurable. Recall, too, that only one infectious disease contributes to the leading causes of death. But few revel in that
knowledge. Headlines are grabbed and fear generated by the few troublesome
"germs" of the unseen world.

Our global village is currently beset by a series of microbial threats dubbed
"emerging infections," abetted by a seditious menace, bioterrorism. Both
require elucidation and comprehension. Emerging infectious diseases are
"infections that have newly appeared in a population, or have existed previously but are rapidly increasing in incidence or geographic range" [1].

With that as guidance, several questions are thrust upon us: how they are
transmitted, whether we are passive recipients without protective defenses,
and perhaps most important, why now—a core epidemiologic question that
requires pursuit. The assault of pathogenic diseases is ageless, and illness a fact
of life. Has anyone not had a cold, or the "flu"? Yet, with all the ailments we

are heir to, we have thrived and multiplied, which indicates a tolerable relationship with our pathogenic consorts.

Communicable diseases that currently afflict people in other parts of the world have become threats to the United States because distance and time have literally collapsed. Modern transportation, commerce, and widely shifting populations place us in contact with microorganisms never before seen here. Infectious organisms know no borders. Consequently, we become vulnerable, as our immune defenses have had little to no experience with these newcomers. Over the past decade, the threat has grown; 10 years ago we had not heard of West Nile encephalitis; Marburg, Ebola, or Lassa hemorrhagic fevers; severe acute respiratory syndrome (SARS); and avian (Asian) flu. These previously arcane afflictions have not only become household words but also borne fear in their wake, as it is entirely possible for these diseases to develop into widespread outbreaks with awful consequences. The Centers for Disease Control (CDC) recently estimated that should an influenza epidemic reach the United States, it could result in an estimated 300,000–700,000 additional hospitalizations with 89,000–200,000 deaths, and associated costs of from $71 to $167 billion [2].

In addition to naturally occurring disease outbreaks, we are now faced with the real possibility of infectious pathogens used as instruments of urban terror. In either case, natural or terrorist, no one knows, can know, when or where a new affliction may appear. Furthermore, since 1995, nearly 70% of emerging infections have been zoonotic—illnesses transmitted from animals to humans—which creates additional complications, but also, as we shall see, can provide links for interdiction.

How will or could these invisible microbes penetrate our borders? One has only to let the mind roam free. Who would have believed that a passenger on an Austrian Airlines flight from Vienna to Brussels would have smuggled birds into the passenger cabin in hand luggage? These birds, diagnosed as having been infected with avian flu virus, were illegally brought from Thailand to Brussels, where they were confiscated by European Union inspectors, but not before exposing passengers to the virus—passengers who may well have taken flights to other countries, thereby continuing to spread the virus [3]. Truth remains stranger than fiction.

Although polio (poliomyelitis) is not a new disease, it could reemerge in a susceptible population, and offers instruction on yet another mode of disease transmission. Cases of polio reached Mecca in January 2005, just before 2 million Muslims arrived for the hajj pilgrimage. Polio, nearly eradicated by 2003, began spreading from Nigeria across Africa after Islamic clerics and officials condemned the vaccine by spreading the word that the vaccine could make women sterile, could transmit AIDS, and was made of pork products. New cases skip along the highways used by pilgrims on their way to Mecca, in Saudi Arabia. And as polio symptoms take as long as 35 days to emerge, pilgrims traveling by bus or boat can take weeks to get home, all the while mixing with thousands of fellow pilgrims. The virus resides in the intestines and spreads via fecal–oral transmission, so changing a diaper, sharing food, and/or bathing in contaminated water can readily pass it to others. Polio vaccination is not a requirement for pilgrims. Thousands arrive illegally, and thousands arrive with forged immunization documents [4]. Could a microbe wish for more?

During the week of the hajj, vast crowds, literally hundreds of thousands, pressed tightly together circle the grand mosque and stone pillars of Satan. At week's end, the 2 million will be returning home with many asymptomatic individuals carrying the virus with them, with the potential for spreading it worldwide, a nightmare for international health officials.

What of international trade in exotic animals? Monkeypox, a close relative of smallpox, caught Illinois, Indiana, Michigan, Wisconsin, and Ohio by surprise in May 2003. Shock would be more like it, as monkeypox had never before been seen outside Africa. On April 9, 2003, 5–10-lb giant Gambian pouched rats, with monkeypox in their secretions, landed in Texas via Ghana, destined for Phil's Pocket Pets in suburban Chicago [5]. In fact, the shipment contained approximately 800 small mammals of nine different species including six genera of African rodents, bushtailed porcupines, striped mice, and tree squirrels. House pets all. But it was the prairie dogs in close association with the huge rats at Phil's that became infected with monkeypox virus carried and passed along by the rats. Prairie dogs were growing in popularity as house pets until the pox descended on 82 unsuspecting people in the Midwest.

Monkey pox, like its cousin smallpox, can be highly contagious, and was known to have a high death rate in Africa. The symptoms of a 10-year-old girl, a patient at Swedish-American Hospital, serves as a cautionary tale: "After admittance Rebecca McLester's fever rose. Pox sores were erupting all over her body, including inside her mouth and throat, on the palms of her hands, and the soles of her feet. She couldn't swallow and was having vivid, choking nightmares of choking to death" [6]. She survived only because several physicians and nurses were not afraid to care for her.

The United States has exported thousands of prairie dogs to Japan, where they are not naturally found. The Japanese adore these round, cuddly rodents that can harbor bubonic plague fleas. Public health officials in Japan are in a swivet over this trade that could erupt into a nasty epidemic.

The fact is, viruses are moving around our global village faster than ever before, and our behavior fuels this movement, and business takes precedence over health. Although the federal government banned the sale of prairie dogs along with the importation of Gambian rats, the illegal trade in exotic beasts continues and amounts to several billion dollars annually [7].

Mosquitoes, the vectors of malaria, dengue, and yellow fever, and a range of Encephalotides, often reside in pools of water inside discarded tires. While a desultory tire in a backyard here and there can host mosquitoes, and contribute to local infections, it is the wholesale trade, export, and import of millions of scrap tires that moves tires containing mosquitoes from the Far East to the United States, where literally multimillions of stockpiled tires sit around cities and towns, waiting for disposal that rarely if ever comes, become permanent harborages for mosquitoes and rats. In her most recent book, *Irresistible Empire: Americas Advance through Twentieth Century Europe*, Victoria de Grazia, Professor of History, Columbia University, maintains that American capitalism has changed the continent forever. She suggests that "the supermarkets relentless pressure for cheap food and industrial agriculture all pushed farmers to feed processed meat to cattle, leading to "mad cow" disease" [8].

Between March and May 2005, 399 cases of Marburg virus (an African RNA virus that causes green monkey disease) infection had been recorded with 350 deaths in Uige, Angola. Angola is one of 10 mainland countries in southern Africa, and Uige (pronounced Weezh) is its northernmost province, bordering Zaire and Congo. International health officials believe that this fearsome epidemic, if not stopped, could spread through Angola, and beyond its borders. Given Angola's long border on the Atlantic Ocean, the virus could easily be carried to other countries via ship or plane. Once contracted, Marburg has one of the highest mortality rates of any known disease. Fatal cases have occurred from mere contact with body fluids—blood, urine, saliva, which can easily occur during preparation of a body for burial, as has occurred so often among family members. Symptoms appear in less than a week: headache, muscle ache, fever, diarrhea, and vomiting, followed by massive hemorrhage from all body orifices, including the eyes. Death ensues from shock as blood pressure sharply drops, as fluid leaks from blood vessels. Although it is a highly contagious illness, it does require close proximity for transmission, which offers means of control. Nevertheless, 14 nurses and 2 physicians died while caring for patients. This outbreak is highly unusual as its first victims were children in the regional hospital's pediatric ward. How the first child became infected remains a mystery [9].

The Western world first learned of Marburg in 1967, when 32 scientists in Marburg and Frankfurt, Germany, and Belgrade, Yugoslavia, fell seriously ill, and seven died. Postmortem exams found collapsed liver function and a curious combination of hemorrhage and blood clots. These illnesses and deaths were decisively traced to a batch of fresh monkey cells that the microbiologists in these cities had used to grow polio virus. The cells had been taken from the kidneys of recently imported Ugandan green monkeys that had been naturally infected with this lethal virus—one of the most virulent known. The virus has since carried the name of the German city, Marburg, which first experienced this pernicious virus. Although monkeys can harbor and spread the virus, monkeys were not the source in Uige. The wild-animal reservoir in Uige has yet to be found. Until it is, additional outbreaks are to be expected. Marburg is a rare disease, but decidedly dramatic when it occurs, and with a high fatality rate, usually due to mismanagement by unprepared and often superstitious populations. While it is true that no one knew that the monkeys imported to Germany and Yugoslavia were carrying the virus, those monkeys from the African bush were purposely brought in for medical research without regard for their provenance. Could this happen again? More than likely. Consider monkeypox.

West Nile virus is naturally transmitted among birds by mosquitoes. Until 1999, West Nile virus was never seen in North America. In 1999, crows and blue jays began falling from trees and the sky in New York City. Shortly after the birds fell, people—especially the elderly and the immune-suppressed—become ill with neurological symptoms. West Nile encephalitis had reached North America, and between 1999 and 2004, had spread across the country to California, infecting some 12,000 people in 40 states, and taking over 300 lives [10]. How did West Nile cross the Atlantic? The suspects are imported exotic birds harboring the virus, an infected asymptomatic individual traveling between the Middle East and New York, or mosquitoes harboring the virus and hitching a plane ride in luggage. These three seemingly best possibilities involve peripetetic microbes readily moving about our globalized world. Which one is actually the culprit we'll never know.

When severe acute respiratory syndrome (SARS), an atypical pneumonia in which the lungs fill with fluid, swept out of southern China, it was a total mystery. Over 4 months in 2003, SARS sickened over 8000 people worldwide, killed some 800, and took a fierce economic toll. The World Health Organization has called the possible resurgence of SARS, now known to be a new member of the Coronavirus family, *SARS CoV*, one of the most pressing issues facing public health agencies around the world. Because of its flulike symptoms, it had created a diagnostic nightmare for hospitals, where it was often mistaken for influenza. Hunting down its origins, its reservoirs, is essential. The belief is that China's wild-animal trade, its wild-animal markets, provided the means for SARS to attack human populations. "The Chinese commonly eat exotic wild animals. The virus involved in SARS was found in the masked Palm Civet, a domestic-cat-sized mammal resembling a skunk, given white stripe from head to snout. Several of the early SARS cases were in chefs, and all had handled and eaten civet before becoming ill" [5].

In July 2005, scientists at the Institute of Molecular Biology of the Austrian Academy of Sciences reported that the SARS virus interferes with the crucial renin–angiotensin enzyme pathway that regulates body fluid balance and blood pressure. By blocking the enzyme angiotensin-converting enzyme 2 in the lungs, fluid leaks into the alveoli, allowing the lungs to fill with fluid, as well as curtailing the transfer of carbon dioxide and oxygen. This new piece of the puzzle, a salient biochemical key, was found in mice, but as both men and mice possess this enzymatic pathway, there is high hope of direct transfer to humans. Intravenous infusions of the enzyme could prevent the viruses' deadly effects. If this proves out, a new therapy for SARS and other viral respiratory infections could become a reality in the near term [11]. Because SARS spreads via respiratory secretions, its droplets should be readily managed and controlled which is not the case with Asian flu, also spread by respiratory droplets. Droplets containing avian flu virus are finer and far lighter and can be carried via air currents among groups of people and throughout a building.

Asian flu also poses special problems as the virus has an uncanny ability to mutate and jump from birds to humans, or as it begins to appear, spread to a range of animals such as pigs and cats that live in close proximity to people. Should this species jump occur, with a mutation permitting human–human spread, a worldwide pandemic could became a dreaded reality. So far, the avian (Asian) flu has not been found in the United States. But with only hours of flight time between the Orient and the United States, this needs to be considered a real threat to the United States. It is also worth considering that some 70% of the 1400 species of potential human pathogens are transmitted by animals, but as we shall see, epidemic diseases are primarily the result of microbes directly transmissible by human–human contact. Furthermore, a sense of proportion is required. As of April, 2007, 58 countries have reported a total of 170 deaths.

If avian flu, A(H5N1), hits the United States, lots of people are prepared to stop eating poultry because they do not know that the heat of cooking kills the virus. It would be extremely unusual and inordinately difficult to become infected from eating cooked chicken. One of the benefits of chickens raised indoors, as opposed to free-range chickens, is that the Former do not contact infected wild birds. Ergo, that is another layer of protection and safety. It surely is worth thinking about.

The question raised time and again is How is it that H5N1 is so lethal, yet so difficult to spread? From recent studies, virologists at the University of Wisconsin, and Erasmus University, Rotterdam, have found that unlike human influenza virus, A(H5N1) "preferentially infects cells in the lower respiratory tract making them difficult to expel by coughing and sneezing, the usual route of spread." Both research groups found that while human flu virus binds to alpha-2,6 galactose receptors that are spread throughout the upper respiratory tract, H5N1 binds preferentially to α-2,3 galactose receptors, common in birds, but also found around human alveoli deep in the lower respiratory tract. It is also known that H5N1 originated in China and escaped to Europe in migrating birds. This Qinghai strain contains the mutation PB2 E 627K, which in molecular biologic notation informs us that at position 627 on polymerase basic protein 2, glutamic acid (E), has been replaced by lysine (K). This single amino acid substitution allows the virus to replicate in the nose which is far cooler than birds intestines, which also suggests that this virus has spent time in humans—a worrisome fact [12]. Furthermore, it does appear that the countries of the world will be free of avian flu this year. In fall 2005, health officials nationally and internationally predicted that flocks of migrating birds would be flying south to Africa and back in spring 2006, and would spread the virus across Africa and Europe. With the arrival of the flocks in May 2006, it was clear that the birds did not carry the virus or spread it during their annual migration. In fact, not a single sample obtained from over 7500 birds in Africa, contained A(H5N1). Are we lucky, or do we lack significant information about this virus? Will it return in 2007, or 2008? At the moment no one has answers.

MEANS OF PATHOGEN TRANSMISSION

Pathogens are fragile. They need moisture, warmth, food, and a place to reproduce. To obtain these essentials, they must locate a suitable host animal or
human. A diversity of routes are available.

Airborne Transmission

Air is not an appropriate growth medium for any pathogen. A pathogen in
the air had to come from someone or somewhere—food, water, soil, an animal,
or a human. In the air a pathogen is riding a magic carpet—a dried mucus
carpet; a dry fecal flake or dust particle, and can be inhaled into human lungs.
Some pathogens are hardier than others. *Mycobacterium tuberculosis* can
remain desiccated for months after expulsion from an infected individual and
remain viable, infecting new individuals once inhaled. Coughing, sneezing,
talking, and laughing release moist droplets into the air. With a sneeze, secretions from the nose and mouth propelled from behind clenched teeth can be
ejected with the force and speed of a bullet sailing through air at the rate of
50–100 meters per second (m/s). Figure 3.1 demonstrates the explosiveness of
a sneeze, with eyes closed, teeth clenched, lips open, and an aerosol forcefully
ejected. A sneeze can eject as many as 10^6 droplets from 10 microns to a millimeter in diameter. And note how far from the source droplets are ejected.
The larger, heavier droplets fall colse to the source and are less of a hazard.

Figure 3.1. High-speed photograph of an aerosol generated by an unstifled sneeze. Note
the distance the smallest droplets traveled from point of ejection. Also note the sizes of
droplets, which consist of virus and bacteria-laden saliva and mucus. If inhaled by a susceptible person, a new infection can occur. (Figure adapted from *The Major Ascana. The
Female Picture Galleries Deck 2*, VIII.)

The smaller, lighter particles evaporate rapidly, leaving behind droplet nuclei which can remain suspended indefinitely, with the consequent accumulation of large numbers of potentially infective organisms. Respiratory transmission via droplet nuclei is the preferred means of transmission of most human infectious diseases—a highly efficient route for passing organisms among crowds, and a highly efficient means for pathogens to locate new venues, which they must do to survive.

Direct Contact

The hemorrhagic fevers are excellent examples of direct contact transmission. Susceptible, healthy individuals need but touch secretions or excretions, blood, saliva, urine or feces of an ill individual to pick up the microbe. Its inordinate mortality rate bespeaks uncommon virulence, and minscule immunity within the host.

Sexually transmitted diseases such as HIV/AIDS, gonorrhea, and syphilis are also members of this highly select route of transmission. Mucous membrane–mucous membrane is the direct route sexual transmission requires. These microbes are so fragile that they cannot survive outside the human body and must again be rapidly deposited within a new host. Kissing, and the exchange of saliva between individuals, is another example of direct transmission.

Fomites, inanimate objects, offer yet another means of passage for certain organisms. Another name for this route is *vehicle transmission*, and common vehicles include water and food. Food poisoning at picnics and church suppers are often the result of common vehicle transmission in which hundreds of people can eat an egg, tuna, or chicken salad containing pathogenic bacteria or their toxins, and become ill with gastroenteritis in 4–8 hours. Fomites encompass bedding, drinking cups, food utensils, surgical instruments, catheters, and clothing, which have all been implicated in disease transmission. Campylobacteriosis, listeriosis, leptospirosis, and salmonellosis fit comfortably in this mode of transmission. Given the fact that fomites are an unsuitable growth medium for most microbes, contact with a fomite is usually a local phenomenon.

These groupings can often aid in the diagnosis of disease, as all microbes fit into one category or another, and can be transmitted only by a specific route. Influenza virus cannot be transmitted by food or water, nor can the gonococcus be transmitted by droplet nuclei, or the treponema of syphilis by such an inanimate object as a toilet seat.

Vectorborne disease is the province of arthropods—insects, mites, ticks, and fleas, as well as the vertebrates—dogs, cats, bats, rats, skunks, and the like that bite, penetrating the skin, a major protective barrier—and depositing pathogens they harbor into tissue, muscle, and blood. Included in this route are rabies, Lyme disease, Rocky Mountain Spotted Fever, West Nile encephalitis, and its relatives, St. Louis, Japanese, and Venezuelan encephalitis. Rabies cannot be transmitted by fomites, nor can West Nile be passed by droplet nuclei.

Fecal-Oral Transmission

Traveler's diarrhea, the Inca half-step, Montezuma's revenge, salmonella, and staphylococci food poisoning, typhoid fever, all travel by means of the fecal–oral route, which requires fecal material in food or water, contributed by a symptomatic or asymptomatic individual, be ingested, and within hours an explosion occurs as the body, in its wise attemps to expel toxins produced by the ingested organisms. Almost everyone is susceptible. Immunity plays no part in these episodes.

Parenteral transmission is the preferred route of intravenous drug users who often share contaminated hypodermic syringes and needles that deposit both "recreational" drugs and microbes directly into blood and/or muscle. Wounds created by gunshot and knives can also carry infectious organisms into deep tissue, where anaerobic organisms such as the clostridia can initiate tetanus and gas gangrene.

With such an array of routes into our bodies available to the 1400+ pathogens, ostensibly to wreak havoc on us, we have not only survived but flourished. Obviously we possess efficient and effective protective defenses working full time for us.

Our continuing success depends on an elaborate, dynamic, and regulatory communications network of hundreds of millions of specialized calls passing information among them. This sensitive system produces a rapid, specific, and effective immune response.

A vast network of specialized cells and organs, the lymphoid system, extends throughout our bodies, and is responsible for producing a specific type of white cell, lymphocytes, especially B and T lymphocytes, which will do the heavy lifting in warding off infection. But the body must know and recognize friend from foe. Ergo, every cell contains specific molecules that identify it as self. This is essential as we surely do not want our defense system not knowing who the enemy is. It must be able to distinguish self from nonself. Not knowing the difference could mean attacking its own tissues. Unfortunately this does occur at times, and results in what is known as *autoimmune diseases*. Rheumatoid arthritis and lupus erythematosus are two such mistakes.

Any protein capable of triggering an immune response is called an *antigen*. It can be a virus, a bacterium, or even tissue or cells from another individual— as in transplants—that can trigger rejection. An antigen heralds its presence, its foreigness by special shapes, epitopes, receptors that protrude from its surface. Some have several; others, hundreds; and some will be more effective than others in evoking an antibody response. Enter the B and T cells ("B" for bone marrow and "T" for thymus, where each undergoes maturation and multiplication). They are now capable of producing an immune response and capable of distinguishing self from nonself. The vital business of B cells is to secrete antibody into the body's fluids. Antibodies will interact with circulating antigens. Each B cell is programmed to produce one specific antibody. So, for example, one B cell will produce an antibody that will recognize and block the rhinovirus from attaching to cells and eliciting symptoms of the common cold. Yet another B cell will produce antibodies that attack the pneumococcus, blocking its attempt to induce pneumonia.

All microbes have a collection of antigens on their surface. The uniqueness and specificity of these surface antigens, and the remarkable ability of lymphocytes to recognize these antigens, forms the basis of our immune defense. The exquisiteness of the immune response lies in the fact that each of us is born with millions of B lymphocytes—each capable of detecting a different surface antigen. Before entering the world and encountering foreign proteins, we come outfitted with a full complement of B cells and their antibodies that have a designated target. B cells may have as many as 10,000 surface—specific antigen receptor sites. Thus, each person's total mature B-cell population carries receptors specific for literally millions of different antigens.

Antibodies are immunoglobulins, large protein molecules that can coat microorganisms, making it easier for phagocytic cells to engulf and destroy them. Nine chemically distinct immunoglobulins (Ig) have been identified: four kinds of IgG; two IgAs, an IgM, an IgE, and an IgD, each with a specific defensive role.

T cells are the first to be activated in any immune response. They contribute by secreting cytokines, chemicals that can attack an antigen directly, but also have a range of other functions, including directing cellular traffic and activating macrophages, phagocytic scavenger cells that consume dead microbes and other foreign debris. One of the first cytokines discovered was interferon, a protein with antiviral properties. T-cells that attack microorganisms directly are designated Tc for cytotoxic. With B and T cells we have both humoral (B-cell) immunity, also referred to as *antibody-mediated immunity*, and cell-mediated immunity, the province of T cells. There is also complement, or the complement system, a group of 30 different proteins that function together as a nonspecific defense against infection, to complement the action of antibodies in destroying germs.

Complement proteins can lyse bacteria and can also amplify the effects of antibodies. Specific complement chemicals can neutralize or inactivate bacterial toxins. Immunity to a disease like diphtheria depends on the production of antibodies, antitoxins, which inactivate the toxins produced by the corynebacterium. Viral neutralization can also occur when antibodies such as IgG, IgM, and IgA bind to a virus, preventing it from attaching to a target cell. Complement proteins circulate inactively in blood until triggered by an antigen–antibody reaction. This trigger activates a complement cascade—a set of regulated steps in which the product of the first reaction activates the next until potential infections are cut short or disabled [13].

As an additional defense, we also have acquired immunity. Whenever B and T cells are activated, some become "memory" cells. The next time a person is exposed to the same antigen, the immune system is primed to destroy it quickly. This can be accomplished by either active or passive immunity. Active immunity is naturally acquired by encountering microbes during our lifetime—becoming ill and recovering, or being exposed to a microbe without experiencing an actual illness, but still developing antibodies and developing immunity. Most of us have encountered *Mycobacterium tuberculosis*, but do not have full-blown tuberculosis. Nevertheless, tests can demonstrate our level of antibody to the mycobacterium antigen. We have solid protection. Similarly with polio; we've encountered polio, and developed antibodies that protect us against its full-blown effects—paralysis.

Passive immunity is the immunity we receive via vaccination, which provides sufficient, but weakened antigen (of a specific infections disease) to stimulate antibody production and protection. DPT immunization confers immunity to diphtheria, pertussis, and tetanus. Influenza vaccine (a mixture of several viral types) confers immunity to those types, and Pneumovax confers immunity to pneumonia for about 5 years.

Passive acquired immunity can also be obtained by inoculations of serum containing antibody to a specific infection. Antibodies from a donor can be injected into an individual, or gammaglobulin, a collection of antibodies from the pooled serum of many donors, can be used to provide temporary protection. Tetanus immunoglobulin contains high concentrations of antitetanus antibodies against tetanus toxin. Protection usually lasts 5 or more years. Should a wound infection occur, a booster dose of antibodies will send the body's level of tetanus antibodies into overdrive, ensuring added protection.

Active natural vaccination to stimulate the body to produce antibodies can have protective effects beyond the individual. "Herd" immunity is the resistence of a population, a community, to infection and spread of a pathogen. The greater the number of immune individuals in a community, the lower the chance of transmission as an infected person will have few susceptibles available. Ergo, the group or herd, will have high resistance and protection. That's why, and quite properly so, the public is constantly urged to get their "shots". The more immune people in the community, the less chance for a communicable disease to penetrate. The group protects its members, including the non-immune. The high proportion of immune individuals becomes a virtual barrier against infection. Although herd immunity will vary with each disease, it does speak forcefully for greater attention to vaccinations by the public., which can be translated into economic terms. The greater "the herd," the fewer tax dollars need be spent on dealing with sickness. Bear in mind that the immune defenses described defend against those microbes that enter the body via droplets, direct contact, airborne inhalation, and the fecal–oral rate, those pathways that bypass the skin and bring pathogens into the body. Although the immune system puts up a splendid defense against the continuous onslaught of enterprising infections agents, and is more often victor than vanquished, certain pathogens are so highly virulent that the B ond T cells along with the complement system are simply overwhelmed. Marburg, Ebola, rabies, and smallpox are four such virulent malefactors. It is well worth remembering that few have such power.

Resistance to complement is an attribute limited to several viruses and bacteria, but at the end of the day the human organism is seen to have devised sufficient defenses to defeat all comers—for the most part.

VIRUSES

Viruses among Us

We literally live among and contact millions of virus particles daily. Measles, mumps, and chickenpox have been a fact of life of every generation's children—and with the current clutch of emerging infectious diseases, all viral, it is evident that we remain vulnerable. Similarly, however, the fact that for the most part colds last but a week, that most of us recover from a bout of the "flu," and that recovery from exotic infections such as West Nile, SARS, and avian flu is the norm, must also inform us that our system can manage viral infections as well as bacterial.

The natural response to viral infection is the domain and responsibility of powerful proteins such as interferons and interleukins, members of the complement system, and whose signals mobilize its many and diverse specialized cells, which can and do destroy virus particles. The antibody response of the humoral system, with its built-in memory, plays a major role in defending against viral attack. Indeed, it is a constant and continuing struggle. The complement cascade can be activated quickly and begins functioning within hours of viral penetration; the antibody system requires days or weeks to mobilize and enter the fray at maximum strength. Current theory holds that the rapid complement response provides essential information to the antibody system about the immediate hazard. For the most part, the system works well, in conjunction with the phenomenal protection afforded by the 14 currently available antiviral vaccines that stimulate antibody generation and/or immune memory of past viral exposure.

A natural experiment demonstrating the fact of immune memory occurred on the Faroe Islands, a remote speck of land between Iceland and Norway. In 1781, a devastating measles outbreak drastically reduced the island population. During the ensuing 65 years, the population rebuilt, and the islands remained measles-free. In 1846, with the arrival of a ship from Denmark, measles struck again, infecting some 80% of the population with similar mortal affects. Ludwig Panum, a Danish physician investigating the outbreak, astutely observed that "Of the many aged people still living on the Faroes, who had measles in 1781, not one was attacked a second time." He also found that "All the old people who had not gone through with measles in earlier life, were attacked when they were exposed to infection" [14]. Panum learned that immunity to measles was long-lived and that reexposure was not essential for maintaining long-term protective immunity. Interestingly enough, Thucydides, the Greek historian, in describing the plague of Athens in 430 bce, wrote that "the same man was never attacked twice."

The current batch of emerging infections has raised concern worldwide because viruses are more complex biologically than bacteria. To reverse the idiom, the smaller they are, the more difficult to knock down. Accordingly, a brief recapitulation of virological fundamentals can be helpful in following the heightened concerns.

Viruses are simple creatures, but that simplicity is deceiving—fraudulent— as it belies the mischief they can forge. They are nothing more than sacs of genetic material, and as such are obligate parasites totally dependent on living human or animal cells for reproduction—which is what they must do. Even the largest is smaller than the largest bacterium. They range in size from about 25 manometers (nm) to about 300 (10^{-9}–10^{-10}), which places them within the resolving power of the electron microscope, and finds them some 100 times smaller than the smallest bacteria. No matter their size or shape, they follow a common strategy to ensure survival. Viruses can exist in two phases, outside or inside a cell. Outside, they can only exist—but not for long. Inside human or animal cells they can induce their host cells to synthesize additional viral components so that multitudes of new viral particles, virions, are produced and expelled on host cell disruption—to continue the reproductive cycle at the expense of the host.

The basic structure of a virion, a complete virus particle, is a nucleic acid core surrounded by a protein that permits dividing them into two major classes: lipid-enveloped viruses, which possess a lipid coat, and nonenveloped, naked viruses, which are coatless. The lipid coat surrounds the protein structure and genetic material—the genome—and protects it from recognition by the immune system. The core nucleic acids are either deoxyribonucleic acid (DNA) or ribo nucleic acid (RNA). Viruses, then, are either DNA or RNA viruses.

This lipid coat also helps the virus infect host cells by merging, or fusing, the virus coat with the host cell surface, its plasma membrane. It is the epithelial cells lining the respiratory tract, nose, pharynx, bronchi, and lungs that constitute the major portal for most virulent viruses. The plasma membrane has a twofold function: containing the cells' cytoplasm and regulating passage of whatever moves in and out of the cell. Successful entry requires the virus to breech the plasma membrane. But cell membranes are not permeable to particles as large as viruses—even the smallest—so viruses employ their unique talents. The highly versatile virus fuses with the plasma membrane, removing its coat as it does so and spills its nucleic acid, DNA or RNA, into the cell. The host cell is now infected. Once uncoated, the viral nucleic acid is in charge, taking over the cellular machinery—its replication system—and directing the host cell to produce more virus components. With the host cell full of new virions, the cell bursts, releasing the new infectious particles, and the cycle continues anew at the host's expense. But bear in mind that it is the business of the host's immune system to ensure that virus establishment does not occur, or if it does occur, to cut it short, or control it. Herein lies the ongoing and eternal battle. With that as prelude, we can consider the emerging viral infections causing current worldwide discomfort—or worse.

West Nile Encephalitis

West Nile fell from the sky and out of the trees as sickened birds succumbed to this new interloper from Africa and the Middle East. Those crows, sparrows, blackbirds, and blue jays in Manhattan's trees, loaded with virus particles, had provided yet another winged messenger, *Culex pipiens*, a mosquito, with fulfilling virus-containing blood meals. But *Culex*, with options, bird or human blood, prefers human blood when available. So, now full of West Nile virus, *Culex* seeks a human meal. When found, and that's not too difficult in a city as diversely populated as New York's boroughs, the mosquito, active in the evening when the folks are out perambulating, sinks its sharp proboscis into the skin of the unsuspecting, to drink deeply, and in so doing, in goes West Nile with an opportunity to begin its takeover cycle. That's the essence of the bird–mosquito–human relationship, a classic arthropodborne (arbovirus) transmission cycle.

But West Nile is a disease of Africa. How did it get to Manhattan and travel cross-country, engendering hysteria as it moved westward? West Nile was first isolated from the blood of a woman by British virologists in the West Nile region of Uganda in 1937. It has since been identified as an enveloped, single-stranded RNA, flavivirus of the Flaviviridae family that includes Japanese B encephalitis and yellow and denque fevers. In the Middle East, Asia, and Australia, West Nile virus is known as *Kunjin virus*.

Naturally, in the wild, without the preferred human around, the virus is transmitted among birds, the reservoir, by mosquitoes, the vector. Interestingly, most human infections are subclinical, asymptomatic. In New York one in five people who were infected developed the frank disease, and only one in 150 developed meningioencephalitis, West Nile fever, with severe neurologic involvement. Those whose immune system could manage the virus typically developed flulike symptoms along with headache, fever, chills, and backache within 3–14 days, which disappeared without treatment in 3–6 days. The risk of encephalitis increases with age over 50, as the immune system begins to weaken. Currently there is no antiviral treatment—preventive or cure—for West Nile or any member of the Flaviviridae family. Curiously enough, West Nile was given in an unsuccessful experimental treatment for advanced cancer in New York, in the 1950's [15].

Scientists at the U.S. Geological Survey's National Wildlife Health Center maintain that no one really knows how the virus arrived in Manhattan. It could have been brought in by a migrating bird, an imported bird, a mosquito hitching a ride on a plane or boat, or another unknown host. Dr. Emi Sailo of the Geological Survey believes that West Nile Virus (WNV) is here to stay. As the virus needs its natural host to multiply, "The only way to truly eliminate the virus in the U.S. is to kill every bird and every mosquito" [16]. Even if this were possible, which it isn't, WNV would return by one of the routes noted above. Futile. Between 1999 and 2004, there were over 12,000 reported cases and 300 deaths nationwide. Interestingly, the red-breasted robin is the favorite food source of the *Culex pipiens* mosquito, which carries the virus. In the birds' absence, the mosquito turns to us humans. The number of human cases of West Nile in the United States, is highest in August and September, when few, if any, robins are available. No help there.

Lassa Fever

Lassa fever arrived in New Jersey in August 2004. The virus was carried by a businessman who had spent 4 months in Sierra Leone and Liberia before suffering fever, chills, sore throat, diarrhea, and backache. Two days later he left Sierra Leone, traveling by plane to London, then on to Newark, and by train home, where he was hospitalized in Trenton, NJ. After 4 days of unsuccessful diagnosis and treatment, Lassa fever was diagnosed. Before ribavirin, an experimental drug, could be used, the patient died. Postmortem specimens sent to the CDC confirmed the diagnosis. With Lassa fever confirmed, bells went off and the CDC opened an investigation of those individuals possibly exposed to direct contact with the patient's body fluids. Contacts were categorized as of low or high risk. During the period of the patient's infectivity, some 188 people could have been exposed. Five were classified as high-risk; 183 as low. The five were his wife, three children, and brother, who visited him in the hospital. Each of them had unprotected exposure to his body fluids. Among the low-risk group were 139 healthcare workers at the hospital: 42 lab technicians, 32 nurses, and 11 physicians, along with 16 commercial lab technicians who tested body fluids. Nineteen passengers on the flight from London to New York were identified and classified as low-risk. All were interviewed and requested to self-monitor their temperatures and symptoms compatible with Lassa fever [17]. None of either group reported positively. But this was not the first case encountered in the United States. In 1989, a mechanical engineer born in Nigeria, but a US citizen, traveled to Nigeria to attend his mother's funeral. She had died of an acute febrile illness diagnosed as malaria. During his stay, several family members became ill and died, including his father. He quickly returned to Chicago. Two days later a high fever and flulike symptoms, shaking chills, sore throat, and muscle pain brought him to a private clinic, where his condition deteriorated, as misdiagnoses were made. He was hospitalized with a possible diagnosis of hepatitis or yellow fever. Learning of his travel to West Africa, a diagnosis of Lassa fever was confirmed. Respiratory distress occurred, which took his life [18].

These two cases followed similar, if not exact, scenarios. Sierra Leone is known as an area of hyperendemicity, and West Africa generally is a hotspot for viral hemorrhagic fevers. In both cases the victims returned home, where they anticipated appropriate medical care. The Chicago patient's mother had more than likely been misdiagnosed with malaria rather than Lassa fever, where it is seen regularly and frequently. Some 100,000–300,000 cases occur annually with over 5000 deaths. Outside of Africa Lassa is a rare disease; only 20 cases have been reported worldwide. With a single case in 1989, and another in 2004, it would be surprising if any physician or nurse could recognize Lassa fever in its early stages as symptoms are nonspecific. The only clue was travel in West Africa. In both cases influenza, malaria, and typhoid fever were suspected, and before a firm diagnosis was made, both patients experienced respiratory failure and death. Lassa, Marburg, and Ebola share a common pathogenic attribute: the ability to disable their host's immune response by attacking and destroying the cells that normally would initiate the antiviral response. Viruses engulfed by defending macrophages become virus replication factories, and with virus expulsion, the virions spread to regional lymph nodes, liver, and spleen. Infection of liver cells (hepatocytes) impairs function of blood clotting mechanisms; edema ensues with loss of sodium and severe reduction of blood pressure. Currently there is no effective cure once the virus spreads to major systems.

Although Marburg and Ebola are members of the Filoviridae family, and are enveloped with long, filamented, single-stranded RNA and Lassa with an enveloped, single-stranded RNA virus of the Flaviviridae family, all possess highly virulent and artful genomes that can rapidly overwhelm the human immune system.

The reservoir of the Lassa fever virus is known. The carrier of the virus is the multimammate rat, *Mastomys natalensis*, that readily colonizes human habitats and that the people eat. Until the west Africans choose other sources of protein to augment their diets, and refuse to share their abodes with mastomys, many will continue to fall victim to the virus. Unfortunately the reservoir of both Marburg and Ebola remain unknown. Nevertheless, after an initial infection from their respective reservoirs, further transmission is by direct person to person contact of body fluids and fomates. This close personal contact indicates that these viruses can be appropriately and safely managed by barrier techniques and precautions by physicians and nurses in handling blood, stools, body fluids, and patient-contaminated equipment.

Until the animal reservoirs of Marburg and Ebola are identified, these two illnesses can be expected to occur from time to time, but with diminishing frequency, and the rare, individual case may well continue to be misdiagnosed by physicians with little or no experience with these exotic diseases. With appropriate patient management, the list of emerging disease should decline, taking much fear with it.

POTENTIALS FOR BIOTERRORISM, PANDEMICS, AND ENVIRONMENTAL CONTAMINATION

Bioterrorism

We've discussed the good that microbes do; their activities that make life livable and often pleasurable, and with emerging infections we've taken on the bad with their potential for widespread illness, but lacking malicious intent.

The ugly comes from malignant human mischief; loosing microbes that can inflict devastating illness and death that for the most part are rarely, if ever, encountered; are not of public concern; or, in the case of smallpox, a noxious disease reincarnate, that has been eradicated from human populations.

What more evidence do we need than the horrific attacks of September 11, 2001, on the World Trade Center, The Pentagon, and the abortive attempt to obliterate the White House, that the threat of vast networks of highly organized international terrorist groups intent on spawning mass destruction is real—a no-brainer.

Although there is no universally accepted definition of *terrorism*, the *Code of Federal Regulations* defines it as "The unlawful use of force and violence against persons or property to intimidate or coerce a government, the civilian population, or any segment thereof, in furtherance of political or social objectives." That's it in legalese. More succinctly, a biological attack is the intentional release of a pathogen, or its toxin, against people, plants, or animals. An attack against people seeks to cause illness, fear, death societal disruption, and most assuredly, economic havoc. An attack on crops and livestock would cause fear of the food supply with dire nutritional consequences.

A Backward Glance for Perspective

Biological terrorism began with hurling of diseased animal and human cadavers, clothing, bedding, over city walls or into wells to contaminate water supplies, in attempts to subdue an enemy city. The spread of infections disease along with the sixteenth century age of exploration resulted in untold numbers of deaths. It's estimated that the arrival of smallpox in Mexico with the Spanish Conquistadors reduced the Aztec population by 10–15 million, effectively ending the Aztec civilization [19]. The glorious victories attributed to Spanish arms would not have been possible without the devastation wrought by Spanish diseases.

Writing in the *Bulletin of the History of Medicine*, Professor Duffy informs us that [20]

> The colonists were well aware of the potency of smallpox as a weapon against the Indians, and on several occasions deliberate efforts were made to infect the Redmen. One of the instances occurred during the Pontiac Conspiracy in 1763. The British Commander, Sir Jeffrey Amherst, added the following postscript to a letter to Col. Henry Bouquet, "could it not be contrived to send the smallpox among those disaffected tribes of Indians? We must on this occasion use every strategy in our power to reduce them." Bouquet replied on July 13, 1763, "I will try to inoculate the . . . with some blankets that may fall in their hands, and take care not to get the disease myself." Just how successful Bouquet's experiment was is not known

During World War I, Germany successfully infected allied livestock and animal feed with the bacilli of anthrax and Glanders, to infect mules and horses of the French Calvary. Argentinean livestock intended for allied troops were also infected with anthrax. As many as 15 million fleas per city were released by the Japanese over 11 Chinese cities during World War II, to initiate epidemics of bubonic plague. Indeed, bioterrorism is an old microbial story—well before microbes were known—with today's new microbial twists.

Biological weapons are characterized by invisibility, high potency, ready accessibility, and ease of delivery. Biological agents can be carried in easily in disguisable vials, tubes, or an envelope in a jacket, vest, or back pocket undetectable as a person passes through metal detectors, odor detectors, or pat-downs. They are the real stealth bombers, and with their ability to sow fear and chaos, they can be the real weapons of mass destruction. Being invisible, odorless, and tasteless, no one would know that an attack was underway. Their concealment, transportation, and dissemination are exceptionally easy. Unlike nuclear weapons or missiles, delivery systems are not required for bacteria, viruses, or fungi. In addition, as aerosolization is a major means of dissemination, such low-tech methods as agricultural crop dusters, backpack sprayers, and purse-size atomizers can spread fine powders effectively. Small groups of knaves, modestly financed, and with basic biological and engineering skills, can produce effective weapons. That's scary. To ratchet up this potential, recipes for making bioweapons can be downloaded from the Internet. These simple qualifications must suggest that we are vulnerable.

The obvious question then is if bioweapons are so potent, are so easy to come by, and can easily fall into the wrong hands, why has this not been of greater concern to our political leaders? It is reasonable to believe that because such weapons have never been used on a grand scale against civilian populations, they would never be. It may also be understood that if such weapons were used, nuclear retaliation would be swift. But who could be retaliated against after 9/11, when the attackers were not a declared state, but only a group of terrorists? Nevertheless, it would be reckless and unreasonable to assume that biological weapons would not be used in the future. We must accept the reality that we will not be able to prevent every act of bioterrorism.

A case in point occurred in Oregon in 1984, but not made public until 1997. This event had all the qualifications noted earlier. It was a malicious, malignant, and premeditated attempt to control the outcome of a municipal election by sickening and killing enough voters that the Bhagwanshree Rajneesh Commune (cult) would win the upcoming election and thereby control community zoning and other municipal issues.

The Rajneesh religious commune had its headquarters in the Dalles, Oregon, county seat of Wasco County, a community of some 11,000 people, located near the Columbia River on Interstate 54. Between September 9 and October 10, 1984, 751 adults became ill with a salmonella-induced gastroenteritis. All had eaten at salad bars in some 10 restaurants. It took more than a year to accumulate evidence linking the commune with the epidemic. FBI agents found a vial of *Salmonella typhimurium* (a bacterial culture obtained from the American Type Culture Collection, Baltimore, MD) in the commune's clandestine laboratory. On March 19, 1986, two commune members were indicted for poisoning food in violation of a federal antitampering act, and were sentenced to substantial prison terms. They admitted that they had intended to make enough citizens sick to prevent them from voting. They also admitted sloshing cultures of salmonella on foods in salad bars, and in coffee creamers [21]. As is well known, salad bars are open to all. Who would know? Who could tell? It was so easy, so uncomplicated. That's the beauty of bioterrorism. Given the ease with which the Rajneesh slipped *S. typhimurium* into salads, dissemination of other agents by serious terrorists is foregone. In an open society, the team on the offense has the advantage over the unsuspecting.

The idea of infection due to invisible agents is frightening. "It touches a deep human concern about the risk of being destroyed by a powerful, evil, imperceptible force. These beliefs activate emotions that are extremely difficult to direct with that tools of reason" [22]. Obviously, in addition to the illnesses directly related to a specific microbe, there will be psychiatric casualties. Bioterrorism is an ugly undertaking.

So, where would the microbes be obtained, and what are the desirable characteristics of a biological (warfare) agent, and how would it (they) be dispersed? Microbes can be easily obtained from natural sources. Any microbiological text lists the reservoirs of pathogenic organisms. As all are found worldwide, obtaining them poses little problem. Should the rogues prefer to stay out of the woods, so to speak—well, there are always laboratories, public and private, from which cultures of organisms could be stolen or purchased. Given the nature of humans, a bribe here or there could produce the most lethal organisms.

Only two laboratories in the world, one in the United States and one in Russia (the former Soviet Union), are supposed to house the smallpox virus—and under strict security. If, however, certain countries sequestered the virus and continue to hold stocks of it clandestinely, they could be a source. This is unlikely, but in our uncertain world, one never knows. There is the real possibility that microbes (especially viruses) or their toxins could be synthesized or genetically manipulated in a laboratory, although to do so would require a high degree of expertise along with advance technology; nevertheless, we must assume that there are no bars to acquisition.

Knowing that microbes can be obtained, what would a villain want by way of an effective weapon? It should

- Involve as many people as possible
- Be easy to produce
- Be aerosolizable to within 1–10 μm
- Be able to survive drying, heat, and UV light

- Be capable of causing disabling illness, death, or both
- Be capable of person–person transmission
- Be capable of causing panic
- Be able to reach the target population
- Also, effective treatment must be unavailable or lacking.

Let us also understand that the term *infectious* is different from the terms *contagious* or *communicable*. *Infectious* refers to the number of bacteria, bacterial spores, or viral particles needed to infect an individual. The fewer the number, the more infectious the agent. Agents are contagious if they spread from person to person. Some agents that are highly infectious such as tularemia and the viral hemorrhagic fevers (VHFs) are not contagious. So, which ones do we choose? Which is a terrorist most likely to choose? Category A, set up by the Department of Homeland Security and the Centers for Disease Control, contains the elite five: anthrax, botulism, tularemia, smallpox, and plague. Category A agents are those that pose the greatest possible threat for an adverse effect on public health, may spread across a large area or need public awareness, and need a great deal of planning to protect the public's health. All five fit comfortably in this category. Each is in excellent company.

It's back to the future. The ancient scourges are still those given prime time. Each requires some detail.

As we deal with each of the elite, the "big five," ask yourself whether we are in for a global influenza pandemic that could conceivably remove 100 million of us in rapid fashion, and if so, what more could bioterrorism do? Is bioterrorism as much of a threat as an influenza pandemic, or more so? Is it a different kind of a problem, or in terms of society the same or similar?

Anthrax

Robert Stevens was the first to die from the anthrax attack of September 2001. Robert Stevens worked at *The Sun*, a publication of the AMI Publishing Company of Boca Raton, Florida. AMI also publishes the *National Enquirer* tabloid, and was one of the six media companies that received anthrax-spore-containing letters [23].

The anthrax attack occurred over several weeks beginning on September 18. Letters containing anthrax spores were mailed to six media offices, including the major networks, and two US senators. Twenty-two people became infected. Five died. The letters sent to Senator Tom Daschle (Democrat, South Dakota), then majority leader of the Senate, and Senator Patrick Leahy (Democrat, Vermont), were more potent than the others. Those two contained highly refined dry powder: 1 gram (g) of pure anthrax spores. This highly milled powder is referred to as *weapon-grade* or *weaponized anthrax*.

The postal system was disrupted and even though the mailings were localized to New York, New Jersey, Pennsylvania, Washington, DC, and Florida, the country was reeling and fearful, coming as it did on the heels of 9/11. The fear was not misplaced. Anthrax, in any of its three forms, is nasty. The five people who died, died of inhalational anthrax.

Bacillus anthracis is the etiologic microbe of anthrax. This bacterium can exist in two forms: a vegetative state, which is its infectious form; and its spore stage, in which the vegetative state assumes a dormancy, permitting it to resist environmental stresses and emerge into its vegetative state when conditions are ripe. Soil is its natural reservoir, where the spore can remain viable for decades. Infections can occur as an occupational risk to veterinarians and farmers via direct contact with infected animal tissues at necropsy or when hides, hair or wool are removed from infected carcasses and sold for processing, which has made anthrax poisoning an occupational disease in the hide tanning, gelatin, and animal hair and wool processing industries for hundreds of years. In 1997, the *Journal of the American Medical Association* [24] reprinted an item that had appeared in *JAMA* in 1897. It's pertinence here is evident:

Anthrax Epidemic; Precautions May Avail. August 21, 1897

A news items dated August 12, Clearfield, PA, gives out that the Falls Creek tannery of Dubois, Pa, has closed down, owing to a foreboding anthrax epidemic. Four rather rapid deaths occurred among the employees who had bathed in a stream of water used by the tannery. Several cows suffered to roam at large also drank of the water and likewise died. Hides from China are suspected of being the cause of the unlooked for visitation, and the New York Board of Health having heard that "the Swamp," the great leather exchange of the city, had received some of the importation, made haste to trace out the lines of distribution. Let us have a Department of Public Health!

Although anthrax is a serious disease, since the 1950s less than 10 cases have occurred annually in the United States. These cases were limited to workers who unknowingly handled infected products or inhaled spores from infected products such as sheep wool—for this reason it's called the "wool-sorter's disease." Anthrax is highly infectious but not contagious. Three types are recognized. Human infection can result from a cut or break in the skin producing cutaneous anthrax, with its characteristic black ulcer—hence the name anthrax, from the Greek, meaning coal or carbuncle. Gastrointestinal disease can occur 2–5 days after ingested spores in undercooked meat reach the intestine. Nausea, loss of appetite, bloody diarrhea, vomiting, and fever along with abdominal pain are the usually symptoms, a syndrome that runs its course in about a week. But this is a rare disease. It is, of course, the inhalational form that will occur should anthrax spores be dispersed in a city. One to six days after inhalation, flulike symptoms appear—fever, muscle weakness, sore throat enervation, and/or a nonproductive cough for which most people would not seek medical care, but with these symptoms anthrax would never be suspected. This confusion only compounds the problem as these symptoms are only the first phase. The bacillus is now in its vegetative state and producing toxin that passes into the bloodstream, and symptoms of acute toxicity occur: shortness of breath, profuse sweating, and skin discoloration. If not treated rapidly with appropriate antibiotics, respiratory collapse and death follows.

Curiously enough, the attempts by the Aum Shinriko pseudoreligious sect to sicken and kill large numbers of people in Tokyo, Japan, in 1995 by releasing anthrax spores failed on each of their three attempts. Yet a release of an aerosol of anthrax spores at a military facility in Sverdlovsk, in the Soviet Union in 1979, killed 66 of 77 people [25]. Apparently the Aum Shinriko followers were unable to properly prepare and disperse the anthrax spores. All this does give one to wonder whether the huge number of spores contained in the contaminated letters and the proximity of those who opened the envelopes might account for the difference in infectivity.

Botulism

Botulism toxin appears on everyone's short list of bioterrorist weapons. With good reason. Botulism toxin is an exquisitely toxic protein produced by strains of a group of organisms known as *Clostridium botulinum*. It is estimated that a single gram of the crystalline toxin, if properly distributed, could kill a million people. Given this estimate, botulinus toxin must be considered the world's most poisonous substance.

The clostridia are yet another group of spore-formers found in soils throughout the world, and more than likely the spores can be found on most fresh produce, but these spores are not—repeat, are not—harmful when ingested. The spores enable the microbe to survive in a dormant condition until favorable conditions allow them to morph into vegetative cells that can produce toxin—seven distinct types, designated by letters A–G. Types A, B, E, and F can initiate human disease: muscle paralysis in people of all ages. These toxins inhibit the release of acetylcholine at cholinergic synapses. Flaccid paralysis occurs as a consequence of the interruption of nerve impulses at the myoneural function. No one is immune to these neurotoxins. That's the key!

In addition to paralysis, the four toxins induce double vision, which is almost pathognomonic, as are drooping eyelids (ptosis), slurred speech, and difficulty swallowing. If these symptoms are not treated rapidly, progression to respiratory paralysis, as well as paralysis of arms, legs, and trunk, is assured. Ventilator-assisted breathing can be required for up to 8 weeks before the system is detoxified [26].

Botulism toxin is a highly rated choice as a bioweapon as the early symptoms make it difficult to distinguish from other neurological problems. Difficult and misdiagnoses place a person in jeopardy as treatment begun too late can be fatal. In addition, confirmatory diagnosis can take as long as 4 days, and the tests may not be readily available, further complicating the problem. If that were not problem enough, there is no preventive vaccine for botulism, although scientists at the University of California San Francisco have announced success with a drug that could be cheaply manufactured [27]. The drug, of manufactured antibodies, works quickly and can protect for up to 6 months, as it neutralizes the toxin, preventing nerve endings from sending signals to muscles leading to paralysis.

Whereas botulism's exceptional toxicity and ease of production and transport, appear to make it a potent agent of terrorism, there is continuing debate about it, as the toxin is difficult to weaponize, and it can't remain potent for long in air or in chlorine-treated water supplies.

Normally, about 20 people a year have been paralyzed by eating food containing the preformed toxin. All illnesses have come from eating low-acidity foods made at home and preserved under anaerobic conditions giving the clostridia a perfect opportunity to produce toxin that is ingested along with the food. Here, too, along with soil, are venues for anyone bent on mischief to obtain both organisms and toxin. But the real trick is to prepare and weapon the crystalline protein that would be disseminated.

In June 2005, the National Academy of Sciences published a paper, "Analyzing a bioterror attack on the food supply," by two Stanford University professors, describing how terrorists could use botulism toxin to poison the nation's milk supply. They showed how a small quantity of toxin added to a tanker truck of milk could poison hundreds of thousands of people [28]. Shades of the ice cream problem noted earlier! This publication created a bit of a furor as a number of people questioned the decision to publish as providing a ready-made roadmap for terrorists, and surely not in the best interests of the country. The professors maintain that all the information gathered in their paper can be downloaded from the Internet. Does that make it fitting and proper?

Tularemia

Tularemia, from Tulare County, California, where the disease was first described, is another bacterial zoonosis, induced by *Francisella tularensis*, one of the most virulent bacterial pathogens known. As few as 10 organisms are needed to incite illness, but it is not contagious [29]. We humans are incidentally infected as a result of hunting, trapping, butchering, and eating infected animals, as well as inhaling aerosolized organisms while farming. Mosquito and tick bites and consumption of contaminated water are additional means of transmission. Because of its extreme infectivity, ease of dissemination, and substantial capacity to cause illness and death, it is considered a prime and dangerous potential weapon. Since the 1990s no more than 200 cases per year have been reported. However, it is believed that there is substantial underreporting. Given its diverse and grave symptoms, underreporting seems unlikely; but given its rural setting, that may well be so.

Francisella tularensis has had a documented history as a biological weapon. During 1932–1945, the Japanese used it against the Chinese in Manchuria. International use by both the Germans and Soviets during World War II resulted in thousands of infections among German and Soviet troops. Later, in the 1950s and 1960s, the US military developed weapons to disseminate *F. tularensis* aerosols. In 1969 a World Health Organization Expert Committee estimated that an aerosol dispersal of 50 kilograms (kg) of the organism over an area with 5 million people would cause 250,000 incapacitating casualties and 19,000 fatalities. To add to the effect, if that was necessary, the illness would be expected to last for weeks along with relapses. Vaccinated individuals would be only partially protected. Indeed, this bacterium has more than enough attributes for a terror weapon.

Can terrorists bent on creating panic obtain this organism? Absolutely. *Francisella tularensis* occurs throughout North America and Eurasia and can be found in voles, mice, squirrels, rabbits, hares, and water rats, as well as in contaminated water, soil, and vegetation [30]. Of particular concern, francisella is very small, and can remain airborne for extended periods, which only abets inhalation. Airborne organisms would be expected to cause pneumonialike symptoms, while other exposures could involve the eyes, inducing ocular tularemia, a conjunctivitislike condition. Organisms could also penetrate breaks in the skin, causing glandular disease. Tularemia has earned its bad reputation. The release of this microbe in densely populated areas would be expected to cause abrupt onset of large numbers of acute, nonspecific, febrile illnesses up to 5 days after release, with pneumonias developing in many people over weeks and months. No group would be spared. Both sexes are equally at risk, as are children and adults. Businesses, schools, and government offices could barely function. Hospitals would be hard-pressed to accommodate the sick, and health professionals would be swamped. It would take a while before it was evident that the city had been attacked by a biowarfare agent. But once tularemia was seen in epidemic numbers, it would be clear that an attack had occurred as tularemia is a disease of rural areas, not urban centers. Streptomycin, the antibiotic of choice, would finally tame the outbreak, but not before a substantial number of fatalities occurred. As a bioweapons, Francisella and tularemia both have much to recommend them.

Smallpox

Smallpox comes with a legendary paternity. It was one of the most prevalent of all diseases; it was a massive killer, and created wretched disfigurement. Its very name could petrify an entire community.

But smallpox, as opposed to the "great pox," syphilis, is no more. With the arrival of vaccinia virus vaccination and subsequent widespread immunization, along with the vigorous efforts of the World Health Organization, smallpox was declared eradicated from the world in 1979, the only microbial disease to achieve such destinction.

How could that happen? Eradication was possible because smallpox is so obvious. Thick with pox pustules and scabs, the body is like no other; it can't be missed by searchers, no one would hide a pustule displaying individual, and there are no asymptomatic carriers or animal reservoirs. We humans are the only ones to get smallpox, and it has a short period of infectivity. Because we are the only hosts, its spread was prevented until no new cases developed. Smallpox is gone. Well, almost. The virus is kept frozen under liquid nitrogen in two safe locations: at the CDC in Atlanta and at the Research Institute for Viral Preparations, in Moscow. This maintenance has been the subject of much debate—whether the stocks in both repositories should be incinerated—totally destroyed. One side says destroy, the other says, save, because its genome needs mapping, so the controversy continues. But there is little controversy in microbiology textbooks, as those published since 1990 have little if anything on smallpox. For textbooks and the classroom, smallpox is no more.

If the case on smallpox is closed, and seemingly tightly, why is it on the short list of elite bioweapons? How could would-be terrorists obtain it, let alone use it? Actually—and this is an assumption, a bit of a stretch, but not out of the realm of possibility—as eradication approached, all countries that held small-pox cultures were requested to destroy them or turn them over to WHO. Does that provide a clue? As I say, its only an assumption, but did everyone turn their cultures in, or destroy them? So, do cultures still exist, in viable condition? If so, it is also within the realm of possibility that smallpox is obtainable. The idea is heinous, but in our tormented world, we would be remiss not to prepare our defenses, not to know about this vicious virus. The threat is low, but not zero.

For Europe and the American colonies, the earliest hint of possible control of smallpox arrived with Lady Mary Whortley Montague, wife of the British Ambassador to Turkey. Lady Montague had seen Turkish women inoculating their children, who had mild cases and became immune. On her return to England around 1710, she tried to educate the public about the procedure, but had little success [31].

The year was 1777, and George Washington, Commander of the Continental Army, was in possession of a report indicating that the British planned to infect his troops with smallpox. Taking a bold step, against what was widely decried as ungodly, he ordered his troops inoculated with smallpox-infected material [32]. Edward Jenner had yet to be heard from.

Edward Jenner, a young apprentice to a country doctor, overheard a young girl say that she could not get smallpox because she already had cowpox.

With smallpox, a person would develop a rash of malodorus pus-filled blisters over the body, which would become crusty scabs. If the person lived, the scabs would fall off, leaving deep pocked scars. It could also lead to blindness. Cowpox infected the teats of cows and the hands of milkers, usually young girls. It produced sores and fever but was rapidly transient, with subsequent immunity to smallpox.

Jenner became a physician in Berkeley, England. In May 1796, well after the Revolutionary War, Jenner vaccinated James Phipps, his gardener's 8-year-old son. (Jenner got the term *vaccination*, from the Latin *vacca*, for cow.)

Young James contracted cowpox, but recovered in a few days. Eight weeks later, allowing time for the boy's body to build immunity, Jenner exposed the boy to smallpox. James remained healthy. Today, that kind of experimentation would be considered totally unacceptable, unethical, and immoral. Over the next year Jenner vaccinated another 23 people with similar success. Good thing for all of us that he did this. The results of his work were published in 1798 in "An inquiry into the causes and effects of the variolae vaccinea a disease discovered in some of the western countries of England particularly Gloucestershire and known by the name cowpox." The medical community would have none of it—not until another experiment with cowpox and smallpox in London proved him right [33].

The outcome of his work was seen in reduced numbers of smallpox cases. But the pustular fluid was difficult to obtain and preserve, and the methods of vaccination used by other physicians varied from Jenner's permitting smallpox to continue, but the threat declined.

The preservation of pustular fluid for later use seemed insurmountable until Dr. Xavier de Balmis, a Spanish military physician, ventured yet another ingenious scheme. The Spanish colonies in the new world were being devastated by smallpox. King Charles II ordered Dr. Balmis to bring Jenner's vaccine to the Spanish colonies. But the voyage to South America would take at least a month. How to preserve the cowpox vaccine? De Balmis brought aboard 22 orphan children. He infected one child and waited for about 10 days as pustules formed. He then took the fluid from the lesions and inoculated it into a second child, continuing the cycle with successive immunizations. In this way fresh vaccine reached Mexico, Venezuela, and Puerto Rico. De Balmis continued on to Spain's colonies in Asia, picking up new children along the way and finding homes for those now vaccinated, while an assistant reached Columbia, Peru, and Chile. It is estimated that 100,000–500,000 people could have been vaccinated, marking an end to the epidemic [34].

The variola virus is a double-stranded DNA (dsDNA) member of the genus *Orthopoxvirus* and the Poxviridae family, which includes cowpox, camelpox, and monkeypox. Smallpox is the largest of all viruses and its double-stranded DNA is reported to code for some 200 different proteins, one of the largest viral genomes known, which makes it almost impossible to create a synthetic copy. No effective treatment has ever been devised. Perhaps this is reason enough to maintain the virus—mapping out its genome and seeking to discover both an appropriate preventive and a cure.

When WHO launched its eradication program in 1967, the "ancient scourage" still threatened 60% of the world's population, killed every fourth victim, scarred or blinded most survivors, and eluded any form of treatment. It had to go.

Smallpox had two forms: variola major and variola minor. Both produced similar lesions, but the minor form followed a milder course, and had almost

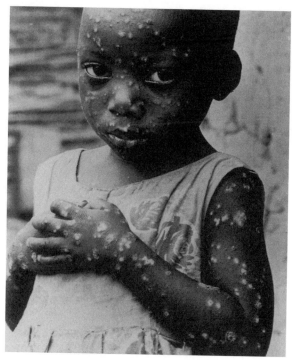

Figure 3.2. The horror of smallpox before it is eradicated. (Figure copyright PAHO/ WHO.)

a negligible fatality rate. Smallpox has an inoculation period of 12–14 days, during which time there is no shedding of virus, and the infected person looks and feels well and is not infectious—the lull before the storm. Suddenly a cascade of symptoms occurs: fever, headache, depression, fatigue, severe back and abdominal pain, and vomiting. Two or three days later, the fever abates and the person feels somewhat better, but now a rash appears on the face, hands, and forearms, progressing to the trunk. Lesions develop in the nose and mouth and ulcerate, releasing huge amounts of virus into the month and throat. This is the time when coughing, sneezing, and talking expel virus particles into the air, and when others are at great risk of breathing in virus-containing droplets or aerosols. Now the lesions are changing from macules (flat spots) to papules (raised spots) to vesicles to pustules and onto scabs, which, when dry, fall off, leaving disfiguring scars on healing; this repulsive affliction is seen in Figure 3.2. In the absence of immunity, recall that the virus has not been circulating in the United States for more than 50 years, so that few people age 50 and under have any natural or acquired immunity. Those over 50 may or may not have antibodies. It's a real concern today, and it is widely accepted that there is universal susceptibility—an excellent reason for smallpox to be on the short list. Contaminated clothes and bedding are also sources of infection, but much less of a risk.

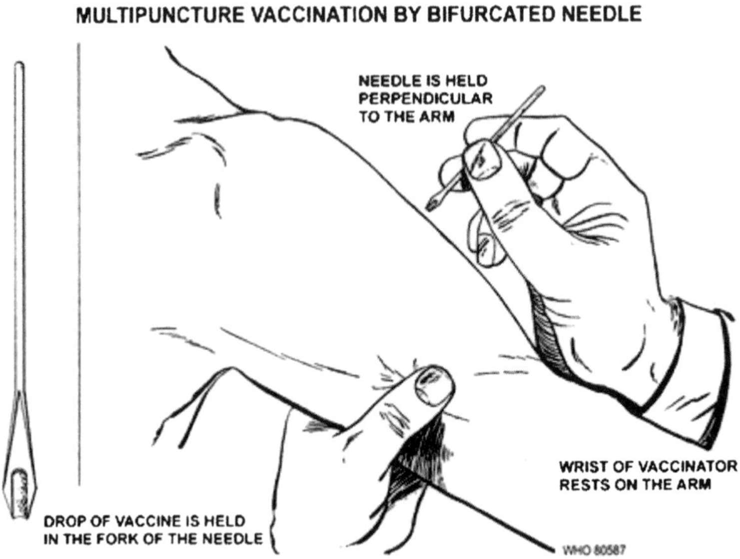

MULTIPUNCTURE VACCINATION BY BIFURCATED NEEDLE

NEEDLE IS HELD
PERPENDICULAR
TO THE ARM

WRIST OF VACCINATOR
RESTS ON THE ARM

DROP OF VACCINE IS HELD
IN THE FORK OF THE NEEDLE

WHO 80587

Figure 3.3. Multiple puncture vaccination for smallpox by bifurcated needle. (Figure copyright WHO, Geneva.)

Epidemics appear to develop relatively slowly; weeks between new cases. Experience gained during the eradication campaign indicates that given the presence of a strong surveillance system, rapid containment can break the drain of infectivity and halt an outbreak in a short time. Containment includes efficient detection, isolation (this is essential—but difficult to ensure in an open society), and vaccination of all known contacts. The public would be told, in the strongest terms, to avoid crowded places and to follow precautions for personal protection. This will, of course, have a chilling effect on business, education, and entertainment or sporting events, movies, theater, and schools. The stock exchanges could be curtailed for months until surveillance indicates a peak and certain decline in new cases. Smallpox surveillance is easier than for any other disease given the distintive rash on face, hands, and feet. Terrorists know all this, and that's their target—to disrupt the economy and society generally, which must not be allowed to happen. Emphasis must be placed on preventing epidemic spread. Bear in mind that immunity develops rapidly after vaccination. Figure 3.3 depicts the vaccination process. It's quick and painless, and postvaccination complications are rare. The best estimates of number and type derive from a 1968 study involving 14 million people. Progressive vaccinia, in which the local lesion failed to heal, occurred only in those with immune deficiency. It occurred in 11 people, resulting in four deaths. Generalized vaccinia, occurring 6–9 days postvaccination, produces a rash that disappears over weeks. That occurred in 143 of the 14 million vaccinated. Eczema vaccinatum occurred in 74 people who had a history of eczema. In these cases an eruption occurred at sites on the body that were affected by eczema. Symptoms were severe. Encephalitis, the most serious complication, occurred 16 times with 4 deaths [35].

From this study we see that approximately one death per million resulted from complications following primary vaccination, and one death per 4 million following revaccination. Whether one death per million is too high a price to pay for a protective smallpox vaccination is a personal decision. But one that must be weighed against the well-known ravages of the disease. Dr. Jenner believed that a successful vaccination produced lifetime immunity to smallpox. That may have been true in the eighteenth and nineteenth centuries, when average life expectancy was under 50, but in the twenty-first century, with life expectancy approaching 80, that is no longer the case. Revaccination will be required for protection in the event smallpox should return.

Consideration of smallpox would not be complete without noting its furious passage among the Indians of the high plains in the 1830s, which was "completely devastating." In fact, it was one of the most catastrophic epidemics recorded on the North American continent.

Smallpox had swept up from Mexico and across the plains in the eighteenth century, decimating the Mandans, Ojibway, and Pawnees, but by the 1830s the acquired immunity of their elders had waned, and none of their younger people had any immunity to smallpox. When the steamboat *St. Peters* from St. Louis arrived at Fort Clark with the first thaw of spring, there was great jubilation, until it was learned that several passengers had contracted variola major. Three weeks after the ship departed, Indians began dying. Many committed suicide by leaping off cliffs and stabbing themselves with knives and arrows, the pain of the pox was so excruciating. Along the upper reaches of the Missouri River, traders bungled an attempt to inoculate Indian women with scabs taken from smallpox cases. Dozens died. Seven months after the first Mandans died, the tribe had been reduced from 1600 to 31, and many other tribes approached annihilation. The Great Plains had been converted into a graveyard. As many as 2000 plains Indians had died of the hideous pocking disease [36]. Today, all manner of Americans have arrived at the same dilemma—wholly susceptible, few vaccinated. All unsuspecting.

Plague

On August 2, 1996, a young man, age 18, with fever, pain, diarrhea, and tenderness in his left groin, was seen at a local outpatient clinic in Flagstaff, Arizona. He was treated with a nonsteroidal antiinflammatory medication and sent home. The next day he had difficulty breathing and collapsed while taking a shower. Brought to a hospital emergency room, he was pronounced dead on arrival. On August 8, cultures of blood samples taken in the emergency department were positive for *Yersinia pestis*, the plague bacillus.

On August 17, 1996, a 16-year-old girl in western Colorado had pain followed by numbness in her left armpit. Fever, chills, and vomiting occurred over the following 2 days. At a local hospital she was treated for chest pain as she said she had fallen from a trampoline. An appointment with a neurologist was scheduled, and she was sent home. On August 21, she was found semiconscious at home and was taken back to the hospital. Within an hour of arrival she went into respiratory arrest, and was transferred to another hospital with a diagnosis of respiratory distress syndrome and possible meningitis. Her condition rapidly deteriorated. She died later that day. On August 23, blood and spinal fluid cultures obtained on August 21, were positive for *Y. pestis* [37].

Five cases of human plague were reported in the United States in 1996, two of which resulted in deaths. On average, 10–20 cases per year are the expectation, occurring mostly in young people under 20, with highest rates are among the Navajo. Others at risk are hunters and veterinarians. Although an ancient affliction, plague remains one of the world's most feared diseases, and appears to be too complex to be eradicated. This zoonotic disease has reservoirs on nearly every continent and has the uncanny ability to overwhelm host immune systems. Obviously plague is a perfect choice as a bioweapon, even though its spread is preventable and the illness curable if diagnosed rapidly and appropriate treatment instituted. Nevertheless, the devastation attributed to bubonic and pneumonic plagues dwarf that of most other diseases. Estimates put the death toll at 200 million.

Plague is primarily a disease conveyed among wild rodents. We humans can become infected when entering rodent territory, as illustrated in Figure 3.4. The infection is easily transmitted from rodent to human by the bite of an infected flea that picked up the bacterium from its rodent host, while taking a blood meal and then taking another blood meal from a human and disgorging bacteria into the bite site. Intimately involved in the affair are rodent, flea,

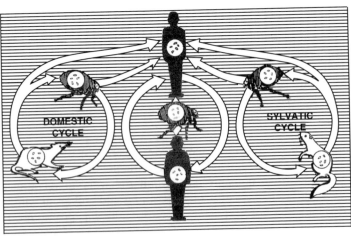

Figure 3.4. Cycles of transmission and sequence of events leading to infection and possible outbreaks of bubonic and pneumonic plagues. (Figure shows fleas carrying *Yersinia pestis* bacteria, obtained from wild rodents, disgorging the bacteria while taking a blood meal from an unsuspecting individual.)

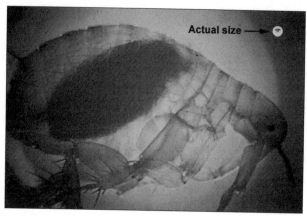

Figure 3.5. Male *Xenopsylla cheopis* (oriental rat flea) engorged with blood. This flea is the primary vector of plague in most large plague epidemics in Asia, Africa, and South America. Both male and female fleas can transmit the infection. (Figure adapted from the Centers for Disease Control, Atlanta.)

bacterium, and human—far too complex to even contemplate eradication. Plague can also be acquired by direct contact with infected animal carcasses or by inhalation of contaminated airborne droplets.

Plague exists in three forms: bubonic, pneumonic, and septicemic. *Bubonic* is the classic and most common type, that produced the infamous Black Death of the fourteenth and seventeenth centuries. After a flea, most often the oriental rat flea, *Xenopsylla cheopis* (Fig. 3.5), ingests *Yersinia pestis* from its rodent host, yersinia produces a chemical, coagulase, that clots ingested blood in the flea's proventriculus, an organ between its esophagus and stomach, which blocks passage of the next blood meal into the flea's stomach. With this blockage, fleas requrgitate *Y. pestis* into the wound site of the bite they made while attempting to feed. It's quite ingenious. Between 25,000 and 100,000 bacteria are sent into the wound site. Those bacteria now migrate to regional lymph nodes, where many are engulfed and destroyed by polymorphonuclear leukocytes—a type of phagocyte. But those that remain viable in the lymph nodes induce inflammation and edema, which cause painful, swollen buboes (the lymph nodes) in armpits and groin, hence bubonic (from the Greek *boubon*, meaning groin). Swollen buboes are timeless indicators of bubonic plague. No other infectious disease produces that manifestation. The "blackness" of the Black Death refers to, and is the aftereffect of, subcutaneous hemorrhages forming dark patches on the skin as blood oxidizes and turns purple-brown to black. But that is not the end of it. The bacteria can also enter the bloodstream, initiating septicemia and systemic shock, which, if not treated promptly and appropriately, is deadly.

Pneumonic plague is a rare but deadly form of plague that is spread via respiratory droplets in close contact (between 2 and 5 feet) with an infected person. The illness progresses rapidly from fever and flulike symptoms to an overwhelming pneumonia with heavy coughing and production of bloody sputum. Especially serious is the fact that the majority of cases in the United States were spread by droplets from infected domestic cats. Veterinarians have been at unusually high risk [38].

Yersinia pestis is so highly infective that no more than 100–500 microbial cells are required to produce illness; and, of course, it is contagious. Edema and respiratory lung congestion are common symptoms followed by respiratory collapse, if not rapidly treated with antibiotics. Mortality approaches 50%. There is little doubt that this form of plague would be a terrorist's choice.

Most human plague occurs as a result of contact between rodents (many kinds) and people. However, human outbreaks are often preceded by epizootics, with large numbers of deaths among susceptible animals, which can signal an alert that human involvement could follow, which is often the case with teenagers who walk through rodent-infested areas, get bitten or scratched, or inhale airborne organisms.

As a bioweapon in today's world, the choice would not be fleas, even though they were used effectively by the Japanese during World War II. Weaponized, airborne *Y. pestis* would be the choice. During the 1950s and 1960s the United States and Russia developed procedures for aerosolizing *Y. pestis*, further enhancing its infectivity and virulence. Among its many attributes is the fact that in the early stages of dispersal and consequent illness it would be misdiagnosed, leading to a high and rapid death rate. The fact that it is communicable would sharply increase infections. Perhaps of even greater concern, plague can trigger fear and panic that would be difficult to contain. As a bioweapon, physicians would be faced with severe and rapidly progressing pneumonia. Many similar cases would occur over several days. Buboes, of course, would be absent, and the illness would be occurring in urban rather than in rural areas, and there would be no indication of prior epizootics. Together these factors could catch people unaware and unprepared, permitting numbers of deaths before a correct diagnosis was made and appropriate treatment instituted. Indeed, here again, surprise is on the side of the bad guys.

Protection of Agriculture and the Food Supply

In his October 2001 executive order establishing the Office of Homeland Security, President Bush added agriculture and the food industries to the list of critical infrastructure sectors requiring protection, and acknowledged that the agriculture sector and the food supply are vulnerable to bioterrorists. The Secretaries of both Agriculture and Health and Human Services have publicly recognized that the US food supply is susceptible to deliberate contamination. Attacks could be directed at livestock, crops, and food products.

Our agriculture and food sectors have features making them vulnerable; our livestock industry is highly concentrated, as is the centralized nature of our food processing industry, which suggests that infectious pathogens and chemicals could be intentionally added at a number of points along the farm-to-table continuum. Even the threat of an attack could adversely affect the public is confidence in the safety of the food supply and destabilize the export market, as recently occurred with a dozen countries refusing to import our beef because of the widely reported finding of a single "mad" cow in the State of Washington.

An excellent case in point is the 2001 foot-in-month epidemic that swept through England. On February 20, 2001, the presence of foot-and-month disease was confirmed in Essex, England. The virus, a single-stranded RNA rhinovirus, spread rapidly, eventually resulting in over 4 million cattle being slaughtered to control its further spread.

Terrorists seeking ways to harm the United States, could deliberately introduce foreign animal diseases into the country. Accepting that idea, the USDA recently calculated that a foot-and-mouth outbreak could spread to 25 states in as little as 5 days. A simulation by the National Defense University in June 2002 predicted that a foot-and-mouth outbreak could spread to more than one-third of the nation's cattle herds [39]. When it does, vesicles appear in the mouth and on the upper lip, on the coronary band of the foot, and on the mammary glands and areas of fine skin. Foot-and-mouth disease (FMD) is a visual misery, and can't be missed. These vesicles burst, leaving painful, red erosions. Secondary bacterial infections of the open vesicles can also occur in the mouth and on lips. For a terrorist, the fact of virus elimination via all secretions and excretions offers added reward. Large quantities of virus are eliminated in saliva, which is responsible for both environmental and airborne contamination. Virus can then be spread widely by drinking water, breathing in aerosols, and eating contaminated food. The appearance of FMD on a farm signals turmoil. Farmers know that they and all other people on the farm along with their dogs and cats can become transmitting sources (vectors), carrying the virus on boots, wheelbarrows, and truck tires. If infected animals aren't spotted quickly and culled from the herd, spread and ruin are inevitable [40]. In Britain, the epidemic caused suicides among farmers who saw their livelihoods go up in smoke. Add to that the fear of air and water pollution from animal carcass disposal. And as the epidemic occurred in August, it totally wiped out tourism as the tens of thousands of tourists who had planned to arrive canceled their reservations. FMD followed hard on "mad cow" disease, which the public worldwide had been led to believe was both a human and animal disease, that humans could get from animals, and that eating the famous English beef was unsafe. That was more than enough; people headed in other directions. Bioterrorism has many faces—all ugly.

The U.S. General Accounting Office, looking at measures for preventing FAD and "mad cow" from entering the United States, found that because of the sheer magnitude of international passengers and cargo entering the country every day, the available inspection resources are simply unable to ensure preventing entrance. They simply can not inspect everything. They also found that

FMD can be carried on the shoes of international passengers who visited farms, and on packages they carry. The virus can be transmitted in international mail and in garbage from international ships and airlines. The FMD virus is a hardy bit of protein that can survive for considerable periods in diverse environments. It can remain viable in salted bacon for 3 months, and in air-dried hides and skins for 6 weeks. If a person steps in infected manure, the virus can endure for 2 months in summer and 3 in winter. Such a delivery spectrum can be deliberate or benign. Mad cow disease, benign spongiform encephalitis (BSE), has been more sound than fury, and does not lend itself to terrorism. But could an animal be slipped in among a herd coming into the country, to be discovered later? Indeed it could, and that "discovery" could trigger a ruinous economic boycott. Figure 3.6, illustratres the layers of governmental involvement that provides substantial and reassuring protection against such a possibility.

Food and food processing is generally considered the midpoint of the farm-to-table continuum, and it is generally agreed that terrorists could use food as vehicles for introducing bioagents into the food supply. A food poisoning epidemic in 1994 that sickened 224,000, people can serve as a model and cautionary tale, even though it was an inadvertent accident. The fact is, it highlights the vulnerability of our open society.

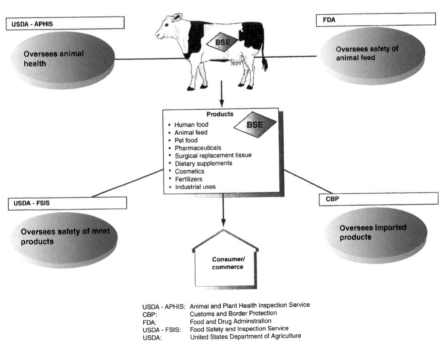

Figure 3.6. Federal government agencies involved in bovine spongiform encephalopathy (BSE) oversight. (Figure adapted from General Accountability Office, GAO-04-588T.)

Salmonella enterididis, a known food poisoning organism, was the microbe that produced the largest salmonellosis outbreak and the largest food poisoning outbreak in US history—and, of all things, it did so in America's favorite food, ice cream. The centralized nature of our nation's food distribution networks lends itself to endangerment of huge numbers of consumers. Commercial ice cream producers use only pasteurized milk, and pasteurization kills all salmonellae. It's the homemade ice cream, often made with unpasteurized milk, that allows salmonella to grow and produce its toxin in ice cream. How, then, could ice cream become the vehicle of the gastroenteritis that laid low so many thousands of ice cream lovers.?

Tanker trailers hauling thousands of gallons of ice cream made from base premix had hauled nonpasteurized eggs to processing plants before taking on the ice cream mix. That would have been okay had the tanker drivers not broken established work rules, which required tankers hauling unpasteurized liquid eggs to be thoroughly washed and sanitized before loading ice cream premix. The salmonella-containing liquid eggs contaminated the premix, which was not repasteurized after transportation, and 224,000 vanilla ice cream devotees became surprised and accelerated half-steppers [41]. Could this happen again? Yes. Deliberately? Again, yes.

The way to prevent it is to take control away from drivers and maintenance crews. Either liquid eggs cannot be delivered in trucks that also carry premix, or trucks that have dischared a load of liquid eggs must be automatically sealed when their contents are emptied and should not be opened until they are in a cleaning station, or some such unit. But to evade the system shouldn't be difficult. Again, it's the human equation. A bribe might be all it takes to avert eyes as a culture of tularemia or plague is dropped into a tanker. Yes, it surely could happen again, and it could sicken and kill thousands. This is exactly what the two Stanford University professors had in mind with their idea of botulinum toxin in milk.

Unfortunately Congress has not yet moved to give the FDA authority to require food processors to take safety measures to improve security against deliberate attempts at compromising the country's food supply.

Food crops—especially the cereal grains, wheat, barley, and rye, along with the most widely eaten staples, corn, rice and potatoes—are recognized targets of terrorism. Disruption of basic food supplies has caused famine and can cripple a country. Characteristically, fungal diseases of wheat and rice, and other cereals, are spread by fungal spores that can withstand environmental stresses and distruction. These spores infect the aerial parts of plants, causing diseases that have the capacity to spread rapidly, attaining epidemic (epiphytic) proportions during a single growing season. For a notable, and unforgettable, natural fungal devastation of a crop, we need only glance backward to the Irish potato famine of 1845–1848, a famine of unimaginable proportions, the result of infection of the potato crop by *Phytophthora infestans*, which continues on current short lists of potential anticrop weapons, and has been given the codename, "LO."

Intense potato cultivation in Ireland had accommodated a remarkable 70% population increase between 1745 and 1841. By 1840, potatoes were the sole food of 30% of the people, and essential to another 30%. Potatoes were eaten in place of bread, and the two main meals of a laborer's day consisted of potatoes.

Potato Late Blight first occurred in autumn 1845. In August 1845, John Lindley, the newly appointed professor of botany at the University of London, and editor of the *Gardeners Chronicle* and *Horticultural Gazette*, wrote: "A fearful malady has broken out among the potato crop. On all sides we hear of destruction. In Belgium the fields are said to be completely desolate. There is hardly a sound sample in Convent Garden market. . . . As for cure for this distemper, there is none." He continued: "We are visited by a great calamity which we must bear." The desperate nature of the devastation was announced on September 13: "We stop the press with very great regret to announce that the potato Murrain has unequivocally declared itself in Ireland. The crops about Dublin are suddenly perishing. . . . Where will Ireland be in the event of a universal potato rot?" [42]. The rot was all-encompassing. Starvation was widespread. A wholesale exodus followed in which hundreds of thousands of people fled the country. Between 1845 and 1851, the population of Ireland decreased by more than 2%. A mold, a fungus, changed the course of Irish history.

Molds, rusts, and smuts are fungi. Although they often look like plants, they aren't. Plants are green, and contain chlorophyll, which does not occur in fungi. Consequently fungi can not produce their own food and must obtain nourishment by digesting plants, living or dead. Fungi secrete digestive enzymes into the plant they parasitize, and absorb their nutrients. Fungi can destroy crops by causing diseases such as leaf spot, blights, smuts, rusts, and blasts. Trees can also be victimized. Dutch Elm disease, the worst such outbreak in the United States in recent decades, was the work of *Ceratocystis ulmi*.

Anticrop agents considered well qualified include stem rust of wheat, the result of infection with *Puccinia graminis*, codenamed TX. *Puccinia* is the largest genus of rust fungi, and *P. graminis* is the most destructive fungus of all cereal grains. Black stem rust of cereals is a pathetic sight. Pustules of spores erupt through leaf surfaces looking for all the world like smallpox of plants. As the plant matures, the pustules blacken, then break open, releasing masses of brown spores that are now freee to infect surrounding plants.

Ustilago, the principal genus of the smut fungi, includes *U. maydis* on corn and *U. nuda* on wheat and barley. The ears of barley infected with *Ustilago* look as if dusted with soot; hence they are called "smut" fungi. Interestingly enough, wheat cover smut, a strain with crop-destroying potential, was evaluated for use as a weapon. Research done at Salmon Pak, outside of Baghdad, Iraq, "in 1985, demonstrated that wheat cover smut spores sprayed over immature wheat plants would be lethal to crops." In his discussion of Iraq's biological weapons, Raymond A. Zilinskas informs us that "In 1988, young wheat plants growing in large fields near the town of Mosul were infected with this agent: the infected wheat subsequently was harvested and moved to Fudaliyah for storage. The Iraqi's claim that no attempt was made to recover fungus from the harvest and that the infected crop was destroyed in 1990." He goes on to say that" the investigation of wheat smut implies that Iraqi leaders know that biological weapons were more than anti-person weapons; they could also be used against crops as part of economic warfare" [43].

Magnaporthe grisea, codenamed L4, the causal organism of rice blast and the most destructive worldwide, is also on the short list of potential anticrop agents. Clearly these fungi could easily be used to adversely affect the world's must economically and socially significant food and cash crops: cereals, potatoes, and rice, the world's major food crops. Attacks on cereals could seriously affect bread production, and coffee berry disease due to *Colleotrihium coffeanaum*, which seriously reduces yields and kills coffee plants could surely disrupt supplies of coffee beans. With over 13,000 fungal species, the possibilities appear endless, and that a given crop species such as wheat can be destroyed by over 200 different diseases worldwide, indicates that terrorists have unlimited choices at their disposal.

The potential of dual-use capabilities of weed and pest control chemicals is also worth considering. The fact these could go either way poses yet another serious threat that they could be used for malign purposes. The immensity of the security needs faced by leaders at all levels of government can be daunting.

In the United States, use of biopesticdes and biocontrol agents is increasing, replacing chemical pesticides, which can be a double-edged sword. *Bacillus thuringiensis* produces a protein toxic to lepidoptera (caterpillars), diptera (flies), and coleoptera (weevils and beetles). Freely available scientific literature provides detailed descriptions of fermentation techniques used in the rapid and large-scale production of a range of organisms, or their toxic products. If that were not information enough, genetic manipulation of organisms to enhance their effectiveness is available to anyone with a computer and Internet connection, which means just about anyone.

Biological and Toxic Weapons Convention

In 1975, in response to concerns of many governments, the Biological and Toxic Weapons Convention (BTWC) came into being as the most important tool against the use and development of biological weapons. Nevertheless, since 1975, there have been confirmed cases of states breaching the Convention. As of 2004, efforts to strengthen the BTWC by a supplementary legally binding protocol have failed. The BTWC was set up to regulate the behavior of its signatories. But a number of countries now believe that nonnation terrorist groups pose the greatest threat for use of bioweapons and have suggested that the BTWC can no longer ensure security. The ability of the BTWC to ensure the security of its member states is also being challenged by rapid developments in biotechnology and genetic engineering with its promises of improving quality of life, could also be converted for hostile purposes. A sixth BTWC Review Conference in 2006 to address these issues is in the planning stage. Failure to address and fix these issues would surely undermine the relevance of the BTWC. To deal with the challenge of the rapid pace of scientific progress, a seventh Review Conference is planned for 2011. Be that as it may, the BTWC has been in existence for 30 years, yet in all that time it has had no effective means of verifying that any of its member states have lived up to its treaty obligations. In fact, some countries are opposed to a verification system. It doesn't require a high degree of cynicism to believe that conventions and treaties are only so much paper [44].

It must be evident that our global village can tolerate neither terrorist behavior nor the threat of such behavior, and must strenuously counter this challenge. Terrorism must be relegated to the dust bin of history.

Atlantic Storm

In the meantime nations must prepare for the worst, which raises at least two questions: What would countries do if faced with bioweapon attacks on their cities, and how would they deal with a rapidly spreading epidemic?

In January 2005, a tabletop exercise codenamed Atlantic Storm simulated a smallpox attack on countries of the transatlantic community. Atlantic storm was the brainchild and presentation of the Center for Biosecurity of the University of Pittsburgh Medical Center, the Center for Transatlantic Relations of Johns Hopkins University, and the Transatlantic Biosecurity Network.

During the exercise, held in Washington, DC, former prime ministers and other senior government officials played the roles of the heads of government of their respective countries. Madeleine Albright, former U.S. Secretary of State, played the part of the US President. Other participants represented the European Union as well as the United Kingdom, France, Germany, Italy, The Netherlands, Poland, Norway, and Sweden in a mock summit.

The scenario presented was the simultaneous outbreak of smallpox in Istanbul, Frankfurt, and Rotterdam, with smallpox detected in the United States later in the day. Early in the exercise the participants were told that the disease had been unleashed deliberately, and that a terrorist group claimed responsibility. Heated discussion included availability of vaccine; did every nation have enough to vaccinate everyone, or should vaccine be limited to those infected, and what about sharing or not sharing vaccine, and why are borders being closed? Polish citizens are streaming into Germany demanding vaccine because there is none in Poland. Should they be given vaccine from their limited supply? As the discussion continued, numbers of smallpox cases increased and the number of countries involved also increased. Cases were reported in Mexico and Canada and throughout Europe. How can further spread be curtailed or prevented?

At the end of the day they found that neither the UN, the EU, nor NATO had the ability to be a rapid responder, and WHO had no budget to cover the cost of the vaccine. Atlantic Storm had shown how critical it was for leaders to be prepared to respond to such surprise attacks. They also found that there was insufficient awareness at the highest levels of government of the possibility and consequences of an attack. Atlantic Storm was the right idea at the right time. We can only hope that what was learned will not be allowed to languish [45]. There ought to be more of them. As surprise attacks become less of a surprise, terrorists may get the idea that there is nothing to be gained.

Chronic Disease Microbiology: Possible Emergence?

Illness has been an intimate part of life for as long *Homo sapiens sapiens* walked the earth. We have wondered why. Hippocrates wondered, and believed that how and where people lived determined their state of health. For him, health or illness was the result of environmental risk factors. He was persuasive [46]:

> Whoever wishes to investigate medicine properly, should proceed thus: in the first place to consider the seasons of the year, and what effect each of them produces (for they are not at all a like, but differ much themselves in regard to their changes). Then the winds, the hot and the cold, especially such as are common to all countries, and then such as are peculiar to each locality. We must also consider the qualities of the waters, for as they differ from one another in taste and weight, so also do they differ much in their qualities. In the same manner, when one comes into a city to which he is a stranger, he ought to consider this situation, how it lies as to the winds and the rising of the sun; for its influence is not the same whether it lies to the north or the south, to the rising or to the setting sun. These things one ought to consider most attentively, and concerning the waters which the inhabitants use, whether they be marshy and soft, or hard, and running from elevated and rocky situations, and then saltish and unfit for cooking; and the ground, whether it be naked and deficient in water, or wooded and well-watered, and whether it be naked and deficient in water, or wooded and well-watered, and whether it lies in a hollow, and confined situation, or is elevated and cold; and the mode in which the inhabitants live, and what are their pursuits, whether they are fond of drinking and eating to excess, and given to indolence, or are fond of exercise and labor, and not given to excess in eating and drinking.
>
> From these things he must proceed to investigate everything else. For if one knows all these things well, or at least the greater part of them, he cannot miss knowing, when he comes into a strange

city, either the diseases peculiar to the place, or the particular nature of common diseases, so that he will not be in doubt as to the treatment of the disease, or commit mistakes, as is likely to be the case provided one had not previously considered these matters.

This appraisal was conceived 2500 years ago, on the Greek Island of Cos, where Hippocrates practiced medicine, and it remained the prevailing and widely accepted view for 2000 years. But there were stirrings. Ignatz Semmelweis, an obstetrician in 1840 Vienna, wondered why maternity clinics were "houses of death," where childbed fever was rampant and fatal in more than 50% of cases. Physicians couldn't possibly be the problem. It was the environment—obviously. But why was his division, consisting of himself, other obstetricians, and his medical students, having far higher maternal mortality rates than the midwives' division? Explanations—and there were many—were unavailing until the day a colleague died of a wound inflicted during the dissection of a cadaver. To his utter shock and astonishment, Semmelweis recognized that the collected symptoms were the same as those of the women who had died of childbed fever. He realized that he, his students, and the obstetricians had been autopsying cadavers just before examining women in labor. Semmelweis had his epiphany. They were all infecting the women. Midwives had no contact with cadavers. He also found that mortality among the women rose and fell with the number of autopsies performed. He knew what to do. All physicians and students coming from the autopsy room were required to wash their hands with soap followed by rinsing in a chlorine solution before entering the maternity wards. The death rate plummeted. Semmelweis tells us that "the cause of Prof. Kolletschka's untimely death was known: it was the wound by the autopsy knife that had been contaminated by cadaverous particles. Not the wound, but contamination of the wound by cadaverous particles that caused his death" [47]. He could not know what those particles were, but disinfection with a chlorine solution could remove them. It had to be something in the environment, not the environment itself.

In 1854, cholera returned to London. Dr. John Snow (1813–1858) was a thinking man's physician. When he was a young medical apprentice he was sent to help during an outbreak of cholera among coalminers. His observations convinced him that the disease was usually spread by unwashed hands and shared food, not by "bad air." In 1854, cholera struck London. The city's water supply came from two companies—the Southark & Vauxhall Company and the Lambeth Company. Snow interviewed cholera patients and found that most of them purchased their drinking water from the Southark & Vauxhall Company. He also learned that this company took its water from the Thames River below locations where Londoners discharged their sewage. By contrast, the Lambeth Company's intake pipes were in the upper reaches of the Thames well before the water reached London. His data showed that those households receiving water from the Lambeth Company had a mortality rate 5 times less than those purchasing water from Southark & Vauxhall. Snow inferred the existence of a "cholera poison" transmitted by polluted water. He also noted that the cause of the disease must be able to multiply in water, and he believed that consecutive cases suggested a means of transmission, more so than could be accounted for by coincidence. He tells us that "diseases which are communicated from person to person are caused by some material which passes from the sick to the healthy." Of course, he was right, but again, he couldn't know why. But he was close. As he put it, "the morbid material is ingested through the mouth, multiplies in the gut, and is excreted with feces" [48]. Stirrings, indeed. Ferment. New ideas were circulating. Louis Pasteur in France discovered that fermentation of wine was controlled by microscopic yeast cells that converted sugar in grapes to alcohol and carbon dioxide. When vats of wine became contaminated with other microbes, the wine would turn sour; acid was produced, instead of alcohol. Pasteur developed a procedure in which liquids are heated for a short time to kill the microbes without affecting the flavor of the liquid—pasteurization.

Joseph Lister, a surgeon in England, learning of both Semmelweis' disinfection techniques and Pasteur's heat treatment, began heating and soaking surgical instruments and dressings in carbolic acid (phenol). This procedure proved highly successful in reducing surgically related infections, and it provided strong but indirect proof of the role of microbes in disease because phenol, which killed bacteria, also prevented would infections.

It remained, however, for Robert Koch, a young country physician in Wollstein, Germany, to demonstrate unequivocally the direct role bacteria played in human disease. Using criteria suggested by Jacob Henle, a former professor and friend, he established a direct causal relationship between *Bacillus anthracis* and anthrax. Koch injected healthy mice with the blood of diseased sheep. The mice sickened. Placing bits of infected spleen from ill mice into sterile beef serum, he obtained luxuriant growth of the bacillus. Koch them inoculated the new growth of bacillus into healthy mice, which also fell ill, and he again obtained the bacillus from these newly ill mice.

This was the first direct proof that infectious illness was a function of microorganisms, and he set forth a series of criteria to establish that connection. He's criteria required that:

1. The suspect organism be present in every case of the disease, but absent in healthy animals.
2. The suspect organism be isolated from an infected animal and grown in pure culture.
3. The disease be reproduced when the isolated organism is inoculated into healthy animals.
4. The same organism be isolated from ill animals.

Koch's proof of the relationship between *B. anthracis* and anthrax was then independently verified by Pasteur, which effectively ushered in the "golden age of bacteriology," with its germ theory of disease, and its underlying principle of one agent, one disease—which ensured the decline of the environment.

By 1882 Koch had used his techniques to discover and isolate the tubercle bacillus. The ensuing 40 years was a period of feverish activity during which the majority of human bacterial pathogens were discovered. The principle of one agent, one disease was in full bloom [49].

But what of the chronic, noncommunicable diseases—were they microbially related? As we have seen, the role of microorganisms in human illness is based on the ability to isolate and grow the organisms in laboratory systems and subject them to the Koch (Henle) postulates. "The causes of Crohn's disease, ulcerative colitis, Wegener's granulomatosis, rheumatoid arthritis, tropical sprue, lupus erythematosis, Kawasaki's disease and other chronic illnesses remain unknown despite features that are suggestive of infections etiolgy" [50]. One microbe, one disease may still operate, but that one microbe remains elusive. Crohn's disease is a severe inflammation of the intestines. It is chronic (persistent and long-term) and recurrent, with fever, weight loss, abdominal pain, and diarrhea. Fever, pain, inflammation (swelling), and redness, which usually accompany inflammation, are the traditional criteria of an infectious process, so it is reasonable to consider an organism or combination of organisms as contributing to Crohn's disease. But the intestines are the seat of hundreds of bacterial species. To run these down would be an enormous and possibly fruitless undertaking. Ulcerative colitis, an inflammation of the colon, has similar symptoms, and arthritis is marked by inflammation and swelling of joints. Considering that traditional microbiological approaches have been unavailing, a totally new approach is required. Laser-capture microdissection could fill that need by microscopically focusing on suspect areas or cells within tissue sections. Such anatomic targeting could get beyond the normally heavy microbial density of tissues associated with intestinal diseases. This microscopic technique could also overcome the difficulty of culturing viruses, which could be shown to be a triggering or causative agent.

No one considered peptic ulcer a microbial condition. Far from it. For generations, gastroenterologists lectured medical students that stress causes the stomach to produce excess acid and that the corrosive acid produces ulcers. Everyone knew it. As far back as 400 bc, Hippocrates wrote that "a spare diet and water agree with ulcers, and with the more recent rather than the older" [51]. In 1596, Edward Spenser in Faerie Queen wrote "that my entrails flow with poisonous gore, and the ulcer growth deadly more and more" [51].

In 1920, Karl Schwarz, at the Merciful Brothers Hospital in Agram, Germany, wrote this mantra: "Keine Saur, Keine Geschwur" (no acid, no ulcer) [52]. The words stuck and became received medical wisdom for the next six decades, until 1984, when two Australian pathologists in Perth found that it was a bacterium, *Helicobacter pylori (H. pylori)*, that produces the characteristic inflammation, acidity, and chronic superficial gastritis. That revelation sent shockwaves through the medical world. Although it has low mortality, ulcer disease is high on the list of human suffering. For the millions of adults who have ulcers, a microbial etiology came as welcome news, as the great majority of cases can be alleviated by appropriate antibiotics. Dr. Marshall also informs us that "one of the main difficulties with the H. pylori theory was our inability to develop an animal model for the disease. After failing to infect rats, mice and pigs, I decided to infect myself with H. pylori to see if chronic gastritis developed. In July, 1984, I drank a pure culture of H. pylori (109 organisms)." He goes onto say that he became ill, but after 14 days the illness resolved, "but culture and histologic diagnosis demonstrated severe, acute gastritis and many H. pylori organisms" [53]. Dr. Koch would be smiling.

In his recently published and delightful little book, Dr. Marshall has collected the personal thoughts of those researchers around the world who, after learning of his initial publication in *The Lancet*, realized that they too, had seen the stubby bacterium, but hadn't realized they had discovered the cause of gastric ulcer [54].

It was eye-opening to learn that the stubby microbe had been "discovered" in Italy (1892), Greece (1958), Boston (1967), China (1972), Russia (1974), and England in the 1970s. For all of them, the stubby organism resembled campylobacter, a well-known food poisoning microbe, and so it was first named *Campylobacter pyloridis*, then *C. pylori*, and finally by Marshall and Warren, *H. pylori*. It's a wonderful and heroic tale in which we can take comfort in the fact that highly competent scientists are working on similar studies and that victory over disease will come from one country or another, as no one country has a lock on talent, imagination, or creativity.

The recently published report *Microbial Triggers of Chronic Human Illness*, by the American Academy of Microbiology [55a], is the outcome of 3 days of impressive deliberation by microbiologists, immunologists, and virologists who addressed the question of how microbes might trigger chronic illness. A product of their ruminations was a list of chronic diseases that may have microbial components.

Table 3.1 does catch one's breath. The idea that mesothelioma, a malignant tumor of the tissue covering the lungs, long held to be due to asbestos fibers, could play a role along with a virus, is stunning. And what of Alzheimer's disease, which is considered a problem of protein misfolding; could chlamydia or other microbes trigger the misfolding? The idea does grab the imagination. That Bell's palsy, the most common cause of acute neuromuscular paralysis, could have a viral etiology, will offer hope to hundreds of thousands who suffer with this palsy. Additionally, if prostate cancer turned out to be a virus-related disease, antiviral medications could well do away with current surgery that offers men two options—incontinence and/or impotence—both wicked! As if to dot that "i," at the 2006 Prostate Cancer Symposium, it was reported that men with a genetic mutation in the human prostate cancer 1 (HPC1) gene harbor a virus that may cause the cancer. The HPC1 gene encodes an antiviral protein (RNaseL) activated by a viral infection. Men who have two copies of a mutated HPC1 gene have twice the risk of developing prostate cancer compared to men with two normal versions of the gene. Further testing is in

TABLE 3.1. Chronic Diseases for Which There Is Suspicion of an Infectious Etiology

Disease	Suspected Agent(s), if Any
Primary biliary cirrhosis	Helicobacter pylori, retrovirus
Mesothelioma	Simian virus 40
Multiple sclerosis	Epstein–Barr virus
Tics and obsessive–compulsive disorder	Group A *Streptococcus agalactiae*
Obsessive–compulsive disorder	Group A *S. agalactiae*
Crohn's disease	*Mycobacterium paratuberculosis* and others[a]
Alzheimer's disease	*Chlamydia pneumoniae*
Diabetes	Enteroviruses
Sjogren's disease	*H. pylori*
Sarcoidosis	*Mycobacterium* species
Atherosclerosis	*C. pneumoniae*, CMV
Bell's palsy	Herpes simplex virus
Schizophrenia	Intrauterine exposure to influenza
ALS	Prions
Chronic fatigue	HTLV-1; EBV
Prostate cancer	BK virus

[a] *Clostridium, Campylobacter jejunji, Campylobacter faecalis, Listeria monocytogenes, Brucella abortus, Yersinia pseudotuberculosis, Yersinia entercolitica, Klebsiella* spp., *Chlamydia* spp., *Eubacterium* spp., *Peptostreptococcus* spp., *Bacteriodes* fragilis, *Enterococcus* faecalis, and *Escherichia coli.*

Source: Reprinted with permission of the American Academy of Microbiology, *Microbial Triggers of Chronic Human Illness*, 2006, Table 2, p 9.

progress to determine how this virus initiates tumor development, and how it may be shut down [55b].

The prospect that these 11 distasteful conditions could all be microbial, holds out tremendous hope. If only one or two prove so, that, too, would be a welcome victory. With these intriguing possibilities, microbiologists have cause to burn the midnight oil.

Prion Diseases—Proteins Gone Awry

So we begin with proteins, the molecules that do the body's heavy lifting, making possible all activity. But proteins begin with amino acids. According to instruction written in gene DNA, cells synthesize proteins from 20 essential amino acids, the building blocks, each with its own size, shape, and properties. With 20, the variety of combinations can be almost—but not quite—infinite. Nevertheless, there are over 100,000 different proteins, each consisting of from 50–5000 amino acid combinations. Each protein has its unique sequence of amino acids. These amino acid chains are not arranged in straight lines; they twist, fold, and buckle. When the chains settle into their final shape, it becomes active. Protein folding is both a complex and difficult process given the number of molecules crowded into a cell. Some proteins fold rapidly; others fold over several days, and they do so in groups of four or more. Chaperone molecules help proteins fold into their correct three-dimensional shape or conformation. If a protein becomes unfolded, a chaperone refolds it. A range of cellular stresses, in addition to mutations, can cause misfolding. Too many defective molecules in a cell could clog the system, causing an accumulation of imperfect, misfolded proteins. Current thinking holds that accumulation and aggregation of misfolded proteins are responsible for such neurodegenerative diseases as Alzheimer's, Parkinson's, amyotrophic lateral sclerosis, and the prion diseases. In each one, proteins convert from normal, soluble configurations to insoluble, sticky fibers known as *amyloids*. Two such are B amyloid of Alzheimer's and the L (gamma) synuclein of Parkinson's [56]. These proteins coalesce into fibrillary aggregates—tangles—with characteristic and identifiable structures.

Prions (from proteinaceous and infectious) are infectious proteins believed to be composed exclusively of the protease—resistant prion protein that replicates in the body by inducing the misfolding of the normal cellular prion protein PrPc. Compelling evidence supports this hypothesis [57].

Prion diseases (pronounced preeon) and referred to as transmissible spongiform encephalopathies (TGFs), are causative agents of infectious disorders affecting the brains of both human and animals that are invariably fatal, by means of an entirely novel mechanism, but all involve modification of the prion protein.

Prions, an entirely new type of disease causing agent are distinct from viruses in that PrP is encoded by a chromosomal gene, PRNP, located on the short arm of chromosome-20, and they lack a nucleic acid genome. Prion can also exist in multiple molecular forms; viruses exist in only a single form and have a consistent structure and elicit an immune response. Prions do not have a consistent structure and do not elicit immune responses. They are also resistant to both radiation and ultraviolet light treatment. Currently, an accepted and trim definition of a prion is a proteinaceous particle that lacks nucleic acid.

Prion diseases are associated with the conversion of the normally occurring prion protein PrP^c (found in nerve cells where it appears to help maintain neuronal functioning) into an abnormal, twisted, insoluble PrP^{sc} that is believed to be infections. PrP^{sc} makes more of itself by converting normal prion protein into copies of itself. PrP^{sc} can force PrP^c to refold and make more PrP^{sc} [57]. Although cells have the ability to destroy and clear misfolded proteins, conversion may occur faster than clearing, with consequent buildup of abnormally folded PrP^{sc} producing the characteristic spongelike holes in brain tissue, and ensuing progressive and fatal encephalopathy. Why this occurs remains a mystery. Most often this misfolding occurs for no apparent reason. Sporadic Creutzfeldt—Jacob disease (CJD) is the most common—but rare—human prion disease, with approximately a single case occurring in a million people.

Disease may also result from a mutation in the gene that codes for the PrP, or from ingesting distorted (infectious) bovine protein, in which case vCJD, varient Creutzfeldt–Jacob disease, ensues. Humans have always been exposed to CJD, but because its transmission requires either direct ingestion or injection of infected tissue, CJD has remained sporadic. Recent studies indicate that there are two forms of vCJD: one that results in a subclinical carrier state, which has public health officials in the United Kingdom biting their nails that these dormant cases may become overt. But bear in mind that between 1980 and 2005, a total of 156 cases of vCJD have been documented worldwide. That's 156 in 25 years—6 cases a year; 146 of these occurred in the United Kingdom and 10 were distributed among six countries in the European Union.

Current worldwide concern about prion diseases is a concern about food of animal origin. Four animal diseases are recognized: (1) scrapie (of sheep), (2) bovine spongiform encephalopathy ("mad cow"), (3) chronic wasting disease (of elk and deer), and (4) transmissible mink encephalopathy (TMB), which is of concern to mink ranchers who harvest mink for their pelts.

Since the 1700s, farmer's in Great Britain have described uncommon behavioral and physical changes in their sheep that clearly match the current description of clinically affected sheep, which suggests that scrapie has existed for at least 300 years [58].

Scrapie's most obvious manifestation is the severe rubbing against fence posts that sheep undergo in an attempt to relieve their intense itching. They stumble, tremble, lose their appetite. Death is certain. The French call it *la tremblante*, trembling; the Germans, *trabukrankheit*, the trotting disease. Each country named it for a symptom, but the Scottish moniker, *scrapie*, stuck. Scrapie is a degenerative disease of the central nervous system, and most importantly, the brain always exhibits massive collections of protein and holes in the tissue resembling nothing so much as a sponge.

In the 1950s scrapie was thought to be due to a "slow virus," as bacteria could not be cultured from brain, body tissue, or fluids, nor were any visible under the microscope. An organism that took unusually long to produce a disease could reasonably be referred to as a *virus*, and a "slow" one at that.

In the late 1950s-early 1960s 200 people a year were dying of kuru, a rare, progressive encephalopathy dubbed the "laughing death," in the highland region of Papua, New Guinea. Kuru affected predominantly women and children who were ritualistically consuming (cannibalizing) their dead relatives as a mark of respect. They ate everything—brain, entrails, and bone, and in fact what they didn't eat they smeared on their skin. The men, especially older men, had fewer cases as they received only leftovers. Dr. Carlton Gajusek, of the National Institutes of Health, first described this illness in 1957 [59]. The victims developed muscular weakness, loss of coordination, tremors, staggering gait, and inability to speak, and died with a kind of grimacing smile.

An astute observation of the pathological similarity between kuru and scrapie impelled the inoculation of kuru brain tissue into monkeys and chimpanzees. These infected primates displayed symptoms similar to those exhibited by the Fore- people, and their brain tissues showed the now classic spongy degeneration. Transmissibility was proven. The question now arose as to whether kuru and Cruetzfeldt–Jacob diseases could be similar. In 1968, the transmission of CJD to chimpanzees was documented [60]. Scrapie, kuru, and CJD were alike; all displayed the spongiform degeneration of the central nervous system, and microorganisms could not be found in tissues or body fluids. But transmissibility had been established. Now, with the decline in cannibalism, kuru has almost disappeared, providing additional evidence of the presence of an infectious process.

For Stanley B. Pruisiner, 1997 was a splendid year. The Nobel Prize in Physiology or Medicine was awarded to Dr. Pruisiner for his discovery of prions, and for explaining and clarifying the principles underlying their mode of action; and, as Dr. Pruisiner noted in his Nobel Lecture, "Prions are unprecedented infections pathogens that cause a group of invariably fatal neurodegenerative diseases of an entirely novel mechanism" [61]. Scrapie, kuru, and CJD are a group of neurodegenerative animal diseases now known as *transmissible spongiform encephalopathies* (TSEs), which are the result of modification of the prion protein.

Back to the future. Intense worldwide concern is not with CJD, but with vCJD, the human variant of BSE, bovine spongiform encephalopathy, "mad cow" disease. Concern and worry proceed from the thought—threat—of possible ingestion of animal products—beef, lamb, elk, and deer, but primarily beef, which may contain prions. Just how possible is this, and what has occurred to provoke this concern, and is worry appropriate?

BSE is an always fatal animal disease that has been found in cattle in 26 countries since it was first identified in Great Britain in 1986. Cattle contract BSE by eating feed derived from the remains of BSE-infected animals, a common nutritional practice of adding protein to promote growth. That vCJD, the human variant of BSE, is difficult to contract is suggested by the fact that since 1986, only 156 people worldwide have died from it.

Available evidence indicates that recycling the remains of diseased animals, specifically scrapie-infected sheep, as meat and bonemeal, into feed for livestock was responsible for the emergence and spread of BSE in the United Kingdom. A change in the rendering process used to produce the meat and bonemeal entailed discontinuation of a hydrocarbon solvent/steam treatment to extract fat for the meal. Apparently solvents and steam reduce prion infectivity. The new process used only high heat in the rendering. High heat can eliminate the usual microorganisms, but not PrP^{sc}, which continued to infect cattle. In turn, infected cattle became food for other cattle, and BSE spread with it until the process was banned in 1996 [62]. Since the ban, the number of new BSE-infected animals declined markedly from a high of 37,000 in 1992, to about 600 in 2003, and 343 in 2004. But "mad cow" PrP^{sc} is now believed to have infected humans, causing vCJD. In 1992, the first human vCJD cases were reported in England. By 1996, 11 young adults had contracted and died of the disease. The pattern of their brain degeneration differed from that of CJD, but still showed the typical spongy holes.

It has also become known that PrP^{sc} can enter the bloodstream and be transmitted by blood donation. In, May 2005, the father of Matthew Middleton, 18, who died of vCJD in March 1997, one of a cluster of four from a Yorkshire Village, accused medical authorities of engaging in a 7-year coverup to hide the fact that his son had been a blood donor. In 1998, Mr. Middleton was told that seven people had received Matthew's blood. Those seven were not told that they had his blood until September 2004.

Similarly, in 1997, there was grave concern in Ireland that as many as 3000 people might have been infected by blood products, taken from another donor who had died of vCJD [63]. This further suggests that although by 2005, 146 people in England had died of vCJD, the full reckoning is yet to occur. Nevertheless, the vCJD outbreak has peaked, and should be on its way out of England. To further ensure elimination, two compelling developments have occurred that should help alleviate further blood and surgical transmission. Pall Medical Life Seances, a British Company, plans to market a new filter, the "leucotrap affinity prion reduction filter," which removes infections prions from red blood cells, without damaging the cells. As prions are often found inside white blood cells, the filter also removes white cells, the leukocytes, along with the free prions, reducing, according to Pall Medical, prion infectivity by 99% [64]. With a name like "leucotrap affinity prion reduction filter," one can only anticipate remarkable results.

Researchers at University College, London, have formulated a detergent wash that reduces the possibility of CJD transmission during surgical procedures. When people who unknowingly harbor CJD undergo surgery, prions could be passed to others via surgical instruments. With the new disinfectant, surgical instruments will be required to soak for only an hour in the detergent [65].

Now, what of the United States, where vCJD has not occurred? What protections are in place to ensure the continued absence of this illness? The U.S. Department of Agriculture (USDA) is responsible for detecting disease in cattle, and the Department of Health and Human Service's Food and Drug Administration (FDA) is responsible for preventing its introduction and spread through animal feed. To protect both cattle and consumers, these three agencies have erected two firewalls:

> Controls over imports—which prohibits the importation of live cattle and certain cattle products where BSE exists. In 1992, the FDA began identifying medical products and other FDA-regulated foods and cattle-derived products from countries with BSE. In cooperation with the Department of Homeland Security's Custom and Border Protection, shipments of such products are screened.
>
> Animal surveilliance—To detect BSE, the USDA has been testing brain tissue from cattle that exhibit neurological symptoms, as well as adult cattle that die of unknown causes, and those slaughtered for meat.

Ruminant Meat Ban

The United States and Canada have banned feeding ruminant meat and bone-meal to other cattle, and have removed high-risk tissues from meat for human consumption. These are important steps to prevent the spread of BSE. To detect prohibited material in cattle feed, the FDA uses a "feed microscopy" test, which checks samples under a microscope for the presence of animal tissue, hair, and/or bone particles. Currently the FDA is evaluating a more sensitive test, the polymerase chain reaction, which detects animal DNA, and can distinguish ruminant DNA [66].

The question that remains open is, how many animals should be tested? Japan requires all animals going to slaughter to be tested for prions. The United States slaughters 30–35 million cattle annually. Should all be tested? Can 3 million head of cattle be tested monthly? Over a 3-year period Japan tested 1.9 million cattle and found 9 BSE-infected animals. Clearly, a reasonable sample number requires determination.

Over the past decade, two cows in the United States have tested positive for prions. Neither entered the food supply. The last one occurred in June 2005. This cow, unlike the first, was not imported. It was born in Texas, making it the first domestic case of the disease. This was a 12-year-old animal that never left the ranch. Given its age, it is likely that it was infected prior to the 1997 ban on feeding ruminant proteins to cattle. The USDA is now tracing all animals born on the ranch in the same year, or in years before or after, as well as any offspring of the cow, born in the last 2 years. All will be tested for BSE [67]. And by 2006, we shall know the numbers and types of tests that the USDA will initiate to determine the actual prion state of US cattle.

It may also be useful to know that the three most prevalent human neuro-degenerative diseases—Alzheimer's, Parkinson's, and amyotrophic lateral sclerosis, although the consequence of protein misfolding—are neither transmissible nor prion diseases, and have no connection to the food supply.

Chronic Wasting Disease

Hundreds of pounds of venison, the sweet-tasting meat of deer and elk, packed into frozen-food lockers is a hunter's dream. Unfortunately chronic wasting disease (CWD)—a scrapielike prion infection, has cast a pall over thousands of square miles throughout the midwestern states, and up into Alberta and Saskatchewan, Canada, where thousands of hunters track the tasty cervids, and where thousands of game farms and ranchers raise additional hundreds of thousands, the question is being raised as to whether CWD can be passed to humans. For the moment, and for some years to come, the only answer is, no one knows. It might pose a similar risk, it might not. As beef is far more popular than venison, CWD doesn't pose anything like the threat beef does. Nevertheless, the Centers for Disease Control investigated the recent deaths of several young people, 30 and under, and another 6 who were known to have eaten venison and later died at these tender ages. No connection to CWD could be made. While reassuring, it is still too early to be certain as the incubation period between initial infection and frank illness can be as long as 35–40 years, and CWD had only started to spread a decade ago. The question is also asked, and rightly so, can CWD spread to cattle? That question is uppermost in many minds because efforts to seal our borders from infected foreign cattle could be nullified and meaningless if our domestic herds succumbed to BSE from domestic deer and elk. That could be catastrophic. Thus far, however, cattle have not contracted CWD. Cows have been penned with infected cervids for over 5 years and have remained healthy [68].

Conclusion

From the best available evidence, it is fair and reasonable to say that thus far, US cattle appear to be safe, and that our country is now far more secure than it was on 9/11.

REFERENCES

1. Morse, S. S., Factors in the emergence of infectious diseases, *Emerg. Infect. Dis.* **1**(1):7–15 (1995).

2. *Emerging Infectious Diseases. Review of State and Federal Disease Surveillance Efforts*, U.S. Gov't. Accountability Office, GAO-04-877, Washington, DC, Sept. 2004.

3. Sick birds smuggled on Austrian flight, *Wall Street Journal* D3 (Oct. 26, 2004).

4. McNeil, D. G., Jr., Mecca pilgrims may be spreading polio, experts say, *New York Times*. (Jan. 11, 2005).

5. Kahn, L. H., Viral trade and global public health, *Issues Sci. Technol.* 57–62 (Winter 2004).

6. Reynolds, G., Why were doctors afraid to treat Rebecca McLester? *New York Times Mag.* 30–35, 76, 82–86 (April 18, 2004).

7. Control of communicable diseases: Restrictions on African rodents, prairie dogs, and certain other animals, *Fed. Reg.* **68**(213), (Nov. 4, 2003), FDA, Dept. Health and Human Services.

8. DeGrazia, V., *Irresistible Empire: America's Advance through Twentieth-Century Europe*, Harvard Univ. Press, Cambridge, MA, 2005.

9. La Franiere, S., and Grady, D., Stalking a deadly virus, battling a town's fears, *New York Times*, 1, 12 (April 17, 2005).

10. *West Nile Encephalitis: Cases and Deaths*, CDC, Div. Vector-borne Diseases, Atlanta, July 11, 2005.

11. Kuba, K. (and 23 others) A crucial role of angiotensin converting enzyme 2 (ACE 2) in SARS coronavirus-induced lung injury, *Nature Med.* **11**(8):875–879 (2005).

12. Normile, D., Avian influenza: Studies suggest why few humans catch the H5N1 virus, *Science* **311**:1692 (2006).

13. Shindler, L. W., *Understanding the Immune System*, NIH Publication 93–529, U.S. Dept. Health and Human Services, Bethesda, 1993.

14. Ahmed, R., and Gray, D., Immunological memory and protective immunity: Understanding their relation, *Science* **272**:54–59 (1996).

15. Solomon, T., Ooi, M., Beasley, W. C., and Mallewa, M., West Nile encephalitis. *Br. Med. J.* **326**:865–869 (2003).

16. Here to stay, *FDA Consumer* 23 (Jan./Feb. 2003).

17. Imported Lassa Fever—New Jersey, 2004, *JAMA* **292**(23):2828–2830 (2004).

18. Holmes, G. P., McCormick, J. B., Trock, S. C. et al., Lassa fever in the U.S., *NEJM* **323**(16):1120–1123 (1990).

19. The Viruses, in *Microbiology*, 3rd ed., Prescott, L. M., Harley, J. P., and Klein, D. A., eds., Wm. C. Brown, 1996, Chap. 16.

20. Duffy, J., Smallpox and the Indians in the American colonies, *Bull. Hist. Med.* **25**:324–341 (1951).

21. Torok, T. J., Tauxe, R. V., Wise, R. P. (and 8 others), A large community outbreak of salmonellosis caused by intentional contamination of restaurant salad bars, *JAMA* **278**(5):389–395 (1997).

22. Holloway, H. C., Norwood, A. E., Fullerton, C. S. et al., The threat of biological weapons: Prophylaxis and mitigation of psychological and social consequences, *JAMA* **278**(5):425–427 (1997).

23. Stevens, R., *Death from Anthrax* (www.crimelibrary.com; accessed July 2005).

24. Anthrax, 1897, *JAMA* **278**(5):382 (1997).

25. Meselson, M., Guillerman, J., Hugh-Jones, M. et al., The Sverdlovsk anthrax outbreak of 1979, *Science* **266**:1202–1208 (1994).

26. Arnon, S. S., Schechter, R., Inglesby, T. V. et al., Botulism toxin as a biological weapon: Medical and public health management, *JAMA* **285**(8):1059–1070 (2001).

27. Ostrov, B. F., *Botulism Vaccine Created at UCSF. Mercury News*, UCLA, Dept. Epidemiology, School of Public Health Website, 2002 (accessed, June 2005).

28. Wein, L. M., and Liu, Y., Analyzing a bioterror attack on the food supply: The case of botulinum toxin, *Proc. Natl. Acad. Sci.* USA **102**(28):9984–9989 (2005).

29. Dennis, D. T., Tulavemia, in *Diseases Transmitted Primarily from Animals to Man*, 14th ed., Public Health and Preventive Medicine (series), Wallace, R. B. ed., Appleton & Lange, Stamford, CT, 1993, pp. 354–357.

30. Dennis, D. T., Inglesby, T. V., Henderson, D. A. et al., Tularemia as a biological weapon: Medical and public health management, *JAMA* **285**(21):2763–2773 (2001).

31. Lady Mary Pierpont Wortley Montague: Letters from the Levant, during the Embassy to Constantinople. 1716–1718, J. A. St. John, London, 1838 (www.questra.com/pm.qst? accessed June 2005).

32. Healy, B., A medical battalion, *U.S. News World Report* 76 (Feb. 10, 2003).

33. Edward Jenner and the Discovery of Vaccination (www.sciedu/library/spcoll/nathist/Jenner.htm).

34. Need to know: Orphan drug, *Sci. Am.* **292**(5):25 (2005).

35. *WHO Fact Sheet on Smallpox. Communicable Disease Surveillance and Response*, WHO, Geneva, 2001.

36. Jones, L. Y., Tribal fever, *Smithsonian* 92–97. (May, 2005).

37. Fatal human plague: Arizona and Colorado, 1996. *JAMA* 278(5):380–382 (1997).

38. Perry, R. D., and Felthaston, J. D., *Yersinia pestis*—etiologic agent of plague, *Clin. Microbiol. Rev.* **10**(2):35–63 (1997).

39. *Bioterrorism; a Threat to Agriculture and the Food Supply*, General Accountability Office, GAO-04-259T, Washington DC, Nov. 19, 2003.

40. Foot-and-mouth disease, in *Zoonoses and Communicable Diseases*, Acha, P. N., and Szyfres, B., eds., Sci. Publ. 354, pp. 226–231, Pan American Health Org., WHO, Washington DC, 1980.

41. Hennessey, T. W., Hedberg, C. W. et al., A national outbreak of Salmonella enteridis infections from ice-cream, *NEJM* **334**(2):1281–1286 (1996).

42. Woodham-Smith, C., *The Great Hunger*, Harper & Row, New York, 1962.

43. Zilinskas, R. A., Iraq's biological weapons: The past as future, *JAMA* **278**(5):418–424 (1997).

44. Zanders, J. P., *Bioweapons Report 2004*, Bioweapons Prevention Project, June 17, 2005. (www.bwpp.org).

45. *Atlantic Storm: Exercise Illuminates Trans Atlantic Leaders Reactions to Bioterror Attack*, The Center for Biosecurity of Univ. Pittsburgh Medical Center (www.atlantic-storm.org), Jan. 17, 2005.

46. *Hippocrates, the Genuine Works of Hippocrates* (transl. Francis Adams), Williams & Wilkins, Baltimore, 1939, pp. 19–42.

47. Codell Carter, K., and Carter, B. R., *Childbed Fever: A Scientific Biography of Ignaz Semmelweis*, Greenwood Press, Westport, CT, 1994.

48. Snow, J., On the mode of communication of cholera, in *Cholera*, Snow, J., ed., The Commonwealth Fund, New York, 1936.

49. Prescott, L. M., Harley, J. P., and Klein, D. A., *Microbiology*, 3rd ed., Wm. Brown, 1996, p. 9.

50. Relman, D. A., The search for unrecognized pathogens, *Science* **284**:1308–1310 (1999).

51. Graber, M. A., and Nugent, A., Peptic ulcer disease: Presentation, treatment and prevention, *Emerg. Med.* **31**:66–70 (1999).

52. Schwarz, K., Uber Penetrierende Magen and Jujunalgeschure, *Beitrage fur Klinischen Chirugie* **67**:96–101 (1910).

53. Marshall, B. J., Helicobacter pylori. The etiologic agent for peptic ulcer, *JAMA* **274**(913):1064–1065 (1995).

54. Marshall, B. J., *Helicobacter Pioneers: First Hand Accounts from the Scientists Who Discovered Helicobacters. 1892–1982*, Blackwell Publishing, Victoria, 2002.

55. (a) Carbone, K. M., Luftig, R. B., and Buckley, M. R., *Microbial Triggers of Chronic Human Illness*, American Academy of Microbiology, Washington DC, 2005; (b) New virus linked to prostate cancer, *JAMA* **295**(13):1503 (2006).

56. Conn, P. M., and Janorick, J. A., A new understanding of protein mutation unfolds, *Am. Sci.* **93**:314–321 (2005).

57. Prusiner, S. B. ed. *Prion Biology and Diseases*, 2nd ed., Cold Spring Harbor Lab. Press, Cold Spring Harbor, NY, 2004.

58. Carp, R. I., and Kascap, R. J., Taking aim at the transmissible spongiform encephalopathie's infectious agents, in *Prions and Mad Cow Disease*, Nunnelly, B. K., and Krull, I. S., eds., Marcel Dekker, New York, 2004.

59. Gajdusek, D. C., and Zigas, V., Degenerative disease of the central nervous system in New Guinea: The endemic occurrence of "kuru" in the native population, *NEJM* **257**:974–978 (1957).

60. Gajdusek, D. C., Gibbs, C. J., Jr., Asher, D. M., and David, E., Transmission of experimental kuru to the spider monkey (Ateles geoffreyi), *Science* **162**:693–694 (1968).

61. Prusiner, S., Prions, *Acad. Sci. USA* **95**:13363–13383 (1998).

62. Brown, P., On the origin of BSE, *Lancet* **352**:252–253 (1998).

63. Rice, D., and Knowsley, J., Health chiefs knew dead CJD boy was blood donor, and ordered cover up, *Sunday Mail* 11 (May 11, 2005).

64. Prion busting filter to reach Europe in spring (PharmaTechnologist.com), Feb. 27, 2005.

65. Breakthrough in cutting CJD risk, *BBC News* (http://news.bbc.co.uk/2/hi/uk_news/4296467.stm), Feb. 25, 2005.

66. *Mad Cow Disease: FDA's Management of the Feed Ban Has Improved, but Oversight Weaknesses Continue to Limit Program Effectiveness*, GAO Report GAO-05-101, Washington, DC, 2005.

67. McNeil, D. G., Jr., Case of mad cow in Texas is first to originate in U.S., *New York Times* (June 30, 2005).

68. Yam, P., Shoot that deer: Chronic wasting disease, *Sci. Am.* **288**(6):39–43 (2003).

4

FOOD UNCOVERED

> You are entitled to your our own opinions,
> but you are not entitled to your own facts.
> —*Daniel P. Moynihan*

With a splitting headache, numb mouth, deafness in her left ear, and burning sensations shooting through her body, Wendy Bannerat lay curled quietly in the darkened confines of a bunk below deck of her 41-foot aluminum centerboard sloop Elan. Wendy was in terrible pain and needed expert medical help. But she and her husband Scott were anchored off Vaka'eitu, an island in the VaVa'u Group, Tonga—in the far Pacific east of Australia and New Zealand.

Airlifted to Bishop Auckland General Hospital in Auckland, New Zealand, where a battery of tests were done, CAT scans, MRIs and lumbar puntures were negative for bacterial and viral meningitis. Additional tests finally indicated that Wendy had parasitic meningitis—rat lung worm meningitis. She had ingested the rat lung worms while eating coleslaw made from raw vegetables. Eating insufficiently cooked escargot, or ingesting raw crabs, freshwater shrimp, fish, or frog legs have all been culprits. Wendy got hers from cabbage, insufficiently washed and disinfected cabbage that she used for coleslaw.

She was not alone. On the island of Jamaica, 12 of 23 Chicago medical students on spring break became infected to varying degrees from a single shared meal that included raw vegetables. In 1993, the first case of parasitic meningitis was diagnosed in North America, when, on a dare, an 11-year-old boy from

Our Precarious Habitat . . . It's In Your Hands, Fourth Edition. By Melvin A. Benarde
Copyright © 2007 John Wiley & Sons, Inc.

New Orleans ate a raw snail, became severely ill, but luckily recovered fully in a week. How did these cases occur?

Adult lung worms reside in rats' pulmonary arteries. These worms lay eggs that circulate to lung capillaries; hatch; and migrate to the trachea, down the esophagus, and into the gastrointestinal tract; and are eliminated in rat feces. These feces-encrusted larvae are picked up by land and aquatic snails, crabs, and/or slugs, which crawl over fruits and veggies. Wendy, the medical students, and the young boy ate raw vegetables and an uncooked snail, picking up *Angiostrongylus cantonensis*, a worm that readily penetrates the gastrointestinal wall, making its way to central nervous system tissue—usually getting into the fluid-filled spinal canal. Now the real trouble starts. Worms begin to die, and the body musters its immune defenses against a foreign protein. Cerebrospinal fluid pressure increases on tissues and nerves, causing great pain.

As a consequence of immune-mediated tissue injury, fever, nausea, blindness in one eye (until surgical removal of the worm), photophobia partial paralysis, vertigo, and stiff neck—a nasty cascade—can occur, and in the more serious cases can continue for months or years [1]. There is one simple way to prevent the rat lungworm—cook vegetables!

Pomatomus saltatrix (bluefish) is seldom suspect. But for the five physicians who sat down to lunch at a New Hampshire Inn, that was cold comfort. Although such nonscombroid fish as amberjack and mahimahi have been implicated in poisonings typical of the Scombridae family (tuna, mackerel, bonito), bluefish have not. Nevertheless, within 4 hours of eating they developed headache, flushing, redness of the upper body, diarrhea, abdominal pain, and pounding heart. Bacterial growth and metabolism on the surface of unrefrigerated fish decomposes histidine to histamine in muscle tissue. Ingestion of histamine causes the symptoms noted as well as tightness in the throat, vomiting, and metallic or peppery taste in the mouth, depending on the amount ingested. The fish look, taste, and smell just fine [2].

After Martha Jefferson (not her real name), a 34-year-old devotee of sushi, a trendy Japanese dish often containing raw fish, experienced continued pain, she relented and went for X rays. Embedded in the mucosal lining of her stomach was an inch-long third-stage larva of *Anisakis simplex*. Raw salmon is considered the most likely source of this parasitic worm.

In New York State during an 8-month period, over 1000 people became ill with typical symptoms of gastroenteritis as a result of eating bacterially contaminated clams and oysters.

Forty-one employees of a department store in southern Taiwan were hospitalized after eating white-tipped mackerel. Dizziness, burning mouth, numbness of the lips, flushing, headache, and abdominal cramping were the clear signs of scombroid poisoning.

At Princeton University, it was nothing so sophisticated. For three consecutive, years, students and nonstudents were afflicted with salmonellosis—one of the most common forms of bacterial food poisoning. One of the outbreaks involved 80 students and six administrators. Improper food handling practices appeared to be the precipitating cause in all three events.

In another incident, 1700 employees of a large pharmaceutical plant and their families gathered for a picnic lunch. Food was served at about 11:30 A.M. By 1 P.M., the first cases of food poisoning (gastroenteritis) occurred. By 4 P.M., over 800 people had become ill. By 9 P.M., the Indiana Department of Health had recorded 1000 victims.

Since "forewarned" is usually "forearmed," planners of a picnic the following Saturday, by an electrical parts manufacturing firm, should have been alerted by the previous week's tragic events. Apparently, it was not a learning experience. It was a particularly warm day: 1813 people turned out for the festivities. Food was served at 4:30 P.M. By 7 P.M., nausea, vomiting, abdominal cramps, and diarrhea were evident in 25–30 people. An additional 100 were ill by 10:30 P.M. In all, 216 cases were reported.

These are a representative sampling of the unfortunate but tremendous number of foodborne illnesses that occur annually in the United States and around the world. Foodpoisoning. Foodborne illness is totally inclusive, having little respect for national or other geographic boundaries, seasons, age groups, or gender. We are all susceptible, anytime, given the appropriate conditions.

Although the American food supply is remarkably safe, microbes can contaminate our food and cause illness. In fact, yearly, there are an estimated 76 million foodborne illnesses; most are mild, lasting only a day or two. Nevertheless, the Centers for Disease Control (CDC) in Atlanta, has documented 325,000 hospitalizations and 5000 deaths. Although these are sizable numbers, 50% and more are easily preventable. While this is a sad commentary on our willingness to live with such unnecessary risks, it is also an indictment of our food safety system.

But it is also an indictment of such current practices as eating seared tuna, salmon, and other fish, all the rage today in upscale restaurants. *Searing* refers to the clashing fact of flash-heating the sides of the fish without heat penetration. The "in" crowd today, the very chic, are into raw or seared fish, which now takes its place alongside steak tartar. Are we betting against the odds, or is it denial, or just old-fashioned foolishness? Indeed, one need not look too far afield for answers to the why of foodborne illness.

It is currently well established, and I'm certain this will come as a surprise, perhaps even a shock, that seafood, fin and shellfish—crustaceans—crabs and lobsters, and shellfish—oysters, clams, mussels, and scallops, account for some 15% of all foodborne illness. That's a greater percentage than either meat or poultry even though we eat 8 times more meat and 6 times more poultry. Why is this? Simple. More than 80% of all seafood consumed in this country is imported, coming from over 13,000 foreign suppliers, in some 160 countries. In 2002, the last year for which data are available, the FDA was able to inspect only 100 of those 13,000 seafood suppliers [3]. It is just this lack of US inspectors and inspections in China, added to the total lack of regulations and standards by the Chinese government, that permitted melamine ($C_3H_6N_6$, 1,3,5, triazine, 2,4,6 triamine), a nonnutritional chemical, high in nitrogen content, but a fake protein to be added to pet food sold in the US in 2007, and responsible for widespread illness and death of cats and dogs. Overall, of 8.9 million shipments of food to our ports from around the world, 20,662 inspections were made—0.23% of all shipments were inspected and sampled. Until a better system of inspection is in place, and the suppliers made to perform at US standards, foodborne illness from seafood, let alone meats and poultry, can be expected to continue, because food poisoning can be likened to a stealth bomber; you don't know it's there. The food gives no warning signals, no hint of trouble, no sign that means stop—don't eat me! This is a world of difference from food spoilage, which is not harmful, does not cause illness, but does provide the senses with decisionmaking information.

MICROBES AND BACTERIAL INFECTIONS

The microbes of the unseen world can be segregated into three lists. The A list includes the spoilers; the B list, the sickeners; and the C list, the benefactors, those that make our lives livable and enjoyable. The A list contains such members as the genuses *Pseudamonas*, *Serratia*, *Aeromonas*, *Leuconostoc*, *Fusarium*, *Flavobacterium*, *Shewanella*, and *Microccus*.

On the B list we have *Campylobacter*, *Listeria*, *Escherichia coli* O157:H7, *Yersinia*, *hepatitis A and C*, *Clostridium*, *Salmonella*, *Staphylococci*, and *Mycobacterium*.

Our C list includes such stalwarts as *Saccaromyces*, *Lactobacillus*, *Penicillium*, *Acetobacter*, *Brevibacterium*, *Pediococcus*, *Bifidobacterium*, *Proprionobacterium*, and *Lactococcus*.

These, of course, are not exhaustive lists, and are presented here only to emphasize that each is an easily distinguishable group with little to no crossover.

It is, of course, the wee beasties of the A list that constantly challenge us in the race to the food. They, as we, covet the fats, carbohydrates, proteins, and minerals that make food tasty and nutritious, and offer an ideal environment for growth. It would be hard to imagine that anyone has not experienced, from time to time, slimy meat, fish, or poultry and noxious odors, off-tastes, and texture and color changes in any number of foods.

Microbes, especially bacteria, require moisture, so the more water in a food, the more susceptible it will be to bacterial metabolism. Foods can be divided into acidic, neutral, and nonacidic or basic. Citrus fruits and tomatoes are least susceptible even with their naturally high moisture content, as bacteria shun acid. On the other hand, vegetables such as lettuce, cucumber, bell peppers, and celery, which are acid-neutral, and with high moisture content, are uniquely susceptible, as are the proteinaceous meats, poultry, and fish.

Bread is relatively dry, but not as dry as sugar and starch. Bread is attacked by mold, not bacteria, while sugar and starch are not at all susceptible.

Milk, an almost perfect food, would succumb quickly to a host of bacteria if not pasteurized. Whether it is the old-fashioned way, at 140°F for 30 minutes, or flash-pasteurized at 160°F for 15 seconds, it is not sterile—devoid of all life. The heat of pasteurization is meant to remove one of the sickeners on the "B" list, the tubercle bacillus, *Mycobacterium*, responsible for tuberculosis. While doing so, pasteurization also removes most spoilers, which can reside on the cow's hide, and inadvertently enter milk collectors. However, pasteurization does not remove all bacteria. It is not meant to do so. It is calculated to remove the more fragile pathogens, the sickeners. By keeping milk at refrigeration temperatures of 35–40°F (1.6–4.5°C), milk can last up to 12 days, possibly more, before turning sour and developing off-odors and clotting. The sourness and off-odors are chemicals contributed by the multiplying and metabolizing spoilers. But—and this is a crucial but—those organisms that sour and clot the milk for its nutrients are not harmful—not illness-producing. Drinking the off-tasting and odorous milk may be distasteful, repugnant, but nothing more. This souring is not to be confused with the taste of buttermilk, which is a controlled bacterial fermentation, by members of the B list, producing lactic acid. The bacterial conversion of citrate in milk produces diacetyl(2,3-diketobutane).

After a week or more at refrigeration temperatures a unique odor begins to emanate from raw fish. That's the noxious and highly distinctive odor of trimethylamine, produced by bacteria in fish muscle, as well as those contributed by the catching, storage, and preparation—between landing and retailing. Again, it is uninviting, but it is not a poison. Not harmful. Not the problem of histamine noted earlier, which gives no warning. Do not forget the refrigerator's freezer unit. Although freezing does not kill bacteria, maintaining the freezer at from 0–5°F (−17.8–15°C) will slow bacterial growth markedly, which translates into longer life for frozen meat and poultry.

What about fruit? Peaches, pears, grapes, plums, pineapple, cherries, strawberries, and blueberries all contain reasonably high levels of sugar, a nutrient that both bacteria and yeasts adore. All they require is a minuscule cut or bruise to enter, digest, and multiply. As they do, they produce alcohol. Should a container of cooked fruit remain in the fridge for 2–3 weeks, gas bubbles begin rising in the juice—a sign that the fruit and liquid are fermenting and producing alcohol. Too many people are too quick to trash the lot. I and others gobble it up. Remember, we call grape alcohol, "wine." Pineapple, cherries, peaches, plums, all produce their own distinctive-tasting wine. Of course, fermentation is also a means of preservation. Isn't that how people the world over preserved cow, horse, goat, and sheep milk—creating their special brands of yogurt and clabbered milk? "Clabber" is an old Irish word meaning cuddle or thickening, and that's what was needed when refrigeration was lacking.

Sliced or cut bell peppers (red, green, and/or yellow), head (iceberg) lettuce, sliced cucumbers, and sliced onions often develop a slimy softening at the cut surfaces. These acid—neutral veggies with high moisture content are highly susceptible to bacterial attack, but harmful, again, no.

Spoiled food is a problem of aesthetics—looks, feel, and taste. In fact, if it smells or tastes bad, it's safe. Can there be any better examples than Limburger, bleu, and Camembert cheeses? Can anything smell worse than Limburger? But few would trash it. Need I remind you that people pay exorbitant prices for these noisome foods. Spoiled? Yes, for some; for others, no. But harmful? No, no, no.

As to when a food is unfit to eat, an appropriate response might be that a food is inedible when its look and odor so testify. Fortunately, or unfortunately, depending on individual preferences, little universal agreement exists as to what is or is not fit to eat. Some people like their meat "high." Most Americans would call such meat spoiled and not fit to eat. "High" or "gamy" meat usually has a strong odor and taste, both produced by bacterial action. Titmuck, fish buried to allow bacterial fermentation to occur, is eaten by Eskimos as a delicacy. Eskimo dogs, however, refuse to eat this semiliquid, foul-smelling delight. Most Americans would call it putrid. Yet it is putrefaction that gives Limburger cheese its gourmet qualities!

Having mentioned that foul word, we must deal with it. Putrefaction is nothing more than the bacterial splitting of proteins by anaerobic metabolism producing foul-smelling, incompletely oxidized chemicals such as mercaptans and alkaloids. Foul, yes. Illness-producing, no.

Serratia, a bacterium on our "A" list, has the distinction of playing a prominent role in religious history. From the twelfth to the nineteenth centuries, bloodlike, often liquidy red-spotting of starchy foods, has been observed and described—especially when the blood-red liquid appeared on sacramental bread or wafers. Twenty-five bloodlike appearances on bread, chicken, meat, and cake, but most often on the Host, were recorded between 1171 and 1874, in Italy, France, Germany, Belgium, and Poland. The most famous appearance has come to be known as the "miracle of Bolsena," which occurred in the Church of Santa Christina near Bolsena, Italy in 1263. Here a German priest on his way to Rome, stopped to celebrate mass. As he bent over the Host during consecration, "the blood that Christ had sweated in his agony in Gethsemone oozed from the host and dripped down upon the linen of the alter" [4]. In memory of this miracle, Pope Urban IV issued a Bull, Transitaurus de hoc mundo, which instituted the feast of Corpus Christi.

During the eighteenth and nineteenth centuries, in a number of Italian villages red spots appeared on Polenta causing civil disturbances. Explication arrived in 1879, when Bartolomeo Bizio conducted a series of tests and found that the reddish matter of the polenta is produced in a very damp and warm atmosphere, and he also noted that the reddening of the polenta had "manifested itself at other times in the warm season." He also showed that the cornmeal and other cereal grains carried "seeds," which reproduced themselves; Bizio named this organism *Serratia*, in honor of Serafino Serratia, an Italian physicist, and he added a species name, *marcescens*, to the new genus *marcescens*, from the Latin *marcescere*, to wither or decay, because, "as it decays, it dissolves into a fluid and viscous matter which has a mucilaginous appearance". Today we know that the miracle of Bolsena and the mystery of the red polenta were due to the growth of the bacterium *Serratia marcescens*, which produces a bright red pigment [5].

Now the "B" list. We would not knowingly invite them to our table. They are invisible in every way. That's another key. The B list contains the "sickeners," which give no indication of their presence. The food looks and tastes just fine. That's why so many people, men, women, and children, at church suppers, picnics, and banquets become ill with explosive gastroenteritis, 4, 6, 8, 10 hours after eating the toxin-laced salads and meats as the body attempts to rid itself of the bacterial poisons—the result of preparing large batches of food under less then ideal conditions, but most important, not properly refrigerated, or left unrefrigerated, permitting bacteria to grow and produces their odorless, tasteless, toxins. Illness-producing microbes do not spoil food, and the spoilers do not produce toxins and illness. So, you might want to think again before chucking out what you thought was harmful food.

Before considering the members of the "B" list, take a glance at the "C" list and notice the very different and unfamiliar names, and the many tasty dishes they bring to the table to make life enjoyable. Note, too, that the three lists bear no resemblance to one another. There is a high degree of specificity in the microbial world. They know what they are meant to do, and don't get confused. Something to dwell upon.

Campylobacter Infection

Campylobacter heads the B list. That stubby little bacterium is the most common cause of bacterial gastroenteritis in the Western world, and responsible for over 2 million illness in the United States annually. Although microbiologists encountered campylobacter long years ago, only relatively recently has it been tagged as responsible for scads of foodborne illnesses via three prime routes. Researchers at the University of Wales College of Medicine and the Public Health Laboratory Service Communicable Disease Surveillance Center, now inform us that bottled water, prepackaged veggie salads, and chicken are campylobacter's major vectors. As for veggies, we readily recall the recent outbreak of *E. coli* O157:H7 induced gastroenteritis during September, October, and November 2006, that felled over 100 people who had eaten prepackaged raw spinach and green onions. More about this in the discussion of food irradiation.

Would you believe that we consume about 2 billion gallons of bottled water a year? That number is rising, and we are eating a lot more prepackaged salad veggies, and chicken. All three can be readily, and naturally, contaminated by campylobacter, and it remains available in these items because of inadequate heating in the case of water and chicken, and inadequate washing and disinfecting of veggies prior to packaging [6].

Neither eating salad vegetables nor drinking bottled water had been recognized as risk factors Contaminated soil and/or contaminated water during harvesting are ready sources. Additionally, salad vegetables are imported from countries with low sanitation standards, where food handlers infected with campylobacter, transferring them to vegetables while packaging them—and what of cross-contamination during preparation of vegetables? Tomatoes and cucumbers diced on chopping boards can contaminate other vegetables. Washing and disinfecting of tomatoes and cucumbers prior to dicing can also be inadequate.

The most consistent finding in etiologic studies of Campylobacter infection has been chicken. This has been documented and reported from the United Kingdom, Sweden, Switzerland, New Zealand, the Netherlands, and the United States. The relationship with chicken is most commonly with undercooked chicken, and eating chicken away from home.

Studies of raw poultry found extensive campylobacter contamination, which can be readily dealt with by adequate cooking. And as will become evident, the key to prevention of all foodborne members of the infamous B group is adequate hand washing before food handling, and adequate cooking of raw foods. Foodborne illness should have little to no incidence in our twenty-first-century world.

Listeriosis

Listeria and its illness, listeriosis, is another affiliate of the B group. *Listeria monocytogenes* is responsible for at least 2500 illnesses annually and 500 deaths. Pregnant women are not only at increased risk but are also 20 times more likely to become infected than women generally. In fact, 30% of all listeria infections occur during pregnancy. But it's the newborn infant that suffers the effects, not the mother. In addition, immunosuppressed individuals, especially those with AIDS, are 300 times more likely to become ill. Fever, muscle aches, headache, stiff neck, confusion, and loss of balance—the predominant symptoms—suggest central nervous system involvement.

Listeria monocytogenes is present in soil and water. Vegetables become contaminated from soil as well as manure used as fertilizer. Animals can carry the microbe without appearing ill, and can contaminate meats and dairy products. Unpasteurized milk can surely contain it, as well as cheese made with unpasteurized milk. Listeria are easily killed by the heat of pasteurization and cooking.

Yersinosis

Although *Yersinia enterocolitica* causes some 100,000 infections a year, it appears to have a preference for children, who develop fever, abdominal pain, and bloody diarrhea up to a week after exposure. Preparation of raw pork intestines, chitterlings, is a prime source of acquiring the bug, and infants can become infected when their nannies handle chitterlings and fail to wash their hands before attending to the baby. Eating raw and undercooked pork is the major risk factor, abetted by oysters, fish, and raw milk.

Salmonellosis

Nontyphoidal salmonellosis, particularly induced by *Salmonella enterididis*, with 1.5 million new cases a year, is on the list perennially. Egg-associated salmonellosis is an ongoing public health problem. Contaminated eggs appear normal. Nevertheless, 12–72 hours after sucking up raw eggs, fever, stomach cramps, vomiting, and diarrhea occur as the body tries to expel the toxin produced by the organism. Symptoms can last up to a week, and the diarrhea can be severe enough to require hospitalization. Salmonella is one of those microbes that can produce toxin and symptoms on ingestion of relatively few cells. "But the eggs were washed and disinfected" is the lament of the recovered consumer. Of course, they were; current egg cleaning procedures has curbed the fecally contaminated eggs of the past. But our stealthy salmonella infects the hen's ovaries and is present before the shell forms, which means that washing and disinfecting eggs shells will not deter the bugs inside.

But that is not the end of it. Foods can readily be contaminated by food handlers who don't wash their hands before leaving the restroom—even when a sign cautioning "wash hands before leaving," appears in the restroom where they wash their hands.

Pets are yet another common source of infection, especially turtles and other reptiles, which are notorious for harboring salmonella. Hand washing after handling these animals is not optional. Children must be taught to do this regularly.

Hepatitis A

Hepatitis, an inflammation of the liver, comes in three forms: A, B, and C. A is on our "B" list as it is unique among the picornaviruses (single-stranded RNA viruses) to be foodborne. Forms B and C are transmitted by illegal drug use and unsafe sexual practices, and are not on our B list.

Hepatitis A is the most common form of hepatitis, with some 25,000 cases reported annually; the actual incidence is believed to be 10 times higher because of underreporting. It is a highly infections virus, requiring only 10–100 virions to initiate a frank illness with a touch of jaundice. Most US cases are seen in the western and southwestern states. Children age 5–14 have the highest incidence. Day-care centers contribute some 8% of cases and household contacts, another 14%. The prime source of infection appears to be contamination of fresh produce during harvesting, processing, and packaging. Hepatitis is a classic example of the fecal–oral route of microbial transmission, which means that a person unknowingly places in his/her mouth food that has been contaminated with fecal material of an infected individual. That's why hand washing is not an option for food handlers. A single infected food handler can pass the virus on to dozens, even hundreds, of unsuspecting people [7].

Two recent hepatitis A outbreaks, one in Tennessee–Georgia–North Carolina and another in Pennsylvania, which engulfed 500 people who had eaten in a restaurant, were linked to scallions grown in Mexico. Viruses isolated from affected individuals in Pennsylvania were genetically similar to the viruses isolated from patients in the Tennessee–Georgia–North Carolina group. Epidemiologists from the CDC believe that at harvesting, a hepatitis A excreter handled the scallions, stripping off the outer layers, bundling them with rubberbands, and contaminating all of them, leading to an outbreak. Should the contaminated scallions have contacted other foods or been used in another popular recipe, the outbreak could have been huge [8].

Of the 76 million foodborne illness events in the United States, 37% appear to be due to viruses. Between them, the Norwalk-like virus (noroviruses), rotavirus, and astrovirus account for 28 million estimated gastroenteritis cases. These are self-limiting, lasting 2–10 days with watery stools, abdominal pain, fever, and headache. These inflammations of the stomach and intestine are often referred to as "stomach flu," but are unrelated to the influena virus. These viruses inhabit raw fruits, vegetables, raw shellfish, and undercooked foods. They are refractory beasties capable of surviving 140–150°F cooking temperatures, and the rotavirus can survive on vegetables for days at 4°C—the temperature of refrigerator freezers [9]. Nevertheless, the quarter part of foodborne illness is preventable, and should be.

Ptomaines

To this point, nothing has been said about ptomaines or ptomaine poisoning, an old concept predating our knowledge of the microbial involvement in food poisoning. But ptomaines (from the Greek *ptoma*, meaning corpse or cadaver) are a group of amines, foul-smelling chemicals with names to match. Putrescine, an aliphatic diamine, is formed by the decarboxylation of the amino acid ornithine, and the ptomaine cadaverine occurs when lysine loses a CO_2 moiety. With the bacterial degradation of tissue protein amino acids to amines and ammonia, obvious odor changes occur. These alterations are elements of food spoilage and do not enter into the food poisoning equation. But raw food does. Between 2001 and 2004, a 15-month-old infant died and some 35 people—22 adults and 13 children—contracted Tuberculosis by eating cheese made from raw milk. Queso fresco, an unripened soft cheese imported from Mexico, contained *Mycobacterium bovis*, an organism that can cause tuberculosis. The City of New York's Department of Health has expressed concern about paqueterias, courier services that bring food to and from Mexico skirting border inspections [10].

Eating raw fish, raw meat, and raw eggs and drinking unpasteurized milk are needless risks. The cows, pigs, chickens, sheep, goats, and fish that provide these foods are loaded with microbes, which the heat of cooking and pasteurization can easily remove. Eating these raw foods is akin to playing Russian roulette with your health.

There is a further consideration. To forestall spoilage, refrigerators should be maintained at 35–40°F (1.5–4°C). It's worth checking every few months. Place a thermometer on shelves at front, rear, top, and bottom. Allow the Fridge to equilibrate for 20–30 minutes. If temperatures are above or below, reset the thermostat. The freezer should be maintained at 0–5°F.

The larger question is: How can we protect ourselves against a range of food poisoning microbes, and forestall spoilage at the same time? Today, the appropriate and optimum solution is food irradiation.

FOOD IRRADIATION

On July 1, 2005, the population of the United States stood at 296,660,955. It is anticipated that for 2005, the number of foodborne illnesses will be 76 million; similar to 2004, along with several hundred thousand hospitalizations and 5000 deaths. That's approximately 30% of our total population. One in every 3 or 4 of us can expect to become a statistic—a case of gastroenteritis, or worse. It's an unimaginagle number. And there is an economic toll in medical treatment costs and lost productivity, which the U.S. Department of Agriculture pegs at $37 billion. Is this yearly magnitude of illness and economic loss acceptable? Foodborne illness is one of the most preventable conditions in the country. Assuredly, there is no magic bullet or wonder drug to eliminate foodborne disease, but there surely is a better way than current food safety practices. Food irradiation is a food protection–preservation process whose time has not only come, but is long past due. Its benefits clearly out weigh any risks. Irradiation of food can effectively protect the health of the public.

Just what is food irradiation? What does it accomplish, and is it safe? Before dealing with these questions, a pause for clarification. At this moment in history "radiation" is a much encumbered word, saddled, loaded with distracting and misleading undercurrents and associations. To say that it is maligned is not an overstatement. Ergo, cleansing is required.

The sun delivers a tremendous amount of energy daily, most of which travels to the earth as radiation. Radiation simply means radiant energy. It has nothing—repeat, nothing—to do with radioactivity. It refers only to the transmission and absorption of radiant energy that moves through space as invisible electromagnetic waves with differing energy levels. Consequently different types of radiant energy, visible light, microwaves, radio and TV waves, infrared, ultraviolet, X rays, and gamma rays can be distinguished from one another by their wavelengths. At the ends of the electromagnetic radiation spectrum, we have cosmic, gamma, and X rays with the greatest penetrating power; and FM, radio, TV, and microwaves, with the lowest energy levels and least penetrating power at the other end. Visible light, the only energetic wave we can see, exists as a narrow band between the ultraviolet and infrared. It is the waves of radiant energy that bring radio and TV broadcasts into our homes. At this end of the energy spectrum we use low energy levels to toast our bread and cook our food.

The radiant energy used in food preservation is called *ionizing radiation* or *irradiation*. These are among the shorter wavelengths, at the more penetrating end of the spectrum, and are capable of damaging the type of microorganisms that bring us foodborne disease, and spoil food. As with the heat of pasteurization of milk, the irradiation process substantially reduces, but does not eliminate, all spoilage organisms. It does remove pathogens.

Irradiation exposes food to short-bursts of high electromagnetic energy. When microbes, present as they are in food, are irradiated, the energy from the wave is transferred to the water and other molecules in the bacteria, molds, and insects. The energy creates temporary, short-lived, reactive chemicals that damage the pests DNA, causing defects in their genetic instructions. Unless the damage can be repaired, the pests will die when attempting to duplicate themselves. Disease-causing organisms differ in their sensitivity to irradiation. The larger the organism, the larger the available DNA ,the easier it is to kill. The smaller the organism, such as a virus, the more difficult to remove, and the greater the need for penetrating energy. Also, because the doses are so small, there is never—repeat, never—any chance of the food becoming radioactive. Furthermore, the radiant energy sent into the food during irradiation dissipates as it passes through the food. None remains in the food So, how are foods irradiated? Two things are needed for irradiation: a source of radiant energy and a place to confine the energy. Three sources, three irradiation processes are currently avaialable. Each uses a different tyupe of energy.

Radioactive Isotopes

The oldest form uses the radiation emitted by a radioactive substance, such as cobalt-60 made by neutron bombardment of the natural isotope[59] Co- or cesium-137 (a naturally occurring isotope), which emits high-energy photons, gamma rays that can readily penetrate food. These photons do not produce neutrons, which means that nothing in their path can become radioactive. That's an essential key. Recall, too, that people who have radiation treatments for a medical condition and receive doses of radiation higher than those ever received by food, do not become radioactive. The astronauts, whose food is sterilized by radiation, to remove all signs of life, receive 45 times the amount of radiation used for commercial food, have eaten irradiated food since the 1970s, and have never experienced adverse reactions from their highly irradiated foods. These are the types of response needed for those who perversely claim that eating irradiated food can induce radioactivity.

Energy given off by radioactive isotopes has been used for over 30 years to sterilize medical, dental, and household products, and of course for radiation treatment of cancer. As radioactive isotopes continually emit gamma rays (no off/on switch), when not in use, the source is stored in a pool of water that completely and harmlessly absorbs the radiation. To irradiate food, or other products, the source is raised out of the pool and into a room with massive concrete walls that contain the radiation. Food to be irradiated is wheeled into the room and exposed to the radiation for a defined period. After use, the source is returned to the pool, and the food is sent on its way.

Electron Beams

Beams of electrons or "e-beams" are streams of high energy electrons propelled out of an electron gun, the same—but larger—version of the device at the back of the household TV tube (before flat screens came on the scene) that propels electrons into the TV screen at the front of the tube, making it light up. The e-beam generator works on an on/off switch, and no radioactive substances are involved. Massive concrete shielding is unnecessary as gamma rays are not involved. Electrons, however, with their far less penetrating energy, can go no more than an inch into a food so that irradiation by electron beam requires thin sections—or the process must use two beams, one above, and one below to irradiate thicker portions. The electrically charged particles emerge from a linear accelerator at tremendously high speeds, with energies ranging from 3–10 million electronvolts (MeV). Figures 4.1a and 4.1b display a 10-MeV Rhodotron, a recirculating accelerator that operates on the principle that electrons gain energy when crossing an existing electric field. This makes it possible for the unit to be operated continuously, which has the advantage of high electrical efficiency and low maintenance costs. Its electron beam travels a rose-shaped path; hence its name Rhodotron, from Rhodos, Greek, meaning rose. Figure 4.1b displays the process, showing the accelerator in a shielded section above the the food irradiation area. The arrow points to a scanning horn that delivers the deeply penetrating beam to food packages on a conveyor belt. As the electrons flash through the foods, they create secondary particles, ions, and free radicals that disrupt microbial and insect DNA chains, destroying most but not all of them. This type of radiation is called *ionizing radiation* because the energy emitted is high enough to dislodge electrons from atoms and molecules and convert them to electrically charged particles, called ions.

X Rays

Machines used for food processing are more powerful versions of those used in hospitals and dental offices to take X ray pictures. To produce the X rays, a beam of electrons is directed at a thin gold plate or other metal, producing a stream of X rays beneath the metal plate, which then penetrate the food being processed. X rays can also penetrate deeply and can be switched on and

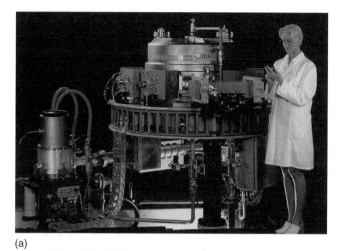

(a)

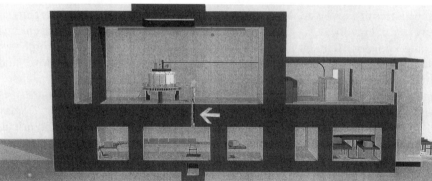

(b)

<u>Figure 4.1.</u> (a) The Rhodotron TT100, a 10-MeV/35-kW recirculating electron beam accelerator, showing the circular area where electrons are generated and accelerated; (b) general overview of a suggested layout of a food irradiation plant employing a rhodatron accelerator—the arrow points to the radiation vault where electrons scan back and forth across a conveyor system carrying the products to be irradiated. (Both photographs courtesy of Ion Beam Applications, Lourain-La-Neave, Belgium.)

off, and although they require heavy shielding, radioactive substances are not involved. In the newer X ray machines, an electron beam collides with a tungsten target plate, creating X rays capable of completely penetrating food products.

We need to give these irradiation processes an opportunity to work on our behalf, and to do so we address two questions: Why food irradiation, and how does it affect food? The "why" is straightforward; it's protective, preserving, and preventive.

TABLE 4.1. Irradiation Doses Permitted in Foods

Product	Dose (kGy)
White Potatoes	0.05–0.15
Fruit	1 maximum
Fresh vegetables	1 maximum
Poultry, fresh or frozen	3 maximum
Shell eggs	3 maximum
Meat, uncooked, chilled	4.5 maximum
Meat, uncooked, frozen	7.0 maximum
Seeds for sprouting	8 maximum
Spices	30 maximum

In meat and poultry it kills salmonella, *E. coli* O157:H7, listeria, campylobacter, toxoplasma, and other bacteria at levels of 99.9%. Irradication also

- Readily removes *Yersinia* and the *Trichinella* worm in pork.
- Prevents sprouting of potatoes and mold on strawberries.
- Kills insect larvae in grains, fruits, and vegetables and kills insects and bacteria in spices.
- Will delay ripening of fruit, increasing shelf life of many products that are closer to their fresh state.

With all this, what can we expect of the irradiated food?

The many studies done over the past 40 years indicate that irradiation produces no greater nutrient loss than does canning. Some foods may taste slightly different, just as pasteurized milk tastes slightly different than unpasteurized milk.

If irradiation is done at low temperatures, vitamin loss is minimal. Thiamine, a B-complex vitamin, is more sensitive to cooking temperatures than it is to radiation. Up to the maximum 10 kGy dose, there is no significant loss of proteins, fats, or carbohydrates, and minerals are not at all affected by irradiation.

Irradiation energy is measured in units of grays (Gy), the amount of energy transferred to the food. Irradiation of food is delivered in kilograys. One thousand (1000) grays equals one kilogray (kGy). A single chest X ray provides a dose of 0.5 mGy, a thousandth of a gray unit (Gy). By way of comparison, the FDA has approved the use of 3 kGy for irradiation of raw chicken to kill salmonella. Table 4.1 provides the approved dose levels for a sampling of foods approved for irradiation.

Irradiation can increase shelf life, but it cannot improve spoiled food Off-odors and off-tastes cannot be undone. Nor does irradication destroy preformed bacterial toxins already in food, or viruses that may also be present. Irradiated foods require refrigeration and must be handled and cooked safely. Now, what of safety?

Evidence from numerous studies conducted worldwide over the past 50 years informs us that chemicals formed in irradiated food are generally the same as those produced during canning, cooking, and other forms of food preservation, and do not put consumers at risk. The standard procedure in those studies is to feed laboratory animals the irradiated food and look for impacts on longevity, reproductive capacity, and tumor incidence [11]. A recently published study provides some guidance. This experimental study sought to determine whether 2-alkylcyclobutanone (2ACB), a radiolytic derivative of triglyceride, promoted colon cancer. In this study, one group of rats were fed 2ACB, while another group received a 2ACB-free diet for 6 months. Both groups received injections of a known chemical carcinogen. The question was whether 2ACB would enhance tumor growth. The researchers found that the 2ACB-treated animals had more tumors than did the control group. However, the authors inform us that the 2ACB-treated rats received 1000 times more 2ACB than would be found in any irradiated food eaten by humans. They also indicate that irradiated foods may contain several components that may reduce the bioavailability of 2ACB. Furthermore, they say that the benefits of food irradiation are becoming increasingly recognized [12].

It has been suggested that there is a potential for chromosomal damage to those who eat irradiated food. This, of course, refers back to the notion of induced radioactivity in such foods. According to the International Consultative Group on Food Irradiation (an FAO/WHO/IAEA affiliate), irradiation cannot induce radioactivity no matter how long the food is exposed to a radiation source, or how much of the energy is absorbed [13]. It's worth recalling that photons do not release neutrons, so that without radioactivity, chromosomal damage cannot occur.

Yet another concern raised about food safety has been the notion that irradiation could increase the virulence, the pathogenicity of bacteria. The FDA has found no evidence of such a response. In fact, it has been found that pathogenic bacteria that survive irradiation are destroyed at lower cooking temperatures than are unirradiated pathogens. It is also worth considering that because the FDA defined irradiation as an additive rather than a process per se, it had to undergo extensive testing to ensure its safety. Additionally, the U.S. General Accounting Office reviewed 6000 published articles and concluded that "the benefits out weigh the risks" [14].

Labeling and Recordkeeping

As *irradiation* is defined as an additive, all irradiated food must be labeled in conformity with FDA and USDA regulations. Both agencies require that irradiated foods be labeled with the international food irradiation symbol, the Radura: a stylized flower, shown in Figure 4.2. In addition, the label must state that the product has been intentionally irradiated. The Department of Agriculture's Food Safety and Inspection Service (FSIS) also requires that meat and poultry ingredients in multiingredient meat and poultry products be iden-

Figure 4.2. The Radura, an internationally accepted symbol required on all packages and containers of irradiated food. (*Source:* USDA.)

tified on the ingredient list. And food processors are permitted to add information explaining why irradiation is used. Their products can state, "treated with radiation to inhibit spoilage," or "treated with irradiation instead of chemicals to control insect infestation." And recordkeeping is essential. The FDA and USDA require processors to maintain accurate records that verify when irradiation occurred and the level received.

The Controversy

In the conclusion to their review of food irradiation, the U.S. General Accounting Office stated that [11]

> Despite the benefits of irradiation, the widespread use of irradiated food hinges largely on consumer confidence in the safety and the wholesomeness of these products. The cumulative evidence from over four decades of research—carried out in laboratories in the United States, Europe, and other countries worldwide—indicates that irradiated food is safe to eat. The food is not radioactive; there is no evidence of toxic substances resulting from irradiation; and there is no evidence or reason to expect that irradiation produces more virulent pathogens among those that survive irradiation treatment. In addition, nutritional losses from food irradiation are similar to other forms of food processing and would not adversely affect a food's nutritional value.

Additional meritorious reasons to support irradiation proceed from the fact that we are in an unrelenting race with a bewildering array of ravenous microbes and other parasites for our food supply. As much as one-fourth of the world's food supply is lost annually to microbial and pestilential spoilage; literally billions of dollars a year are lost. Irradiation can easily cut that in half. Furthermore, irradiation is an energy-efficient form of food preservation.

That's another plus, as is the fact that it totally precludes the use of chemical preservatives. The irradiation process occurs after the food is packaged, which means that the risk of recontamination is essentially zero. Adding all this to the fact of retention of freshness, retention of original nutritional status, little loss of flavor or appearance, and a return to medium and medium–rare hamburgers, we are left to conclude from such a calculus that we are indeed in thrall to a band of vocal, misguided Luddites opposed to the procedure. Change is overdue. But how is this to happen?

The American Medical Association's Council on Scientific Affairs published its Report No. 4, *Irradiation of Food*, in 1993. After a lengthy discussion of the process of food irradiation, it informed its membership and journal readers that irradiated food was not only safe and efficacious but also a desirable addition to the pantheon of food preservation processes. It concluded by stating: "The Council on Scientific Affairs recommends that the AMA affirm food irradiation as a safe and effective process that increases the safety of food when applied according to governing regulations" [13]. Similarly, the American Dietetic Association, a national and international professional association of thousands of dieticians and nutritionists, published its position on irradiated foods. It had this to say: "The ADA encourages the government, food manufacturers, food commodity groups, and qualified food and nutrition professionals to work together to educate consumers about this additional food safety tool and to make their choice available in the marketplace" [14].

In addition, the Food and Agriculture Organization and the World Health Organization, both United Nations affiliate organizations, have endorsed food irradiation without reservation, as has the American Gastroenterology Association, thousands of physicians with professional expertise in the gastrointestinal tract. The list of endorsements is long and honorable:

US government agencies
Food and Drug Administration
Public Health Service
Centers for Disease Control and Prevention
US scientific organizations
American Veterinary Medical Association
Council for Agricultural Science and Technology
Institute of Food Technologists
National Association of State Departments of Agriculture
International organizations
    International Atomic Energy Agency
    Codex Alimentarious Commission
    Scientific Committee of the European Union
    United Nations

All must be given high marks for their efforts. They're on the record. Unfortunately, that is not nearly enough. All of these efforts are passive. Not one has brought its recommendations to the public, where it would have an effect.

Unless and until each of these organizations or combinations of them actively reach out and bring their message to the people, little will change; illness and death will continue to pile up, needlessly.

While I gave high marks to those national organizations who went on record in support of food irradiation, I now rescind those high marks and in their place award "F"s for failure. Their on-the-record statements go nowhere, accomplish nothing. Has anyone ever read or heard of them? Has the media given voice to them? To regain their positions of excellence, they must become active and inform consumers—directly, by taking ads in newspapers across the country, to reinforce the message or irradiation's safety and nutritional merit.

It is also time for the thousands of members of professional organizations to realize that they, too, are citizens and voters, that bad decisions affect them as adversely as everyone else. Passivity is no longer acceptable. The means to deal effectively with microbial pathogens, to rid our food of them, is at hand. Given the recent food irradiation approvals, food distributors and retailers will need all the help they can get to ensure that items are there to be purchased. For example, on July 20, 2000, the FDA announced that it had approved the use of ionizing radiation for fresh shell eggs to reduce their level of salmonella. The new regulation permits a treatment dose of up to 3 kGy [15a], and the eggs will bear the required Radura. Given the huge numbers of eggs consumed in the United States, and considering that fresh eggs are among the top three vehicles for transmission of salmonella and human salmonellosis, this federal approval will add a needed level of safety to the food supply. Furthermore, there was also good news for alfalfa sprouts. On November 3, 2000, the FDA approved the use of irradiation on seeds for sprouting, including alfalfa, to reduce its levels of pathogenic bacteria such as salmonella and *E. coli* O157:H7. Here, too, a considerable reduction in numbers of microorganisms will be achieved by irradiation of up to 5 kGy. Finally, the FDA has also been petitioned to approve a large number of additional foods, such as preprocessed meats and poultry, and fruits and vegetables. When approved, the numbers of foods having obtained clearance will have more than doubled.

All the above notwithstanding, at the end of the day, it is the public that must embrace food irradiation. Enthusiastic praise from professional organizations and governmental agencies will not suffice. I'll tell you why. Three years ago, a major supermarket chain in our area announced that it would begin selling irradiated, fresh, ground beef. That was good news for us as we cared not for well-done meatloaf and burgers, nor did we want to worry about *E. coli* O157:H7 and other potential pathogens. Over 18 months we ate irradiated ground beef. In the spring of 2004, we received a card from the supermarket informing us that the SureBeam Corporation that provided the electronic irradiation of the ground beef was going out of business for lack of funding. There simply was inadequate support for irradiation. We are back to eating well-done beef, and wondering whether we are taking undue risks. It is difficult to believe that the public prefers to accept the risks of foodborne illness rather than safe, irradiated food. Consequently it is clear to me that lacking confidence, the public will continue to shun food irradiation. Too many people associate radiation, irradiation, with radioactivity, atom bombs, Hiroshima, and Three Mile Island. Confidence building would help clear away these flawed notions. Confidence building cannot begin too soon.

We now return to the prepackaged raw vegetable fiasco. Apparently the findings and recommendations of food scientists at the University of Georgia, that raw vegetables prepackaged in an atmosphere of 3% oxygen and 97% nitrogen had no effect on the growth of *E. coli* O157:H7, was ignored or fell on deaf ears [15b].

With the recent country-wide outbreak of gastroenteritis among many who ate raw spinach, lettuce, and green onions, food packers and fast-food restaurant chains, such as Taco Bell, are scrambling to determine the appropriate dose of irradiation that will kill off *E. coli* and campylobacter, yet not wilt raw leafy vegetables. By my estimate, irradiated prepackaged foods could be in supermarkets by early 2009.

However, the anti-irradiation forces have blocked the use of food irradiation for more than 30 years, preferring illness and death to irradiated foods. They've worked zealously to ensure that the word irradiation is deemed an epithet, and they've been successful.

Now comes the FDA in April 2007, announcing that it will consider approving the term *pasteurized* on labels of irradiated foods. The designation *irradiated* would appear only on foods whose texture, taste, and odor have been altered. The FDA will have a 90-day public comment period. You can bet that the anti's will be out protesting such a change. Should *pasteurized* prepackaged foods make it to market shelves, the public's purchase of them will be proof that after years of blind rejection of food irradiation, they have opted for the comfort and safety offered by prevention of foodborne illness. I'm betting on them.

NUTRITION AND HEALTH

Oxidants/Antioxidants and the Fruit–Vegetable Connection

A new label on cans of V8 vegetable juice announces in bold white letters on a blue background "essential antioxidants" and on another banner, yellow letters on a green background, trumpets, "100% A,C,E, vitamin-rich." Also with lycopene: 17 mg per serving. The rear of the label reads "essential antioxidants"; thus V8 makes a good thing even better by combining the powerful antioxidants lycopene and vitamins A, C, and E to help protect against the harmful free radicals that can damage cells.

As a V8 drinker for decades, this piece of news was of more than passing interest for me. It did raise questions. The primary question was: What are the facts about free radicals and cell damage, and how does it play out in human health or illness? Are the folks at Campbell's correct, or is their marketing ahead of the science? So, a look at oxidants and antioxidants seems like a good place to begin. We shall pursue this with an in-depth look at the evidence for antioxidant protection against cancer and heart disease.

But first a brief overview of the biochemical fundamentals. Oxidative stress has been implicated in a range of pathologic processes—and the oxidative stress theory of aging appears to be increasingly accepted as a major function of the aging process.

Free radicals (FRs) generated in human cells are derived from molecular oxygen (O_2) and are referred to as *reactive–oxygen species* (ROSs) and *reactive nitrogen species* (RNS), which are highly reactive because of their lack of an electron. Those FRs are generated in the body as inhaled oxygen is incompletely reduced, and are also known as *oxidants*, which are capable of removing electrons from any biological material in their path. Superoxide (the ion O_2), hydrogen peroxide (H_2O_2), and the hydroxyl radical (OH) are continuously generated in human cells during normal cell metabolism—but especially so during mitochondrial respiration. When produced in excess, these three radicals appear to produce harmful effects by damaging the large DNA, protein, and lipid molecules. (Direct detection of ROS and RNS radicals are difficult because they are so short-lived and so reactive, seeking to gain an electron and become stable.)

This oxidative stress is seen as playing a role in human cancer and neurodegenerative disease such as Alzheimer's and Parkinson's, as well as stroke and atherogenesis.

To control the balance between production and removal of oxygen radicals, cells contain antioxidant systems such as superoxide dismutase (SOD), catalase (CAT), and glutathione peroxidose. And to further prevent free-radical accumulation, this defensive system also includes metal-containing proteins that scavenge free radicals, as well as repair machinery for correcting ROS—induced DNA damage.

Diets rich in fruit and vegetables contain a mixture of the antioxidants ascorbic acid (vitamins C), carotenoids, vitamin E, and phenolics such as the flavinoids catechin, quercetin, and naringin. Current theory holds that vitamins C and E and the carotenes are major sources of the cells' antioxidant defenses, responsible for protecting the gastrointestinal tract from oxidative assault and damage, and forestalling cancers of the alimentary tract [16–19].

Theory is fine as far as it goes, but is it supported by evidence from human studies? And what types of studies are available from which to draw conclusions?

Epidemiologic Studies

As time marches on, our vision becomes clearer; not that aging increases visual acquity, but rather that long-term studies begin to yield significant new data that refute short-term studies.

We learn, for example, from a pooled analysis of eight prospective studies that included 430,281 men and women followed for up to 6–16 years across studies, that elevated consumption of fruits and vegetables was associated with "a modest reduction in lung cancer risk, mostly attributable to fruit, not vegetable intake." But the 20 researchers from five countries who participated in this substantial study tell us that, "The modest inverse association observed for fruit intake, and the absence of a reduction with vegetable intake, reinforces the public health message that the primary focus for reducing lung cancer incidence and mortality should be in smoking prevention and cessation" [20].

From another study of fruit and vegetable intake conducted by 10 members of the Harvard University School of Public Health, and the National Sun Yat-Sen University of Taiwan, in which prospective cohorts of 71,910 nurses and 37,725 male health professionals were followed for 15 years, the investigators found that increased fruit and vegetable consumption was "associated with a modest, although not statistically significant reduction in cardiovascular disease and not cancer" [21].

The European Prospective Investigation into cancer and Nutrition, referred to as EPIC, which involved 43 researchers from 10 countries, not including the United States, and enrolling 285,526 women between the ages of 25 and 70 years, were followed for >5 years to determine both the total and specific vegetable and fruit intake on the incidence of breast cancer. This formidable study found that neither the total nor the specific vegetable or fruit intake was associated with risk for breast cancer, and that the absence of a protective association was observed among almost all of the participating countries. They make the additional point that the duration of follow-up is relatively short and that, "we cannot exclude that associations will be found after more years of follow-up." They also note that it is of interest and concern that a protective effect is supported by a vast number of case–control studies [22].

This concern is being reiterated time again as differences of considerable magnitude in vital issues are published. Contrary results between observational, case—control studies and prospective cohort studies, and randomized clinical trials is an ongoing epidemiologic problem. For the most part, case–control studies show favorable responses to increased consumption of fruits and vegetables, while cohort studies and clinical trials are equivocal to negative. It should be recalled that because case–control studies are observational, relying on the memory of those who have had cancer or heart disease, bias can mislead when cases are compared to healthy control volunteers, who may be health-conscious and eat more fruits and vegetables.

Concerned about this continuing problem, and the fact that observational studies consistently show that individuals who use large doses of vitamin obtained cardiovascular benefits, researchers from the University of Miami Medical Center and Yale University Medical School conducted a metaanalysis of seven large-scale randomized controlled clinical trials of the effectiveness of vitamin E in the treatment and prevention of cardiovascular disease. Data on myocardial infarction, stroke, and cardiovascular death were included. They found that six of the seven trials "showed no significant effect of vitamin E on cardiovascular disease," and they also state that "vitamin E had neither a statistically significant nor a clinically important effect on any important cardiovascular event" [23]. They pointed out that the importance of this conclusion is enhanced by a recent survey indicating that 24% of adults in the United States are taking vitamin E supplements, which in effect precludes their use of known beneficial drugs.

At this point we should recall both the HOPE and HOPE-TOO trials discussed in Chapter 2, which orchestrated a large andomized, double-blind, placebo-controlled international trial: the initial heart outcomes prevention evaluation. Recall, too, that this trial enrolled 9541 individuals at 267 centers around the world, that HOPE-TOO enrolled 7030 individuals at 174 centers, and that after a median of 7 years of follow-up, the investigators found an increase in the risk of heart failure and concluded that "Vitamin E supplements should not be used in patients with vascular disease or Diabetes mellitus. Furthermore, their data clearly showed that vitamin E supplementation does not prevent cancer or major cardiovascular events [24]. And while back at Chapter 2, revisit the discussion of Miller et al. (Ref. 28 in Chapter 2), on high-dose vitamin E.

In July 2005, in yet another study, this one a randomized, clinical trial, conducted between 1992 and 2004, in which 39,876 healthy US women aged at least 45, were randomly assigned to receive vitamin E or placebo, and aspirin or placebo, were followed for an average of 10.1 years. The study found that 600 IU (international units) of vitamin E taken every other day "provided no overall benefit for major cardiovascular events or cancer, did not affect total mortality, nor decreased cardiovascular mortality in healthy women" [25]. Low-dose aspirin (100 mg) taken every other day had no effect on overall cancer incidence of mortality.

Three investigators, two from Canada, one from Spain, collaborated on a study to determine whether consumption of tomato products and lycopene reduces the risk of prostate cancer. From the results of their metaanalysis of 23 published studies that met their multiple criteria, they reported that "The existing evidence is not overwhelming enough to recommend the use of lycopene supplements in the prevention of prostate cancer" [26].

What can we glean from published studies specifically concerned with antioxidant supplementation and its effects on chronic medical illnesses? A recent trial, the Alpha-Tocopherol, Beta-Carotene Cancer Prevention study (ATBC), assessed the effects of supplemental α-tocopherol only, β-carotene only, α-tocopherol plus β-carotene, or placebo on the incidence and death from lung and other cancers among 29,133 male smokers aged 50–69. After a median of 6 years of follow-up, this randomized, placebo-controlled clinical trial revealed that lung cancer risk and mortality increased among those who received β-carotene. Prostate cancer incidence decreased among those receiving α-tocopherol. However, both the beneficial and adverse effects of supplementation disappeared during postintervention follow-up. The Finnish researchers indicated that the preventive effects of α-tocopherol on prostate cancer requires confirmation, and that smokers should avoid β-carotene supplements [27].

A team of investigators from Denmark, Serbia/Montenegro, and Italy sought to determine whether antioxidant supplements prevented gastrointestinal cancers. They identified 14 randomized trials (the "gold standard" of epidemiologic studies), which included 170,525 men and women. Their systematic review found that β-carotene, vitamin A, vitamin C, and vitamin E supplements alone or in combination, had no effect on the prevention of gastrointestinal cancers. But most disconcerting was their finding that these supplements "seem to increase overall mortality" None of the supplements protected against esophageal, gastric, colorectal, or pancreatic cancer [28].

In yet another metaanalysis using the Cochrane Controlled Trials Registry and Medline, this one from the Oregon Health and Science University, Portland, the researchers addressed the issue as to whether supplementation with vitamins A, C, and E; β-carotene; or a multivitamin reduce cardiovascular death, all-causes mortality, or cardiovascular events in the general adult population. After reviewing over 40 studies, they found that randomized controlled trials of specific supplements failed to demonstrate a consistent or significant effect of any single vitamin or combination of vitamins on the incidence of or death from cardiovascular disease [29].

A metaanalysis of randomized controlled clinical trials of nonvertebral fracture prevention with vitamin D supplementation was conducted by a group from the Harvard University School of Public Health and Boston University School of Dental Medicine. They began with the idea that "given the high prevalence, severity and cost of osteoporotic fractures, prevention strategies that are effective, low cost, and well tolerated are needed." Their 12 randomized control trials included 9294 men and women with hip fracture and 9820 for nonvertebral fracture risk. They found that an oral vitamin D dose of 400 IU per day (IU/d) was not sufficient for fracture prevention, but that 700–800 IU/d appeared "to reduce the risk of hip and any non-vertebral fractures in ambulatory and institutionalized elderly persons" [30].

This overview of what can reasonably be considered the optimum studies currently available reveals generally disappointing results for the efficacy of fruits and vegetables to accomplish what was expected of them by way of chronic diseases. The new prospective cohort studies and randomized clinical trials not only do not support the results of observational case–control studies; they contradict them. Also, given the fact that both the prospective studies and clinical trials are far more definitive and reliable, we are compelled to consider them seriously and favorably—that fruits, but not vegetables, may provide some small benefit in reducing risks of heart disease. But neither fruit nor vegetables appear to mitigate cancer risks. The idea of "five a day" for better health may have stalled in front of an insurmountable obstacle on the pathway to better health.

This is not the way it was supposed to be. In one of the most widely quoted and respected publications, "Oxidants, antioxidants, and the degenerative

Figure 4.3. Marketing of blueberries with the benediction of university research.

diseases of aging," Professor Bruce Ames, the lead author, contended that dietary fruits and vegetables are the principal source of ascorbate and carotenoids and are one source of tocopherol. He states: "Low dietary intake of fruits and vegetables doubles the risk of most types of cancer as compared to high intake and also markedly increases the risk of heart disease and cataracts. Since only 9% of Americans eat the recommended five servings of fruits and vegetables per day, the opportunity for improving health by improving diet is great" [31].

That has been the canon, the dogma, for a quarter of a century. Fruits and vegetables were to be the clear path to health, based primarily on biochemical studies and case–control epidemiologic studies. These new prospective studies and clinical trials will not go down easily.

Nevertheless, if anything has been evident, it is that the advertising and marketing of antioxidant vitamins has been excessive, and that the hype (hyperbole), the claims of benefit for antioxidant-filled food, has obtained ludicrous proportions. What a surprise it was to pick up a container of the most wonderful New Jersey blueberries and read the label, shown in Figure 4.3, and learn that in addition to their luscious taste, Tufts University Research has found blueberries beneficial for their antioxidant power, urinary tract health, reduction of eye strain and fatigue, and reduction of buildup of "bad" cholesterol. In their search for an edge, the berry producers use whatever is available.

And what are we to do about the "French paradox," which holds that the antioxidant polyphenols in red wine are protective, allowing the French to eat boldly of fatty cheeses and buttery sauces and have less heart disease than Americans? Or the "Japanese paradox," based on soy and green tea, and the Mediterranean diet heavy on olive oil? Is there any strong evidence for any of these?

The oxidant/antioxidant story is far from over; questions persist. For example, is there no limit to the amount of antioxidant we can use or need? Does anyone know the essential level of antioxidants that we humans require, if there is one? Is there an optimum level? Does it vary weekly, monthly, or yearly, and require checking? Can a healthy individual benefit from additional amounts of antioxidants? Is a person's antioxidant level a measure of health? I'd certainly like to see these questions answered before throwing out the baby with the bathwater. Haste in eliminating vegetables from the diet is uncalled for as new research indicates other beneficial effects. Vegetables are the main dietary source of nitrate (NO_3^-), which accounts for 60–80% of our daily nitrate intake. This new research indicates that the previous view that dietary nitrate has only harmful effects requires reconsideration. For example, in the stomach's acidic environment, following a meal with high nitrate concentration (lettuce), the levels of nitrate and nitrite increase markedly, killing most enteric pathogens within an hour, as well as increasing gastric mucosal blood flow [32].

Having raised questions and concerns about the future of antioxidants, it is now necessary to raise yet another question that brings the entire question of oxidant cell damage and consequent aging into sharper focus.

A newly published study appears to call into question the longstanding and widely held belief that reactive oxygen species are important in the aging process. The new claim is that apoptosis, programmed cell death, which selectively eliminate cells with damaged DNA, not oxidative damage from free radicals, drives the aging and degenerative disease process. This is the work of 18 investigators from five American universities: Wisconsin, Florida, Brown, Vanderbilt, and Texas, along with the University of Tokyo.

These researchers found that mice accumulating mtDNA mutations display features of accelerated aging, and that such accumulation was not associated with oxidative stress. They conclude by stating that "accumulation of mtDNA (mitochondrial) mutations that promote apoptosis may be a central mechanism driving mammalian aging" [33]. This will not be received lightly. We can expect a flurry of new research to either support or refute this new paradigm. But that is what science is all about. We do need to know. If found to be true, we may be a step closer to the fountain of youth. Is that something to look forward to, or is it another one of Pandora's boxes?

Foodborne Illness with a Tangled Web

In the previous edition of *OPH*, amyotrophic lateral sclerosis–Parkinson's dementia complex (ALS/PDC), a progressively fatal neurodogenerative disease, clinically and histologically similar to Alzheimer's disease and Parkinson's dementia, was examined and believed to be due to consumption of the lemon-sized seeds of the cycad tree, *Cycas circinalis*. At the time, ecologists and anthropologists were mystified by the fact that among the Chamorro people of Guam, the incidence of ALS/PDC was 50–100 times greater than in the continental United States. It was during World War II, with tens of thousands American troops stationed on Pacific Islands, that the alarming numbers of ALS cases came to light. By the 1960s the disease had been characterized, and suspicion had fallen on the cycad's seeds, which grow in clusters just beneath the palm fronds. The cycad is actually a false sago palm, currently known as *Cycas micronesica* Hill. Chemical analysis of the seeds revealed the presence of cycasin, a glycoside, and an unusual nonprotein amino acid, β-*N*-methylamino-L-alanine (BMAA), a known neurotoxin, and when fed to macaque monkeys, ALS symptoms appeared. It was also found that in preparing the cycad flour, from which a number of foods were made, the BMAA content was substantially reduced—too low to produce the typical ALS symptoms.

Fifteen years would pass before further progress was made, and the suggestion advanced that exposure to BMAA might well occur with consumption of *Pteropus marianus marianus* (of the Marianas Islands), the Marianas flying fox—actually a large bat, that fed on cycad seeds. As the investigation proceeded, it was learned that flying fox are a prized food item of the Chamorros, especially among the older men, who boil them in coconut cream and eat them whole. Chamorros who died of ALS/PDC had a substantial level of BMAA in their brain tissues, and flying fox museum specimens were analyzed and found to contain large amounts of BMAA. The investigators postulated that biomagnification could readily occur, resulting in toxin concentration in the body of the fox. But how did BMAA become a component of cycad seeds? Dr. Sandra Banack of California State University recently laid bare the mystery—put the pieces of the puzzle together. As she recounted it, the inaugural source of this unique amino acid are nitrogen-fixing, photosynthetic, chlorophyll-containing cyanobacteria of the genus *Nostoc* that grow symbiotically on cycad roots. When unrelated to cycad roots, cyanobacteria produce minimal amounts of BMAA, but in close association, the level surges 100-fold. The levels are further concentrated in the seeds, and concentrated yet again in flying foxes. So, it is now known that the web of causation includes a primitive bacterium, a palm tree and its seeds a flying fox, and finally the ultimate consumer, we humans [34–36]. But a question remains—why do cyanobacteria produce BMAA? Where does it fit in its metabolic pathway, or is it a protective, defensive chemical? Is it evolutionarily beneficial? Stay tuned.

From the discussion of cancer and its supporting statistical data (Chapter 1), it is evident that not only is there no cancer epidemic in this country but also that in fact cancer incidence and cancer mortality (lung cancer excluded) have been declining for the past 20 years. The contributing risk factors for cancer—smoking, dietary imbalances, alcohol consumption, microbial and occupational risks—leave little room for pesticides in foods, yet from the level of coverage given pesticides as a potential (more likely actual) risk factor, an objective observer could reasonably assume that pesticides in the foods we eat were at the top of the risk list.

Clearly pesticides have no place in our food supply as they provide no nutritional or other enhancing value. So, why pesticides—are they a hazard, but most importantly, is our food supply safe?

The reality is that no plot of land is immune to attack by every manner of pest, and that we are in an actual race against insects, microbes, worms, rodents, and weeds, to see who gets to the food supply first. Yearly losses run over 25%. Pesticides are needed to ensure a consistent, wholesome supply at reasonable prices. Pesticides are required, and residues will be present. Nevertheless, a major contributor to health in the twentieth century were human-made (anthropogenic), synthetic pesticides, which markedly decreased the cost of food production and ensured that most of the corps planted would be eaten by us humans rather than the realm of pests [37].

Fine, but crops have been treated with chemicals—chemicals we call "pesticides," which can be either natural or synthetic and are used to kill off pests attacking crops. How are they removed, and how are they tested to determine residual pesticide content?

Dietary exposure to a pesticide depends on both the actual residue in or on foods and food consumption patterns—as in how much ingested, over a specific period of time.

To ensure safety, the EPA sets safety standards that limit the amount of pesticide residue that legally may remain on or in any foods sold in this country. Both domestic and imported foods are monitored by the FDA and USDA to ensure compliance.

Of course, the highest residue levels would be found on crops in the field, where levels decline naturally as a consequence of disruption via ultraviolet light and removal by rain. Nevertheless, these crops can go two ways: to markets as fresh produce, and/or to processing plants for canning, freezing, juicing, milling, baking, brewing, or winemaking, each involving mechanical washing of the food product several times over. Processed foods must also be tested, even though their residual pesticide levels may be minuscule. Concern must cast a spotlight on fresh fruits and vegetables that are washed before reaching markets, and where washing can occur again. At point of sale is where it begins for consumers, we have "total diet studies," otherwise known as "market basket studies," to guide and protect us.

In fact protection begins before crops are planted. In setting tolerance limits, the EPA requires that maximum levels to be found on food crops must be determined by the pesticide manufacturer who must plant crops, apply the pesticide, and determine residues. Food producers must also submit data on residues in their finished products. The EPA also requires toxicity testing of every pesticide used. Acute, short-term feeding studies, and chronic, long-term studies are done with laboratory animals to determine risks that may occur to infants, children, and adults. These include developmental studies that determine risks to developing fetuses that could result from a mother's exposure to pesticides during pregnancy.

From these battery of studies a reference dose (RFD) is ascertained. This RFD is defined as "an estimate of a daily exposure level for the human population, including sensitive subpopulations, that is likely to be without appreciable risk of deleterious effects during a lifetime—without appreciable risk," is understood to mean a safe level of dietary intake. A safe level is not the same as zero level, which is neither expected, achievable, nor necessary. Just how safe is safe enough? This is both an individual assessment and a question for the ages, as each generation will make its own decisions. But in each generation, levels are set. Thus, the EPA sets levels of permissible residues, but it is the FDA's responsibility to test foods for the presence of pesticides and ensure that tolerances are not exceeded.

Total diet studies are used to monitor pesticide levels as well as nutritional concerns. To determine whether radioactive fallout from atmospheric nuclear weapons tests resulted in elevated levels of radionuclides (see Chapter 6) in foods, total diet studies to monitor these and other chemical contaminants were begun by the FDA in 1965. Since then, there has been an ongoing testing program. Four times a year FDA personnel shop in supermarkets and grocery stores, once in each of four geographic regions of the country. Shopping in three cities from each region, they purchase the same 234 foods (including meat) selected from nationwide dietary survey data to typify the American diet. These foods are called "market baskets."

These market basket foods are then prepared as a consumer would prepare them. Using a standard recipe, for example, beef and vegetable stew is made from the collected foods, and is then analyzed for pesticide residues, and the results, together with USDA consumption data, are used to estimate the dietary intakes of pesticide residues for eight age–sex groups ranging from infants to senior citizens.

For their most recent report [38], the FDA researchers included results from 27 market baskets. Included were 33 different infant foods (both strained and junior), 10 adult foods eaten by infants and children, and four types of milk. The infant foods included cereals, combination meat–poultry dinners, vegetables, desserts, fruit and fruit juices, and infant formulas. The adult foods included apples, oranges, pears, and bananas; apple, grape, and orange juices; applesauce; grape jelly; and peanut butter. Milks were chocolate, evaporated, 2% low fat, and whole. What did they find? No residues were found in infant formulas, and no residues over EPA tolerances or FDA action levels were found in any of the "total diet study" (TDS) foods. Low levels of malathion were found in some cereals because malathion is widely used on grains both before and after harvest. Low levels of thiabendazole, a postharvest fungicide used on fruits, were found on some fruits and fruit products.

The low residue levels found in the TDSs demonstrate how food processing and preparation for table consumption can dramatically reduce residue levels. Washing fruits and vegetables at home further reduces residues. The highest thiabendazole levels on raw apples was 2 parts per million (ppm). EPA tolerance level is 10 ppm: 0.08 in apple juice and 0.06 in applesauce.

Since its origination in the United States, TDSs have become standard practice in other countries. Canada reported on its latest market basket study in 2004 [37]. The intention in this study was to determine whether pesticide residue levels were different in northern Canadian communities compared to previous determinations in the more populated southern communities. Foods were purchased in local markets and prepared for consumption following community standards. They found that residue levels in the northern territories were similar to those in the major southern cities and that here, too, levels were well below established tolerances. As in the United States, malathion was the most frequently detected compound. Not only were no dietary intakes above the accepted daily intakes for any age–sex category; their data compared favorably with residue levels in Japan and Italy, two countries that are especially fastidious about extraneous chemicals in foods, which raises another vital concern—the safety of imported food of which huge amounts, as noted earlier, enter the United States annually.

The FDA randomly tests imported foods for pesticides. Samples are collected at points of entry, and must meet the same tolerances as domestically produced foods. Should imported foods contain excessive levels, the FDA can stop all future shipments if there is reason to believe that the problem will persist. About 5% of imported foods have been impounded. Many imported foods come from Mexico and can contain a pesticide not registered in the United States, or they may contain a pesticide for which no tolerance has been established. One aim of the North Atlantic Free Trade Agreement (NAFTA) seeks to ensure that pesticide use in Mexico meets US requirements [38].

From time to time, the question is raised about the safety of pesticides given the fact that single-pesticide ingestion studies may not represent reality, that multiple pesticides occurring together as they do in our diets offer a more plausible scenario, and that safety might be better judged on ingestion of multiple pesticides simultaneously. Well, fair enough.

Japanese investigators took up this question and fed a mixture of 20 pesticides, all approved for use in Japan—19 organophosphorous compounds and one organochlorine pesticide. Two feeding levels were conducted: one, at the World Health Organizations Acceptable Daily Intake level, and one at 100-fold the ADI. Five groups of rats were fed these pesticides for 8 weeks. They found that feeding the 20 pesticides a the WHO ADIs, "did not enhance rat liver preneoplastic lesion development. In contrast, a mixture 100 times ADI significantly increased the number and area of lesions" [39].

The 100-fold increase in pesticide level fed to animals to ensure development of tumors has long been a point of concern. It raises the fair question as to interpretation of how we translate this to human ingestion and safety. As Professor Bruce Ames contends, the toxicological data on synthetic chemicals as a significant risk to human health is simply not supportable: "When you look at the huge dose you need to give cancer to a rat, the levels humans are ingesting don't make much sense as a being a significant cause of concern" [40]. Furthermore, 99.99% of the chemicals we eat are natural. The amounts of synthetic pesticide residues in fruits, vegetables, and cereal grains are insignificant compared to natural pesticides produced by the crops. "Of all dietary pesticides that humans eat, 99.99% are natural. These are chemicals produced by plants to defend themselves against fungi, insects and other animal predators" [40]. Ames goes on to inform us that "On average, Americans ingest roughly 5000 to 10,000 different natural pesticides and their breakdown products. Americans eat about 1500 mg of natural pesticides per person per day, which is about 10,000 times more than they consume of synthetic pesticide residues." Is this a tempest in a teapot—are we shooting butterflies with rifles? An exemplary case in point befell us in 1989 when CBS's Ed Bradley delivered an expose on "60 Minutes," and the celebrated actress Meryl Streep spoke out against the pesticide Alar (daminozide) as the "most potent cancer causing agent in our food supply." Nationwide hysteria followed, and apples became "the great satan." As children were thought to be at exceptional risk, apples were removed from school lunches in 50 states. The root of the affair lay in the finding that UDMH, unsymmetric dimethylhydrazine, a trace contaminant in Alar, and believed by some to be a carcinogen, appears when Alar is hydralyzed in the body. In fact, there is more hydrazine in a portion of mushrooms than anyone, adult or child, would get from apples or apple juice. The *Washington Post* calculated that to get a harmful dose of hydrazine, a person would have to drink 19,000 liters of apple juice. A statement issued jointly by the FDA, EPA, and the Department of Agriculture, declared that: "Recently, the Natural Resources Defense Council (NRDC) has claimed that children face a massive public health problem from pesticide residues in food. Data used by NRDC, which claims cancer risks from Alar are 100 times higher than the Environmental Protection Agency estimates, were rejected in 1985 by an independent scientific advisory board created by Congress." They noted too, that, "risk estimates for Alar and other pesticides based on aminal testing are rough estimates and are not precise predictions of human disease. Because of conservative assumptions used by EPA, actual risks maybe lower or even zero" [41].

In his book, *But Is It True: a Citizen's Guide to Environmental Health and Safety Issues*, Aaron Wildavsky appears to have it right. "We should be guided," he writes, by the probability and extent of harm not its mere possibility. The search for possibilities in endless, and it trivializes the subject. There is bound to be great diversion of resources without reducing substantial sources of harm. Consternation is created but health is not enhanced" [42].

Genetically Modified Food—Genetically Modified Organisms

> As for the future, your task is not to foresee it, but to enable it.
> —*Antonie de Saint-Exupery, 1948, The Wisdom of the Sands*

In his chapter on orthobiosis, Metchnikoff (Ref. 42 in Chapter 2) lauds Luther Burbank, an outstanding American botanist, for his improvement of useful plants. "Burbank," he tells us, "cultivated great numbers of fruit trees, flowers, and all kinds of plants, with the object of increasing their utility." He goes on to say that

> He has modified the nature of plants to such an extent that he has cactus plants and branches without thorns. The succulent leaves of the former provide an excellent food for cattle . . . and to obtain such results much knowledge and a long period of time were necessary. To frame the new ideal of the plant it was necessary not only to have an exact conception of what was wanted, but to find out if the qualities of the plants in question furnished any hope of realizing it. Indeed, results could never be predicted; only hoped for.

Random genetic variation occurs naturally in all living things and is the basis of evolution via natural selection. Well before its scientific basis was understood, farmers took advantage of natural variation by selectively breeding wild plants and animals to produce variants better suited to their needs. But this selective breeding involved the transfer of unknown numbers and types of genes between organisms of the same species that rendered outcomes haphazard and unpredictable, and could take decades. Specificity, predictability, assurance, and timeliness had to await recombinant DNA technology, also known as *genetic engineering, genetic manipulation, gene technology,* and *genetic modification,* all of which can be defined as the introduction of genes from one organism into another so that the recipient organism acquires genes that encode for unique new traits that will be expressed. An era of directed genetic change is at hand.

Biologists are simply using DNA—the inherited chemicals of all living things—from viruses to humans, from one plant or animal species and inserting it into the DNA of another, which means that the only thing being transferred are four nitrogenous bases: adenine, guanine, cytosine, and thymine, which are the same chemicals in flies, fleas, fish, frogs, flowers, and my friend Fred, which translates to—a chemical, is a chemical, is a chemical; adenine is adenine no matter where it is found. Understand this, and genetic modification becomes a no-brainer.

The total, the complete four-letter alphabet rendered by adenine, cytosine, guanine, and thymine, a, c, g and t, in different sequences, repeated over, and over constitutes the remarkable and astonishing "alphabet" in which the genetic code is "written." The words of the code, the instructions, govern and direct, how and when the literally hundreds of thousands of proteins will be made. The "words" are determined by the order of the bases along the DNA molecule. Bear in mind that all cells in our bodies (except red blood cells, which do not have a nucleus and have no DNA), all cells in plant and animal tissue contain the DNA sequences that make each organism unique; that is our genome, a plant's or animal's genome.

Each sequence of bases actually spells out an amino acid. The code specifying an amino acid must consist of at least three bases. These three-letter words are called *codons*, of which there are 64, and can spell out all 20 essential amino acids. So, for example, TTT codes for lysine; CAT, for methionine; GCC, for alanine; and ACC, for tryptophane. Thus the arrangement of the four bases spells out all amino acid sequences. Amino acids join to form polypeptide chains, and polypeptide chains join to form proteins. The various combinations of 20 amino acids can produce millions of proteins. Again, the takeaway message in all this is, a chemical, is a chemical, is a chemical, no matter what its initial origin. Since all species of living things have the same four-letter genetic alphabet, A,G,C,T, they share the same genetic language and thus any sequence of bases can be "read" and understood as an instruction to make something. So, bacteria, for example, can have a gene for human insulin production inserted into its genome, and the bacterium will produce (express) human insulin—unmistakable human insulin. In a word, the bacterium becomes a human insulin producing factory. Similarly, a set or sets of bases can be inserted into plants that can prevent freezing, permit growth under low moisture conditions, or tolerate high salt content, prevent formation of toxic chemicals, prevent specific microbes from initiating an infection, increase the size of fish, permit more rapid growth, and increase vitamin content. The fact is, it is impossible to identify a gene as belonging to a turkey, tomato, or trout—a gene is a gene, is a gene.

With these fundamentals as prelude, the question then is, why genetically modified foods? What are their advantages and potential benefits? Changes to food as a consequence of gene transfer are essentially no different from those that occur naturally as a result of selective breeding, except that in the gene transfer procedure, a selected few genes are involve, drastically reducing the time to achieve a specific trait, and totally removing the random assortment of all other genes. In addition, genetic modification makes it possible to transfer genes between different species. This is where Luddites raise the spector of Frankenfoods. Such transfers are truly revolutionary and bring with them the potential for a range of benefits, but they have also brought with them such bones of contention as safety, ethics, environmental impact, and consumer choice.

Given these concerns, it may be essential at this point to inquire as to how it began and how gene transfers are made. It began in Japan, in 1901. *Bombyx mori*, the silkworms, were dying. Shigatone Ishiwata, a microbiologist–entomologist at the Japanese National Institute of Sericultural and Entomological Science, isolated the bacterium that was infecting, softening, and killing the

Lipidoptera larvae. He called the silkworm disease *sotto* and the culprit, *Bacillus sotto* [43]. In 1915, in Germany, Ernst Berliner was investigating a disease of the Mediterranean flour moth, *Angasta kueniella*. These infected moths were obtained from a flour mill in Thuringia. In his published report he described the isolation of a spore-forming pathogenic bacterium from the dead and dying insects, which he dubbed *Bacillus thuringiensis* [44]. As it turned out, *B. sotto* and *B. thuringiensis* were the same gram-positive soil bacterium. But *B. thuringiensis* persevered. This bacterium possesses the unique ability to produce a parasporal crystalline protein during sporulation that is not only insecticidal but also exquisitely selective, with over 20,000 strains found in soils around the world, having the ability to produce seemingly inexhaustible insect destroying proteins. Insects are not all that different from we humans in their susceptibility to microbial infection and death. Microbes are one of nature's ways of limiting insect populations. *Bacillus thuringiensis* has been performing this function for eons. From Ishiwata to Berliner, during 1900–1915, the number of insect orders that *B. thuringiensis* was found to infect and destroy jumped from the single Lipidoptera to currently nine additional orders that now encompasses flies, beetles, mosquitoes, protozoa, worms, flatworms, mites, and ants. The delta(δ)-endotoxins possess the impressive attributes of our ideal biopesticide:

- Easily grown and made.
- Petrochemicals are superfluous.
- Nontoxic to vertebrates: mammals, birds, and fish.
- Resistance to it has not developed over the past 17 years. (Since the toxic crystal protein consists of a string of amino acids, it is reasonable to believe that insect genes will mutate and resistance may develop.)

Bacillus thuringiensis (Bt)'s insecticidal promise took off in 1981 when H. Ernest Schnepf and H. R. Whitely of the University of Washington, Seattle, cloned and sequenced the Bt toxin gene; all the words of the four-letter alphabet became known. In 1987, scientists inserted Bt genes in cotton plants. A year later the first cotton plants containing Bt genes, with its insecticidal crystalline protein expressed, were harvested. In 1995, the Bollgard gene, a Monsanto product, became commercially available throughout all cotton-producing US states, as well as worldwide. For the first time in human history, there was a crop that could defend itself against the voracious bollweevil. That's the brief history of Bt: traditional, careful, creative science at work. Knowledge building on knowledge. And the cotton plants defend themselves the way all crops containing the Bt crystal do. The process is remarkably straightforward. Ingestion of the CRY toxin (cry for crystal) results initially in paralysis of the gut and mouthparts. When swallowed, the protein is released at specific receptor sites in the insect's stomach. The protein opens a channel in the insect's stomach, flows in, and dissolves the intracellular cement. As the gut liquid diffuses between the deteriorating epithelial cells, paralysis occurs, followed by bacterial invasion and subsequent insect death. Each Bt protein has its narrow insect host bounds—and "insect" is the operative locution. Bt has no effect on birds, fish, or mammals, including we humans. Recall that our stomachs, and those of all vertebrates, are acidic, unlike those of arthropods, which are alkaline, and since the Bt crystalline protein is alkaline, it can function effectively at the higher pH ranges. Receptor sites for this protein are lacking in acid environments. Ergo, Bt is harmless to all but insects. That is another cardinal takeaway message. That is the point to retrieve when the Luddites raise the spector of safety.

Currently three techniques or processes are available for placing a new gene into cells normally foreign to it: a physical method, a splicing procedure, and a microbial messenger. But it is also necessary to have isolated a gene with the traits to be imparted. The vehicle of insertion will be one of the above, and of course the host plant, animal, or fish DNA to be modified is needed. Perhaps most important, we need a way to determine that the gene we believe has been successfully inserted, has in fact been placed in, and is working. This requires a marker gene that must be readily detectable in the modified DNA. These marker genes must be attached to or in close association with the gene carrying the desired traits, which may not be visible. There is yet another consideration that goes hand-in-hand with the gene that we wish to impart. Where do we obtain this gene, or any gene? Currently the world is fortunate to have a gene bank that goes by the name Genbank and is an integral part of the National Center for Biotechnology Information (NCBI) established in 1988 (at the National Library of Medicine, Bethesda, MD) as a national resource for molecular biology information. It contains an annotated collection of all publicly available DNA sequences, via the humanitarian and public-spirited inclinations of our academic scientists, as well as scientists the world over, who freely deposit the results of their researches. As of February 2004, the Genbank had on deposit 37,893,844,733 bases in 32,549,400 sequences, which are also freely available to anyone. NCBI can be accessed at http://www.mcbi.nlm.gov/Web/GenbankOverview.htm.

So, with the specific DNA in hand, the physical procedure in question requires that it be mixed with gold or tungsten pellets, 1 μm in size, which are placed in position with the target cells so that a mighty blast of compressed helium gas, from the helium gene gun, sends the DNA-coated pellets into the plant cells, which migrate to the cell's nucleus. Some remain in, some exit. But as there are huge numbers of cells, it is fairly certain there will be "takes"—cells that receive the new DNA and survive. When it works, the cell regenerates into a new plant, and its new genome will express the newly acquired trait.

A second procedure for depositing a gene into a cell is *splicing*. Using readily available biochemical techniques, DNA can be removed from cells and, using restriction endonuclease enzymes, to cut the DNA strands into segments and insert a foreign gene. Not unlike splicing a segment of tape into a reel of cassette tape.

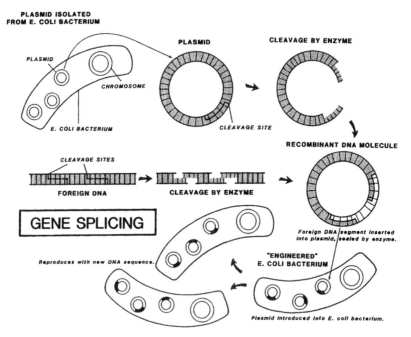

Figure 4.4. Follow the arrows to comprehend the gene splicing process.

Once the restriction enzymes slice DNA and the new gene is inserted, a ligase enzyme, acting as a genetic "glue," reattaches the cut segments and the splicing is complete. Thus, bacteria that divide and double every 30 minutes are virtual chemical manufacturing plants. These engineered microorganisms can be made to produce or metabolize almost anything.

In a test tube, using detergent-type chemicals, the cell membrane is dissolved, allowing the cell contents, including the small plasmids, to spill out. These are the most easily modified. The segregated plasmids are mixed with restriction enzymes that cleave it at various points, opening it up and "stretching" it out. Conveniently, sticky ends are produced. The solution also contains the ligase that cements the sticky ends together and the new gene into place. The result is a new plasmid loop. The complete process is shown in Figure 4.4 New plasmids are placed in a solution of cold calcium chloride, which contains normal untreated bacteria. This solution is heat-shocked. As a result, the membranes become permeable, allowing the new plasmids to pass through and become part of a microbes' new genome. When the bacteria reproduce, as part of their normal metabolism, they will produce—express—the new information now contained in their genome.

Plant and animal genomes can be modified using the common soil bacterium *Agrobacterium tumefaciens* as a vehicle. *A. tumefaciens* contains a recombinant plasmid tumor-inducing (Ti) and regularly enters and exits plant roots.

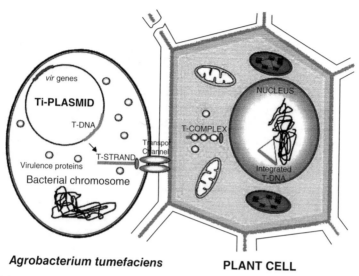

Figure 4.5. Gene transfer from *Agrobacterium tumefaciens* to a plant cell. (Figure courtesy of Prof. Stanton B Gelvin, Dept. Biological Science, Purdue University.)

Infection of normal plant cells by this bacterium transforms them into tumor cells. As a consequence of the insertion of genes into the plant's genome, not only a natural phenomenon, but possibly the only natural example of breeding between two different species, which has been ongoing since root–nodule bacteria began fixing nitrogen, takes place. Microbiologists soon realized that this organism could be used as a vehicle for transferring recombinant DNA genes into plant chromosomes. The Ti plasmid is now used frequently to produce successful transfers, but new strains of agrobacterium have been developed that no longer contain the pathogenic tumor-inducing plasmid, but can still transfer genes. The process is clearly shown in Figure 4.5.

Australian scientists from the Commonwealth Scientifiic and Industrial Research Organization (CSIRO) recently used an Agrobacterium to transfer five genes from Algae (which produce the omega-3 fatty acids that we get from eating the fish that eat the algae) into a crop plant, and demonstrated the production of the omega-3 in the plant, which will provide an alternative source of omega-3s for people who don't care for fish.

To transfer a gene to a plant or animal cell is one thing; identifying whether, and where, the genes have landed is quite another. Indeed, it remains a kind of hit-or-miss undertaking. I say "kind of," as there are literally millions of cells involved so that there is reasonable expectation of a take. Furthermore, as the newly selected gene is copied millions of times over, the original DNA in the new genome is trifling. Hence marker genes are essential. Another takeaway message is that the newly inserted genetic material is present in minuscule amounts. Yet another safety consideration.

The current procedure links the trait with a marker that expresses readily identifiable characteristics. The kanamycin resistance gene, a widely used marker, produces an enzyme that inactivates the antibiotics kanamycin and neomycin. Plant cells that do not express the kanamycin gene (do not contain it) are killed by these antibiotics, providing a means of rapid screening. The kanamycin gene is used as a marker because the chemical produced by the gene is easily digested in the stomach even under low-acidity conditions. It's harmless and doesn't affect antimicrobial resistance. Nevertheless, it raised a question: Could these marker genes be transferred from the modified plant to intestinal bacteria and spread resistance to therapeutic antibiotics? An FAO/WHO expert panel was convened to consider the implications that a maker gene might pose. They concluded that the presence of marker genes in food did not compromise its safety for either human or animal consumption. Following that panel, the FDA noted that marker genes held no threat of toxicity or allergenicity [45]. In February 2004, the British Society for Anitimicrobial Chemotherapy stated that "There are no objective scientific grounds to believe that bacterial antibiotic resistance genes will migrate to bacteria to create new clinical problems." They went on to say that "the argument that occasional transfer of these particular resistance genes from GM [genetically modified] plants to bacteria would pose an unacceptable risk to human or animal health has little substance. We conclude that the risk of transfer of antibiotic resistance genes from GM plants to bacteria is remote, and that the hazard arising from any such gene transfer is at worst, slight" [46].

With this concern for the antibiotic resistance gene attended to (and there will be other safety issues to deal with), we return to the question, Why genetically modified foods? What can they do for the people of the world that traditional foods cannot do?

In a comprehensive article in the *Atlantic Monthly*, Jonathan Rauch [47] made a significant case for genetic modification of foods. "Over the next half century," he wrote, "genetic engineering could feed humanity and solve a raft of economic ills, if only environmentalists would let it." Hard to imagine as it may be, but no less true, that a small but highly organized and vocal group has managed to intimidate governments and keep genetically modified foods from markets the world over. This hubris requires comment, as shall be done, but let us concentrate on the need for GM foods. In his cogent essay Rauch affirms the United Nations estimates that global population will rise from its current 6.4 billion to approximately 9 billion by midcentury, 2050. But he also notes that across the world people must have their pet dogs and cats to the tune of another billion mouths that not only must be fed, but when people move beyond a subsistence lifestyle, they will want to be provided with "the increasingly protein-rich diets that an increasingly rich world will expect—doing all that will require food output to at least double, and possibly triple" [47]. He continued:

If in 2050 crop yields are still increasing, if most of the world is economically developed, and if population pressures are declining or even reversing, then the human species may at long last be able to feed itself, year in and year out, without putting any additional net stress on the environment. We might even be able to grow everything we need returning crop land to wilderness, repairing damaged soils, restoring ecosystems. In other words human agriculture might be placed on a sustainable footing forever. The great problem then, is to get through the next four or five decades with as little environmental damage as possible.

And that, he maintains, is where genetically modified food comes in.

On the basis of pest pressure and current crop protection, the biggest yield gains are expected in South and Southeast Asia and Sub-Saharan Africa [48]:

> South and southeast Asia and sub-Saharan Africa are also the regions with highest population growth, so increases in agricultural output per unit area are vital for poverty alleviation and food security "Bt [*Bacillus thuringiensis*] cotton, Bt maize, and Bt potatoes, which have already been commercialized in some countries, have direct relevance to the developing world. Bt rice, Bt sweet potatoes and a number of food crops with other pest-resistance mechanisms will further broaden the portfolio in the near future.

Pest resistance, while a substantial benefit, is but one among others. So, for example, GM holds out such advantages as

- Allowing a wide selection of traits for improvement such as nutritional, taste, and visual improvements.
- Results obtained rapidly and at lower cost
- Greater precision in selecting traits

These advantages can lead to the following benefits:

- Improved yields with less labor and overall costs.
- Reduced use of herbicides and pesticides.
- Benefits to the soil by no-till farming, which foregoes ploughing and allows underground ecosystems to return which reduces erosion and runoff. Worms do the ploughing, which saves the farmer fuel, which in turn saves energy and reduces pollution. But no-till farming depends on GM crops.
- Crops and grow and flourish in previously in hospitable environments— drought, salinity, flooding, extremes of temperature—which translates into increased yields.
- Improved flavor texture.
- Removal of allergens and toxic components such as cyanide in cassava.

Several pertinent and apropos examples are readily at hand. A more nutritious version of golden rice offers a practical solution to vitamin A deficiency. Initially, golden rice was genetically engineered to produce β-carotene in its seeds, but the low levels of β-carotene in the kennels of this transgenic crop raised questions about its nutritional value. Recall that the desire to develop rice enriched with β-carotene arose in response to the high level of vitamin A deficiency throughout the developing world—especially where rice represents a substantial portion of the diet. Vitamin A deficiency results in blindness and susceptibility to infections via a depressed immune system. Plants do not make vitamin A. Biotechnology changed that by inserting a gene from the daffodil (*Narcissus pseudonarcissus*) to convert geranyl geranyl diphosphate to phytoene, and a second gene from the soil bacterium *Erwinia*, to further convert phytoene to lycopene, which is itself converted to carotene by enzymes in rice. Quite a testament to, and evidence of transgenic legerdemain (sleight of hand). But the level of β-carotene was insufficient to make a difference. Another upgrade was needed. Researchers at Syngenta took up the challenge, and found a gene, phytoene synthase (psy) in maize, that substantially increased carotenoid accumulation and developed Golden Rice 2, which increased the carotenoid level 23-fold, and can now be delivered to children. A child given 60–70 grams of this rice, less than a quarter of a pound (<0.25 lb), per day, can now obtain their full complement of vitamin A [49]. This has to be seen as a splendid accomplishment for science, for agriculture, for humanity.

Mastitis in dairy cows has been, and is, an unwanted consequence of selection for improved milk production. Dairy cows are quintessential examples of hundreds of years of imbreeding for increased yield and quality of milk. But this has had unintended consequences: mammary gland susceptibility to a staphylococcal infection that is refractory to cure, and is the dairy industry's most costly veterinary condition, the world over, as well as a major cause of premature animal death. Elimination of chronically infected cows appears to be an efficient way to prevent the spread of infection. Staphylococci in the milk of infected cows can cause foodborne human illness. To deal with this stubborn problem, researchers at the U.S. Department of Agriculture's Bovine Functional Genomics Laboratory, Beltsville, Maryland, ventured to make cows resistant to *S. aureus* by inducing mammary gland cells to produce the enzyme lysostaphin, an antistaphylococcal protein. They obtained the lysostaphin-producing gene from another staphylococcal strain. The transgenic cows they created now secrete an antibiotic of bacterial origin in their milk, and staphylococci could not be recovered from the milk, as they were being lysed so rapidly that an infection could not occur [50]. Although they were successful, the question remains, Will the presence of the lysostaphin transgene in milk be acceptable to the public? Milk is a special food, carrying a symbolic emotional dimension that could complicate an already complex issue. Nevertheless, the USDA researchers clearly demonstrated the feasibility of introducing disease-resistant genes in cattle to confer protection against a specific disorder.

Although the unconscionable resistance to GM foods has stalled and delayed its benefits to countries that need them, there is light at the end of this troubled tunnel. India offers a shining example with its recent approval of three varieties of Bt cotton, which they have found lets farmers use less pesticide—typically one or two sprayings per harvest as opposed to three to four for conventional cotton plants. This makes it cheaper and more environmentally friendly. Reduction of pesticide use not only saves money, but far more importantly, saves lives, as was well demonstrated in China. The Chinese cultivate tens of millions of acres of cotton and had been spraying the plants with tons of organophosphate pesticides to kill the bollweevil grubs on them that feed so voraciously. Organophosphates contain chemicals similar to those in the nerve gases Sarin and Taben, which means that Chinese farmers and their families placed themselves at undue risk. The death rate among them has been so high that the government does not disclose the number. Bt cotton had to be sprayed 20–30 times between May and September, the height of the growing season, because the bollweevil worms become even more resistant to the pesticides. Use of Bt cotton now saves lives, prevents illness, and increases crop yields by as much as 50% [51]. In India, average yields exceeded those of non-Bt plants by as much as 60%. According to scientists at the Indian Institute of Science at Bangalore, "Farmers have bought it (the Bt cotton) left and right ... farmers are cleverer than the activists or the companies. They won't buy things if they do not work." Also, there is an expectation among researchers that opposition to GM crops will melt away once their home-grown research begins to deliver tangible results. India's farmers are already voting for Bt cotton by buying the seed. GM crops that are "made in India" can only get more popular [52].

Mexico is also opening up to genetically modified food with President Vincente Fox signing a Bill providing a regulatory framework for gene altered crops [53]. Mexican agriculturalists heartily approve the new law as they have been cut off from the new technology and can't compete with the United States and Canada under the North American Free Trade Agreement (NAFTA). They want the freedom to decide whether the technology is worth it. Who could argue with that?

A major research effort in Mexico that political activists have almost entirely shut down is the attempt to develop transgenic corn and rice tolerant of the high levels of aluminum naturally present in soils in many areas of Mexico, that simply stunt the growth of both crops. The new law will give the science an opportunity to flourish, and farmers an opportunity to harvest high yielding crops.

Until March 2005, Brazil was one of the last of the world's major agricultural producers not to have granted approval to plant genetically modified crops. President Inacio Lula da Silva signed the new law that pitted scientists and farmers against environmental and religious groups [54]. Be that as it may, it is estimated that about 30% of Brazil's soy crop is already grown with genetically engineered seeds brought in clandestinely from Argentina.

Despite the fact that GM food is booming in Asia, where people do not have the luxury of denial they have in Europe, perhaps the brightest light in the tunnel comes from Europe, where in May 2004, after a 6-year moratorium, the European Union, the Common Market, finally approved the importation of genetically engineered sweet corn (Bt11) developed by the Swiss Company, Syngenta. Bt11 is resistant to both the corn borer and the corn ear worm, and is also resistant to the herbicide glyphosate. But that good news comes wrapped in not-so-good news. Under European rules the corn must be labeled as genetically modified, which, given current widespread consumer resistance, would discourage potential buyers from purchasing, and food companies from even offering, it for sale. He that giveth, also taketh away. Nevertheless, this is seen as a significant breakthrough even though Friends of the Earth, a group vehemently opposed to crop biotechnology who maintain that there is simply no market for GM foods in Europe as consumers have overwhelmingly rejected them [55]. A self-fullfilling prophesy if ever there was one.

Of course, rejection has been at the top of their agenda for the past 20 years, and they have successfully achieved their goal—so far.

Perhaps the experimental field trials with transgenic rice currently underway in Watson, Missouri will help bring even more light into the GM tunnel. Jason Garst, a 35-year-old, sixth-generation farmer here, has planted 12 varieties of rice plants engineered to produce the proteins found in milk, saliva, and tears. When isolated, dried, and rendered into powder, it is anticipated that these proteins will become a healing ingredient in Granola bars and nutritional drinks, preventing infant deaths from diarrhea in third-world countries. Clearly, the transgenic possibilities are limited only by our creativity and imaginations. This project brings together in a cooperative effort, a local university, Northwest Missouri State, a small bioscience company, and the State of Missouri, to attempt to reverse the long decline in northwest Missouri's farm economy [56]. Here we have a noble humanitarian project, without a corporate CEO in sight and the critics already swinging into action trying to kill the project—aborning, hammering as they do on the health issue—that transferring a gene could imperil human health. In fact, after two decades of searching, there is a pronounced absence of any solid evidence that GM crops are in any way harmful to human health. Nevertheless, the mantra of the environmentalists, the political activists, continues to inform their constituents and the media, who continue to provide them prime coverage, that safety is the vital issue.

Before delving into the safety of GM foods, let us hear from Stewart Brand, the quintessence of environmental activism and founder of the *Whole Earth Catalog*. Musing out loud, so to speak, in a recent issue of MIT's *Technology Review*, Brand informs his acolytes, to their chagrin, of four environmental heresies, reversals of fortune, that they will not only have to accept but also support: population growth, urbanization, genetically engineered organisms, and nuclear power. Heresies, indeed! We shall deal with two: genetic modification and nuclear power. Brand maintains that "The success of the environmental movement is driven by two powerful forces; romanticism and science—often in opposition." He compares the two this way: "The romantics are moralistic, rebellious against the perceived dominant power, and combative against any who appear to stray from the true path. They hate to admit mistakes or change direction." As for scientists, "They are ethicalistic, rebellious against any perceived dominant paradigm, and combative against each other. For them, admitting mistakes is what science is." He is quite correct in stating that there are more romantics, environmentalists, than scientists, which translates into environmentalists having their say and way. Listen to this: "It means that scientific perceptions are always a minority view easily ignored, suppressed, or demonized if they don't fit the consensus story line." Brand is honest and has the self-confidence to say this out loud, even though his environmentalists have held us all in thrall these many contentious years. When they are good and ready to change things, they'll change. Not before; because they own the cat bird seat. It's galling, but he does have the troops. Why else would government agencies refer to them, acquiesce to them, request their romantic views?

Although nuclear power is some distance away, I present his views here along with biotech as they constitute the warp and woof of the total fabric of his beliefs. For Brand, climate change is the most profound environmental issue of our time. It is a disaster so profound that romantics, environmentalists must give up that ghost and genuflect at the alter of nuclear power—a reversal of fortune if even there was one. Recall that at the dawning of nuclear power, environmentalists were its strong supporters given its promise to decarbonize the atmosphere, as it produces abundant energy at high yields and at low cost—cleanly. As Brand spells it out, power derived from nuclear reactors can slow the destabilization of our planet's climate. The fact that the country, the world, has lost 40 years to the carbon of fossil fuels and its direct negative effect on climate causes no pain. It didn't have to be this way. Brand too often paints with a broad brush, placing, for example, the reactors at Three Mile Island and Chernobyl in the same class, which they absolutely are not. But it is this type of aggregation that supports the romantic and denies the science. He also speaks glibly about readily transportable nuclear plant waste, which has long been the hallmark of the environmentalists' antinuclear power efforts; moving spent nuclear fuel by truck or rail through towns and cities could, for them, be catastrophic. Now, with but a sentence, that rhetoric disappears, as it no longer serves their purpose. When James Lovelock, guru of the Greens, and an internationally recognized chemist, who by the way was the discoverer of chloroflurocarbons (CFCs) in the atmosphere, spoke up in favor of nuclear produced energy, he was suppressed, given the silent treatment—ignored, as was Green Peace cofounder Patrick Moore and Friends of the Earth Hugh Montefiore. Here again, Brand's honesty is breathtaking. The man has no shame Listen: "Public excoriation (of these outspoken environmentalists) however, would invite public debate, which so far has not been welcome." They had their agenda and they would stick with it no matter; the rest of us be damned.

His approach to biotechnology is as brazen as his new attitude about nuclear power. One area of biotech, he informs his cohorts, "with huge promise and some drawbacks is genetic engineering, so far rejected by the environmentalist movement." Can he expect any other response? Environmentalist leaders and their organizations have led an unrelenting and vicious crusade against genetic engineering. That rejection he states is a mistake. One could expect the heavens to open and a bolt of high energy to strike. These are the people who have impeded GM crop development for a decade in the most needy countries, simply because GM crops were the offspring of big corporations, and environmentalists would rather swallow hemlock than approve anything corporate. Brand has the temerity to tell his fellows to ignore the corporate and fix on the technology, given the facts that GM crops are more efficient, produce higher yields on less, and often hard-scrabble land, with less pesticide use—facts that the Amish have clearly seen, understand, and hue to, their traditional avoidance of technology notwithstanding.

Possibly the most shocking of his pronouncements pertain to the "scare stories that go around have as much substance as urban legends about toxic rat urine on coke can lids." Obviously environmentalists cannot stoop low enough. Anything goes to forward their agenda. But that is not the end of it. He is quite forthright in telling readers that many leading biologists, who double as environmentalists, have no concerns about genetically engineered organisms, but that they don't say so publicly because, "they feel that entering the GM debate would strain relations with allies and would distract from their main focus which is to research and defend biodiversity" [57]. A pox on both their houses!

We now delve into the safety of GM foods. The safety of the marker gene, the antibiotic resistance gene, as we have seen, has been given a clear bill of health. However, the Organization for Economic Cooperation and development and FDA suggest that GM crops may pose three human risks as they potentially contain allergens, toxins, or antinutrients, although these potential risks are not unique to GM foods. We humans have consumed food containing allergens, toxins, and antinutrients throughout human history. Foods generally cannot be guaranteed to pose zero risk. The concern with genetic modification is the potential for introduction of a new allergen, an enhanced toxin or antinutrient in an otherwise safe food.

Allergic Reactions

Allergic reactions are abnormal responses of the body's immune system to an otherwise safe food. Allergic reactions run the gamut from mild skin rashes and gastrointestinal upset to life-threatening anaphylactic shock. Although all food allergens are proteins, natural proteins are rarely allergenic.

To avoid introducing or enhancing an allergen in an otherwise safe food, such as soybeans, which commonly contain allergens, a GM soybean must be tested and compared to conventional soybeans. Tests require blood serum from individuals allergic to soybeans to be tested against GM soybeans. Similar reactions would except GM soybeans.

Scientists at Pioneer Hi-Bred International, the world's largest seed company, transferred the methionine-rich 2S albumin gene from Brazil nuts to soybeans, which are deficient in methionine. Brazil nuts are a known allergen for susceptible individuals—1 in 100,000. Nevertheless, the response can be severe. Pioneer developed the methionine—rich soybean for use in animal feed, but it is difficult to separate soybeans destined for animals from those for human consumption. Soy protein, which is less allergenic than milk protein, is used in infant formulas, baked products, and dairy substitutes.

Pioneer learned of the potential allergenicity of the Brazil nut gene and took the matter to the University of Nebraska's Department of Food Science and Technology. Their question was: "Would eating transgenic soy protein containing the Brazil nut gene induce allergic responses, that is, release histamine in susceptible individuals, with classic symptoms."

Working with human volunteers, departmental researchers used skin pinprick extracts of Brazil nut, soybean, and transgenic soybean to challenge them. Serum from volunteers with known allergy to Brazil nuts reacted strongly to both Brazil nut extract and the methionine-enhanced soybean. A full accounting of this research was published in the *New England Journal of Medicine* [58]. For Pioneer, the issue was whether to proceed with the enhanced soybean or discontinue the project. Available data showed less than two deaths per year from Brazil nuts. The entire project was terminated and the decision made public. Rather than praise Pioneer for their decision, environmentalists and the media used the information to dramatize the danger of biotech and corporate recklessness [59].

Toxic Reactions

Unlike allergic reactions, all humans are susceptible to toxic, poisonous substances. If a GM food contains toxic components greater than the natural range of its traditional twin, the GM food is unacceptable. To date, GM foods have proved to be no different from their conventional counterparts with respect to toxicity. In fact, in some instances there is more confidence in the safety of GM foods because naturally occurring toxins that are overlooked or brushed aside in conventional foods are measured in GM food premarket safety assessments.

Tomatine, a naturally occurring toxin in tomatoes, was largely ignored until a GM tomato was developed. The FDA considered it important to measure the potential changes in tomatine. Analyzing both the traditional and the GM tomato, they found no differences in the levels of tomatine. GM tomatoes have yet to be approved for human use.

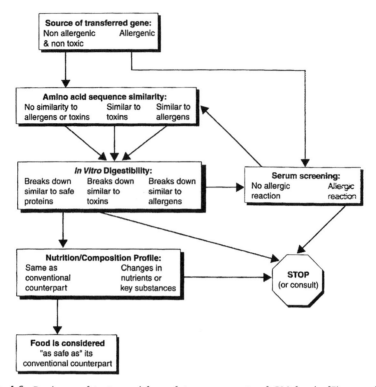

<u>Figure 4.6.</u> Regimen of tests used for safety assessments of GM foods. [Figure adapted from GAO-02-566, *Genetically Modified Foods* (GAO analysis of FDA documents).]

Antinutrients

Antinutrients are naturally occurring compounds that interfere with absorption of essential nutrients. A GM canola oil was submitted to the FDA for approval. The genetic modification had altered its fatty acid composition. To determine whether the modification had produced an unintended antinutrient effect, rendering the oil unsafe, its antinutrient composition was compared to that of traditional canola oil. Here again, the levels were similarly low.

To ensure that GM foods do not have decreased nutritional levels, nutrient composition or the nutrition profile of GM foods is determined. Tests measure amino acid level, oils, fatty acids, and vitamin content. Figure 4.6, displays the battery of tests that a GM food must traverse before obtaining FDA approval. The tests provide evidence at key decision points as to what additional tests must be performed. Tests on the source of the newly expressed protein, amino acid sequence similarity, and digestibility are typical for both allergenicity and toxicity assessments, while serum screening is used only for allergenicity assessment.

Figure 4.6 represents typical tests undertaken by a company in the safety assessment of a GM food. The figure is not meant to be a comprehensive illustration that is used in every safety assessment. Antinutrients are tested as a subset of toxicity. In addition, they are often measured with a simple nutrition/composition profile. If a company transfers genetic material from an allergenic source and undertakes serum screening tests, it does not have to go though serum screening again if in vitro digestibility tests uncover a similarity to an allergen. At such a point, it would be assessed by amino acid sequence similarity and in vitro digestibility tests for potential toxicity.

If a food company reaches the "stop (or consult)" decision point, there are clearly allergenicity or toxicity issues. This means that further tests will be discontinued and the food item removed from development. If a company transfers a gene from a nonallergen source, the amino acid sequence must also be determined. If the sequence is found to be similar to a known allergen, additional specific allergenicity testing is required, as are digestibility tests to ascertain whether the GM protein was broken down in simulated digestive fluids. Should there be concerns about the rate with which the GM protein was split, serum screening tests would be called for to support or refute the results of the digestibility tests. If the serum screening results show that the GM protein does not react with antibodies in serum, it can be concluded that the GM protein does not raise allergenicity concerns.

Checking the source of the transferred genetic material is the safety assessment starting point. Two principles undergird allergenicity testing: (1) avoid transfer of known allergenic proteins, and (2) assume that all genes transferred from allergenic sources create new allergens until proven otherwise. With both the FDA and the food industry agreeing on these principles, the probability of introducing a new allergen, enhancing a toxin or antinutrient, is exceedingly small.

The second step requires the comparison of amino acid sequences of transferred proteins and those of known allergens, toxins, or antinutrients. If an amino acid sequence in a GM food is the same as or similar to one in an allergen or toxin, there's a reasonable likelihood that the GM food poses a risk. Sequence similarity tests can be fruitful in eliminating proteins of concern.

In vitro digestibility tests are a primary component of GM food safety tests. Rapid breakdown of a GM protein in simulated human gastric fluids indicate a high likelihood that the protein is neither allergic nor toxic. Safe dietary proteins are rapidly digested; allergens and toxins are not.

If a gene raises allergenicity issues, serum screening, used only for allergenicity assessment, is required. Serum screening evaluates the reactivity of antibodies in the blood of individuals with known allergies to the plant that was the source of the gene. Serum screenings are valuable because they can expose allergens whose presence was only surmised in amino acid sequence similarity tests. Finally, nutritional and compositional profiles of GM foods are created to assess whether unexpected changes in nutrients, vitamins, proteins, fibers, starches, sugars, minerals, or fats have occurred as a consequence of genetic modification. While changes in these substances do not pose a health risk, obtaining a nutritional profile ensures that a GM food is comparable to its conventional peer.

As for long-term monitoring of potential health risks, the U.S. General Accountability Office, in its report on genetically modified foods, states that, "such monitoring is unnecessary because there is no scientific evidence, or even a hypothesis, suggesting that long-term harm such as increased cancer rates results from these foods" [59].

These many and varied tests notwithstanding, the question persists, are GM foods safe? That question appears to have been reasonably put to rest. For example, Professor Robert Paarlberg of Wellesley College reminds us that "Even in Europe, the epicenter of skepticism about genetic modification, the Research Directorate General of the European Union in 2001 released a summary of 81 separate scientific studies conducted over a 15-year period (all financed by the EU rather than private industry) finding no scientific evidence of added harm to humans or to the environment from any approved GM crops or food" [58]. Between 2002 and 2004, a clutch of European scientific organizations lined up in favor of GM foods. Professor Paarlberg informs us that "the French Academies of Sciences and Medicine drew a similar conclusion, as did the French Food Safety Agency." In May 2003, the Royal Society in London and the British Medical Association joined this consensus, followed by the Union of German Academies of Science and Humanities. Then in May 2004, the Food and Agriculture Organization (FAO) of the United Nations issued a 106-page report summarizing the evidence "that the environmental effect of the GM crops approved so far have been similar to those of conventional agricultural crops." Paarlberg goes on to say that, "as for food safety, the FAO concluded in 2004 that, 'to date no verifiable untoward toxic or nutritionally deleterious effects resulting from the consumption of foods derived from genetically modified foods have been discovered anywhere in the world'" [60].

The U.S. National Academy of Sciences and The Institute of Medicine added their voices to the growing chorus in 2004. In their report, *Safety of Genetically Engineered Foods: Approaches to Assessing Unintended Health Effects*, they state unequivocally that "The most important message from this report is that it's the product that matters, not the system you are using to produce it" [61]. They also noted that special food safety regulations are not needed just because foods are genetically engineered. Genetically modified foods have passed every test demanded of them. There is nothing Frankensteinish or malevolent about them.

What more surely can be provided? Not a great deal. Field trials and both human and animal testing will continue, and experience as with traditional food will provide continuing assurance—or it will not.

Are human health and food safety the real issues? I think not. For the Greens of Europe and the United States, the issue is political. Health and safety are smoke and mirrors, a way of keeping the public nervous, and the issue in the media. The Greens continue hoping to resolve their political concern, which is about—as they see it—nothing less than control of the world food supply. If plants are patentable, farmers could no longer save seeds from season to season. They would be forced to buy new seeds each year at increasing costs, if seeds were available to them. Farmers could be told when and what to plant. They would be captives of the seed owners. Furthermore, central Mexico has long been the world center of maize's genetic diversity. If wild Mexican maize becomes contaminated with modified genes and farmers in other countries who freely trade varieties, obtain modified maize from Mexico, will they be forced to pay royalties to the corporation that holds the patent on that modified gene, now in the maize? This patenting of genes could force farmers to lose control of their farms. In addition, many advances in gene modification were made at public universities with public funds and then given over to large corporations. For opponents of patents on DNA, the issue is, Who will control the food supply?—which is worth thinking about [62].

So, again, the biotech brouhaha is a political contretemps, not a scientific or safety issue. For the public, this cloak of invisibility would be better removed and the real case made on its merits—which appear to be considerable. Little is to be gained by frightening the public with fraudulent horror stories about the food supply. In fact, just what does the public know about genetically modified foods?

For several years the Rutgers University Food Policy Institute sought to take the pulse, as it were, of the American public's knowledge of GM foods. Sample surveys were conducted in 2001, 2003, and 2004. Here is a summary of their findings [63]:

> Most Americans have heard or read little about it, are not aware of its prevalence in their lives, and are confused as to which types of GM products are available. Respondents struggled with the factual questions related to GM food and the science behind it; could not recall news stories related to the topic, and were not very knowledgeable about laws regarding the labeling and testing of GM food. Americans are also unsure of their opinions about GM food and split in their assessments of the technology when forced to take a position.

Not a terribly encouraging scenario. Furthermore, we learn that the responses changed little between 2003 and 2004, although there has been a small but significant increase in awareness since 2001. The numbers are dismaying: 44% of respondents had heard or read nothing about GM foods, while another 42% had read something, which suggests that 86% are poorly informed. This relates to the fact that 79% believed that GM tomatoes were available for purchase in supermarkets. Of course, the first and only GM tomatoes, the Flarsavr, were taken off the market in 1997. Yet 8 years later the belief remains. More disturbing was the fact that 32% believed that, or were uncertain as to whether, by eating a GM fruit, a person's genes could also become modified. And 40% either believed or were unsure as to whether tomatoes modified by a catfish gene would taste fishy. No, not at all encouraging. If the Rutgers survey is representative of the nation, it is readily understandable why environmentalist groups have such success purveying their brand of nonsense.

For too long the public has relied on others to make weighty decisions that affect them. It is time for the people to assume that responsibility and arm themselves with the knowledge vital for decisionmaking.

What kind of reception will the public accord cultured meat? Tissue engineering to produce edible meat may or may not be a form of genetic modification, but it is bioengineering, and is currently being developed at several universities. A collaborative team of researchers from the University of Maryland, South Dakota State, The Medical University of South Carolina, and Wageningen University of The Netherlands are pursuing an unimaginably new means of food production; mass-producing meat. These researchers from departments of cell biology and agriculture are using stem cells, undifferentiated cells, from animals or tissue specimens, and allowing them to grow in nutrient solutions in a rotating bioreactor, where they develop into muscle cells that form skeletal muscle tissue, myofibers. This tissue is then harvested and processed into a variety of meat products. If this method of producing sausages, meat patties, and beef and pork loins is perfected, waste would vanish, as would foodborne disease. World hunger would decrease markedly, and there would no longer be a need to slaughter cattle. "Remarkable" is an understatement. But it will require 10–15 years of further development to move from the laboratory to the marketplace [64]. Will the public accept so nontraditional a departure in a food? That's arguable, but what is not, is the fact of widespread scientific creativity and the availability of biologic and genetic engineering tools. Down the road expect a clutch of novel foods—to feed an escalating population.

REFERENCES

1. Bannerot, S., and Bannerot, W., Health underway: An unexpected intruder, *Cruising World* 94–102 (Sept. 2001).
2. Becker, K., Southwick, K., Reardon, J. et al., Histamine poisoning associated with eating tuna burgers, *JAMA* **285**(10):1327–1330 (2001).
3. Dyckman, L. J. D., *Federal Food Safety and Security System*, U.S. General Accountability Office, GAO-04-588T, Washington DC, March 30, 2004.
4. Cullen, J. C., The miracle of Bolsena, *Am. Soc. Microbiol. News* **60**(4):187–191 (1994).
5. Merlino, C. P., Bartolomeo Bizio's letter to the Most Eminent Priest, Angelo Bellani, concerning the pheneomenon of the red colored polenta (1823), *J. Bacteriol.* **9**(6):527–543 (1924).
6. Evans, M. R., Ribiero, C. D., and Salmon, R. L., Hazards of healthy living: Bottled water and salad vegetables as risk factors for Campylobacter infection, *Emerg. Infect. Dis.* **9**(10):1–13 (2003).

7. Fiore, A. E., Hepatitis A: transmission by food, *Clin. Infect. Dis.* **38**:705–715 (2004).

8. Grady, D., Produce items are vulnerable to biological contamination, *New York Times* 3 (Nov. 18, 2003).

9. Woteki, C. E., and Kineman, B. D., Challenges and approaches to reducing food-borne illness, *Annu. Rev. Nutr.* **23**:315–344 (2003).

10. Santora, M., Tuberculosis cases prompt warning on raw-milk cheese, *New York Times* B4 (March 16, 2005).

11. U.S. General Accountability Office, *Food Irradiation. Available Research Indicates that Benefits Outweigh Risks*, GAO-RCED-00-217, Washington, DC, Aug. 24, 2000.

12. Raul, F., Gosse, F., Delincee, H. et al., Food-borne radiolytic compounds (2-alkylcyclobutanones) may promote experimental colon carcinogenesis, *Nutr. Cancer* **44**(2):188–191 (2002).

13. *Irradiation of Food*, Report 4 (I-93), American Medical Association, Council on Scientific Affairs (Ref. Comm. E), Chicago, adapted 1992.

14. Position paper: Food irradiation, *J. Am. Diet. Assoc.* **100**(2):246–253 (2000).

15. (a) Irradiation in the production, processing and handling of food: Safe use of ionizing radiation for the reduction of salmonella in fresh shell eggs, *Fed. Reg.* **65**(141) (July 21, 2000). (b) Abdul-Raouf, U. M., Beuchat, L. R., and Ammar, M. S., Survival and growth of *Escherichia coli* O157:H7 on salad vegetables, *Appl. Environ. Micro.* **59**(7):1999–2006 (1993).

16. Caudill Seed Co., Inc., Safe use to control pathogens in alfalfa (filing for food additive petition). *Fed. Reg.* **65** (64605–07) October 30, 2000.

17. Cejas, P., Casado, E., Belda-Iniesta, C. et al., Implications of oxidative stress and cell membrane lipid peroxidation in human cancer (Spain), *Cancer Causes Control* **15**:707–719 (2004).

18. Hore, A., The cost of free radicals, *Sci. Culture* **70**(7–8):249–254 (2004).

19. Prior, R. L., Biochemical measures of antioxidant status, *Top. Clin. Nutr.* **19**(3):226–238 (2004).

20. Smith-Warner, S. A., Spiegelman, D., Yaun, S. S. et al., Fruits vegetables and lung cancer: A pooled analysis of cohort studies, *Intl. J. Cancer* **107**:1001–1011 (2003).

21. Hung, H. C., Joshipura, K. J., Jiang, R. et al., Fruit and vegetable intake and risk of major chronic disease, *J. Natl Cancer Inst.* **96**(21):1577–1584 (2004).

22. Van Gils, C. H., Peeters, P. H. M., and 41 others, Consumption of vegetables and fruits and risk of breast cancer, *JAMA* **293**(2):183–193 (2005).

23. Edelman, R. S., Hollar, D., and Hebert, P. R., Randonized trials of vitamin E in the treatment and prevention of cardiovascular disease, *Arch. Intern. Med.* **164**:1552–1556 (2004).

24. Lonn, E., Bosch, J., Yusuf, S. et al., Effects of long-term vitamin E supplementation on cardiovascular events and cancer, *JAMA* **293**(911):1338–1347 (2005).

25. Cook, N. R., Gaziano, J. M., Lee, I. M., et al., Vitamin E in the primary prevention of cardiovascular disease and cancer, *JAMA* **294**(1):56–65 (2005).

26. Etminan, M., Takkouche, B., and Caammo-Isorna, F., The role of tomato products and lycopene in the prevention of prostate cancer: A meta-analysis of observational studies, *Cancer Epidemiol. Biomark. Prevent* **13**(3):340–345 (2004).

27. Virtamo, J., and seven others, Incidence of cancer and mortality following alpha-tocopherol and beta-carotene supplementation, *JAMA* **290**(4):476–485 (2003).

28. Bjelakovic, G., Nikolova, D., Simonetti, R. G., and Sluud, C., Antioxidant supplements for prevention of gastrointestinal cancers, *Lancet* **364**:1219–1228 (2004).

29. Morris, C. D., and Carson, S., Routine vitamin supplementation to prevent cardiovascular disease, *Ann. Intern. Med.* **139**:56–70 (2003).

30. Bischoff-Ferrari, H. A., Willet, W. C., Wong, J. B. et al. Fracture prevention with vitamin D supplementation, *JAMA* **293**(18):2257–2264 (2005).

31. Ames, B. N., Shigenaga, M. K., and Hagen, T. M., Oxidants, antioxidants, and the degenerative diseases of aging, *Proc. Natl. Acad. Sci. USA* **90**:7915–7922 (1993).

32. Lundberg. J. O., Weitzberg, E., Cole, J. A., and Benjamin, N., Nitrate, bacteria and human health, *Nature Rev. Microbiol.* **2**:593–600 (2004).

33. Kujoth, G. C., Hiona, H., Pugh, T. D. (and 15 others), Mitochondrial DNA mutations, oxidative stress, and apoptosis in mammalian aging, *Science* **309**:481–484 (2005).

34. Murch, S. J., Cox, P. A., Banack, S. A. et al., Occurrence of B-methylamino-L-alanine (BMAA) in ALS/PDC patients from Guam, *Acta Neurol. Scand.* **110**:267–269 (2004).

35. Banack, S. A., and Cox, P. A., Biomagnification of cycad neurotoxins in flying foxes, *Neurology* **61**:387–389 (2003).

36. Banack, S. A., and Cox, P. A., Distribution of the neurotoxic non-protein amino acid BMAA in Cycas micronesica, *Botan. J. Linnean Soc.* **143**:165–168 (2003).

37. Rawn, D. F. K., Cao, X.-L., Doucet, J. et al., Canadian total diet study in 1998: Pesticide levels in foods from White Horse, Yukon, Canada, and corresponding dietary intake estimates, *Food Addit. Contam.* **21**(3):232–250 (2004).

38. *Questions about Pesticides in Food. Extonet. FAQ's*, Jan. 1998 (at http://extonet.orst.edu/taqs/pesticide/pestfood.htm, accessed Feb. 10, 2005).

39. Ito, N., Hasegawa, R., Imaida, K. et al., Effect of ingestion of 20 pesticides in combination at acceptable daily intake levels on rat liver carcinogenesis, *Food Chem. Tox.* **33**(2):159–163 (1995).

40. Ames, B. N., and Gold, I. T., Environmental pollution, pesticides, and the prevention of cancer, misconceptions, *FASEB J.* **11**:1041–1052 (1997).

41. Media Coverage of Controversies Regarding pesticides—Alar. www.entomology.cornell.edu/faculty—staff/scolt/370 accessed 4/28/07. See also Alar use on apples. FDA. www.fda.gov/bbs/topics/news/new00128.html.

42. Wildavsky. A., *But Is It True: A Citizen's Guide to Environmental Health and Safety Issues*, Harvard Univ. Press, Cambridge, MA, March 1997.

43. Ishiwata, S., On a severe softening of silk-worms (Sotto disease), *Dainihan Sanbishi Kaiho* **9**(114):1–5 (1901).

44. Berliner, E., Uber die Schlasfsucht der Mehlmottenraupe (Ephestia kuhniella Zelli) und ihren Erreger Bacillus thuringiensis n.sp., *Z. Angewan. Entomologie* **2**:29–56 (1915).

45. Goldman, K. A., Bioengineered food—safety and labeling, *Science* **290**:457–459 (2000).

46. *Genetic Modification and Food*, Working Party, British Society for Antimicrobial Chemotherapy, Feb. 2004. (http://www.bsac.org.uk/).

47. Rauch, J., Will Frankenfood save the planet? *Atlantic Monthly*103–108 (Oct. 2003).

48. Qaim, M., and Zilberman, D., Yield effects of genetically modified crops in developing countries, *Science* **299**:900–902 (2003).

49. Paine, J. A., Shipton, C. A., Chaggar, S. et al., Improving the nutritional value of golden rice through increased pro-vitamin A content, *Nature Biotechnol.* **23**(4):482–487 (2005).

50. Wall, R. J., Powell, A. M., Paape, M. J. et al., Genetically enhanced cows resist intra mammary Staphylococcus aureus infection, *Nature Biotechnol.* **23**(4):445–451 (2005).

51. Smith, C. S., China rushes to adopt genetically modified crops, *New York Times* A3 (Oct. 7, 2003).

52. Randerson, J., By the people, for the people, *New Sci.* 36–37 (Feb. 19, 2005).

53. Malkin, E., Science vs. culture in Mexico's corn staple, *New York Times* 10 (March 27, 2005).

54. Benson, T., Brazil passes law allowing crops with modified genes, *New York Times* (March 4, 2005); see also A harvest in peril, *New York Times* C6 (Jan. 6, 2005).

55. Meller, P., and Pollack, A., Europeans appear ready to approve a biotech corn, *New York Times* D1–D3 (May 15, 2004).

56. Barrioneuvo, A., Fields of bio-engineered dreams, *New York Times* C1 (Aug. 16, 2005).

57. Brand, S., Environmental heresies, *Technol. Rev.* 60–63 (May, 2005).

58. Nordlee, J. H., Taylor, S. L., Townsend, J. A. et al., Identification of a Brazil-nut allergen in transgenic soybeans, *NEJM* **334**(11):688–692 (1996).

59. *Genetically Modified Foods*, U.S. Government Office of Accountability, GAO-02-566, Washington DC, May 2002.

60. Paarlberg, R., Genetically modified crops, *Issues Sci. Technol.* 30–32 (spring 2005); A response to: Agricultural biotechnology: Overregulated and under appreciated. (winter 2005).

61. *Safety of Genetically Engineered Foods: Approaches to Assessing Unintended Health Effects*, National Research Council and Institute of Medicine. National Academy Press, Washington, DC, July 2004.

62. Cayford, J., Breeding sanity into the GM food debate, *Issues Sci. Technol.* 49–56 (winter 2004).

63. Hallman, W. K., Hebden, C. W., Cuite, C. L. et al., *Americans and GM Foods: Knowledge, Opinion and Interest in 2004*, Publication RR-1104-007, New Brunswick, NJ, Food Policy Institute, Cook College, Rutgers—State Univ. New Jersey.

64. Edelman, P. D., McFarland, D. C., Mironov, V. A., and Matheny, J. G., In-vitro cultured meat production, *Tissue Eng.* **11**(5/6):659–662 (2005).

5

CLIMATE CHANGE AND GLOBAL WARMING: WILL THE GREENHOUSE BECOME A HOTHOUSE?

> To know that we know what we know, and to know that we do not
> know what we do not know, that is true knowledge.
> —*Copernicus*

The spring and summer of 2005 saw the worst drought in 20 years in states of the American Midwest. That year (2005) appears to have set a record that might just bump 1998 from its position of prominence as the hottest year on record.

The 10 hottest years on record have occurred since 1991; 1998 was the hottest, with 2002, 2003, and 2004 in second, third, and fourth positions, respectively, since systematic temperature measurements were initiated in the mid-1800s. The summer of 2003 will not soon be forgotten in Europe. Some 15,000 people died from the sweltering heat in France, and London was scorched under multiple hundred plus degree days, which hadn't occurred since 1500. It is now well documented that over the past 1000 years, temperatures have not been as high as those of the past century [1].

Is our normally stable, salubrious climate ending? Are we moving into a warmer world? A key conclusion of the United Nation's Intergovernmental Panel on Climate Change's (IPCC) *Second Annual Report* was that "the increase in temperature in the 20th century is likely to have been the largest in any century during the past 1000 years." And they also noted that, "the balance of evidence suggests that there is a discernable human influence on global climate" [2]. These conclusions raise yet another issue: whether human activity can perturb the forces of nature. This issue is at once pressing, and reasonable given the dire consequences for humankind should we possess such power.

The inimical effects of human activity were not lost on nineteenth-century observers cognizant of the human condition. George Perkins Marsh (1801–1882), author of *The Earth as Modified by Human Activity* (1874), was certain that where man is concerned, "there is scarcely any assignable limit to his present and prospective voluntary controlling power over terrestrial nature." And it was the Congregational Minister, Horace Busnhell (1802–1876), who thundered from his pulpit that "not all the works, and storms, and earthquakes and seas, and seasons of the world have done so much to revolutionize the earth as man has done since the day he came upon it, and received dominion over it." Indeed, we have dug deeply, re-routed and dammed rivers, sent men to the moon, and spaceships toward the heavens, but altering climate? Do we possess such might? If the IPCC is correct, we are unimaginably powerful. But the consequences of a far warmer world signal that we may have gone too far. Can we reverse the shift? These and several additional questions require pursuing. What, in fact, is climate change, or as some prefer to call it, global warming? How did it come to pass after 12,000 stable years, and how will a warmer world affect our lives?

GREENHOUSE AND OTHER ATMOSPHERIC GASES

A Beneficent Greenhouse

Since the retreat of the last ice age, some 12,000 years ago, we have enjoyed the benevolence of a virtual greenhouse. A greenhouse or hothouse is a glass-enclosed structure devoted to the cultivation of flowers and plants—usually out of season—that's the key, out of season, because greenhouses are warm places, all year long. The glass of a greenhouse permits passage of the visible, shortwave radiation coming from the sun, but absorbs the reradiated longer infrared radiation, which warms the otherwise cool space. The incoming light (heat/energy) falls on the objects inside, warming them and raising their temperature. Our natural greenhouse, or greenhouse effect, is a highly salutary condition found only around planet Earth. Venus is a hothouse, a very hothouse, and Mars is a veritable icehouse.

Radiation, light, from the sun hurtling through space penetrates a blanket of gases before striking the earth. For some 300 miles from ground zero, the earth is sheathed in an envelope of gases, an actual protective cocoon, its atmosphere. This atmosphere consists of the radioactively active gases carbon dioxide (CO_2), methane (CH_4), nitrous oxide (N_2O), and water vapor, the so-called greenhouse gases, and is transparent to the incoming visible, shortwave solar radiation. For a comprehensive understanding of the greenhouse effect, the earth's energy budget must be known. This budget, like any budget, will have its credit and debit sides—energy in, energy out. The energy, heat, coming in from our closest star, the sun, heats the earth such that each square meter receives an average of 342 watts per square meter (W/m^2) of solar radiation, but 30% is immediately reflected back as the longer, less energetic infrared radiation. The remaining $235\,W/m^2$ is absorbed by the gaseous atmospheric "blanket"; $168\,W/m^2$ warms the earth's land and ocean. Much of the solar radiation is absorbed by both the organic (living) and inorganic (nonliving) materials, resulting in a transformation into thermal energy that increases the temperature of the absorbing surfaces as well as the air in contact with these surfaces. For a stable climate, a balance is required between incoming and outgoing radiation. Consequently the earth must reradiate an average of $235\,W/m^2$ back into space. Figure 5.1 describes this energy balance, with incoming energy on the right, and outgoing on the left.

Although the greenhouse gases (GHGs) are transparent to the incoming visible light, they are opaque to the reflected infrared. In effect, the greenhouse is actually a retention of heat energy by the earth's atmosphere, but the GHGs rereflect a portion of this retained heat earthward, while another portion of the infrared escapes back into space. This is and has been the natural order of the earth's balanced heat budget, and has worked in our favor for thousands of years until recently, when the balance began tilting in the direction of more heat being received than being reflected back into space.

The unusual buildup of carbon dioxide, as well as the other non–carbon dioxide GHGs in the atmosphere traps more of the infrared that would normally escape into space, rereflecting it back to earth. This additional heat, this

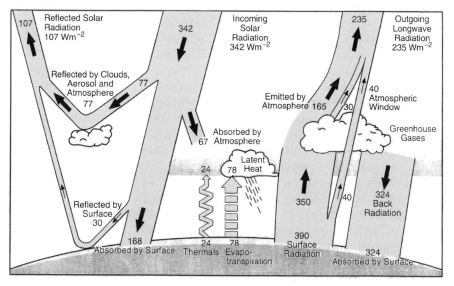

Figure 5.1. The earth's annual and global mean energy balance (follow the arrows).

CO_2 forcing has been occurring over the past 200 years, but much more so over the past 100 years, and the earth's average temperature has risen along with it.

Dr. James Hansen, Director of NASA's Goddard Institute of Space Sciences, and his team have calculated that "these human-made gases were (are) heating the earth's surface at a rate of almost two watts per square meter. A miniature Christmas tree bulb delivers about one watt. So it is as if we humans had placed two of these tiny bulbs over every square meter of the earth's surface, glowing 24/7" [3]. Was it—is it—possible that such minuscule amounts of heat could tweak—or more to the point, destabilize—the awesome forces of nature? Indeed, they could. We now know that small forces maintained over time can induce climate change. Again, Hansen informs us that "the earth is now absorbing nearly 1 watt per square meter more energy from the sun than it is emitting to space, portending further warming even if GHG levels were immediately stabilized" [4].

The greenhouse has been our benefactor these past 12,000 years, maintaining a nicely balanced energy budget—equal amounts of heat coming in and going out. If CO_2 were absent, the energy balance would tilt to the cold side and our earth's temperature would plunge some 70°F (35°C) below our current worldwide average of 14–15°C (58–59°F), which would bring us to a noninhabitable –18°C (–0.4°F), well below the freezing point of water. On the other hand, if the blanket of GHGs swaddling us grew so thick, so deep, preventing all infrared from passing into space, our planet would be another Venus, another red-hot planet, far too hot for life as we have come to enjoy it. We have been fortunate. The quartet of GHGs have been nature's thermostat, automatically set for a splendid balmy temperature—on average. It is difficult to conceive that we are prepared to shut it down. The numbers suggest that we are.

Prior to 1850, the concentration of CO_2 in the atmosphere was approximately 280 ppm. Currently it is 380 ppm—an increase of 35% in 156 years. To edge this a bit more starkly, in 1958 it was 315 ppm, an absolute increase of 65 ppm, or 17% in 48 years. Carbon dioxide is increasing, continuously increasing; that's the rub. But this should not come as a surprise. Who knew? A number of people knew. The sequence of events in which the shorter, visible-light radiation is absorbed by the earth and reradiated as the longer, less energetic infrared, trapped by the naturally occurring GHGs, was referred to as a "greenhouse effect" in 1827 by Jean-Baptiste Joseph Fourier, a French mathematician. But it was from a series of ingenious experiments conducted by John Tyndall, an Irish physicist, that water vapor was found to have a great absorptive capacity for solar radiation. Assigning a value of 1 to dried air, he calculated the absorption of ambient air containing water vapor and carbon dioxide as 15. Deducting the effect of CO_2, he showed the absorption of water vapor to be 13 times greater than air. In his published work Tyndall noted that [5]

> If as the above experiments indicate, the chief influence be exercised by the aqueous vapor every variation of this constituent must produce a change of climate. Similar remarks would apply to the

carbonic acid diffused through the air . . . a slight change in fact may have produced all the mutations of climate which the researches of geologists reveal . . . and they establish the existence of enormous differences among gases and vapors as to their action upon radiant heat."

While not referring to it by name, Tyndall had revealed the existence of a greenhouse 146 years ago. Humankind however, can be forgiven for taking little notice of this esoteric discovery. Not so with the Swedish Chemist Svante August Arrhenius. By 1896, Arrhenius recognized that carbon dioxide allows solar radiation to pass unimpeded through the atmosphere but captures a portion of energy upon its re-radiation from earth [6]. Without benefit of a computer he calculated that doubling CO_2 would produce a 6°F (3.3°C) increase in mean global temperature.

Laying out his calculations in the *Quarterly Journal of the Royal Meteorological Society* for 1938, George S. Callender, a British engineer, who believed that the world was warming, estimated, again without benefit of a computer, a 2°F (1.1°C) increase in air temperature for a doubling of carbon dioxide, and identifying the burning of fossil fuels as the culprit [7]. But concern was elsewhere. Two decades later Gilbert N. Plass of Johns Hopkins University took up the question and categorically established CO_2 as the major greenhouse gas. Listen to him [8]: "If the total CO_2 amount is reduced 50 percent or less of its present value, then a permanent period of glaciation results until the total CO_2 amount again increases." He also contended that "The burning of fossil fuel . . . had greatly disturbed the CO_2 balance. If all this additional CO_2 remains in the atmosphere there will be 30 percent more CO_2 in the atmosphere at the end of the 20th century than at the beginning. Man's activities are increasing the average temperature by 1.1°C per century" [8]. Plass was way ahead of the pack. But there was no urgency. Plass fingered both fossil fuel and those who were burning it. Unfortunately there was no critical mess prompting him to act on it—until 1988, when James E. Hansen sounded the tocsin. Here was a call to arms, and it came at a propitious moment. It was hot. Weeks of drought had parched the American grain belt, and the north and southeast were in the throes of an extended hot spell. Hansen, speaking with a force and authority rarely encountered in an established scientist, not only stated that the world was getting warmer, but that "he was 99 percent certain that the accumulation of greenhouse gases was responsible for the warming trend." Hansen had the heat working for him. The media embraced his account, and the threat of an overheated greenhouse raced around the world. Global warming moved from academic journals to the fast lanes of the electronic media, and "greenhouse" took on a new meaning—and there was a splendid curve—a trendline to back up the caution that atmospheric temperature was rising along with CO_2 on the greenhouse.

In 1958, Charles D. Keeling and Robert B. Bacastow of Scripps Institute of Oceanography, together with the National Oceanic and Atmospheric Administration (NOAA), established a CO_2 monitoring station on the summit of

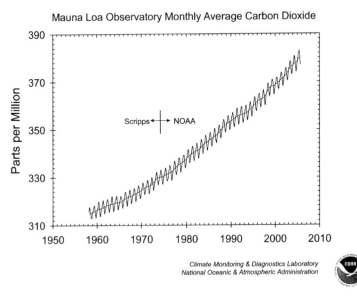

Mauna Loa Observatory Monthly Average Carbon Dioxide

Climate Monitoring & Diagnostics Laboratory
National Oceanic & Atmospheric Administration

Figure 5.2. Monthly average global carbon dioxide levels, 1979–2005. (Figure courtesy of Climate Monitoring & Diagnostics Laboratory, National Oceanic & Atmospheric Administration.)

Mauna Loa, a dormant volcano on the island of Hawaii, well isolated from industrial areas and local air pollution. From 1955 on, Keelings' group collected air samples and analyzed them for carbon dioxide. Figure 5.2 depicts the rendering of monthly mean CO_2 levels and clearly shows the steady, upward trend. In fact, each of the curve's data points is the residual of thousands of analyses. This sawtoothed projection reflects the fact of the warmth of spring and summer with plants and trees absorbing CO_2 and bursting into bloom as CO_2 is removed from the atmosphere during their photosynthetic cycle. In fall and winter, respiration occurs; leaves fall, flowers wither and die; fruit disappears, as all green things exude CO_2. Each leg of the graph presents a half-year's reading. The descending leg represents the photosynthetic absorption and sequestering of CO_2, while the ascending leg represents respiratory activity.

At the outset (in 1958), readings of 315 ppm were obtained. Twenty years later they had arrived at 330 ppm, and by 2004, at 380. Obviously, and unfortunately, an increasing rate of increase at work. Atmospheric CO_2 is increasing faster than ever, which parallels the increased use of fossil fuels around the world, especially in China, India, and the United States. The Mauna Loa curve has been corroborated and verified by an additional station in Antarctica. The manometers have been running continuously for 47 years. The data are solid. Along with these worrisome, steadily increasing CO_2 levels, the IPCC has documented a 0.6°C rise in temperature at the surface of the earth. And

Dr. Hansen recently remarked that "It became clear very quickly that his [Keeling's] measured CO_2 increase was proportional to fossil fuel emissions and that humans were the source of the change. He altered our perspective about the degree to which the earth can absorb the human assault." Dr. Keeling will be missed. He died at age 77, on June 20, 2005. His work goes on [9].

Clearly, CO_2 has been closely monitored, but what of the other non-CO_2 GHGs that interrupt the passage of infrared? What is their contribution, and what of their levels? Are they rising along with CO_2? If they are, it would mean that they, too, would be adding to radiative forcing, and any factor that alters the radiation from the sun can affect climate. A change in the net radiative energy available is referred to as a "forcing." Positive forcings warm the earth; negative, cool it. Let us then consider each of the half-dozen gases that collectively appear to be enhancing the greenhouse.

Carbon Dioxide

Life as we know it would cease to exist without the carbon that is a constituent of all things living. When the fossil fuels coal, oil, peat, and natural gas (all carbonaceous substances) are burned, the major product of their combustion is carbon dioxide, which is at the care of the warming wrangle. Over the past 100 years, disproportionate amounts of CO_2 from tailpipe emissions, smokestacks, cement production, and fires, have been pouring into the atmosphere, and returning to it the carbon that had been removed by trees and plants hundreds of millions of years ago. Suddenly, over two generations, our beneficent greenhouse is being sorely taxed by vast quantities of a strong IR absorber. Recall that an increase in the level of atmospheric CO_2 tends to close the infrared "window" as there are now far more molecules of CO_2 absorbing the radiation, thereby blocking its escape into space. Furthermore, the rate of accumulation is increasing far faster than before. So, for example, in the year 2000, the level of CO_2 was 366 ppm, but by 2004 had risen to 376, and 380 in 2005. Climate forcing via CO_2 is the largest of all forcings, and the tremendous worldwide demand and use of coal and oil are becoming equal sources of CO_2 emissions [10]. Coal use is increasing in the United States, China, and India along with increases in population and international commerce. But for number's sake, comprehensible numbers, consider the contribution to the greenhouse the next time you take a friend for a joy ride. Your car's tailpipe emits 19–20 lb of CO_2 per gallon of gas. With a 20-gallon gastank, using a tankful a week, you're emitting 400 lb of CO_2 weekly. At 50 tankfuls per year, each of our cars is pumping out 20,000 lb per year. Multiply 20,000 by the 219 million registered vehicles currently on the road, and the atmosphere becomes loaded with 4,380,000,000,000 pounds, or 2,190,000,000 tons of CO_2 annually, and that's only for the United States. On a worldwide basis, especially now with ever more cars and trucks in use in China, Indonesia, Vietnam, Korea, and India, the excess is nothing short of Brobdingnagian. Think atto—that's

10 accompanied by 18 zeros. Of course, this does not include the contribution of coal-fired electric power generating plants, more than likely the heaviest polluters of all. But we're not yet done. In their reports, the IPCC speaks of three major CO_2 contributors: fossil fuels, cement production, and land-use conversion. Cement? Can so dense a material contribute to enhancement of the greenhouse? What is cement? Cement is no more than a fine gray powder made of a mixture of limestone, sand, and water. So, what's the problem here? Limestone is a sedimentary rock consisting primarily of calcium carbonate. Recalling a tad of high school chemistry, and decomposition reactions, we know that given the right amount of heat, calcium carbonate decomposes to calcium oxide and CO_2. For cement and limestone, calcining, heating limestone to 1100–1600°F (600–900°C), drives off moisture and produces calcium oxide and CO_2—lots of it in the huge, worldwide cement and concrete industry; gigatons of it. Given the amount of sedimentary rock available, limestone is probably inexhaustible. It is often remarked that a country's progress is measured in tons of cement used. Imagine the ensuing bedlam should it even be hinted that the use of cement and concrete be trimmed.

Land-Use Conversion

The earth's great carbon dioxide stonehouses, its tropical forests, stretch in a belt along the equator from Brazil and the Amazon, through Zaire, to the Democratic Republic of the Congo, Malaysia, and Indonesia. The areas involved are immense. The term "land-use conversion" is a circumlocution, a euphemism, for *deforestation*: the replacement of natural forests with roads, cattle ranches, cropland, and lumber production. As with limestone, these forests are yet another of our great CO_2 repositories—ripe for the ravaging and release of CO_2. What does it mean to say that tens of thousands of square miles of forest disappear annually?

In Peru, 500,000 acres of forest, 780 square miles, 12 times the size of Washington, DC, were cleared for cocoa crops to satisfy the world's craving for cocaine. Deforestation contributes to additional warming in several ways. The cleared areas become more reflecting of infrared back into the greenhouse, and there is a concomitant loss of CO_2 absorptive capacity with the loss of trees. Cutting trees also removes moisture, which causes clouds to develop, playing a major role in the earth's radiation budget. With moisture and cloud cover gone over large areas, these losses could mean adding additional weeks or months to already dry seasons. The large contribution of CO_2 to the greenhouse notwithstanding, the combination of other gases may be equal to, if not greater than, a cause of observed global warming. So, for example, methane (CH_4), the simplest carbon-containing molecule, can now be measured and its concentration accurately determined by the "scanning imaging absorption spectrophotometer for atmospheric chartography" (SCIAMACHY) on board the European Space Agency's research satellite, ENVISAT [11].

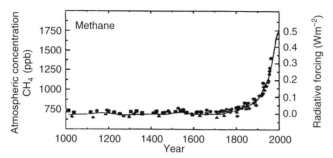

Figure 5.3. Atmospheric methane concentration during the years 1000–2000. [Figure reproduced with the kind permission of the Intergovernmental Panel on Climate Change (IPCC).]

Although the simplest carbon-containing molecule, it is the second most important anthropogenic GHG. More than 50% of current global methane emissions are the work of human pursuits. Curiously enough, although the presence of CO_2 became known by the late eighteenth century, methane was not detected until 1948. Like CO_2, methane, "marsh gas," or, as miners refer to it, "fire damp," methane is not only a strong IR absorber but is also, in fact, 20 times more absorbing than CO_2. But for its presence in the atmosphere at a minuscule 1.7–2.0%, it would be the major IR absorber. From its first analytic measurements in the late 1960s to the most current, it has been found, as shown in Figure 5.3, to be rising steadily. Records in the Antarctic ice cores indicate that since 1750 its concentration in the atmosphere has increased over 150%, and as the trendline so clearly shows, its rise has been almost exponential since the 1920s. Rice cultivation, the prime contributor, has more than doubled since 1950, and will surely continue to increase as the demand for rice keeps pace with population growth. The anaerobic conditions indigenous to rice paddy flooding results in the formation of methane. More than 60% of all the world's rice paddies are found in China and India, where unfortunately data on emissions are unavailable.

In addition to rice production, anaerobic microbial metabolism in the gut of ruminant cattle, sheep, goats, buffalo, and camel, of which there are well over a billion in the world, are heavy emitters of methane. Management of ruminant diets is a high-priority item for those concerned with cutting greenhouse emissions, but, as you may suspect, it is a complex and uncertain undertaking. Yet other contributions come from termites, whose gut is also filled with methane-producing bacteria. Land-use changes, including deforestation, increases land available for incursion by termites, whose populations have also been increasing. In addition, methane emissions arise from landfills, coal mining, and oil and gas extraction. If that were not enough, methane magnifies the effect of CO_2 when chemically reacting with the hydroxyl radical OH, to yield CO_2 and water. Between 1999 and 2002, CH_4 appears to have been increasing at a steady state of 1751 parts per billion (ppb). Indeed, methane remains a prime climate forcer even with modifications in rice farming (cultivar choice, irrigation management, and fertilization) as well as efforts to capture methane at landfills and in mining operations [10].

Moreover, as we shall see, methane, along with the other non-CO_2 GHGs, may emerge as the primary cause of global warming. However, reductions in these gases will require international cooperation at a level that may be extremely unlikely, unless warming ratches up a notch or two and uncertainty recedes. Even if international cooperation were to occur, there is a methane source that, if emitted, could nullify calculations and all reduction efforts.

Methane trapped in marine sediments as hydrates can become an immense energy reservoir, and also a warming time bomb if suddenly released by natural forces [12].

Gas hydrate is a crystalline solid consisting of methane gas surrounded by water molecules, looking much like water ice, and is stable in ocean floor sediments at depths around 1000 feet. The global levels of carbon bound in gas hydrates is conservatively estimated at twice the total amount of carbon in all known fossil fuels on earth. Colossal would not be an overstatement. Consequently, extraction of methane from hydrates could provide an enormous energy resource. Furthermore, methane bound in hydrates appears to be some 3000 times the volume of methane in the atmosphere. There's the rub. There's a great lack of information to judge what geologic processes might affect the stability of hydrates and the possible release of methane to the atmosphere. Methane released as a result of landslides caused by sea-level fall could easily and seriously warm the earth, as would methane released in arctic sediments as they become warmed during sea-level rise. Such releases could exacerbate climatic warming and surely destabilize the climate.

Nitrous Oxide

Nitrous oxide (N_2O) is produced both naturally and via human intervention. Natural sources of N_2O result primarily from bacterial activity in soils. In the process of denitrification, the stripping of oxygen from nitrate, bacteria such as *Pseudamonas denitrificans* and *Nitrobacter* release both nitrous oxide and nitrogen gas to the atmosphere. Tropical soils, far more productive of N_2O than temperate soils, add about 4 million tons of nitrous oxide to the atmosphere annually. Synthetic (human-made) emissions, many and varied, account for some 40% of the total of 15 million tons added each year. During fossil fuel combustion, oxygen and nitrogen combine to produce N_2O; and during the production of nylon (the generic term for all synthetic polyamides fabricated into all manner of tubes, pipes, filaments, and coatings), adipic acid is produced, as is N_2O, as a byproduct from the oxidation of nitric acid and cyclohexanol. Approximately 0.3 metric tons of N_2O are released per metric ton of adipic acid produced. Considering that some 5 billion pounds are manufactured worldwide, N_2O is emitted at appreciable levels.

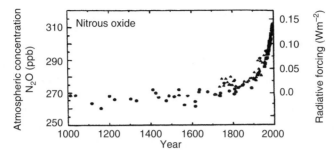

Figure 5.4. Atmospheric nitrous oxide concentrations during the years 1000–2000. [Figure reproduced with the kind permission of the Intergovernmental Panel on Climate Change (IPCC).]

The increasing application of nitrogen-based fertilizers is giving rise to ballooning amounts of N_2O, as is the increasing production of nitric acid, a major ingredient in explosives and nylon, which is manufactured by the catalytic oxidation of ammonia, a process that produces N_2O as a byproduct. The application of fertilizers and livestock manure to pasture and cropland adds significant levels of nitrogen to soils that will be converted to N_2O via bacterial metabolism. Also, although nitrous oxide has a short retention time in the atmosphere as it breaks down readily in the presence of sunlight, its continued and increasing abundance makes it a substantial climate forcer. As shown in Figure 5.4, it, too, has had a substantial increase since the mid-1950s. Perhaps of most concern is the fact that although currently N_2O contributes about $0.1\,\text{W/m}^2$ to the total forcing, it is a strong IR absorber—200 times more absorbing than CO_2. It is its additive contribution together with that of methane, ozone, and the halocarbons that provides the largest climate forcing [4].

Halocarbons

Halocarbons are compounds containing carbon and halogen atoms—chlorine, bromine, iodine, and fluorine. The most notorious of the halocarbons are the chlorofluorocarbons (CFCs), of which there are a large family. If any synthetic chemical could have been said to be perfect, it would be the CFCs. They were tailormade: a response to the need to replace ammonia, sulfur dioxide, and methylchloride, then (in 1928) common household refrigerants, with a far less explosive, corrosive, and nontoxic substance. Frigidaire (at the time a subsidiary of General Motors) rose to the challenge. In 2 years they solved the problem and delivered chloroflurocarbon—chemically stable, inert, nonflammable, nonexplosive, and nontoxic. Who could ask for anything more? Also, their use in refrigerators substantially reduced electric power demand, which meant less combustion of fossil fuel. These Freons, the name DuPont dubbed them, made automobile air conditioning possible, as well as aerosol spray cans, solvents, and blowing agents for packaging materials and foams. There was good reason for scientists to hail CFCs as the perfect solution to the world's cooling and propellant needs. By any set of criteria, they were a success—until James Lovelock discovered trichlorofluoromethane (TCFM) in the air over Adrigale, County Cork, Ireland, in 1971 [13]. The inertness and relative water insolubility of the dozen or so CFCs were being viewed as a bonus. They weren't. They were in fact a calamitous flaw. Because CFCs are not destroyed in the lower atmosphere, they drift into the stratosphere, which is surprising, as CFC molecules are heavier than air. The molecular weight of TCFM is 137; of air, only 29. In the stratosphere they contact a layer of intense, highly energetic ultraviolet (UV) radiation (electromagnetic radiation), which splits the TCFM, releasing chlorine. The free chlorine atoms attack ozone molecules, forming chlorine oxide and releasing a molecule of oxygen—result: loss of ozone. The three reactions are:

$$CfCl_3 + UV \rightarrow CfCl_2 + Cl$$

$$Cl + O_3 \rightarrow ClO + O_2$$

$$ClO + O \rightarrow Cl + O_2$$

It takes about a decade for CFCs to reach the stratosphere. Once there, they can react with ozone for as long as 100 years, with a single CFC molecule destroying 100,000 molecules of ozone. This is the mechanism that thins the protective ozone layer and produces the holes in the ozone that permits harmful UV radiation to reach the citizens below in the troposphere—at the earth's surface. CFCs contribute to over 80% of ozone depletion.

Regretfully, the almost perfect CFCs had yet another inherent imperfection. They are tremendous absorbers of IR radiation—20,000 times more absorbing than CO_2. Fortunately, they are present in concentrations of only parts per trillion. Nevertheless, it had become a new and significant greenhouse gas, a major climate forcer. I say, had become, because of the Montreal Protocol. In 1987, 57 nations established the Montreal Protocol, which set a target for reducing the global production of CFCs by 50% by 1998. In 1992, it was further agreed to phase out production of CFCs in developed countries by 1996, and in developing countries by 2010. Now, with 68 countries on board, we are witnessing a decline in atmospheric CFC levels. However, as noted earlier, CFCs are long-lived and can remain aloft for years, adversely affecting the ozone layer.

Too often, the air, like the ocean, is viewed as a bottomless pit: the recipient of all manner of industrial effluents. William T. Sturgis, an atmospheric chemist at the University of East Anglia, Norwich, England, discovered trifluoromethylsulfurpentafluoride (Sf_5Cf_3)—a rare halocarbon, 18,000 times more IR-absorbing than CO_2, residing in the atmosphere, apparently a byproduct of industrial processes using fluorine [14]. More recently another synthetic intense IR absorber, sulfur hexafluoride (SF_6), was discovered and identified [15]. So much for human forcings, but what of natural?

One of the most extraordinary discoveries was the recent finding that certain types of common fungi can produce ozone-destroying methyl halide gases. For years, the source of these gases in the atmosphere had eluded scientists until Kathleen K. Tresedu, a biogeochemist at the University of Pennsylvania, studied gases produced by four types of ectomycorrhizal fungi, and found that each produced methylchloride, methylbromide, and methyliodide. These ectomycorrhizal fungi envelope roots of trees, forming symbiotic relationships. The fungi remove nitrogen and phosphorus from the soil and pass a portion along to the trees. In return, the trees provide carbohydrate for the fungi, which are nonphotosynthetic. As these fungi are found in forests the world over, they make up as much as 15% of the soil's organic matter. Between them, methylchloride and methylbromide appear to be responsible for some 20% of the current ozone destruction. On the face of it, these fungal emissions do not appear preventable or manageable [16].

Water Vapor

Water vapor is the most absorbent gas of the earth's atmosphere. Fritz Moller of the University of Munich, Germany, pioneered the hypothesis that water vapor could act as an amplifying positive feedback mechanism. That implied that if increases in CO_2 caused additional warming, then increased evaporation from rivers, oceans, reservoirs, and such, according to his calculations, would increase atmospheric water vapor, which, in turn, would absorb longwave IR radiation and drive temperatures still higher [17]. Testing this thesis had to await the field trials of Raval and Ramanathan of the University of Chicago's Department of Geophysical Sciences. They studied the oceans, where water vapor feedback could be expected to be most intense. From their temperature readings of specific areas of the oceans, they collected the heat emitted, and by using satellite-gathered data on the level of humidity escaping into space, they were able to determine the amount blocked by a greenhouse effect. Their data showed that the greenhouse effect increased along with sea-surface temperatures. This verified an hypothesis of immense impact on climate change, while forcing a revision among those who held that water vapor acts as a negative feedback with increasing temperature. Furthermore, their study lent greater credence to climate models that had correctly predicted the magnitude of the feedback [18].

From the IPCC's most recent analysis (2001), "an increase in water vapor reduces the escape of IR only if it occurs at an altitude where the temperature is lower than the ground temperature. And the impact grows sharply as the temperature difference increases. This is now referred to as a positive water vapor feedback" [2]. They also inform us that "the strength of the water vapor feedback is consistent among models, despite considerable differences in the treatment of convection and microphysics." Nonetheless, there is uncertainty here, because as water vapor increases, more of it will condense into clouds, which readily reflect incoming solar radiation—allowing less radiation to reach the earth's surface and heat it. Clearly, the matter is complex, providing skeptics opportunity to cast doubt on the concern for global warming.

Aerosols

Aerosols, dispersions of solid or liquid microscopic-size particles in a gas, consists of dust, soot, ash, volcanic emissions, sea salt, and products of plant and animal decay—the result of both human and natural activity, which is increasing, but comes with a good deal of complexity and uncertainty as to its influence on radiative forcing. The direct effect is the scattering of part of the incoming solar radiation back into space, causing a negative forcing that may partially offset the enhanced greenhouse effect. However, some aerosols such as soot (black aerosols), the result of incomplete combustion of biomass and fossil fuel burning, absorb solar radiation directly, leading to local atmospheric heating, or absorb and emit infrared radiation, adding to an enhanced greenhouse effect. Soot from biomass burning drifts to the North Pole; a third drifts in from south Asia, and a third comes from Russia, Europe, and North America. This black soot on arctic ice may be warming the area, inducing the melting of ice and snow and adding to a warmer atmosphere. Here is the IPCC again. Aerosols, they say, "may also affect the number, density and size of cloud droplets." This, they say, "may change the amount and optical properties of clouds, and hence their reflection and absorption" [2]. Hansen and colleagues maintain that aerosols "may be the largest source of uncertainty about future climate change."

It is the "white" aerosols, the airborne sulfates, arising from sulfur in fossil fuels that are so strongly reflective and thereby reduce heating of the earth. Whether the forcings are opposite but equal, or unequal, and if so in which direction, remains to be sorted out. This will take time, as the amounts of these aerosols is not known. In the mid-1990s, time was not the thief it currently has become.

Ozone

Ozone is another trace atmospheric gas located in a layer between 15 and 50 kilometers (km) (9–30 miles) above the earth's surface in the stratosphere. Figure 5.5 displays the various layers of the atmosphere as well as the position of the ozone layer. It is in this highly chemically energetic area that ozone is constantly destroyed and regenerated via a series of photochemically catalyzed reactions in which oxygen molecules migrating up from the troposphere react with ultraviolet light, splitting it into highly reactive oxygen atoms that combine with neaby oxygen molecules to form ozone and heat. With solar radiation strongest above the equator, the photodissociation of oxygen is strongest there. It is from there that the newly formed ozone is carried around

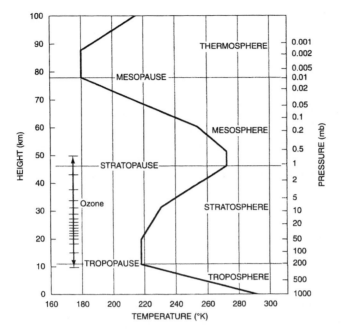

Figure 5.5. Layers of the earth's atmosphere. (Figure reproduced from *Climate Change and Human Health*, with the kind permission of the World Health Organization, Geneva.)

the earth and across the poles by stratospheric winds. Given the normal chemistry of the atmosphere, the level of ozone would be expected to remain constant. As we have seen, however, unique synthetic chlorine-containing chemicals have altered that. For more than two decades we have been witnessing, during a specific period each year, a severe decline in ozone levels. The "hole" in the ozone layer is produced each year during a 4–6-week period beginning in late September, which is the beginning of spring in the Southern Hemisphere. The area of the Antarctic ozone hole, about the size of the continental United States, has grown larger each year during this period.

Four to six weeks after the appearance of the Antarctic ozone hole, ozone from the Southern Hemisphere midlatitudes is carried to the South Pole by the atmosphere's general circulation patterns and replenishes the absent gas. The following spring the cycle is repeated. Ozone depletion due to the CFCs is not a theory; it is a demonstrated fact.

The synthetic halocarbon chemicals, thought to be especially environmentally friendly, drifted into the upper atmosphere, where their chlorine atoms played hob with the ozone layer. Chlorine atoms destroyed stratospheric ozone, and were destroying them faster than they could be replaced. This was not a good omen for the folks beneath, who were and continue to be at increased risk of adverse health effects.

The 2005 Springtime Antarctic Ozone Depletion, the ozone hole, began developing over the South Pole in late August. In early August, surface temperatures were $-110.7°F$ ($-79°C$), the coldest temperature recorded since the late 1990s. By September, ozone concentrations above the South Pole were 20% below amounts measured in August. If this loss rate continues through October, the severity and size of the 2005 ozone hole will approach that of 2003, when the largest ozone hole previously measured opened over the Antarctic [19].

During March 2005, the ozone layer over Britain was reduced to half its normal thickness, a clear loss of shielding against UVA and UVB. The combination of the coldest arctic winter and the high pressure weather system over the North Atlantic had created ideal conditions for ozone loss. The ozone layer is usually 4–5 mm thick. In March it was down to 2.5 mm. Anything below 2 mm is considered a hole. A layer half as thick as it would normally be, will allow 4 times as much UV radiation to penetrate earthward [20, 21].

With the Montreal Protocol in effect, there is reason to be encouraged that continued decline in chlorine atoms in the upper atmosphere will prove effective and protective.

These half-dozen trace gases are just that, traces—minuscule amounts, yet they pack tremendous clout in that the temperature of our planet is controlled by them. So, yes, it does appear that humankind, with its addition to, and effect on, these gases does have the power to affect the forces of nature. As we shall now see, models are needed to simulate and quantify the climatic response to current and future human activity with respect to these trace gases.

But first, how do we know that we humans are in fact the culprits of the current and future warming? A salient question. Let us pursue the evidence, which comes from a number of sources.

First, carbon atoms in CO_2 emitted by fossil fuels—coal, oil, peat, and natural gas—differ from the carbon atoms in CO_2 from present-day plant material. As noted earlier, all living things contain carbon. But elemental carbon has three forms, all of which have the same chemical characteristics, but different atomic weights and different numbers of neutrons in their nuclei. Elements with different atomic weights are called *isotopes*. Carbon has three isotopes; carbon-12, -13, and -14. Carbon-12 has six protons and six neutrons; carbon-13 has six protons and seven neutrons; carbon-14 has six protons and eight neutrons. Carbon-12 is the stable isotope, while ^{13}C and ^{14}C are radioactive isotopes; that is, they emit radiation. Cosmic rays in the outer reaches of space bombard our atmosphere producing ^{14}C, replacing the ^{14}C that decays away. Because of this constant replacement, the ratio of ^{12}C to ^{14}C ($^{12}C/^{14}C$) remains constant. As living things exchange carbon dioxide with the air, they take in this constant $^{12}C/^{14}C$ ratio. When living things die, the ^{14}C continues to decay away. Consequently, by determining the $^{12}C/^{14}C$ ratio, it is possible to determine the age of a sample of carbon. Ergo, the $^{12}C/^{14}C$ ratio that fossil fuels contain will be far different than the ratio in present-day carbon-containing materials.

The carbon ratio of the CO_2 trapped in air bubbles of ancient ice cores has been very different from that of the CO_2 trapped in air bubbles since the 1700s. Twentieth-century observational records and measurements on polar-ice-trapped bubbles show a 20–30% increase in CO_2 over preindustrial levels. Conversely, levels of CO_2 in ancient ice cores, over 200,000 years, are 25% less than current levels. Fossil fuels were formed tens to hundreds of millions of years ago, and the fraction of their carbon nuclei that were once radioactive, are no longer so, while CO_2 from relatively recent natural sources remains radioactive. As we have seen in Figure 5.2, atmospheric CO_2 levels have been increasing steadily since recording began in 1958. Moreover, there is more CO_2 in the atmosphere over the Northern Hemisphere than the Southern Hemisphere. Most human activity resides in the Northern Hemisphere, and it takes about a year for Northern Hemisphere emissions to circulate through the atmosphere and reach the Southern Hemisphere. Also bear in mind that 95% of the total atmospheric CO_2 levels is of natural origin. It is the 3–5% added by human activity that appears sufficient to tilt the energy budget toward a warmer world, with its many predicted dislocations—based on model projections, which raises the question. Are the predictions reliable, and why models?

We first consider the models. *Models* mean simulations by computers because our earth is far too large to bring in to any laboratory, and far too complex, with far too many variables to test singly or even several simultaneously. The many, and more often than not, simultaneously interacting—along with both positive and negative—forcings can be managed only by supercomputers. Models do work, and with good reason. Calculus, the mathematics of continuous change, is the reason. The idea that change or motion can be represented by mathematical equations is the essence of computer modeling of global climate—with its concern for an ever-changing fluidlike atmosphere with a broad range of interactive elements. If climate variables can be represented by an equation, or equations, then calculus can deal with them via differentiation and integration; *differentiation* computes the rate at which one variable changes with respect to another at any instant, while *integration* takes an equation in terms of rate of change and converts it into an equation in terms of the variables that do the changing. Evidence of continuously varying change, if required, is at hand: motor vehicles moving at changing velocities; birds in flight, wheeling and soaring; breezes on a balmy day; or a baseball player preparing to steal a base. In the words of the ancient Greeks, "All things flow."

The atmosphere and oceans are in constant motion, continually changing. If the foregoing is correct, the elements of climate, those that are known, should be reducible to mathematical statements and their solutions, providing descriptions of climate over time. It makes sense. Calculus has worked for over 250 years. It needs no defense. But it is worth remarking that if, say, a model's description of drag forces on a plane's wing, or the stability of a car ferry running in high seas proves less than satisfactory during trial runs, then reruns are not a problem. But climate cannot be studied under controlled conditions in the field or in a lab. It will not hold still for appraisal. Nevertheless, questions must be posed, and an essential one is this: Is the nature of climate, the process, sufficiently well understood to reduce to an appropriate set of equations? Do we need to know it all? How much comprehension is necessary to obtain credible answers? An answer is not elusive. But it must be understood that this vast natural phenomenon of planetary climate can be studied only by computer simulation. There is no other way. It's all we've got. Over the past 40+ years, since the first simulations were seen in 1963, models have improved and have become far more sensitive, and reliable. But because relationships can be expressed in the language of mathematics, this does not mean that they are without fault or error. Nevertheless, models are also based on established physical laws, including the laws of gravity and conservation of energy, momentum, and mass.

It is this reliance on basic physical laws that lends credence to the predication that a buildup of greenhouse gases will lead to an alteration in the earth's climate. Components, or coupled combinations of components, of the climate system can be represented by models of varying complexity. The most complex atmosphere and ocean models are referred to as *general circulation models* (GCMs) or air–ocean GCMs, and, as noted, are expressed mathmatically. Current models are solved spatially on a three-dimensional grid of points on the globe as shown in Figure 5.6, with a horizontal resolution of 250 km and 10–30 levels or boxes, vertically. A typical ocean model has a horizontal resolution of 125–250 km and a resolution of 200–400 m vertically.

Consider a grid covering the surface of the earth and having a number of vertical layers or boxes, a network of points. The computer must calculate temperature, pressure, wind velocity, humidity, cloudiness, and dozens of other variables at literally millions of points. Each of these calculations must be repeated at timesteps of a few minutes. Typical models could have a horizontal resolution of about 6° longitude × 6° latitude with 10 vertical levels, which gives about 18,000 grid points or fields to be calculated. If there are only 8 variables at each grid point, 144,000 evaluations are needed at each timestep. To obtain a 25-year climate simulation with half-hour timesteps, 438,000 timesteps will be needed. If 100 arithmetical operations are needed to produce one of the fields at a point for each timestep, 6×10^{12} calculations are required. A very fast computer can calculate at a rate of 109 operations per second, which indicates that a climate simulation will require about 6000 seconds or 100 hours of running time, and this provides only climate points with gaps of hundreds of miles. Within each box or vertical, there is only one value representing the climate, within, say, 100,000 square miles. Supercomputers and even more sensitive models are constantly reducing these gaps.

Usually the first step attempts to simulate and quantify the present climate for several decades without any external forcing. The quality of these simulations is assessed by comparing them with the actual past climate. The next step would introduce external forcings by, for example, doubling the concentration

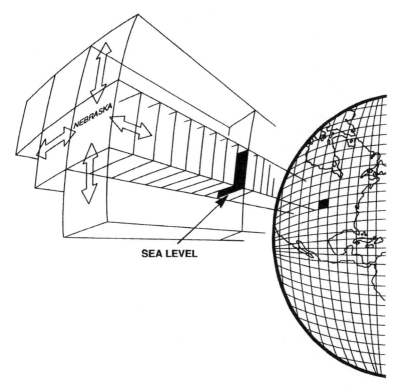

SEA LEVEL

Figure 5.6. Modelers divide the earth's surface and atmosphere into a grid of "boxes" to more effectively manage their data. Here, an exploded view of a single box with its nine levels suggests the degree of complexity of the interacting variables. More than 17,000 boxes or climate units are used to obtain a "picture" of world climate

of CO_2 and running the model to a new equilibrium. The difference between the simulations provides an estimate of the change due to the doubling, and perhaps more importantly, the sensitivity of the climate to a change in radiative forcing. Other simulations would include combinations of GHGs, aerosol additions, and cloud and ocean effects. But these multiple variables do require supercomputers and enough time to achieve another equilibrium climate, which could take weeks or months.

Recently the National Center for Atmospheric Research, in Boulder, Colorado, introduced a powerful new version of a supercomputer-based system to model climate and project global temperature increases. This new system, Community Climate System Model, version 3 (CCSM3), has indicated in early trial runs that the global temperature may rise more than its previous version had projected, if we humans continue to emit large quantities of CO_2 into the atmosphere. CCSM3 has projected a temperature rise of 2.6°C (4.7°F) in a scenario in which atmospheric levels of CO_2 are suddenly doubled [22]. This is more than the 2°C (3.6°F) increase predicted by the earlier version. Although the developers of the new version believe that it is a more accurate model, they do not yet know why it is so much more sensitive to increased CO_2 levels. Be that as it may, the fact that models are based on known physical laws, and can reproduce many features of current and past climates, gives increasing confidence in their reliability for projecting future climate. Although models are far from crystal balls, they have shown that warming would not occur uniformly over the planet. It is expected to be more intense at the higher latitudes, and greater in winter than in summer.

Recently, climate researchers at the Goddard Institute for Space Studies reported that their climate model was validated by actual measurements of the oceans' heat content. From both the model and the measurements, they reported that "the earth was currently receiving 0.85+/–0.15 watts per square meter, more energy from the sun then it is emitting into space." This imbalance they inform us, "is confirmed by precise measurements of increasing ocean heat content over the past ten years" [4]. Indeed, this is the type of validation that will give models greater credibility and reliability.

In 1988, the World Meteorological Organization and the United Nations Environmental Program created the Intergovernmental Panel on Climate Change, the IPCC, an assemblage of the world's foremost climate scientists who would seek to interpret the flow of computer-produced data from around the world. Working Group I was given the task of assessing the climate change issue: Working Group II was to assess the impacts of climate change, and Working Group III was to formulate response strategies. Over the ensuing years, three major reports have been published presenting their best estimates of warming scenarios. The most recent, the *Third Annual Report*, was published in 2001. The Group I assessment projected three possible scenarios as shown in Figure 5.7, which are based on levels of CO_2 emitted, and the level of warming that each could engender. A fourth report is a work in progress and is expected to be published in 2007. However, indications of what could be expected were discussed at a recent IPCC meeting in Paris. Over the decade of the three annual reports, the range of the likely climate sensitivity to be expected by the year 2100 was 1.5–4.5°C (34–40.1°F). Certain model predictions indicated a modest warming, while others found that temperatures could rise by a scorching 4.5°C (40.1°F). Currently, scientist/modelers, using more powerful computers and equipped with a better understanding of atmospheric processes, are beginning to reduce uncertainty, and appear to be reaching a consensus for a moderately strong rise in temperature. Evidence seems to converge on a 3°C (37.4°F) rise for a doubling of CO_2, with a range of 2.6–4.0°C. With a strong consensus, disadvantages, as we shall shortly see, far outweigh advantages of a warmer world. When published in 2007, this will not be welcome news. But it may well shake up the skeptics [23].

The British government, for example, is fully convinced that rising global temperatures are not in its best interests. For them, global warming is approaching a critical point of no return. Beyond that point they are certain that drought, crop failure, and rising sea levels will be irreversible. The American

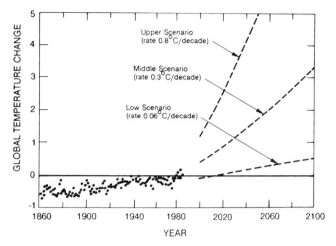

Figure 5.7. Three projected estimations of the earth's surface temperature based on levels of carbon dioxide emitted to the atmosphere. (Figure reproduced from *Climate Change and Human Health*, with the kind permission of the World Health Organization, Geneva.)

position, championed by President George W. Bush, argues that cuts in carbon emissions would sorely damage the US economy.

David A. King, Chief Scientific Advisor to the British government, recently scoffed at that idea. As he put it, "It's a myth that reducing carbon emissions necessarily makes us poorer. "Taking action," he said, "to tackle climate change can create economic opportunities and higher living standards." And he spelled it out with numbers. "Between 1990 and 2000," he observed, "Great Britain's economy grew by 30%, employment increased by 48% and our green house gas emissions fell by 12 percent" [24].

I recently asked James Hansen of GISS whether he was still optimistic that global warming could be avoided. He was optimistic "in the sense that it is possible to limit climate change to a level that avoids the most serious problems, but only if concerted actions are taken . . . If they are not, if the U.S., China and India build huge infra-structures of conventional coal-fired power plants, then we are in trouble" [25].

It is this "trouble" that many countries of the world tried to deal with at the First World Climate Conference in 1979—over two decades ago, as it morphed into the Kyoto Protocols—and are still grappling with.

The First World Climate Conference, held in Geneva, Switzerland, presented the initial evidence of the adverse effects of human interference with global climate and the consequences that could ensue. In 1988, the UN's General Assembly enacted resolution 43/53 urging the "Protection of Global Climate for present and future generations of mankind," which recognized climate change as a concern for all people. It was also in 1988, as noted above, that the IPCC was created and 2 years later issued its first annual report.

It was evident that anthropogenic emissions knew no boundaries and that no country would be spared as economic downfalls in one area would lead to negative changes in other areas. A global treaty was negotiated, and in May 1992, The United Nations Framework Convention on Climate Change (UNFCCC) was adopted, and at the earth Summit Conference in Rio de Janeiro, countries signed on. But its provisions were insufficient to deal effectively with climate change. Firmer commitments were adopted in Kyoto, Japan, in 1997, and the Kyoto Protocols were born. But Kyoto's emission reduction requirements have been sticking points for the Bush Administration. What did Kyoto require? In general, parties to the Protocol, which the US government continues to resist, must reduce or limit their emissions commensurate to their 1990 levels. Developed countries must reduce their collective emissions of GHGs by at least 5% by the end of the first commitment period, 2008–2015. But reduction targets do vary among the first world countries; for example, the Russian Federation and Ukraine were especially favored. They are required only to stabilize their emissions, and bear in mind that Russia is the world's second highest emitter after the United States, with approximately 17% of the total, while Australia, Iceland, and Norway could even increase emissions, but the countries of the European Union, Canada, Japan, and the United States were required to cut emissions in order to achieve the group's 5% goal.

As of April 2004, 124 countries have signed the Protocol. However, their total carbon emissions did not reach the 55% of CO_2 emissions required for ratification, until Russia signed on late in 2004. The United States, the world's largest emitter with 36% of the total, remains a holdout, believing there is insufficient evidence warranting a 7% reduction below its 1990 level. "Although it is important to set targets and time tables, the fundamental problem of climate change cannot be settled that simply" [26, 27].

Furthermore, under UNFCCC, the countries of the world pledged to avoid dangerous human interference with the climate. But "dangerous" was never defined, and over the past decade emissions have done nothing but climb. Current CO_2 levels have not been exceeded over the past 420,000 years, and the rate of increase during the twentieth century has been unprecedented during the past 20,000 years. Indeed, this past century has set many unfortunate records.

On February 16, 2005, a stricter addendum to the Kyoto Protocols took effect, but there remains no agreement on what level of cuts would lead to climate stability. Too many people who ought to know better are urging more research, which is easier than making the difficult decisions required. Dr. James Hansen called for a sense of urgency in 1988, and in February 2005, he reiterated his initial urging. "I think that the scientific evidence now warrants a new sense of urgency." At the recent meeting in February of the World Economic Forum, where Tony Blair, Prime Minister of Britain, urged the United States to join the industrialized nations in agreeing to curbs on GHGs, he said, "It would be wrong to say that the evidence of danger is not clearly and persuasively advocated by a very large number of entirely independent and compelling voices" [28].

As the new, stricter accords take effect, there is some resentment. Europeans have set some of the most stringent emission-reducing targets. They bear some resentment that the United States and China resist bearing the extra costs of emission reduction. But more importantly, they know that the goal of curtailing emissions will not be realized because the atmosphere is a global, not a local, problem. The fact that time is not in anyone's favor has not been lost on a number of major international corporations who are changing their behavior regardless of whether they are in or out of Kyoto member countries. They are changing their behavior because they know that career politicians will not take strong positions supporting reductions of emissions, jeopardizing their elected offices, as climate change deals with a world well beyond their lifetimes.

Michael G. Morris, CEO of American Electric Power, had it right when he said that further delay will serve no one, and will be detrimental; this was highlighted in July 2005, when the Attorney Generals of eight states—Connecticut, California, Iowa, New Jersey, New York, Rhode Island, Vermont, and Wisconsin—filed a lawsuit against the country's five largest power plant CO_2 emitters: American Power Company, the Southern Company, TVA, Xcel Energy, and The Cinergy Corporation. These power companies account for about a quarter of the American power industry's CO_2 emissions, and about 10% of the nation's total CO_2 emissions. Their suit, filed under the federal public nuisance doctrine, claims that the emissions clearly damage their states. That doctrine states and maintains that if a company's activities in one state cause harm in another, then the state where the harm occurs may sue to halt the injurious product. This is not a new law. Apparently it has been applied for over 100 years in interstate pollution cases [29]. Whether they win or lose is not the point. More to the point is the fact that global warming, or climate change, for those who prefer the less threatening expression, appears to be moving away from the political arena, where paralysis seems to have set in, to the movers and shakers who may just force the necessary changes because it is in their best interests to do so.

Adding his clout, Jeffrey Immelt, the CEO of General Electric and one of the heavy hitters in the business world, made it clear in May 2005 that GE, the largest company in the United States, would double its investment in energy and environmental technologies that would prepare it for what he sees as a huge global market for products that help other companies—and countries such as China and India—reduce emissions of GHGs. Further, Immelt believes that mandatory controls on CO_2 emissions are necessary and inevitable. Many companies are moving in this direction in spite of the lack of federal emission regulations, which confer a competitive advantage on those who do nothing.

The questions that linger are whether there is observable evidence of a global warming trend, what means are available for cutting emissions, and whether CO_2 can be appropriately collected and disposed of.

EVIDENCE FOR A WARMING TREND

Model predictions indicate that the major impact of global warming will be felt in the arctic. Temperatures in the arctic have risen almost twice as fast as those in the rest of the world. Models have certainly been correct on this. Although most of the sun's energy hits the tropics, the atmosphere and oceans redistribute the tropical energy north and south to the poles. In the tropics, much of the energy ends up as evaporation. In the arctic, the energy warms the atmosphere. In November 2004, the Arctic Council, an organization of eight northern countries, plus indigenous-peoples organizations, issued a report estimating that by late in the century, average arctic temperatures will rise by some 4–10°C over land and 7–10°C over the oceans, leading to severe changes by century's end. But currently the Inuit people of Canada's northern provinces report that shrinking ice cover and a consequent shortened hunting season is responsible for the emaciated look of increasing numbers of polar bears. Many fiords, formerly covered with hard ice between October and July, are now frozen only between December and May. The walrus have changed their hunting grounds, moving further north, where it is colder [30]. These changes are the harbingers, and as the permafrost uncovers, CO_2 and CH_4 are released into the atmosphere, contributing additional GHGs. We may just be seeing the emergence of a feedback loop, which will speed up warming faster than model predictions. Not a good omen. With warmer temperatures, permafrost underlying the soil surface softens and coastal erosion increases. The arctic warms quickly because as ice melts, which normally is highly reflecting, the now darker seawater and land areas reflect less of the incoming solar radiation, causing the remaining ice to melt ever faster.

In Austria, the Pasterze, its largest glacier, is shrinking at the rate of 13–26 feet a year. But all of Austria's 925 glaciers, these rivers of ice, are also shrinking rapidly [31]. Glaciers in Tibet and China are shrinking along with them. In eastern Tibet the normally massive 27-square-mile glacier, Zepu, has lost 300 feet of its thickness in the last three decades because of rising temperatures in the region [32]. Puddles of melted ice water at 20,000 feet in the Himalayas have replaced ice that covered the mountaintops for thousands of years; and in Peru, the Quelccaya ice cap retreated at a rate of more than 600 feet a year between 2002 and 2004, leaving an 80-foot-deep lake where none had existed before. On Mount Kilimanjaro in Kenya, an 11,700-year-old ice cap that measured 43 square miles in 1912, had shrunk to 0.94 square miles by 2000. "When you see the big picture from many sites, the evidence of drastic climate change becomes quite compelling" [33]. Indeed, the evidence from both the Northern and Southern hemispheres and from continents across the planet the evidence speaks clearly to such a conclusion.

Retreat of the glaciers is only part of the concern. Scientists have warned that lakes forming behind glaciers because of melting ice could burst through cracks in the glaciers and cause tsunami-like devastation to towns below. With the shrinkage, towns depending on seasonal glacial melting could lose their natural water supplies [32].

In Antarctica, on the opposite side of the planet, "which is competing with the Yukon for the title of the fastest warming place on the globe," the Larson A ice shelf, the size of Rhode Island, disintegrated in 1995, followed in 1998 by the collapse of the neighboring Wilkins ice shelf. Over a month in January 2002, at the end of the Southern Hemisphere summer, Larsen B, a 1200-square-mile ice shelf of floating sea ice 722 feet thick, suddenly shattered, losing more than 25% of its total mass and over 35 days setting thousands of icebergs adrift in the Weddell Sea [33]. Then, in just a month, an area twice that size crashed into the sea with its 72 billion tons of ice. Clearly, when greenhouse gases go up, the ice sheets go down.

In Europe's intensively fished North Sea, warming waters over the past 25 years have driven fish populations further north and deeper, making it more difficult for commercial fisherman to locate and land their normal catch [34]. British scientists have reported that increasing levels of carbon dioxide are turning the oceans acidic, which will be harmful to marine life and coral reefs. Acidity does not require computer modeling; it is measured on a pH scale in which 1 is the most acidic and 14, the most alkaline. One unit on the pH scale reflects a change by a factor of 10. With the 25 billion metric tons of carbon dioxide going into the air yearly, and a third of that being absorbed by the oceans, producing carbonic acid, it's anticipated that should this level of CO_2 continue, the pH of ocean water would drop to 7.7 by 2100—lower than at any time in the last 420,000 years [35].

Warmer and acidic seas are only part of the concern. An 11-year study by researchers at the International Rice Research Institute in the Philippines, found a 10% drop in rice crop yields for every 1.8°F increase in nighttime temperature. They believe that warmer nights speeds up plant respiration, causing the rice plants to work harder and waste energy. With Asian populations increasing, and yields flattening, prospects of future food shortages are a growing concern [36].

It is not yet known whether Hurricane Katrina is linked to global warming, but the available evidence suggests that it is highly likely that this warming trend is making hurricanes more destructive and of greater frequency. Hurricanes derive their power in part from warm water, and models show future hurricanes becoming more severe as sea-surface temperatures rise. Professor Emmanuel Kerry of MIT informs us that "The large upswing in the last decade is unprecedented, and probably reflects the effect of global warming." He goes on to say that "any results suggest that future warming may lead . . . to a substantial increase in Hurricane-related losses in the 21st century" [37]. A warmer world also makes hurricanes more destructive by raising sea levels, which means more flooding. Paradoxically, developing countries especially island nations that are not major emitters of CO_2 will be the most severely affected. Rising seas will displace hundreds of thousands, perhaps millions, flooded out of their island homes. How will they be dealt with? Recall the upheaval spawned by Hurricanes Katrina and Rita, and the hundreds of thousands of individuals from Louisiana, Mississippi, and Texas who had to be evacuated to other cities and states with all the stresses attendant on leaving their homes, possessions, family, friends, and jobs. Consider now the displacement of millions of people from islands and shore communities around the world who would need to be resettled in foreign countries with different languages, customs, and food. Such shifting of populations is currently considered as leading to conflicts over living space, food and water, which could easily lead to potential disruption of global security and stability requiring military intervention. This is a nightmare scenario. On the other hand, as noted earlier, global warming's effects will not be felt equally around our planet. The effect of shifting climates is exemplified in the accompanying cartoon, Figure 5.8, suggesting that opportunities and benefits may be the obverse side of the global warming coin.

If, as evidence suggests, climate is changing, and at an unprecedented rate, what impacts can be expected on our health, our food supply, and our lives generally?

Food production is a critical and essential renewable resource. No more than 25 food crops feed the world's population. The most definitive factor in food production is climate, and the lack of water is the single greatest impedi-

Figure 5.8. This cartoon speaks directly to the issue of rising sea levels.

ment to crop growth and production. According to the UNFCCC, degrading soils and water resources will place enormous strains on food production for growing populations. They believe that a warming of less than 2.5°C could reduce food supplies and contribute to higher prices. Assuming prediction of a warming of 1.4–5.8°C over the coming 100 years, evaporation and precipitation will increase along with rainfall intensity. Some regions may become wetter; others will suffer loss of soil moisture and increased erosion. Soil moisture is expected to decline in some midlatitude continental regions in summer, while rain and snow, models predict, will increase in high latitudes during the winter [38].

It is widely believed that increased CO_2 emissions will stimulate photosynthesis in wheat, rice, barley, cassava, and potato plants. Experiments based on a 50% increase in CO_2 concentrations have found that "CO_2 fertilization" can increase yields by 15% under optimal conditions. But new studies suggest that increased CO_2 levels may be a mixed blessing, as there appears to be a tradeoff between quantity and quality. Researchers at Ohio State University have found that nutritional quality declines because "while plants produce more seeds under higher CO_2 levels, the seeds contain less nitrogen" [39]. As the quality of food decreases, you've got to eat more of it to obtain the same benefit. Under the rising CO_2 scenario, livestock and humans would have to increase their intake of crops to compensate for the loss. Echoing the work at Ohio State, Chinese scientists at the Nanjing Institute of Soil Sciences, part of the Chinese Academy of Sciences, found that rising levels of CO_2 could result in increased growth rate but decreased nutritional value of rice and wheat. They maintain that the protein level will decrease by 10% by the year 2050, and elements such as iron and zinc will also decline. These scientists spent 3 years studying the impacts of greenhouse gas emissions on the growth of rice and wheat. Providing the plants with a 50% higher level of CO_2 increased the rice crop yield by 15% and 14% for wheat, and both plants grew faster. But protein levels in both dropped by 10%. They also predict that if air temperatures continues to increase by 0.04°C yearly, the amount of organic matter derived from decaying animal and plants in rice fields will decline by 7% by 2050 [40].

Similar results were obtained by scientists at the USDA's Agricultural Research Service. They reported that elevated temperature and CO_2 levels decreased the food quality of soybeans; total oil, carbohydrate, nitrogen (protein), and phosphorous concentrations decreased with increasing temperature [41].

At a conference on climate change impacts on agriculture, held at the University of Florida, a group of researchers from the Universities of South Carolina and Florida also found that increased levels of CO_2 reduce nutritional content of crops by 10–20%. But they also reported that climate change will affect agriculture via its impact on pest management. Cold and freezing weather are natural pest control mechanisms; as temperatures rise, pest central will decrease. A participant from Oregon State University indicated that an increasing temperature may also increase the need for pesticides, may reduce pesticide effectiveness, and may increase residues, which will raise food safety issues [42].

It has also been widely noted, and accepted, that trees would become CO_2 sinks, absorbing and sequestering much of the increased CO_2 emissions. But that has remained an open question as there was little solid evidence supporting such a contention. Recently, however, researchers at the University of Basel's (Switzerland) Institute of Botany, appeared to have answered this question. They set up a 140-foot -tall construction crane in the middle of a forest of oaks, lindens, maples, and pines, and used it to run a network of porous plastic tubes through the forest canopy and pumped CO_2 through it for 4 years. Indeed, the trees grew vigorously, but did not add biomass. For the investigators, this study "does not support expectations of greater carbon bonding in tree biomass in such deciduous trees" [43]. This study suggests that large trees will be of little help in storing excess carbon dioxide. From a purely agricultural view, climate change does not appear to bode well for either human or animal populations.

AND WHAT OF HUMAN HEALTH?

As part of a Congressionally mandated national study of the impacts of climate variability and change in the United States, a team of 12 experts from universities and government agencies across the country assessed the potential impacts that climate change could cause by 2030 and 2100. However, before dealing with each of the five groups of potential adverse health effects, a caveat is in order. The Committee, headed by Professor J. A. Patz, of the Johns Hopkins University School of Hygiene and Public Health, concluded that "the levels of uncertainty preclude any definitive statement on outcomes," which is worth bearing in mind as the potential impacts of climate change on health are rendered [44].

More than likely the direct effects of heat, which currently cause substantial numbers of deaths among vulnerable populations in summer, will increase. From the numbers of deaths in Europe and England in 2003, there is less uncertainty here than with other conditions. In 2003, France suffered the hottest summer of the last 50 years. By the end of the heat wave, the French Institute of Health and Medical research reported that there had been 14,802 deaths related to heat in just 2 weeks. Failure to grasp the severity of the crisis delayed timely decisionmaking by public officials [44]. This lamentable event offers a cautionary tale. During August 2003, France was in the grip of the Tour de France. Total concern was focused on the hardship of the heat on the cyclists. The health services paid little attention to the elderly and the fact that heat leads to fluid loss and exhaustion with consequent heat stroke. Advanced age, excessive sweating, vomiting, and diarrhea (additional fluid loss) predisposes the body to heat stroke; high humidity, exertion, poor ventilation, and heavy clothing contribute to dehydration and exhaustion. Heat stroke occurs when the body's thermoregulatory system ceases functioning and sweating stops. During this sequence of events the person is unaware of what is happening. The skin becomes dry and hot and reddens dramatically; body temperature is up to 104°F, and rising, which can trigger mental confusion. If at this point, a person is not quickly cooled, the progression is usually delirium, convulsions, unconsciousness, and death. Too often these people are alone, so that no one can help or call 911. If the country is ill-prepared for increased numbers of heat waves, which are clearly in the scenario, the death rate can be expected to climb and hospital facilities sorely taxed.

Computer models indicate an increased frequency and intensity of storms and hurricanes. After Hurricane Andrew roared through south Florida, posttraumatic stress syndrome increased dramatically in the area. How the thousands of refugees from Katrina will fare is yet to be seen. Anger at the failure of government at local, state, and federal levels may not have a lasting effect, but it surely will be translated into future preventive modalities as lessons will undoubtedly be learned.

With water purification and sewage treatment systems rendered inoperable, microbial diseases are always anticipated. During Katrina's ravagings, none appeared. Why? It must be borne in mind that for waterborne diseases to spread, individuals in the besieged areas must shed bacteria into the water, which means that they must either be frankly ill with a disease that can be transmitted by water, or they must be healthy, asymptomatic carriers; otherwise the fecally polluted water will be no more than of aesthetic concern. Boiling water to kill potential pathogenic bacteria is widely recommended, as no one knows if anyone is ill, or a carrier.

Walking through floodwaters can be a real concern. Although leptospirosis occurs around the world, 10% of cases occur in the United States. Symptoms of leptospirosis include fever, jaundice, hemorrhage, and liver and renal failure contracted by picking up the spirochetes of the genus *Leptospira*, deposited in water in the urine of animals and rodents. Leptospires gain entry into the body via breaks in skin as well as orally through mucous membranes when contaminated water is accidentally swallowed. Person-to-person spread is uncommon. Fortunately, leptospirosis does not seem to have been one of Katrina's byproducts.

Excessive water can cut both ways. Rainfall can increase mosquito populations by offering puddles, breeding sites. On the other hand, heavy rains and flooding can flush such sites, destroying mosquitoes in their larval stages.

Although not widely known, malaria was prevalent in the American colonies, and by 1850, had spread north to the territories now occupied by Minnesota, Wisconsin, and Michigan. By the 1930s, malaria had disappeared from the north, but was still active in Louisiana, Mississippi, and Alabama, and was considered eradicated in the United States early in the 1970s. Ergo, a warmer world with increased rainfall could see malaria revisiting its previous habitats. However, current public health infrastructure and policies suggest that this is unlikely.

A cluster of enteric diseases appear to exhibit a seasonal pattern suggesting sensitivity to climate. In the tropics, diarrheal diseases usually rise during the rainy season. Both floods and droughts are associated with increased risk of these illnesses as heavy rainfalls can wash fecal matter into the water supply, while arid conditions that reduce availability of freshwater can lead to an increase in hygiene-related diseases. Transmission of enteric diseases may be increased by higher temperatures because of direct effects on the growth of pathogens in the surrounding environment. But again, the uncertainty factor is high, and developed countries will have responses markedly different from those of undeveloped countries.

Yet another question must be; What of animals—not just pet cats and dogs, but cattle, sheep, swine, and poultry—our major food animals and protein sources? How will they respond to a warmer world? Surely widespread drought would result in great losses. Increased heat could adversely affect animals physiologically, impairing reproductive capacity, and egg-laying among poultry could well be compromised. Insect pests could proliferate and could so irritate cattle as to cause loss of appetite and malnourishment with its loss of weight and meat supply. The possibilities are there and require consideration and evaluation before temperature increases occur—if we should fail to act in curtailing CO_2 emissions.

CAN EMISSIONS BE CURTAILED?

Amory B. Lovens, cofounder and CEO of the Rocky Mountain Institute, Snowmass, Colorado, thinks so, but he's been trying unsuccessfully for decades to show us how. Ten years ago Lovens couldn't get arrested; today he may be our deliverer. Lovens wants us to change our profligate ways and become energy-efficient, an idea we have firmly resisted. For Lovens, saving fossil fuels is a lot cheaper than buying them. He maintains that both sides in the warming debate have it wrong. Experts on one side say that burning less fuel to reduce or prevent global warming will increase the cost of all goods and services. Environmentalists counter by saying that the increased costs are moderate, but worth it. Skeptics, those who don't believe that global warming is a problem, including President Bush and his advisers, are certain that the extra expense would crush our economic system. Lovens has other ideas, and people are beginning to listen. If properly done, protecting the climate would reduce costs. The problem as he sees it is this: The energy sector worldwide is totally inefficient. Power plants and buildings waste huge amounts of heat; cars and trucks dissipate most of their fuel energy, and consumer appliances waste much of their power—they even siphon electricity when turned off. If nothing is done, the use of coal and oil will increase, draining hundreds of billions of dollars a year from the economy while worsening the climate and increasing air pollution. Lovens envisions a plan that at its core improves end-use efficiency, which is the fastest and most lucrative way to save energy. His analysis shows that many energy-efficient products cost no more than inefficient ones. Homes and factories that use less power can be cheaper to build than conventional buildings. And reducing the weight of motor vehicles can double fuel economy without compromising safety or cost. Lovens vigorously emphasizes that with improvements in efficiency and with competitive alternative, renewable energy sources, the United States can phase out oil by 2050.

The Rocky Mountain Institute's analysis shows that full adoption of efficient vehicles, buildings, and industries could shrink projected oil use in 2025 by 28 million barrels a day—that's more than half of the country's current use. They further maintain that before 2050, US oil consumption could be phased out totally by substituting alternative fuels [45]. These ideas need full consideration. We shall venture into alternative energy sources in Chapter 7.

Although not in line with the idea of greater efficiency, the northeastern region of the United States—Maine, Vermont, New Hampshire, Massachusetts, Connecticut, Rhode Island, New York, New Jersey, and Delaware— agreed in August 2005 to cap their greenhouse gas emissions at the current level (a typical coal-fired plant emits 3–4 million tons of CO_2 yearly) and then reduce them by 10% by 2020. The northeast is not going to solve the problem of global warming, but its initiative could surely set an example for other states—and, let's face it, the northeast is a major population and power center; cutting their emissions makes a major dent in the amount of CO_2 entering and enhancing the greenhouse [46]. Nevertheless, one would have liked to have seen deeper cuts, and 2020 is some years away. Let us hope that the window remains open. Pushing for, insisting on, action is the easy part. Taking action is difficult.

As human activity is clearly implicated as a major cause of the current rise in greenhouse gases, and consequent average global temperature along with it, it would be well to reflect on the cultures and people that have succumbed to drastic climate changes. Although the disrupted cultures played no part in their demise, the stark consequences and lessons of climate change are writ large, and should give us pause to consider means of avoiding a disastrous warming trend. In April 2007, the United Nations IPCC issued its *Fourth Assessment of Climate Change*. In its most specific depiction of the effects of human driven climate warming, it predicted widespread drought in southern Europe, the Middle East, sub-Saharan Africa, the American Southwest and Mexico, along with flooding of the low lying islands of southern Asia. This new report also noted, that this was no longer model-based data, but empirical data on the ground. Clearly, urgency must now replace doubt and uncertainty on curbing emissions.

Following that report, in May 2007, the IPCC stated that swift reduction of greenhouse gases over the next 20 to 30 years could lead to CO_2 stabilization towards the end of the century, and with it climate stabilization. They were forthright in calling for new technologies for energy production that were carbon dioxide free—especially nuclear and solar power.

In an earlier book, *Global Warning/Global Warming*, I raised the question what would reasonable people do in the face of uncertainty? I believed they would eschew oil. That was in 1992. They didn't. I now raise a modified question: what will the people do given the current level of certainty of severe climate destabilization if greenhouse gas emissions continue unabated? I'm certain they will pull back from the brink by radically reducing use of fossil fuels. If I'm wrong again, well. . . . stay tuned.

REFERENCES

1. Boyer, L., Robitail, S., and Auquier, P., Heatwave in France: How did the media treat the crisis? *Br. Med. J.* **327**:876 (2003); see also: Black, J., Factors contributing to the summer of 2003, *Weather* **59**:217–223 (2003).

2. Intergovernmental Panel on Climate Change, 2001, *The Scientific Basis*, Houghton, J. T. et al., eds., Cambridge Univ. Press, New York, 2001.

3. Hansen, J., Defusing the global warming time bomb, *Sci. Am.* **290**(3):68–77 (2004).

4. Hansen, J., and 14 others, Earth's energy imbalance: Confirmation and implications, *Science* **308**:1431–1434 (2005).

5. Tyndall, J., On the absorbtion and radiation of heat by gases and vapours, *Philosoph. Mag. Ser. 4,* **22**:169–194, 273–285 (1861).

6. Arrhenius, S. A., On the influence of carbonic acid in the air upon the temperature of the ground, *Philos. Mag.* **41**(5):237–276 (1896).

7. Callender, G. S., The artificial production of carbon dioxide and its influence on temperature, *Quart. J. Roy. Meteorol. Soc.* **64**:223–240 (1938).

8. Plass, G. N., The carbon dioxide theory of climate change, *Tellus* **8**:140–153 (1956).

9. Chang, K., Charles D. Keeling, 77, dies; raised global warming issue, *New York Times* C20 (June 23, 2005).

10. Hansen, J., Sato, M., Ruedy, R. et al., Global warming in the twenty-first century: An alternative scenario, *Proc. Natl. Acad. Sci. USA* **97**(18):9875–9880 (2000).

11. Frankenberg, C., Meirink, J. F., Van Weele, M. et al., Assessing methane emissions from global space-borne observations, *Science* **308**:1010–1014 (2005).

12. Dillon, W., *Gas (Methane) Hydrates: A New Frontier*, U.S. Geological Survey—Marine and Coastal Geology Program (http://marine.usgs.gov/fact-sheets/gas-hydrates/title.html; accessed 9/15/05).

13. Lovelock, J. E., Maggs, R. J., and Wade, R. J., Halogenated hydrocarbons in and over the Atlantic, *Nature* **241**:194–196 (1973).

14. Sturgis, W. T., Wallington, T. J. et al., A potent greenhouse gas identified in the atmosphere: SF_5CF_3, *Science* **289**:611–613 (2000).

15. DeVos, C. T., and Vassilou, P., Sulfur Hexafluoride (SF_6): Global environmental effects and toxic by-products, *J. Air Waste Manage. Assoc.* **50**:137–141 (2000).

16. Tresedu, K. K., Forest-soil fungi emit gases that harm ozone layer, *Sci. News* **160**:389 (2001).

17. Moller, F., On the influence of changes in the CO_2 concentration in air on the radiation balance of the earth's surface and the climate, *J. Geophys. Res.* **68**:3877–3886 (1963).

18. Ravel, A., and Ramanathan, V., Observational determination of the greenhouse effect, *Nature* **342**:758–761 (1989).

19. Oltman, S. J., *Antarctic Stratospheric Ozone Depletion*, NOAA Monitoring and Diagnostics Laboratory (CMDL), Sept. 9, 2005 (www.cmdl.moaa.gov/ozwg/ozsondes/spo/).

20. Adam, D., Britain's ozone levels near all-time low, scientist's say, *Guardian UK News*, March 19, 2005 (www.guardian.co.uk/uk_news/story/o).

21. Univ. Cambridge, *Large Ozone Losses over the Arctic*, April 26, 2005 (http://admin.cam.ac.uk/news/press/dpp/2005042601).

22. *New Version of Premier Global Climate Model Released*, U.S. Global Change Research Program National Center for Atmospheric Research (NCAR), Boulder, CO, June 24, 2005 (www.globalchange.gov/#ccsm).

23. Kerr, R. A., Three degrees of consensus, *Science* **305**:932–933 (2004).

24. King, D. A., Climate change science: Adapt, mitigate, or ignore? *Science* **303**:176–177 (2004).

25. Hansen, J., Personal communication, Sept. 8, 2005.

26. Kim, O., *The Kyoto Protocol: Universal Concern for Climate Change*, UN Chronicle No. 3, 2004.

27. Landler, M., Mixed feelings as Kyoto Pact takes effect, *New York Times* (Feb. 16, 2005).

28. Special Address by Tony Blair, Prime Minister of the Untied Kingdom, World Economic Forum, Davos, Switzerland, Jan. 26, 2005.

29. Letter to the Editor, If feds won't fight global warming, we will (signed by eight Attorney Generals) *Wall Street Journal* (Aug. 17, 2004).

30. Kraus, C., Eskimos fret as climate shifts and wildlife changes, *New York Times* (Sept. 6, 2004).

31. Bernstein, R., Melting mountain majesties: Warming in Austrian Alps, *New York Times* (Aug. 8, 2005).

32. French, H. W., A melting glacier in Tibet serves as an example and a warning, *New York Times* (Nov. 9, 2004).

33. Kunzig, R., Turning point, *Discover* 26–28 (Jan. 2005).

34. Inman, M., Fish moved by warming waters, *Science* **308**:937 (2005).

35. Chang, K., British scientists say carbon dioxide is turning the oceans acidic, *New York Times* A5 (July 1, 2005).

36. Kunzig, R., Two-degree rise drops rice yield by 10 percent, *Discover* 29 (Jan. 2005).

37. Kristoff, N. D., The storm next time, *New York Times* A15 (Aug. 8, 2005).

38. Wittwer, S. H., *The Global Environment: It's Good for Food Production* (www. comnett.net/-wit/food/html).

39. Curtis, P., *Increased CO2 Levels are Mixed Blessing for Agriculture*, Ohio State Univ. Research (http://researchnews.osu.edu/archive/CO2plant.htm).

40. Hepeng, J., Rising carbon dioxide could make crops less nutritious, *Sci. Devel. Network*, March 4, 2005 (www.scidev.net/news/index.cfm?fuseaction= readnews&itemid=1969&language=1).

41. Thomas, J., Boote, K., and Hartwell, A. H., Effects of elevated temperature and carbon dioxide on composition and gene expression in soybean feed, *Crop Sci.* **43**:1548–1557 (2003).

42. Lincoln, D., and Hartwell, A. H., Increased atmospheric levels of carbon dioxide and nutritional content of leaves, *Proc. Conf. Climate Change Impacts on Agriculture and Potential Mitigation Schemes*, Gainesville, FL, Sept. 10, 1999, Center for Sustainable Resource Development, Univ. California Berkeley (http://nature. berkeley.edu/csrd/global/flconf/).

43. Korner, C., Asshoff, R., Bignucolo, O. et al., Carbon flux and growth in nature. Deciduous forest trees exposed to elevated CO_2, *Science* **309**:1360–1362 (2005).

44. Patz, J. A., McGeehin, M. A., Bernard, S. M. et al., The potential health impacts of climate variability and change for the U.S., *Environ. Health Perspect.* **108**(4): 367–376 (2000).

45. Lovens, A. B., More profit with less carbon, *Sci. Am.* **293**(3):74–83 (2005).

46. DePalma, A., 9 States in plan to cut emissions by power plants, *New York Times* A1, B7 (August 24, 2005).

6

TAPPING THE ATOM

As for the future, your task is not to forsee it, but to enable it.
—*Antoin de Saint Exupery. 1948 The Wisdom of the Sands*

ELECTRIC POWER FOR THE PEOPLE

Antoine Henri Becquerel (1852–1908) was clever. He had to be to win a Nobel Prize. How did he do it? He wrapped heavy sheets of opaque, black paper around light-sensitive photographic plates and placed them in opaque envelopes. He then placed the envelopes in a desk drawer with lumps of uranium salts sitting on the envelopes, or with metal coins or a metal Maltese cross sitting between the envelope and the uranium salts. For one trial he left the envelope undisturbed in the dark drawer for several days; in another, for 5 hours. After each trial he removed the photographic plates, developed them, and found them either fogged from the lumps of uranium, or with silhouettes of the coins or Maltese cross. He found that all uranium compounds fogged his light-sensitive plates. Obviously invisible rays emanating from the uranium salts (pitchblende) were affecting his plates. Becquerel had discovered natural radiation; spontaneous, penetrating, natural radioactivity. For that discovery, he was awarded half the Nobel Prize in Physics for 1903. The other half went to Pierre and Marie Curie for their work on Becquerel radiation [1].

But "natural" is the relevant term. Radiation is all around us, and has been for the millions of years of our evolution and life on earth. Indeed, our exist-

Our Precarious Habitat . . . It's In Your Hands, Fourth Edition. By Melvin A. Benarde
Copyright © 2007 John Wiley & Sons, Inc.

ence has occurred in concert with the continuous and unrelenting cosmic radiation arriving from the sun and other energy sources in our galaxy. But its density is affected by the earth's magnetic field, which makes it greater nearer the poles than at the equator. The natural doses people receive increases with latitude and longitude.

The earth's crust contains components that are naturally radioactive. Uranium is dispersed throughout rocks, and soil, as are thorium and potassium-40. They emit gamma rays, which irradiate our bodies. Building materials, bricks and cinderblock that originated in the earth, are radioactive so that we are irradiated both in and outdoors. Radon is a naturally occurring radiative gas that results from the decay of uranium-238, in rocks and soil. When radon, as radon-222, enters a building, its concentration increases. Since radioactive materials occur everywhere, it is inevitable that they will be present in our food and water. Potassium-40 is a major source of internal radiation. Foods such as shellfish and Brazil nuts concentrate radioactive particles so that those who consume fair quantities can receive a dose of radiation significantly above average. And our bones contain radioactive potassium, while our body tissues contain radioactive carbon. Radiation is a fact of life that cannot be eliminated or undone. We can only consider radiation as an integral part of our environment.

Because radiation can be precisely measured and controlled, a wide variety of artificial sources of radiation have been developed to improve our quality of life. So, for example, radiation is used in nuclear medicine to diagnose and treat disease, and in dentistry to locate and identify a range of dental defects. The most widely used radioactive substance in nuclear medicine is technetium-99, given to millions of people annually as "tracers," permitting radiologists to see how internal organs are functioning. The procedures are pain-free, and avoid the need for surgery. Radio tracers are also used in basic medical, chemical, and biological research. Food irradiation, as we have seen, keeps food safe by destroying harmful bacteria. Radiation is used in agriculture to protect crops from pests, and is used to ensure the structural integrity of planes, trains, bridges, and pipelines. And not to be overlooked are smoke detectors, TV sets, and nuclear and coal-fired power plants. Natural sources of radiation contribute far more radiation than do all others (Table 6.1). Natural radiation contributes 2.4 millisieverts (mSv; 1 Sv = 100 rems), or 86% of the total; artificial radiation, with 0.407 mSv, or 14.5%, is a meager contributor. Clearly, if there is concern about radiation, it must be with our natural background, which contributes 6 times more radiation than that contributed synthetically (by humans). If not for medical uses, man-made would approach nil. Nevertheless, we know that extremely high radiation doses can cause sickness and death. But we also know that large populations have been subjected to exposures from nuclear weapons testing, nuclear reactor accidents, and occupational exposures, as well as unusually high naturally occurring radiation. How have these people fared? We shall consider each.

Radiation is categorized according to the effects produced in tissues as ionizing and nonionizing. Ionizing includes cosmic rays, X rays, and radiation from

TABLE 6.1. Doses of Radiation from Natural
and Artificial Sources

Source	Dose (mSv)
Natural	
Cosmic	0.4
Gamma rays	0.5
Internal	0.3
Radon	1.2
	2.4
Artificial	
Medical	0.4
Atmospheric nuclear testing	0.005
Chernobyl	0.002
Nuclear Power	0.0002
	0.4072
Total (round) mSv	2.8

radioactive substances. Atoms of different types are known as *elements*. Some of the 100+, such as uranium, radium, and thorium, are unstable; that is, their nuclei are overloaded with neutrons. The nucleus of each atom contains a specific number of protons and neutrons and as such is either stable or unstable. Unstable atoms that want to become stable and must shed neutrons to do so. In the process of shedding neutrons they emit invisible and highly energetic particles or rays. This emittance is known as *ionization* or *ionizing radiation*. In the process of ionizing, an atom can be stripped of an electron, which can alter the chemical composition of living tissue.

Three types of ionizing radiation are alpha and beta particles and gamma rays. Alpha particles are the most energetic and the most massive, some 7000 times that of the beta particle, but despite their energy, they can travel only a few inches in air, losing their energy as soon as they collide with any matter. Ergo, they have weak penetrating power. As shown in Figure 6.1, alpha particles can be stopped by a sheet of paper, or the outer layer of human skin. Beta particles are smaller, high-speed electrons ejected from the nucleus of radioactive atoms, which can penetrate water, or paper, but are stopped by aluminum foil, an inch of wood, and glass, but can penetrate the top layer of human skin. Heavy exposure to beta particles can cause skin burns, and can be hazardous if inhaled or ingested.

Gamma rays are electromagnetic waves emitted from the nucleus of some radioactive atoms, and have more energy and penetrating power than do alpha and beta particles, traveling at the speed of light. This combination of high energy and high penetrating ability makes gamma rays useful in cancer treatment as a means of killing harmful tumor cells. Gamma rays can be shielded by dense barriers of concrete, steel, or lead. Gamma rays and X rays

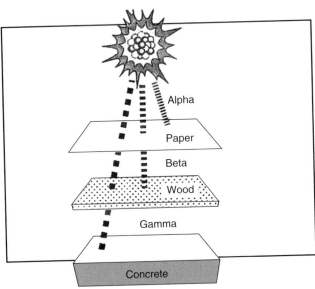

Figure 6.1. Types of ionizing radiation and their penetrating powers. The three main types of ionizing radiation are alpha (α), beta (β), and gamma (γ) rays. Alpha particles are the most energetic but can travel only a few inches in the air. They lose their energy as soon as they collide with matter. Beta particles can pass through paper, but are stopped by wood, glass, or foil. Gamma rays travel at the speed of light and are highly penetrating. A wall of concrete, lead, or steel stops gamma rays. (Courtesy of the U.S. Department of Energy.)

are essentially similar. However, gamma rays are also most dangerous because of their ability to penetrate large thicknesses of matter.

The radioactivity of an unstable element decreases with time, ultimately becoming nonradioactive. This process of decay is referred to as *half-life*, and each radioactive element has its own unique fingerprint or half-life. The half-life is the time it takes for 50% of the element's activity to dissipate or decay away. This information allows us to know exactly how much of a radioactive material still remains. Table 6.2, lists a sampling of half-lives for a dozen radioactive elements, while Figure 6.2 diagrams the activity of two radioactive elements (polonium and radium) versus time. Take note that the time needed for the activity to decrease by a factor of one-half (0.5) is always the same. The figure shows that the time taken for activity to drop by 50% of this value is constant throughout the entire decay process. In one half-life, the activity of each sample decreases by a factor of 2. During the next half-life by a second factor or 2, to one-fourth its initial value. After each additional half-life the activity remaining would be $\frac{1}{8}$th, $\frac{1}{16}$th, and $\frac{1}{32}$nd that of their original values. It takes about 7 half-lives for the original activity to fall below 1% of its initial value. So, for example, strontium-90, a radionuclide always found in fallout

TABLE 6.2. Half-Lives of Radioactive Elements

Isotope	Half-Life	Nuclide
Thorium-223	0.9 second	$^{222}_{90}$th
Nitrogen-16	8 second	$^{16}_{7}$N
Fluorine-17	66 second	$^{17}_{9}$Fl
Bromine-85	3 minues	$^{85}_{35}$Br
Iodine-131	8 days	$^{131}_{53}$Ra
Sodium-22	2.6 years	$^{22}_{11}$Na
Strontium-90	28 years	$^{52}_{38}$Sr
Cesium-137	30 years	$^{82}_{55}$Cs
Radium-226	1620 years	$^{138}_{88}$Ra
Carbon-14	5,630 years	$^{14}_{6}$C
Plutonium-239	24,000 years	$^{145}_{94}$Pu
Uranium-238	4.5×106 years	$^{146}_{92}$U

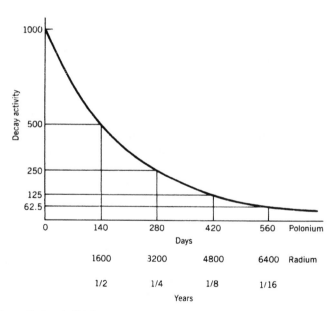

Figure 6.2. Radiation half-life decay rate. Continued activity indicates that nuclei not yet decayed.

from nuclear weapons testing, with a half-life of 30 years, will, after 6 half-lives, 180 years, still be around, albeit in small amounts—$\frac{1}{128}$ th of its original level, but it will not be zero.

The cell is the basic unit of biological tissue, and its nucleus is its control center. About 80% of the cell consists of water; the remaining 20% consists of complex biological and chemical compounds. When ionizing radiation passes through cellular tissue, it produces charged water molecules, which break up into highly reactive free radicals (OH)—reactive oxygen species—that can disrupt proteins, damaging the large DNA molecules by destroying individual bases, particularly thymine, by breaking single- and double-stranded DNA. The chromosomes that carry DNA are also at risk. DNA damage can lead to cancer, birth defects, and death. However, cells have repair mechanisms that can reverse these damaging effects, allowing the body to tolerate low-dose radiation.

BACKGROUND

Isotopes and Half-Lives

All atoms of an element, any element, have a fixed number of protons in their nuclei. This number is the element's atomic number; examples hydrogen are 1; carbon, 6; oxygen, 8; sodium, 11; and uranium, 92. All elements also have neutrons in their nuclei, but some elements can have nuclei with more than one proton/neutron grouping. These different groupings are called *isotopes*. For example, carbon has three isotopes; one with 6 neutrons, one with 7, and one with 8. But their chemical characteristics are identical. Furthermore, all elements have an atomic mass number, the total of all protons and neutrons in their nuclei: carbon-12, carbon-13, and carbon-14.

In chemical notation, isotopes are written as, for example, $^{12}_{6}C$, $^{13}_{6}C$, and $^{14}_{6}C$. The subscript denotes the number of protons, and the superscript informs us that the total number of neutrons and protons is 12, 13, and 14, respectively. One need only subtract the subscript from the superscript to obtain the number of neutrons in any nuclide or isotope. Uranium-238, with 92 protons, has 146 neutrons in its nucleus—and one additional item. Many nuclei are overloaded with neutrons, which makes them unstable and hence radioactive. Isotopes ^{12}C and ^{13}C are stable, but ^{14}C is unstable and radioactive. However, all three carbon isotopes occur naturally and are incorporated in carbon dioxide, and consequently are in all plant and animal tissue. Although ^{12}C and ^{13}C are stable and fixed in concentration, ^{14}C, which is radioactive, decays away over time, as long as the plant or animal remains alive. When the plant or animal dies, ^{14}C becomes fixed. All isotopes have half-lives—a specific period of time during which 50% of their concentration decays away. Because the half-lives of all isotopes are known, ^{14}C, for example, can be used to determine the age of a plant or animal's remains, by determining the amount of ^{14}C in the remains.

Each of the isotopes listed in Table 6.2 is radioactive, and Figure 6.2 shows the trendline for all radioactive isotopes. As shown, one-half, or 50 percent will have decayed away in one half-life. It also means that 50 percent remains. After a second half-life, an additional 50 percent has decayed away, and so on.

For instance, for cesium-137 and strontium-90, with 30- and 28-year half-lives respectively—two isotopes always present after an explosion of an atomic bomb, or radioactive emissions from the explosion at Chernobyl or Three Mile Island (a remote possibility)—the decay rates would be

Number of half-lives:	1	2	3	4	5	6
Amount remaining:	$\frac{1}{2}$	$\frac{1}{4}$	$\frac{1}{8}$	$\frac{1}{16}$	$\frac{1}{32}$	$\frac{1}{64}$
Percent remaining:	50	25	12.5	6.25	3.12	1.56
Years passed						
For strontium-90	30	60	90	120	150	180
For cesium-137	28	56	84	112	140	168

After the passage of 168 and 180 years for cesium and strontium, although both are greatly diminished, small amounts remain. As for ^{238}U, after its decay series of 13 transformations, it arrives at ^{206}Pb, lead—a stable element with no radioactive emissions. But it does require 4.5 million years.

When ionizing radiation slams into body tissue, its energy is transferred to the tissue. The amount of energy absorbed per unit weight of tissue or organ is referred to as the *absorbed dose* and is measured in terms of energy deposited per unit mass of tissue, and is expressed in units of gray (Gy). One gray is equivalent to one joule of radiation energy absorbed per kilogram of organ tissue. In the older metric system, the *rad* was the unit of radiation absorbed dose. Time out for the joule. Although "joule" may be unfamiliar, it is one of the few metric units used in the United States. A measure of energy used everyday is the watt—the energy of our lightbulbs. One watt is the transmission of energy at the rate of one joule per second. One kilowatt-hour on a utility bill means that 3.6 million joules of electrical energy were used—(1000 watts × 60 minutes × 60 joules per minute per watt). A joule, therefore, is a minuscule amount of energy. Nevertheless, a dose of one gray has been shown to cause harmful effects. But equal absorbed doses (equal amounts of deposited energy) may not have equal biological effects.

One gray, in the new International System of Units (SI), is equivalent to 100 rads: 1 mGy = 0.1 rad. Bear in mind that equal doses of the three types of ionizing radiation are not equally harmful. Alpha particles produce greater harm than do beta particles and gamma rays for a given absorbed dose because of the differences in the way in which different types of radiation interact with tissue. To rationalize this difference, radiation dose is expressed as equivalent dose in units of sievert (Sv). The dose in 1 Sv is equal to absorbed dose multiplied by a radiation weighting factor—a dimensionless constant, specific for each type of radiation. Prior to 1990, this weighting factor was referred to as *quality factor* (QF). Equivalent dose is commonly referred to as "dose" in radiation terminology. The old unit of "dose equivalent" or "dose," was the rem (roentgen equivalent in man):

$$Dose (Sv) = absorbed dose (Gy) \times radiation weighting factor$$

In the older metric system

$$Dose (rem) = dose (rad) \times QF$$

($1 Sv = 100 rem$ or $1 mSv = 0.1 rem$). Clearly different doses of radiation have far different effects. For example: a dose of $10 Sv$ can mean risk of death within days or weeks. But a dose of $1 Sv$ may confer a risk of cancer of about 5 in 100, while a $100 mSv$ dose may incur a cancer risk of 5 in 1000 individuals.

Along with the absorbed dose, there is an effective dose (ED), which is the sum of weighted equivalent doses in all organs and tissues:

$$ED = sum of (organ doses \times tissue weighting factor)$$

Let us suppose that a radionuclide induces exposure of the lungs, liver, and bone surfaces. The equivalent doses to the tissues are 100, 70, and $300 mSv$. The tissue weighting factors for these tissues are 0.12, 0.05, and 0.01, respectively. The effective dose, ED, would be calculated as $(100 \times 0.12) + (70 \times 0.05) + (300 \times 0.01) = 12 + 3.5 + 3 = 18.5 mSv$. That $18.5 mSv$ effective dose indicates that the risk of harmful effects from this pattern of exposure will be the same as the risk from $18.5 mSv$ received uniformly throughout the body, as the tissue weighting factor for the whole body is 1.00 [2].

As we have seen, both natural and artificial radionuclides pervade our world. That being the case, we now consider a number of events and studies that can help assess the health effects of radiation.

Health Effects of Radiation Exposure, Natural and Anthropogenic

With the first use of atomic bombs in warfare in 1945, what have we learned about the adverse effects of radiation, over the past 61 years? To assist in understanding, and aid in decisionmaking, we begin with Hiroshima and Nagasaki, then consider two nuclear power plant accidents, and follow that with studies of occupational exposures to nuclear industry workers. Finally, we examine the health of populations living in areas of the world with unusually high natural radiation backgrounds.

Hiroshima and Nagasaki

Monday, August 6, 1945. Three B-29 Superfortress bombers appear over Hiroshima. At 8:15 A.M. the *Enola Gay*'s bomb-bay doors open. "Little Boy," The world's first uranium bomb, plunged earthward. Colonel Paul W. Tibbets turns his plane upward and away from the rising mushroom cloud. Major Charles W. Sweany, piloting the *Great Artiste*, dropped radiation measuring instruments, then followed the *Enola Gay*. Within minutes, three-fifths of Hiroshima vanished. The pressure wave, fire, acute radiation, and thermal burns claimed over 70,000 lives.

Thursday, August 9, 1945, Major Charles W. Sweeny, this time driving *Bockscar*, a borrowed B-29, lifted off the tarmac at Saipan, on its way to Kokura—the site of the second coming of a nuclear bomb. Once over the target city, it was evident that fog and clouds were too thick for a bomb site to pierce. Major Sweeny's orders were to proceed to his second target of opportunity. He turned southeast for Nagasaki. At 11:01 A.M., the bomb-bay doors opened. "Fat Man," a plutonium bomb, fell swiftly. A multicolored mushroom cloud rose faster than at Hiroshima, demolishing the steelworks, arms factory, and thousands of residential buildings, and taking the lives of another 70,000+ people [3].

On the basis of President Harry L. Truman's directive to initiate a long-term and comprehensive epidemiologic and genetic study of the atomic bomb survivors, and there were many, the Atomic Bomb Casualty Commission (ABCC) was established in Hiroshima in 1947 and in Nagasaki in 1948. For 25 years the ABCC studied the effects of atomic radiation until it was replaced in 1975 by the Radiation Effects Research Foundation (RERF), a nonprofit Japanese foundation funded by both the Japanese Ministry of Health and the U.S. Department of Energy [4]. To this day, it remains the foremost international research collaboration, conducting large-scale, systematic studies of the long-term effects of radiation. The RERF has benefited, and continues to benefit, from four groups of individuals: a cohort of 120,000 survivors; the "life span study" (LSS), which has followed this cohort over the past 50 years; by means of a national death certificate retrieval system; and the "adult health study" (AHS), which follows a subsample of 20,000 survivors using twice-yearly health examinations. More recently mortality studies have been augmented by cancer incidence studies using their tumor registries from both Hiroshima and Nagasaki. They have a cohort of some 3000 individuals who were in utero at the time of the bombing.

In a recent publication, Dr. Itsumo Shigematsu, Director of the RERF, informs us that the ABCC's first large-scale program was a genetic study of the first-generation children of survivors. The REFR's genetic investigators, who have taken over the ABCC's studies, have, he wrote, "searched vigorously for heritable effects of radiation in the off-spring of the survivors. Not a single one of the many end points has shown a significant effect. The data suggest that humans are not unusually sensitive to the genetic effects of radiation and, further, are probably not as sensitive as had been initially extrapolated from experiments in mice." [4]—a crucial determination to be borne in mind.

Dr. William J. Shull went to Japan in 1949, at age 29, to head up the ABCC's Department of Genetics. For the ensuing five decades he was a key researcher. His recent book, *Effects of Atomic Radiation: A Half-Century of Studies from Hiroshima and Nagasaki* [5], is an engrossing and sober account of the results of observing, testing, recording, and interpreting the data obtained from the survivors and their children and grandchildren. Much of what follows comes from Dr. Shull's book.

As cancer is a fact of life and an unwelcome condition in every human society, it challenged the ABCC's scientists with a conundrum: Which of the survivor's cancers were radiation-induced, and which were not? No tags or labels offer an insight as to cause. We also know that some organs are more sensitive to radiation than others and that, depending on the site, younger people are at higher risk than older individuals. Over the 47 years from 1950 to 1997, time enough for cancers to express themselves, the LSS, with a cohort of 86,572 survivors, registered 9335 deaths from solid tumors—liver, pancreas, brain, and colon. Of these, 440 were associated with radiation exposure. Perhaps of prime importance was the observation that "there is no direct evidence of radiation effects for doses less than 0.5 Sv. Furthermore, about half of the original LSS cohort was alive at the end of the follow-up period, which suggests that there remains more to be garnered on long-term radiation effects" [6]. Be that as it may, the number of radiation-induced cancer deaths, thus far 440, must be an unimaginable revelation. The common wisdom would have it in the thousands.

The ABCC and RERF mandate was not limited to cancer. Fertility, for example, measured as chance of fertilization, successful reproduction, time between beginning of cohabitation and first pregnancy, and first live-born delivery, were not altered by radiation exposure. And what of chromosomal abnormalities? This, of course, is of critical concern as chromosomes are the vehicle for transmitting genetic information from generation to generation. As malignant tumors and chromosomal abnormalities are frequently associated, the researchers fully expected high numbers. They found no evidence of Down's syndrome or alterations in six ratios. They quite reasonably searched for biochemical changes. Using blood samples obtained from children of survivors, electrophoretic studies were done to detect abnormal proteins. Both enzyme and nonenzyme proteins were checked for 28 proteins of the blood plasma and red cells. After more than a 1.25 million tests, four mutations were found among children whose parents received more than 0.01 Sv. Of the 500 combined pregnancies in both Hiroshima and Nagasaki exposed to more than 0.01 Gy, 25 terminated with an infant with mental retardation, higher than the four or five normally expected, and a number of pregnancies terminated with infants with small heads; that is, two or more standard deviations below the average.

Among 1473 individuals ages 9–19, on whom head sizes were measured, 62 had a small head. Some were mentally retarded; some were not. In utero exposure to ionizing radiation increases the frequency of mental retardation. There was also a loss of some 25 IQ points among those exposed to 1 Gy or more. Prenatal exposure to 1 Gy appears to entail a loss in average school performance scores of about 1.6, which is equivalent to a shift of an average student from a score of 3 to about 1.4, that is, from the middle 50th percentile of a class to the lower 5th or 10th.

Ocular damage was yet another concern. Of some 464 individuals exposed in utero, 309 were examined. Only one, a male, had any degree of opacity.

Shull refers to a major concern of survivors—that of premature aging and dying. This was another challenge as, after all, what are normal signs of aging? Gray hair, loss of hair, wrinkled and coarse skin, and losses of hearing and vision. Other signs are the loss of smell, taste, and memory; spots on the skin; and the inability to walk or run as fast as one used to. But "premature aging" also refers to a time well before what is considered normal. As life expectancy increases, so does the age of premature death, which surely varies by country and time. As many of the survivors are still alive, definitive answers will emerge only as that cohort expires.

From the LSS it has become apparent that radiation at any level is not a health risk. It was also found that congenital abnormalities did not occur following the bombings, nor is there, Shull tells us "evidence that the health and development of children of survivors has been measurably impaired." Although there certainly are excess cancers among the survivors, the numbers are for smaller than the common wisdom reckoned. Dr. Shull's book should be required reading for elected officials at all levels of government, as well as students and teachers in our public and private high schools and universities, and of course for journalists. They would all be well rewarded. By the way, it is of more than passing interest that 2 weeks after dropping the bomb on Nagasaki, Major Sweeny visited Nagasaki and stood on the spot where the bomb was dropped. Major Sweeny died at age 84, on July 16, 2005. Major Sweeny had a good deal of company in Nagasaki. On New Year's Day, 1946, Marines of the 2nd Division played football in the "Atom Bowl" in Nagasaki. The Marines arrived in Nagasaki in September 1945, 6 weeks after "Fat Man," the 22-kiloton bomb had been dropped.

As Christmas approached, the Marines were feeling homesick, and to lift their spirits, Colonel Gerald Sanders organized a football game. Debris was cleared away, and Atomic Athletic Field 2 was prepared. In the game, the Nagasaki Bears, led by Angelo Bertelli, Notre Dame's star quarterback, lost by 14 to 13 to Bullet Bill Osmanski's Isahaya's Tigers. Osmanski, a Holy Cross star, led the NFL in rushing in 1939, as fullback for the Chicago Bears.

Colonel Sanders, 81, the game's last surviving participant, now living in Oxford, Ohio, also organized a Japanese children's choir for the Christmas festivities. None of the participants were concerned about radiation, and after the game the Japanese fans who watched the game left arm in arm with the Marines [7].

Three Mile Island

On March 27, 1979, Metropolitan Edison's Three Mile Island Nuclear Power Plant, near Middletown, Pennsylvania, was running at 97% full power. Like most of the 110 nuclear power plants in the United States, both TMI-1 and TMI-2, were pressurized-water reactors, as shown in Figures 6.3a and 6.3b. In

(a)

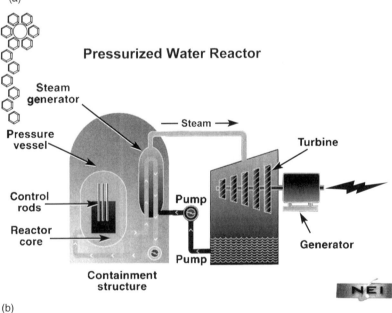

Pressurized Water Reactor

Steam
generator

Pressure
vessel

— Steam —→

Turbine

Control
rods

Pump

Reactor
core

Generator

Pump

Containment
structure

NEI

(b)

Figure 6.3. (a) View of Three Mile Island with its twin cooling towers and reactor building off to the left (courtesy of TMI); (b) the major components of a pressurized-water reactor (courtesy of the Nuclear Energy Institute); (c) schematic diagram showing greater detail within the reactor and the piping leading to the turbine, generator, and cooling towers (courtesy of the Nuclear Regulatory Commission).

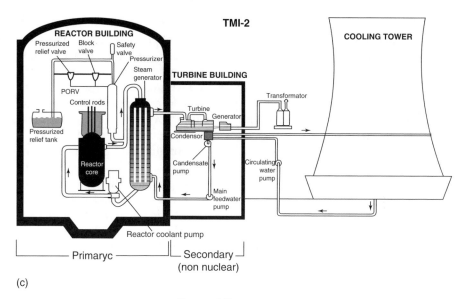

Figure 6.3. *Continued*

PWRs, the water pumped through the pressure vessel, at the rate of 90,000 gallons per minute, is kept under high pressure. As the water passes through the core, it is heated to about 600°F (315°C), and the 2200 pounds per square inch (lb/in.^{2}) of pressure prevents the water from boiling. The pressure vessel that housed the core was 36 feet high and had 9-inch-thick steel walls.

Water circulates through the core, then proceeds to the steam generator, where boiling occurs, creating the steam that drives the turbine in the adjacent building. After transferring its heat, the steam is cooled to liquid water and returns to the core to continue its endless cycle.

As evening slipped into dawn of March 28, trouble was nowhere in sight. TMI-1 was out of service for refueling, and TMI-2 was functioning normally. The staff working the graveyard shift were veterans, highly qualified professionals, having obtained their nuclear training and experience in the U.S. Navy's nuclear submarine program. They had years of experience in both the Navy and at TMI.

At 4 A.M., a pump tripped in the nonnuclear section of the plant. What should have been no more than a minor event escalated into a crisis. The main feedwater pumps stopped running, preventing the steam generators from removing heat. Both the turbine and the reactor automatically shut down, thus increasing the pressure in the nuclear portion of the plant. To relieve the pressure, the pilot-operated relief value (PORV), at the top of the pressurizer (see upper left, above control rods in Fig. 6.3c), opened. It was here at the PORV that, unbeknownst to the staff, trouble mounted. The valve should have closed when pressure decreased, but it failed to do so. Signals at the control room did not show the valve as being open. With the valve open, cooling water poured out of it, causing the core to overheat. This is where confusion ran riot. As there was no instrument to show the water level in the core, the operators judged the level according to the level in the pressurizer, and as it was high, with water pouring out of the core, into the pressurizer and out of the PORV, they judged the core to be adequately covered with cooling water. As alarms rang and warning lights flashed, the operators had no way of knowing that a loss of coolant was occurring in TMI-2. Then it got worse. Believing that there was sufficient coolant in the system, they further reduced the flow. The nuclear fuel now overheated to the point that the zirconium tubing containing the nuclear fuel pellets ruptured, allowing the pellets to melt. This was the meltdown, the so-called worst-case scenario, in which predictions held that a meltdown would breach the reactor's 9-inch-thick steel walls, the steel shielding lining the 36-inch-thick reinforced concrete containment building, and spew radioactive material around the countryside, radiating and killing thousands of people—literally. This was the "China syndrome," and was too vivid for the press to resist. Their steady drumbeat of impending disaster caused thousands to flee the area. Little had been learned since Hiroshima and Nagasaki, in which the press trumpeted the allegations that Hiroshima would be uninhabitable for decades, if not centuries, and would see epidemics of misshapen monsters, both animal and plant. Of course, none of this occurred. In fact, just months after the bombing, reconstruction was under way. Today, both Hirshima and Nagasaki are lovely cities attracting tourists from the world over.

Recall that the accident at TMI occurred only several days after the film *The China Syndrome*, starring Jane Fonda and Jack Lemon, was released. The press had a field day with that. In the film, Jane Fonda, a news anchor at a California TV station doing a series on nuclear energy, is at a nuclear power plant with a cameraman, and is raising the prospect of how unsafe the plant is. In the film, Fonda's character speaks with a safety expert who says that a meltdown could force an area "the size of Pennsylvania" to be evacuated, and the fictional near-accident in the film stems from a plant operator's misunderstanding of the amount of water within the core. Jane Fonda began lobbying against nuclear power. The physicist Edward Teller began lobbying in favor of nuclear power. It was during this period that Teller, then 71, suffered a heart attack, which he blamed on Fonda, noting that "You might say that I was the only one whose health was affected by the reactor near Harrisburg. No, that would be wrong. It was not the reactor. It was Jane Fonda. Reactors are not dangerous" [8].

Given the media circus surrounding the TMI accident, what have we actually learned about adverse health effects of the accident now that almost three decades have passed? Studies of the consequences have been conducted by the Nuclear Regulatory Commission, the EPA, the Department of Health and Human Services, the Department of Energy, and the State of Pennsylvania, as well as independent studies by the University of Pittsburgh, School of Public Health, and Columbia University's School of Public Health. Consensus governmental estimates are that the average dose to about 2 million people in the area was about 1 mrem, or 0.01 mSv. By way of comparison, exposure from a set of chest X rays is approximately 6 mrem (0.06 mSv), compared to the natural background of about 100–125 mrem (1–1.25 mSv) per year; thus the collective dose was minuscule. As for the reactor building, it has become clear that most of the radiation was contained, and the actual release, if there was any, had negligible effects on the population and the physical environment. Perhaps most illuminating was the finding that iodine-131 in the core did not remain in the gaseous state long enough to escape, but had dissolved in water or had attached to metal surfaces of the reactor building [9].

At the request of the Three Mile Island Public Health Fund (created by federal court order in 1981, to supply financial support for analysis of radiation effects in the area), the Division of Epidemiology, School of Public Health, Columbia University, conducted a study to determine whether cancer occurrence following the March 28 accident was related to radiation releases from TMI-2.

New cancer cases from 1975 to 1985, among the 160,000 residents living within a 10-mile radius of TMI-2, were obtained from all hospitals within 30 miles of TMI. During those 11 years, 5493 new cases were diagnosed. How were they interpreted? No associations were seen for leukemia in adults, nor for childhood cancers as a group. In fact, a negative trend was found for adults. Here again, it must be recalled that cancer is a fact of life, and studies must distinguish between which are related to the study question and which are not. Interestingly enough, one of the observations found that rates of leukemia among children in the area were low compared to both regional and national rates, which may speak to both the fact of any accidental emissions as well as living near a nuclear power plant. Also, the authors tell us that they "failed to find definite effects of exposure on cancer types and population subgroups thought to be most susceptible to radiation." They concluded that, "overall, the pattern of results does not provide convincing evidence that radiation releases from the Three Mile Island nuclear facility influenced cancer risk during the limited period of follow-up" [10]. Furthermore, their computer simulations found that projected exposure patterns agreed with the data obtained by TMI dosimeters used at the time of the accident.

Recognizing that there would be wide interest and concern in the results of this study, the editor of the *American Journal of Epidemiology* requested Dr. Colin R. Muirhead of the National Radiology Protection Board, Oxfordshire, UK, to provide an independent evaluation of the publication. Muirhead began by stating that "There are a number of problems associated with performing this study. Doses received as a result of the accident are estimated to be less than the annual background gamma dose of 1 mSv." Too many people who ought to know better have either overlooked that fact or prefer to ignore it. He noted, too, that "consequently it would not be expected to detect any increased risk associated with these emissions," and concluded by stating that "the authors conclusion that their study does not provide convincing evidence that radioactive emissions from Three Mile Island influenced the risk of cancer during the period of follow-up therefore seems to be reasonable" [11].

Continuing concern by residents prompted the University of Pittsburgh, School of Public Health, in cooperation with the Pennsylvania Department of Health, to undertake another study of the residents in the area. This one looked at the mortality experience and specific cancer risks of the 32,135 individuals enrolled in the study. At the outset, it is of paramount importance to know the levels and types of radiation that residents may have been exposed to. From this study we learn that investigators from the Pennsylvania Department of Health determined that the average likely and maximum whole-body gamma doses were 9 mrem (0.09 mSv) and 25 mrem (0.25 mSv), respectively. Compared to 300 mrem (3 mSv) annual effective natural background dose received by the residents of the United States, the emissions from Unit 2 were Lilliputian. As for beta radiation, with its shorter range in air, and poor penetrating power, and given clothing, shelter, and other shielding factors, the impact of beta radiation was substantially reduced, but to an uncommon level. The researchers concluded that "the radiation released from Unit 2 did not provide evidence that low-dose radiation had any impact on the deaths among the cohort studied. And, as the latency period for most cancer is at least 15 years, more likely 20 or more, continued follow-up may provide a more comprehensive description of the mortality experience" [12].

The University of Pittsburgh's first report on health affects from the TMI accident covered the period 1979–1992. Their follow-up study covered an additional 6 years: 1974–1998, and the 32,135 individuals involved in the first study were again available. A major observation was that "overall cancer mortality among this cohort was similar to the local population." Again they remarked that radioactivity released during the accident had no impact on the overall mortality of the residents. Nevertheless, given the timeframe, they suggest four areas for future investigation: the alcohol and smoking consumption of the cohort, continued monitoring of the children, studying the natural background as it relates to cancer rates, and continued follow-up of the overall mortality experience of this cohort.

March 28, 2004 was the 25th anniversary of the accident. It seems fitting that a 30-year follow-up could be useful as children at the time of the accident would then be adults, and young adults at the time would be entering the period when cancer normally exacts a heavy toll. But it is also evident that some people will never believe that TMI was not the cause of their ailments [13]. More than likely one day a journalist will admit that for them, TMI was no more than a media circus. By the way, has anyone inquired as to the adverse health affects acquired by the dozen or more operators who remained on duty during TMI's meltdown? Surely they should have received doses of radiation far in excess of doses received by any nearby residents. In fact, they had no radiation-related injuries—something more to consider, as is the fact that 325,000 cancer deaths from all causes are normally expected in the population of 2 million people living within 50 miles of TMI.

Unfortunately current terrorist concerns have mandated that visits to nuclear power plants be curtailed. Security is the order of the day. The former open-door policy permitted many the luxury of a highly educational guided tour, with questions and answers, from which you came away comfortable in the knowledge that these plants are in capable, professional hands. Worse yet, public opinion was so adamant that Metropolitan Edison decided, after the cleanup, not to reopen Unit 2. It has been standing idle ever since its cleanup, and will be decommissioned shortly. Of course the public has to find its electricity elsewhere. Over the past two decades one would have suspected that public opinion would have changed. But Chernobyl intervened. It will be instructive to see if the most recent data out of Chernobyl have a salutary effect on Pennsylvania residents. So, let us now consider Chernobyl.

Chernobyl

Chernobyl was different from Three Mile Island. Situated some 80 miles north of Kiev, and 12 miles south of the border with Belarus, forming a triangle with Chernigov and Kiev, the Chernobyl nuclear power complex consisted of four nuclear reactors of the RBMK-1000 design. The RBMKs are Soviet designed and built, graphite-moderated pressure-tube-type reactors, using slightly enriched (2% ^{235}U) uranium dioxide fuel. As shown in Figure 6.4, it is a boiling, light-water reactor, with direct steam feed to the turbines, without an intervening heat exchanger. Water pumped to the bottom of the fuel channels boils as it progresses up the pressure tubes, producing steam that feeds two 500-MW electric turbines. The water acts as a coolant and also provides the steam used

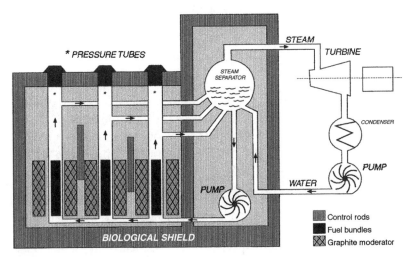

Figure 6.4. A schematic rendition of the Russian RBMK reactor. Figure adapted from *Chernobyl: Assessment of Radiological and Health Impacts*, International Atomic Energy Agency, Vienna.

to drive the turbines. The vertical pressure tubes contain the zirconium-alloy-clad fuel around which the cooling water flows [14].

The Unit 4 reactor was to be shut down on April 25, 1986 for routine maintenance. To take advantage of this shutdown, a test was set in motion. Its aim was to determine whether cooling the core could continue safely if there were loss of power. As the shutdown proceeded, the reactor was operating at about half power. Then the emergency core cooling system was switched off. Standard operating procedure required that a minimum of 30 control rods, of the total 231, were necessary to maintain control of the reactor. In the test, only about 7 were used. There was an increase in coolant flow and a resulting drop in steam pressure. The automatic trip that would have shut down the reactor when steam pressure was low, was bypassed. The reactor became unstable. The loss of cooling water exaggerated the unstable condition by increasing steam production in the cooling channels, and the operators could not prevent an overwhelming power surge.

At 1:23 A.M. Saturday, April 26, 1986, two explosions destroyed the core of Unit 4 and the roof of the building. The two explosions sent fuel, core components, structural items, and burning graphite into the air. The smoke, radioactive fission products, and core debris rose a half-mile into the air, with an unprecedented release of radioactive materials. The lighter materials were carried by the prevailing wind to the northwest of the plant and on into European countries. Fires raged in what remained of Unit 4. But these were not graphite fires. Graphite played no part in the fires as high-purity, nuclear-grade graphite doesn't burn.

Over 100 firefighters were needed, and it was this group that received the highest levels of radiation exposure and sustained the highest losses of personnel. By 5 P.M., the fires were extinguished, but many of the firefighters continued their radiation exposure by remaining on site. The intense heat of the graphite was responsible for the dispersion of radionuclides high into the atmosphere, which continued for about 20 days.

The Chernobyl accident was the result of a poorly designed reactor, with no containment building and poor safety regulations and conditions. The operators performing their test had no idea that what they planned was a recipe for disaster, and they failed to comply with established operational procedures. The combination of poor reactor design and human error created the worst nuclear reactor calamity in history. There were dire predictions that thousands, tens of thousands of people—men, women, and children—in the Soviet Union and Europe might die of radiation-related illnesses. What, in fact was its legacy?

Twenty years after the accident, a clear understanding of its impact had yet to be obtained by the countries and individuals involved. To fill this void, the Chernobyl Forum was established in 2003 by the International Atomic Energy Agency in cooperation with six specialized UN agencies—the World Bank and the governments of Belarus, the Ukraine, and the Russian Federation—as the Soviet Union no longer exists. The IAEA convened an expert working group of scientists from around the world to deal with the environmental effects, while WHO convened a panel of experts to consider the health effects. Their efforts produced a stunning report: *Chernobyl's Legacy: Health, Environmental and Socio-economic Impacts and Recommendations to the Governments of Belarus, The Russian Federation and Ukraine*, published in September 2005 [15]. No sooner had the UN-commissioned report been issued, when it was attacked by environmental groups as a biased attempt to whitewash the potential dangers of nuclear power. The fact that this report was a consensus document prepared by and contributed to by eight agencies and several countries mattered little to these naysayers. But we can let the report speak for itself.

With the exception of the on-site reactor personnel and the emergency workers present near the destroyed reactor, most of the people living in the contaminated territories received relatively low-dose whole-body radiation— comparable to background levels. Some of the reactor staff and emergency workers received doses of external gamma radiation of 2–20 Gy; 28 of them died within the first 4 months from radiation and thermal burns, and another 19 died over the years to 2004. The number of deaths attributable to Chernobyl has been of the greatest import to the public, scientists, the media, and government officials. The claims of the number of deaths that would ensue has, of course, been thoroughly exaggerated. The total number that could have died, or could die in the future due to exposures from Chernobyl in the most contaminated areas, is estimated at about 4000. This includes some 50 emergency workers who died of acute radiation syndrome, 9 children who died of thyroid cancer, and an estimated 3940 that could die from cancer as a result of radiation exposure.

The report makes it clear that confusion about the impact of Chernobyl has arisen because thousands of emergency and recovery operation workers as well as people who lived in contaminated areas died of a range of natural causes not attributable to radiation. Widespread expectations of illness, and a tendency to attribute all health problems to Chernobyl, have led residents to assume that Chernobyl-related fatalities were higher than they actually were. Of additional interest was that fact that among the general population, radiation doses were low, and acute radiation syndrome did not occur.

Now, what of the children? Iodine-131 was one of the main radionuclides released by the explosion. As part of normal metabolism the thyroid gland accumulates iodine from the bloodstream. Consequently fallout of radioactive iodine led to considerable thyroid exposure via inhalation and ingestion of contaminated foods, especially milk. The thyroid gland is one of the most susceptible to cancer induction by radiation, and children are the most vulnerable. During 1992–2000, about 4000 cases of thyroid cancer, were diagnosed among those up to 18 years of age at the time of the accident. The survival rate was 98.8%. Eight children died of thyroid cancer, and 6 died of other causes. There is reasonable certainty that the cancer deaths were due to radiation. The report also makes it clear that "There is no convincing evidence that incidence of leukemia has increased in children or adult residents of the exposed populations."

   Examinations of children's eyes and those of the emergency and recovery operation workers shows that cataracts do develop as a result of exposure. Doses as low as 250 mGy may be cataractogenic. Because of the relatively low dose levels to which the general population was exposed, there is no evidence, nor is there expected to be, of decreased fertility among men and women as a result of exposure. This is similar to the results seen in Hiroshima and Nagasaki. Birth rates may be lower in contaminated areas because of concerns about having children, but there is no discernable evidence of hereditary effects. In fact, the greatest public health hazard has been psychiatric. Because of greatly undue fear of the risks in the area, many have become alcoholic, abuse drugs, or are unemployed, anxious, or unable to function. So, as of mid-2005, no more than 50 deaths were deemed attributable to radiation from the Chernobyl eruption; almost all of these deaths were rescue workers and fire-fighters who had been directly exposed to the radiation resulting from the accident. A total of up to 4000 people may eventually die of radiation in the years ahead, but this is an estimate that could be revised upward or downward. Obviously it is important that the true scale of the accident's consequences become widely known and understood. These data offer means for coping with major releases of radiation whether caused by accident or terrorism and should improve the level of confidence in nuclear reactors as sources of much-needed electricity.

   Paralleling the Chernobyl Forum report, and of singular importance, is the news that in October 2005, farming resumed in the radiation zone. The summer crop of rye, barley, and rapeseed came in at a record 1400 tons, and none tested radioactive, and the winter rye was sprouting green. Families have begun returning to this area of Belarus 150 miles from Chernobyl, which has been among the most contaminated spots on earth. Interestingly enough, among the Chernobyl Forum's recommendations was that authorities in Russia, Ukraine, and Belarus move to reverse the psychological trauma caused by the chaos after the explosion, and encourage investment and redevelopment. Lands where agriculture was banned can be safe for growing crops once again, but farmers cannot sell their produce without a government certificate of approval [16].

   Adding a lighter touch to the increasingly positive developments, curiosity being what it is, Chernobyl has become a tourist attraction; its very name lures people in. Chernobyl may be a dead zone to some, but Chernobylinterinform, the zone's information agency, conducts chaperoned, guided tours that they maintain do not carry health risks. Although tourists are not allowed to wander about on their own, the guided tours are extensive and close-up. One-day group excursions cost $200–$400, including transportation and meals. For many tourists this is an extraordinary photo opportunity, and a bird-watcher's nirvana, cataloging the zones resurgent life [17].

   The investigations by Ronald K. Chesser and Robert J. Baker, both of Texas Tech University, at Chernobyl taught "tough lessons about politics, bias, and the challenges of doing good science" [18]. They found radiation exposure to be less hazardous than generally believed. Shades of Hiroshima and Nagasaki. Their article "Growing Up in Chernobyl" is must read for undergraduate and graduate students planning a career in science. I also recommend this insightful essay to all those trying to negotiate between fact and fiction.

Having considered warfare and nuclear plant accidents, we now turn our attention to environmental and occupational exposures as well as exposures to populations living in areas of unusually high natural backgrounds.

The Hanford Thyroid Disease Study

From 1944 to 1957, large amounts of radioactive substances were released into the atmosphere at Hanford, Washington. The Hanford Nuclear site occupies some 560 square miles in southeastern Washington State, and was the world's first large-scale plant for the production of plutonium for nuclear weapons. Over the course of years, large amounts of radioactive nuclides were released into both the Columbia River and the atmosphere. Iodine-131 was carried by winds to surrounding areas and deposited on vegetation, where it was taken up by grazing cows and goats. Ergo, drinking contaminated milk caused most of the radiation dose to the thyroid of exposed individuals mostly children. Breathing contaminated air and eating fruits and vegetables were additional sources of iodine-131 exposure.

The HTDS was undertaken because public health officials and scientists were concerned about the possible health effects of the Hanford emissions. In response to these concerns Congress directed the CDC to study the effects of Hanford iodine-131 emission on thyroid disease.

Unlike Chernobyl, or medical exposures, Hanford's exposure range was long-term, over years. The CDC's study population consisted of a sampling of people born between 1940 and 1946, who were children at the time of the largest radiation releases, and who lived in seven countries downwind of Hanford. They also represented a range of possible doses of iodine-131. Starting with birth certificates of 5199 individuals, investigators were able to locate 4350 who were still living. Of these, 3440 were willing and able to participate. However, 249 of the 3440 had moved out of the Hanford area before releases began and didn't move back before the end of 1957. They were referred to as "out-of-area participants". All participants received complete medical/physical exams including blood tests, and supplied detailed information about their diets. After 13 years, the study determined that risks of thyroid among participants were about the same regardless of dose received, and that the rates of disease occurrence were the same as in other populations. These findings do not prove that Hanford radiation had any effect on the health of the area population, but they do show that there was no increased risk of thyroid disease from exposure to the iodine-131 releases. If there were an increased risk of thyroid disease, it was too small to detect using epidemiologic methods [19].

PUBLIC HEALTH CONCERNS

Radiation Effects among Workers In the Nuclear Power Industry

Why would anyone work in a nuclear facility? Isn't that throwing caution to the wind, taking inordinate risks to life and limb? Perhaps. Perhaps not.

One of the earliest studies of nuclear power industry workers, and to this day one of the most remarkable epidemiologic studies, was conducted by the Department of Epidemiology, Bloomberg School of Public Health, Johns Hopkins University. In 1980, the U.S. Department of Energy awarded Dr. Genevieve M. Matanoski, the principal investigator, a $10 million contract to determine whether there was an excess risk of leukemia or other cancers associated with exposure to low levels of gamma radiation (from the decay of cobalt-60) to workers overhauling nuclear-powered submarines. This study lasted 8 years, and included over 70,000 shipyard workers from six government and two private bases: Charleston, Groton, Mare Island, Norfolk, Pearl Harbor, Newport News, Portsmouth, and Puget Sound. The workers, aged 16–70, were classified into three groups: nuclear workers with cumulative doses greater than 0.5 rem (0.005 mSv), workers with cumulative doses less than 0.5 mrem, and nonnuclear workers who had had similar jobs and were of the same age. Many of the workers at these eight shipyards had worked for as long as 20 years. The nonexposed population "was over three times larger than the exposed group so there were adequate numbers among the non-exposed for comparison of the effect of radiation." The final report concludes with these findings: "The nuclear worker population does not show a significant increase in risk of any of the cancers studied except for mesothelioma (which is a rare, asbestos-induced cancer) when compared to the general population." They further state that "The data clearly indicate that both nuclear worker groups have a lower mortality from leukemia and lymphobic and hematopoietic cancers than does the non-nuclear group. All three groups have lower rates than the general population." This is an important study, yet since its submission to the DOE in June 1991, it remains on the shelf, not published in any peer-reviewed scientific journal [20]. The fact that the media have yet to discover this naval shipyard workers study and bring its illuminating conclusions to a nuclear-fearing public is a greater mystery. We will have more to say about this study in Chapter 9, when dealing with epidemiology and publication bias.

All too often, official, national, and international agencies charged with assessing cancer risks to workers have extrapolated risks at low doses, using information obtained from groups exposed at high doses, which brings a good deal of uncertainty with it. To overcome this chanciness, a team of researchers from Columbia University's School of Public Health, and Canada's Radiation Protection Bureau, analyzed the mortality experience among Canadian nuclear power industry workers whose working environment exposed them routinely to low doses of whole-body ionizing radiation. Using a cohort of 45,468 workers, they sought to determine risks directly, and to compare them with risks estimated from higher doses. They found little consistency for solid tumors and "the other individual cancers assessed do not appear to offer any meaningful evidence of a positive association" [21].

In collaboration with the International Agency for Research on Cancer, a UN affiliate in Lyon, France, 16 investigators from the United States, Canada, and the United Kingdom mounted a study of the effects of low-dose ionizing radiation among nuclear industry workers in their three countries.

Using data from 95,673 workers monitored for external exposure, for up to 24 years, they reported "no evidence of an association between radiation dose and mortality from all causes or from all cancers" [22]. A more recent study of US nuclear power workers obtained results similar to the Nuclear Shipyard Workers Study, in that both cohorts showed a substantial healthy worker effect, meaning that considerably lower cancer and noncancer mortality than the general population was seen in these workers. This healthy worker effect was also seen in the study of Canadian nuclear industry workers, and may need explication. People who are employed are, on average, healthier than those who are not. Since the general population includes people who are unable to work because of illness or disability, as well as those who are employed, rates of disease and death among the general population are almost always higher than they are for those in the workforce. This can mean that any excess risk associated with a particular occupation will tend to be underestimated when compared to the general population. This "healthy worker effect" many account for a 10–40% decrease below general population rates. On the other hand, some workers will leave their jobs because of job-related ill health. These workers may move to less hazardous employment, creating an unhealthy worker effect on illness or death in these new jobs. However, there is also the real possibility that these healthy workers are intrinsically resistant to low levels of radiation, however protracted [23]. And then there is hormesis, which has been residing in the shadows these many years. Is it possible that small doses of radiation are beneficial? Why should nonnuclear workers doing the same type of work as nuclear workers of the same age, working in the same shipyards, have higher rates of cancer than the nuclear workers? Time out for hormesis.

"In all things there is a poison, and there is nothing without poison. It depends upon the dose whether a poison is poison or not" [24]. These are the words of Phillipus Aureolus Theophrastus Bombastus Von Hohenheim, the sixteenth-century Swiss physician known as Paracelsus. Although hormesis, from the Greek, meaning to excite, has a vulnerable paternity, suggesting that a substance or radiation that can be harmful at high doses may be beneficial at low doses, has been marginalized by scientists generally, and remains highly controversial. Nevertheless, data continue to indicate that chemicals we call "contaminants" may not simply be poisons, but beneficial at low doses. A major proponent of the hypothesis. Dr. Edward J. Calabrese, a toxicologist at the School of Public Health, University of Massachusetts, maintains, in the face of stiff resistance, that hormesis must be considered in any human risk assessment, and that estimates of adverse health effects at low doses linearly down to zero, is flawed science, and that "hormesis is a dose–response phenomenon characterized by a low dose stimulation and a high dose inhibition." Until there is a revolution in toxicology, hormesis may just remain in the shadows [25]. But questions will continually be raised. So, for the time being, the preference for explaining the lower cancer rates in nuclear industry workers is the healthy worker effect—which brings us to populations around the world living in high natural background areas.

Populations Living in Naturally High-Radiation Background Areas

In high natural radiation background areas of the world, as, for example, Brazil, India, China, and Iran, the radiation levels experienced by the local inhabitants are similar to or above the levels to which nuclear industry workers are exposed, as well as doses received by some residents of Three Mile Island.

Studying these populations directly may eschew the need to extrapolate or estimate risks from high doses to those receiving lower doses. If in fact there is a linear relationship between radiation dose and cancer induction, and, as some would have us believe, there is no safe dose, then in areas of high background levels, we should see high or increasing cancer rates, and in low background areas, low cancer rates.

Many areas of the world have natural backgrounds far in excess of those in Denver, the Mile High City, with 50 mrem (0.5 mSv), or in Leadville, Colorado, further up the Rocky Mountains, with 125 mrem (1.25 mSv). In most regions of the world natural backgrounds vary from 0.3 to 1.0 mSv, a variation by a factor of 3, but can exceed that by 10–100 times.

So, for example, the 33-mile-long coastal area of Kerala State, where the Indian subcontinent meets the Indian Ocean and the Arabian Sea, has natural deposits of monazite sands containing radioactive thorium. Kerala has been inhabited for over a thousand years, and its inhabitants have been inhaling and ingesting the alpha and beta particles and gamma rays emitted by the high levels of thorium all their lives—from infancy to adulthood to their dying days. Levels of total radiation have been measured at 70 mGy per year, and on average they are more than 7 times the level monitored in interior areas.

Recently the Regional Cancer Center at Trivandrum studied the population for cancer prevalence, looking for the community's burden of cancer, but "found no evidence that cancer occurrence is consistently higher because of levels of external gamma radiation" [26]. It is of interest to note that the investigators, using portable scintillometers, measured radiation levels in 66,306 houses. Following that, another team from the Cell Biology Division, Bhabha Atomic Research Center at Trombay, screened 36,805 newborn infants for congenital malformations and other pregnancy- and reproduction—related events. We are informed that "The stratification of newborns with malformations, still births, or twinning showed no correlation with the natural radiation levels in the different area. Thus no significant differences were observed in any of the reproductive parameters between the population groups based on the monitoring of 26,151 newborns from high-level natural radiation and 10,654 from the normal-level natural radiation areas of the Kerala coast" [27].

Another team from the Bhabha Atomic Research Center scrutinized lymphocytes of infants for chromosomal aberrations. In fact, this surveillance has been in place since 1986. During this time 10,230 infants have been screened for an array of abnormalities. Comparing 8493 infants from the coastal areas with 1737 infants from normal background areas, they found that "within the limitations of sample size, the frequencies of total autosomal and sex aneuploides as well as structural anomalies were comparable between the high-level and natural radiation areas" [28].

Researchers of the Dental Wing, Medical College at Trivandrum, on the Malabar coast, concerned about the number of cases of oral submucus fibrosis (OSMF), a precancerous condition that they were seeing, studied the area. OSMF results in stiffening and thickening of the oral mucosa and deeper tissues, preventing the mouth from fully opening and causing protrusion of the tongue. This disorder is not uncommon in Southeast Asia, appearing to have a predisposition for the Indian ethnic group, and Kerala has a relatively high percentage of cases. The results of the investigation determined that there was "no significant difference between the observed prevalence of OSMF in the study and control populations." They noted, too, that "It appears highly improbable that the cases of OSMF encountered in the study area were induced by high background radiation alone. Our findings suggest that the observed prevalence of OSMF in the area sampled was on a par with that in other endemic areas; the high natural background radiation cannot be said to have a causal relationship with the disorder" [29].

Of additional interest is the fact that Trivandrum (also known as Thiruvanathapurum), on the ancient Malabar coast and port, as well as the entire coastal area of Kerala to the north, along with Tamil Nadu on the southeastern tip of the continent, sharing a common border with Kerala, have tourism as one of their main industries. Their light sandy beaches and aqua, crystal-clear balmy waters lapping at their shores, attracts thousands of tourists a year from around the world who also inhale and ingest thorium's gamma rays with little, if any, apparent harm. As these studies multiply, a reevaluation of the effects of low-dose radiation exposure warrants consideration.

Researchers from the U.S. National Cancer Institute and the People's Republic of China conducted a study in the industrial city of Shenyang in the northeast of China that is reported to have the world's highest rates of lung cancer in women. Alpha-track radon detectors were placed in homes of 308 women with newly diagnosed lung cancer, while another 356 detectors were placed in homes of randomly selected female "controls." The median household level of radon was 2.3 picocuries per liter of air (pCi/L). Twenty percent of homes had levels greater than 4 pCi/L. After a year of monitoring for radon, levels were no higher in homes of those women who had developed lung cancer than in homes of those who did not. They also found that lung cancer did not rise with increasing radon levels. These results, they indicate, suggest that dose–response exposure relationships may not be as steep as risk models have suggested [30]. Here again, direct observation of exposed populations provides outcomes different from those predicted by extrapolation from high to low doses. Surely worthy of further consideration.

Yet another US/Chinese team investigated thyroid nodularity and chromosomal aberrations among 1001 women aged 50–65 in a high-natural-background area of southern China, where radiation levels are 3 times higher than normal backgrounds. Women in the high background areas were compared to 1005 women of the same age in normal areas. Here again, the results were the same. "Continuous exposure to low-level radiation throughout life is unlikely to appreciably increase the risk of thyroid cancer" [31].

With Mexico City, at 6700 feet above sea level and with some 20 million people; Quito, Equador, at 8500 feet and one million people; and La Paz, Bolivia at well over 11,000 feet and another million people; and with natural radiation backgrounds 3, 4.2, and 7.5 times higher than populations at sea level, there are no indications that such exposures to cosmic radiation have caused radiation-induced cancers beyond normal expectations. What conclusions can be drawn from the panoply of studies of the world's diverse populations? One thing is certain—we live in a world of radiation. It is all around us, and from the many populations studied, it is reasonable to conclude that low-dose radiation, below a certain level, perhaps 60 mSv and possibly higher, is safe. It is also reasonable to venture that many studies of low-dose radiation directly affecting diverse populations are at odds with the assumptions of a linear, no-threshold theory, that extrapolates from verifiable high-level radiation effects back linearly to low levels. In this no-threshold model, adverse effects are anticipated down to zero. The many direct studies show this to be an inappropriate use of the model. Nevertheless, governmental agencies prefer it. How this can be sustained in the face of data to the contrary remains a conundrum.

Of course, high doses of radiation can be harmful but are, as we have seen, exceptionally rare events. Our world, our environment, is a low-dose world, with no need to fear it. Why not determine your level? Figure 6.5 provides an opportunity to calculate the level of natural radiation in which you live. Why not fill it in.

It is also reasonable to wonder, then, whether additional studies are needed to convince a suspicious, fearful, and unaccepting public that support for radiation in the form of nuclear power as a means of generating much-needed electricity is entirely appropriate. However, given the level and type of information gathered over the past three decades that unequivocally shows low-level radiation to be safe, additional studies would be only more of the

Radon gas in our dwellings (average): $\dfrac{200}{26}$

Cosmic radiation that reaches the earth:

Because cosmic radiation is modified by the atmosphere, add 1 for every 100 feet above sea level:

Pittsburgh is 1200 ft., so add 12
Denver is 5300 ft., so add 53
Atlanta is 1050 ft., so add 10
Chicago is 600 ft., so add 6
Coastal cities are at sea level so add 0

If your house is brick, concrete or stone add 7:

Ground radiation (US average): $\dfrac{26}{}$

Water, food, air radiation (US average): $\dfrac{24}{4}$

Nuclear weapons testing fallout:

if you've had a chest x-ray this year add 9 for each one:

If you've had an intestinal x-ray add 210:

For each 1500 miles you've flown in a jet airplane during the year add 1:

If you live within 5 miles of a nuclear or coal-fired power plant add 0.3:

If you sleep with your spouse add 0.1:

Your yearly total

Figure 6.5. Radiation is all around us as part of our natural environment. By completing this chart, you will obtain a realistic value of the amount of radiation you are exposed to annually. On average we are exposed to about 300 mrem per year. How much are you exposed to?

same. A change in attitude, if it is to occur, may have little to do with studies and proof of safety. There is good reason to believe that studies of radiation effects, no matter the number and how well done, are not the issue. Other, unrelated issues are at work. We shall tackle this problem after visiting a nuclear power plant.

THE NUCLEAR POWER INDUSTRY

Generating Electricity via Nuclear Energy

Let us now visit a nuclear power plant. Let us step inside. No, this is not a giant teakettle, but it certainly acts like one. Nuclear power plants and teakettles share a common characteristic. They boil water, and produce steam. With steam, a kettle can whistle to indicate that it's ready. The steam in a nuclear power plant turns a turbine that spins a generator that produces electricity that lights our homes when we flip a switch. That's it. Generation of electricity is the name of the game.

A nuclear power plant (NPP) uses steam to generate electricity, just as coal-, oil-, and gas-fired power plants do. The difference between a fossil fuel power plant and a NPP is the method used to heat the water that produces steam and, of course, the fact that NPPs do not produce greenhouse gases. In a NPP, uranium is the fuel used in place of coal or oil. However, unlike coal or oil, uranium doesn't burn; it fissions. Also, as noted earlier, although atoms of the same element have an equal number of protons in their nuclei, they can have different numbers of neutrons. These different forms of the same element are called *isotopes*. The isotope for reactor fuel is uranium-235, which releases heat via fission: the splitting apart of a heavy atom into two atoms having slightly less mass. Time out for fission.

Enrico Fermi, an Italian physicist, conducted experiments in 1934 showing that neutrons could split atoms. When he bombarded uranium with neutrons, the elements he got were lighter than he expected. Four years later, Otto Hahn and Fritz Strassman, conducting experiments in Germany, fired neutrons into uranium and were similarly surprised to obtain barium, a much lighter element, half the atomic mass of uranium. Lisa Meitner in Copenhagen, working with Niels Bohr and Otto Frish, was quick to realize that the lighter elements that both Fermi and Hahn and Strassman found, fit perfectly with Einstein's mass/energy equality as expressed in his iconic formula, $E = mc^2$. She showed that the lost mass had changed to an unprecedented amount of energy and heat, which proved that fission had occurred, and also was another confirmation of Einstein's work.

Natural uranium consists of 0.72% of ^{235}U, 99.27% ^{238}U, and a trace of ^{234}U. The 0.72% of ^{235}U is not sufficient to produce a self-sustaining critical chain reaction in the type of nuclear reactors used in the United States, which require uranium fuel enriched to 2.5–3.5%.

The 103 operating reactors in the United States are either pressurized-water reactors (PWRs) or boiling-water reactors (BWRs). Other than the difference in pressure, both consist of four major components: nuclear fuel, control rods, coolant (or moderator), and shielding.

The essential factor in fissioning is the enormous amount of energy released as a consequence of the slight difference in mass that Einstein's equation indicates, namely, that mass and energy are interchangeable. The fissioning of a single ^{235}U nucleus releases about 200 million electronvolts (MeV) of energy—far more energy than the reaction of burning coal or oil to heat water. As shown in Figure 6.6, when a neutron strikes an atom of ^{235}U, it releases heat and 2–3 neutrons, each of which is now available for the fission of three more ^{235}U nuclei. In the next stage, about 9 neutrons (3 from each of the three fissioned uranium nuclei) are released. As long as more neutrons are being released, the fissioning continues as a chain reaction. In this way, many billions of ^{235}U nuclei can fission in less than a second, with the release of huge amounts of energy in a short time.

This heat-producing fissioning is generated and controlled in the reactor, which contains the core, control rods, and coolant. The reactor in a NPP performs the same function as a boiler in a fossil fuel plant, but no combustion gases are produced by a NPP.

Boiling-water reactors heat water in the core and allow the water to boil directly into steam, which goes directly to a turbine. A pressurized-water reactor uses water under pressure to cool the reactor and transfer heat. The heated water transfers its heat energy to a secondary system where steam is produced and is then piped to a turbine. The core of the reactor, the heart of a NPP, contains the uranium fuel, which is formed into ceramic-coated pellets

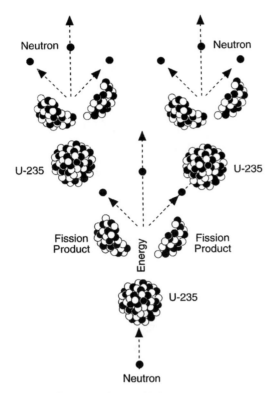

Figure 6.6. The process of nuclear fission; high-energy neutrons can split the nucleus of heavy uranium atoms, releasing lighter atoms and great amounts of energy. (Figure courtesy of the Department of Energy.)

about $\frac{3}{8}$ inch in diameter and a bit more than a $\frac{1}{2}$ inch long. Each uranium oxide (UO_2) pellet releases about the same amount of energy as does a ton of coal or 200 gallons of oil. Indeed, that's one of the great benefits of nuclear power, the tremendous amount of coal and oil *not* needed. These energy-rich pellets are stacked end to end in 12–14-foot-long rectangular arrays, and just under $\frac{1}{2}$-inch-diameter zirconium alloy tubes. These fuel rods, shown in Figure 6.7, are arranged into bundles of 225 or more, making up a fuel assembly. Reactor cores can contain 150–800 assemblies, weighing over half a ton.

Fission occurs within the assemblies, and is controlled by the neutron-absorbing cadmium control rods. As neutrons move swiftly (one-fifth the speed of light), they must be slowed if they are to strike ^{235}U nuclei, rather than simply flying through the rods. The coolant, in this case water, also acts to moderate neutron speed.

When the control rods are raised out of the core, fission increases, producing more heat. As the rods are lowered, fission decreases along with heat production in the coolant flowing between the fuel assembly rods. The coolant pre-

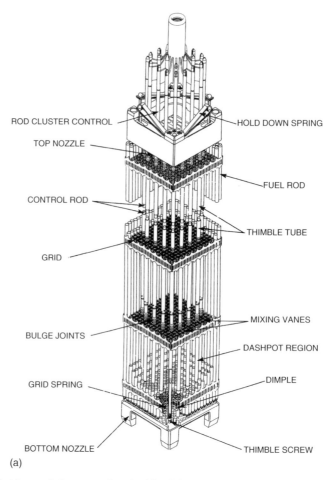

Figure 6.7. Views of the control rods: (a) a labeled schematic diagram showing all the elements; (b) a close-up view showing the size of a bundle; (c) a view showing the tubes in place.

vents the core reactor from becoming too hot and also carries heat away from the reactor to the steam generator.

In a pressurized-water reactor, the piping system that contains the coolant is called the *primary side*. The separate system of piping where steam is produced is the *secondary side*. These systems do not mix. The heated primary-system water flows through the steam generator tubes, which are surrounded by the cooler water of the secondary system. The steam generator is the link between the two systems. In the PWR, a pressurizer maintains the primary side at high pressure to prevent boiling, yet permits water temperatures to reach 600°F (315°C). Since primary-system water is far hotter than secondary-system water, it easily boils the secondary-system water to steam, which is then piped to a turbine whose shaft spins furiously, driving the connected generator,

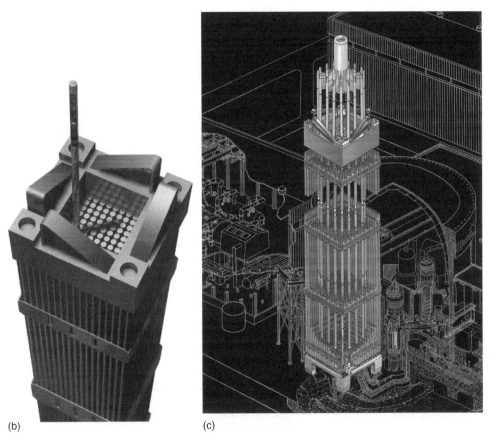

(b)   (c)

Figure 6.7. *Continued*

which churns out electricity. After spinning the turbines, the steam condenses back to water, then returns to the reactor, and the cycle begins again. A reactor coolant pump keeps the primary water circulating in the closed primary system.

Because the fission process is radioactive, barriers are built in to protect against release of radioactivity. As noted, the uranium oxide fuel is formed into ceramic pellets, sealing in the radioactive compound. The fuel pellets packed into the zirconium rods prevent release of fission products.

When a new nuclear power plant starts up, a neutron source such as plutonium is added to provide neutrons to initiate the chain reaction. When a reactor is shut down for refueling, there are neutrons in the reactor that can kick-start the chain reaction. The core, as shown in Figure 6.8, where fission occurs, is placed in a shielded 450-ton, eight-inch-thick steel reactor vessel. The reactor itself is housed in the containment building with its multiple layers of protection noted in the figure (lower right). Although NPPs produce no carbon dioxide or other greenhouse gases, they do produce waste in the form of spent nuclear fuel. Every 12–18 months NPPs are shut down for refueling. One-fourth to one-third of the oldest fuel assemblies are replaced. After 3–4 years in a reactor, the uranium pellets are no longer sufficiently energized to power the chain reaction, but still hot enough to be harmful. This waste, referred to as *spent fuel*, or *high-level waste*, must be managed appropriately, as it contains radioactive cesium, strontium, technetium, and neptunium. Given their various half-lives, some will remain radioactive for several years, while others will be radioactive for millions of years. If we were to take all the spent-fuel assemblies produced to date in the United States by the 103 operating NPPs, which generate 20% of the country's electricity, and stack them side-by-side, end-to-end, the assemblies would cover an area of about the size of a football field to a depth of about 15 feet. Not all that much. But how to store them safely?

When spent fuel is initially removed from a reactor, the assemblies are placed on racks in a 40-foot-deep pool of water contained in a steel-lined concrete basin (e.g., as shown in Fig. 6.8, upper left), where the water cools the fuel. After it has cooled for about a year, the fuel is moved to dry, 18-foot-high storage casks with 9-inch-thick skins of carbon steel, which hold about 18 tons each (similar to those shown in Fig. 6.8, lower left). These casks are either placed upright on concrete pads, or stored horizontally in concrete bunkers awaiting removal to a permanent storage facility. In September 2005, after a protracted struggle, the Nuclear Regulatory Commission voted to issue a

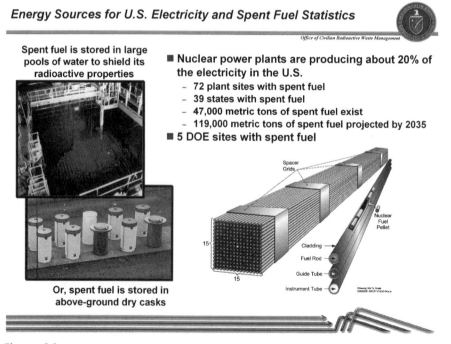

Energy Sources for U.S. Electricity and Spent Fuel Statistics

Office of Civilian Radioactive Waste Management

Spent fuel is stored in large pools of water to shield its radioactive properties

- **Nuclear power plants are producing about 20% of the electricity in the U.S.**
  - **72 plant sites with spent fuel**
  - **39 states with spent fuel**
  - **47,000 metric tons of spent fuel exist**
  - **119,000 metric tons of spent fuel projected by 2035**
- **5 DOE sites with spent fuel**

Or, spent fuel is stored in above-ground dry casks

Spacer Grids

Nuclear Fuel Pellet

15

Cladding
Fuel Rod
Guide Tube
15
Instrument Tube

Figure 6.8. Energy sources for US electricity and spent-fuel statistics. (*Source*: Office of Civilian Radioactive Waste Management.)

license to "private fuel storage," to build and operate a used nuclear fuel storage facility at Skull Valley, Utah, on the Goshute Indian Reservation, which the Goshutes have tried to obtain for the past 10 years. This wind-swept land of sage and scrub, 50 miles west of Salt Lake City, Utah, will become the country's largest bunker for highly radioactive waste, until a final resting place is established—as for example, at Yucca Mountain, Nevada, which we shall deal with shortly [32].

Radioactive waste can also be low-level waste. Approximately 90% of the radioactive waste produced around the world is low-level, but contains no more than 5% of all the radioactivity in low and high levels combined. This type of waste consists of lightly contaminated trash and debris, such as paper, clothing, cleaning materials, metal and glass, and tools used in commercial and medical nuclear industries. Many countries bury their low-level, short-lived waste in protected shallow trenches, or concrete-lined bunkers. Most low-level waste decays away to natural background levels in months or several years. In the United States, low-level waste is sent to disposal sites, such as Barnwell, South Carolina, and Richland, Washington, licensed by the Nuclear Regulatory Commission (NRC). Each state, or group of states, is responsible for disposing of and managing this waste. At these sites, the NRC requires that emissions not exceed an annual dose to any member of the public of 25 mrem to the whole body, 75 mrem to the thyroid, or 25 mrem to any other organ. Actual public exposures are far less than the NRC limits.

Since the mid-1940s spent fuel has accumulated throughout the country, stored in temporary facilities at some 125 sites in 39 states, located in urban, suburban, and rural areas, most near large bodies of water. Over 160 million people live within 75 miles of temporarily stored nuclear waste. Current storage methods are safe, but these above-ground facilities are not meant for long-term storage and will not withstand rain, wind, sun, and other environmental risks for the tens of thousands of years that they are expected to remain hazardous.

The international scientific community has determined that the best option for permanent storage is underground, and that deep, geologic disposal is technically feasible and will protect the public, provide security, and protect the environment. Think Yucca Mountain.

Yucca Mountain, in Nye County, Nevada, 90 miles northwest of Las Vegas, has been studied by the U.S. Department of Energy (DOE) and scientists in university geology departments for 20 years, without agreement. Nevertheless, in 2002, Congress approved President Bush's recommendation of Yucca Mountain as a suitable site for the DOE to construct and operate a geologic repository to safely and permanently dispose of 50,000 metric tons of high-level/spent nuclear waste currently stored at 72 sites across the country. To construct and operate the 1000-foot-deep repository, DOE must obtain a license from the Nuclear Regulatory Commission. As part of the license application, DOE must demonstrate an effective quality assurance program that ensures its safe construction and operation, while protecting public health [33].

DOE must also prove to the EPA's satisfaction that radiation will be safely contained for 10,000. At this time, DOE's plan to have waste shipped to Yucca Mountain by 2010 may be unrealistic, given changes to its plan by the state of Nevada, and the NRC's line-by-line scrutiny of DOE's proposal. However, with pressure mounting for cleanup of the many sites, and for safe, permanent disposal, Congress may force the issue. Safe passage to Yucca Mountain would then become an issue as a safe, dependable transportation system becomes a crucial link. With primary responsibility for regulating the safe transport of radioactive materials in the United States, the Department of Transportation (DOT), has been working with the NRC to set design and performance standards that must be met for a transportation package or container to be certified [34].

Of overriding importance is the fact that during the past 40 years more than 3000 shipments of spent nuclear fuel have been transported over our highways, waterways, and railroads, with a safety record of no fatalities, injuries, or environmental damage caused by the radioactive cargoes. Contributing to this successful record are the rugged, dumb-bell-shaped containers that have been developed. As shown in Figure 6.9, these are heavy, sealed, thick-walled, steel structures that safely confine the spent-fuel assemblies. These casks are considered to be the must robust containers yet developed by the transportation industry. They are also designed to shield a train shipment, on the rails and buffer cars, both fore and aft. Container design and integrity must demonstrate protection against radiologic release under the following hypothetical accidents:

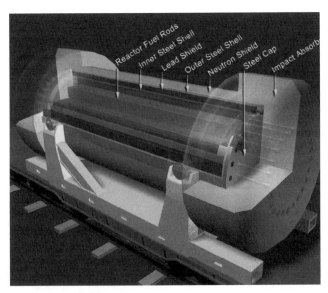

Figure 6.9. The rugged dumbbell-shaped containers used for transporting spent-fuel rods.

- A 30-foot free fall on to an unyielding surface
- A puncture test allowing the container to free-fall 40 inches onto a steel rod, 6 inches in diameter
- A 30-minute fire at 800°C (1475°F)
- An 8-hour immersion under 3 feet of water

For spent fuel, an undamaged cask must be subjected to a one-hour immersion under 660 feet (200 meters) of water [35].

On arrival at Yucca Mountain, waste will be transferred to permanent disposal containers. Management of nuclear waste is an issue that typifies scientific uncertainty and complexity and that can easily polarize people. Its extremely long-term character raises questions of intergenerational concern, and deals with questions of energy sufficiency versus long-term security. As there are no "right" answers to those ethical questions, how can the needs of the current generation be accommodated with those that follow? One option may be to reconsider a 300,000-year repository for a more modest, shorter plan, venturing that the following generation will find better ways of managing the fuel. So, with their anticipated life of a hundred years, temporary storage casks will contain cooler fuel. Recall that after 150 years, 5 half-lives, $\frac{1}{32}$ nd of both cesium and strontium's initial potency will be left. But that 100–150 years will be far from the decay requirements of the heavier isotopes. Nevertheless, time does reduce heat and radiation. Although uranium could be recovered by reprocessing, current reprocessing costs are prohibitive, given the low cost of new uranium. But things could be different in 100 years. If in a 100 years the world has run out of oil, and burning coal would be too climate-damaging, nuclear power would be in high demand, and there would be need for 500–1000 additional NPPs—or, depending on the state of fuel cells and other alternate energy sources, there would be little need for nuclear power. So, why not use the casks for the next 50–75 years, then decide what's next? By that time new storage technologies will surely have been developed, allowing greater confidence in the type of storage that would be suitable for several hundred thousand years. There is even current experimental evidence that transmutation would effectively render high-level waste benign. Particle accelerators, used to produce isotopes for medical use, could be used to fire subatomic particles into high-level waste, changing long-lived radioactive elements to short-lived waste. Creativity is something we excel at. Before too long, the storage problem, too, will be solved [36]—and with the next generation of nuclear reactors, far less waste is anticipated.

The New Generation

As of mid-2005, 440 nuclear power units were operating worldwide. Together, they supply about 16% of the world's electricity. More than half of these reactors are in North America and western Europe. Fewer than 10% are in developing countries, where the greatest growth in energy demand is expected to occur in this century. Of the 31 units currently under construction, 18 are located in India, Japan, South Korea, China, and Taiwan. Twenty of the last 29 to come online are also in the Far East and South Asia.

In addition to new nuclear installations, there has been a steady increase in efficiency and productivity. In 1990, nuclear power plants were generating electricity 70% of the time. By 2002, that had risen to 84%, with an improvement in productivity equal to adding 34 new 1000-MW nuclear plants—at minimal cost.

Yet for all the diversity of new ideas, nuclear power entered the twenty-first century with the baggage of the midtwentieth, when reactors were based primarily on technology that propelled nuclear submarines. However, vital signs within the electrical industry indicate a resurgence in interest in nuclear power because environmental benefits derived from the elimination of greenhouse gas emissions from nuclear power is gaining wide popular acceptance; the economics of nuclear power have become competitive with fossil fuel, and new, improved designs incorporating superior performance and safety features compel a second look.

Several generations of reactors are generally recognized. Generation I reactors were developed in the 1950–1960s; none are in use today. Generation II reactors are embodied in the country's 103 operating units, of two distinct types: the boiling-water reactors (BWRs) and the pressurized-water reactors (PWRs). In BWRs water is heated by nuclear fuel and boils to steam in the reactor vessel, which is then piped directly to a turbine, setting it spinning. These were designed and manufactured by General Electric. In PWRs, designed and built by the Babcock & Wilcox Co. and The Westinghouse Electric Corporation, water is heated by nuclear fuel but kept under pressure to prevent boiling. The hot water is pumped from the reactor to a steam generator. There the water's heat is transferred to a second, separate water supply, which boils to steam and on to a turbine. Although these second-generation units have been found to be safe and reliable, they are being superseded by new, simpler, less costly power plants. The major change in the third-generation units is the incorporation of "passive" or inherent safety features, which do not require active controls or intervention to avoid accidents in the event of malfunction. Both BWRs and PWRs are "active" systems requiring electrical or mechanical operation on command. Fully passive safety systems do not depend on the functioning of pumps or other devices.

Generation III includes General Electric's "advanced boiling-water reactor," four of which are online in Japan, with another one under construction, and one in Taiwan. Four more are planned in Japan and another in the United States. These require less space, have fewer operating components, and are less costly to construct.

Another highly innovative advanced reactor is Westinghouse's AP1000 and its smaller relative the AP600. Both were recently certified by the Nuclear Regulatory Commission [37]. The AP, meaning "advanced passive," has an operating life of 60 years and its passive safety systems require 50% fewer valves, 35% fewer pumps, 80% less piping, and 70% less cable. In this new design, water for emergency cooling is contained above the reactor. Pumps are no longer needed. Should water for emergency cooling be needed, it drops into the reactor. In December 2006, China announced the purchase of four Westinghouse AP1000 pressurized water reactors, as the first stage in its plan to build 30 reactors by 2020, to provide its growing need for electricty.

General Electric has developed yet another Generation III reactor that was recently submitted to the Nuclear Regulatory Commission (NRC) for approval and certification. This one, the "economic simplified boiling-water reactor" (ESBWR), which also employs passive fail-safe design, along with fewer pumps, valves, and piping, also incorporates a newly designed fuel assembly that provides improved fuel utilization.

In September 2005, Grand Gulf Nuclear Station, near Port Gibson, Mississippi, and Bellefonte Nuclear Plant, near Scottsboro, Alabama, were selected as sites for new nuclear reactors. Grand Gulf was designated as the site for GE's ESBWR, and Bellefonte will be the site for Westinghouse's AP1000. However, given the time needed for NRC review, and plant construction, the newest plants in 30 years will not be available until 2014 [38].

Nevertheless, several electrical companies are moving quickly to place orders for new reactors because the energy bill passed by Congress in July 2005 provides financial incentives for a half-dozen. Constellation Energy of Baltimore is planning to build a 1600-MW reactor, half again as large as the current operating systems. They plan to build the EPR, the European pressurized-water reactor, a novel French design, that will have a 60-year life, and provide cheaper power. Two EPRs are currently under construction: one in Olkiluotu, Finland, and another in Flamanville, France. Constellation energy operates two nuclear power plants in Scriba, New York, near Syracuse, and indicates that when the new plant is approved by the NRC, it will be built at either the Scriba site or in Calvert County, Maryland [39]. County officials at each site have already voted in favor of constructing a new reactor in their backyards.

In place of water-cooled reactors, a number of advanced Generation III reactors are gas-cooled. One of the most compelling is the Pebble Bed Modular Reactor, developed by the South African Company PTY, Ltd., in conjunction with the utility company Eskon. Its most captivating features are its small size, simplicity, and low cost. The PBMR is a helium-cooled, graphite-moderated high temperature reactor, in which specks of 9% uranium dioxide (UO_2) are coated in several layers to make particles a millimeter in diameter. About 15,000 particles are mixed with powdered graphite and pressed into spheres 60 mm in diameter—a bit smaller than a tennis ball. Each "pebble" contains 9 grams of uranium. The core of the reactor will contain 456,000 pebbles—equivalent of 4.1 metric tons of uranium.

The helium coolant entering at 500°C (932°F) flows through the pile of pebbles, picking up heat to about 900°C (1652°F), then flows to the turbines directly connected to a generator. The fact that this reactor does not need to be shut down for refueling for about 3 years is another plus. Each day pebbles are taken from the base of the reactor and measured to determine the amount of fissionable material left. Those that are still usable are replaced on top of the pile. This way, there is constant shuffling of the fuel to obtain full utilization. Over the course of 38 months, each pebble passes through the reactor 6 times. There are no fuel assemblies to store, and little waste. This pebble bed design is being further developed by Chinese scientists at Tsinghua University, the Chinese equivalent of MIT.

Yet another desirable feature of the Pebble Bed is its small size—one-sixth the size of traditional reactors; modules are 165 and 200 MW, and are planned to be clustered in units of 2, 3 or 4, to be added as needed. Perhaps of even greater importance, this power plant is meltdown-proof. In the event of a catastrophic cooling-system failure, instead of skyrocketing temperatures, the core climbs to about 1600°C, much below the >2000° melting point in the pebbles, and then drops. Time for decisionmaking is not a problem. There is as much time as needed to fix a problem. It doesn't get much better. The Pebble Bed's fail-safe character was proved in Germany. Germans stopped the flow of coolant through the reactor core and left the control rods withdrawn just as if the reactor were operating normally and generating power. They found that the reactor core shut itself off within a few minutes with no deterioration of any of the components, thus establishing the inherently safe design [40].

Another high-temperature gas-cooled reactor has been developed by General Atomics of San Diego. Their GT-MGR, "gas turbine-modular helium reactor," uses tiny coated hexagonal prisms rather than pebbles. According to General Atomics scientists, they have eliminated the need for safety systems. In the GT-MHR, the worst-case scenario cannot occur. With these new systems, we are a long distance from the nuclear submarine era [41].

What is needed now is for countries to step up and place orders for these units. We certainly are going to need them as demand for electricity grows. But these are not the only new concepts in the pipeline. India has developed an experimental fast-breeder reactor whose fuel is thorium-232, which doesn't fission—but neutrons slamming into thorium-232 produce uranium-233, which does, supporting a chain reaction; and this reactor's coolant is liquid sodium. Other countries are using liquid lead and molten-salt-cooled processes. Creativity is in the air. In water-moderated reactors, water slows down neutrons. Fast breeders use fast neutrons and liquid sodium, lead, or salt as coolants, which do not slow neutrons. These fast breeders also create more plutonium from the uranium-233/plutonium mixture that is used as fuel. They are called "fast reactors," or "fast breeders" because the neutrons are not slowed, but remain at high energy, and because they breed or produce more plutonium than was initially present. Furthermore, their uranium fuel doesn't require enrichment, and the plutonium-239 generated as a consequence of neutron bombardment of uranium-238 can be separated out and reused. This plutonium is wholly unsuitable for atomic bombs [42].

Galena, Alaska needs electricity. It needs cheaper electricity, as it currently pays 3 times the national average for its electricity. Galena doesn't need a great deal of electricity for its 700 residents, but it does get cold up there 550 miles northwest of Anchorage. Toshiba, the Japanese electrical company, would like to help. They have offered Galena a free reactor, if the town would pay the operating costs. Galena's citizens voted to accept the offer. Toshiba's 10-MW 4S reactor (supersafe, small, and simple), using liquid sodium as its coolant, would be installed underground, and in the event of a cooling system failure, heat would be dissipated directly into the ground. The 4S has no control rods to control the movement of neutrons; rather, the reactor has reflector panels around the edge of the core. When the reflectors are removed, the numbers of neutrons becomes too low to sustain the chain reaction. The reactor fuel would be uranium-enriched to 20%, which would allow it to run for 30 years without shutdown for refueling. The NRC, wholly unfamiliar with the 4S design, will review it, but that will take several years. Galena's need for cheaper electricity will not be met soon [43].

Insurance

If nuclear power plants are as safe as they are said to be, why can't they get insurance? This question is frequently asked, and this myth needs exploding. Nuclear power plants are insured; they've always been insured, not by the usual, run-of-the-mill private insurance companies, but by an agency of the federal government. To ensure that funds would be available to settle liability claims in the event of an accident at one of the country's NPPs, resulting in either environmental damage or harm to health, the Price–Anderson Act of 1957 requires licensees for these plants to have primary insurance, currently $300 million per site. The Act also requires secondary coverage in the form of premiums to be contributed by all licensees to cover claims that exceed the $300 million of primary insurance. If these premiums are needed, each licensee's payments are limited to $10 million per year and $95.8 million in total for each of its plants. If claims were to exceed the $10 billion currently available in reserve, the NRC can request additional funds from Congress. No nuclear power plant can be operated without liability insurance as established under the Price–Anderson Act [44].

According to the NRC, the liability insurance policies issued in the United States cover nuclear accidents, including those resulting from theft or sabotage, during transportation of fuel to a reactor site, or following the discharge of radioactive effluent during transportation from a NPP to a storage site. The policy also provides financial assistance for bodily injury, sickness, disease or resulting death, and property damage and loss, as well as reasonable living expenses if evacuation is required. Price–Anderson appears to cover all eventualities. Considering the fact that insurance coverage has been "on the books" for almost 50 years, one can only wonder at the charge that insurance was unavailable, and therefore, surmise that nuclear plants must be inherently unsafe, and uninsurable.

Heat for Homes and Industry

There is yet another dimension to NPPs that is unfamiliar to a great many people—its capability to deliver heat for homes and commercial buildings. Since the early days of nuclear power development, the direct use of heat generated in reactors has been employed by a dozen European countries, India, Japan, and China for district heating in addition to electricity generation. After all, what nuclear plants do is generate heat. But here in the United States, the hot water generated in the reactors has not been tapped for space heating. Technically, over the past decades, a great deal of experience has been gained about using nuclear heat for space heating and industrial processes. Ergo, there is no technical impediment to the application of nuclear reactors as heat sources. Difficulties may arise with transportation of heat via pipelines that require thermal insulation, pumping, and maintenance, especially beyond several miles. Cogeneration has years of use in the United States, but is unrelated to NPPs. As nuclear power plants have been capital-intensive, tapping their hot water for space heating could cover overall costs, as hot-water heating would easily complete with expensive oil, coal, and natural gas; and with an increased number of NPPs, more areas would be served. Penetration into the heating market given current conditions—and for the foreseeable future—should not be difficult. Here, too, both industrial and residential heating contribute substantial greenhouse gases to the atmosphere, which greenhouse-gas-free nuclear-generated hot-water would replace. Locating nuclear cogeneration plants near industrial users, especially the smaller Pebble Bed or GT-MHR-type reactors, should be highly competitive financially, while helping reduce CO_2 and sulfur emissions. Indeed, NPPs are a technology whose time has come—again. Without fear or favor, it can be categorically stated that nuclear power generation, of both electricity and heat, has by far the best safety and environmental record of any technology in general use. Yet the wall of negative perceptions has kept it from working for us. The current window of opportunity must include the nuclear option. To allow it to slam shut without it would be not only irresponsible but also tragic, for as is now evident, demand and competition for fossil fuels has sent prices soaring.

But electricity and space heating are still not the long and short of it. NPPs can be a "triple threat"—in the best sense of the word. NPPs generate very high temperatures, which is what it takes to produce hydrogen—for the hydrogen that wants to replace our carbon–based economy, which means the end of both coal and oil.

At the DOE's Idaho National Engineering and Environmental Lab (INEEL), an innovative nuclear power plant has produced the hydrogen equivalent of 400,000 gallons of gasoline for days running. PBMR (Pty. Ltd), the Pebble Bed folks from South Africa, submitted a proposal for the DOE's billion-dollar hydrogen production project. The hydrogen initiative calls for a plant that can generate electricity and high-temperature heat; with minor modifications, the current Pebble Bed can meet this requirement. The idea is for the heat to liberate hydrogen from water, rather than be piped out to heat homes and businesses. The hydrogen produced would be used to power cars, buses, and trucks, saying goodbye to gasoline. So NPPs can become the unlimited power source for electricity, heat, and hydrogen—surely reason aplenty to get even the most hard-bitten environmentalists to acquiesce. Admittedly, if environmentalists take global warming seriously, they must go for nuclear power.

An Attitude Problem

Earlier, I suggested that a change of attitude vis à vis NPPs, if it is to occur, may have little, if anything, to do with the need for more studies and proof of safety. Considering that there is good reason to believe that unrelated issues are at work in the war against the atom—for that is what it has been this past 25 years, and as the disputes rage, especially over such issues as waste management, the economics and safety of nuclear power compared with other sources of electricity, the possible links with nuclear weapons, and the attitude of the public toward the nuclear industry—decisionmaking is either paralyzed or dominated by those who shout the loudest. As a result, government, industry, and the financial sector have shied away from a coherent energy policy that included nuclear power.

Early on, the dominant voices, organized and led by environmentalists, dubbed NPPs "nukes." For them, nukes, a moniker originally reserved for atom bombs—nuclear bombs—would be seared into the American consciousness to remind us that nuclear power plants were nothing less than glorified atom bombs. It worked, becoming common usage. And simultaneously they have turned a deaf ear and blind eye on the many major developments in nuclear plant, efficiency, and safety. Environmentalists have acquired power far beyond their numbers—holding the country in thrall to their convoluted purposes: Nuclear power plants, no! Fossil fuels, yes. Greenhouse gases, yes, for decades, allowing excess greenhouse gases to exert their climate-changing impact. How could this happen? It happened because so many of us chose to do nothing. We lost our voices and sat on our hands. Allowing other voices to speak for us. Yet the public generally appears agreeable to, and accepting of, nuclear power plants. In April 2004, a survey conducted by Bisconti Research found that 65% of the public favored the use of nuclear energy; 64% said it would be acceptable to add a new nuclear power plant at the site of the nearest operating unit—this was up from 57% in 2003. Furthermore, 54% indicated that new nuclear plants definitely should be built to provide for future electricity needs. Conducting a second poll in 2005, they learned that 83% of Americans living in close proximity to NPPs favor nuclear energy and 76% are willing to have a new reactor built near them.

A survey of 1152 randomly selected nuclear plant neighbors living within 10 miles of 64 NPPs found that these people had a high degree of familiarity and comfort with nuclear energy and would welcome the economic and environmental benefits of new nuclear plants. By a margin of 83% to 16%, plant neighbors said they favored the use of nuclear energy as one of the ways to provide electricity in the United States [45]. But as important and illuminating as these surveys are, they are not as potent as the smoking gun of global warming, which appears to have opened cracks in the environmentalist wall of opposition.

The first crack came in May 2004, when the guru of the Greens, the internationally acclaimed chemist, James Lovelock, published an appeal to drop fossil fuels in favor of nuclear power. Now here was an unmitigated heresy. Lovelock opined that nuclear power was the best way to avert climatic catastrophe. Listen to him: "Opposition to nuclear energy," he wrote, "is based on irrational fear fed by Hollywood-style fiction, the Green lobbies, and the media." He pulled no punches, taking on their most sacrosanct icons. He continued [40]:

> Even if they were right about its dangers, and they are not, the worldwide use as our main source of energy would pose an insignificant threat compared with the dangers of intolerable and lethal heat waves and sea levels rising to drown every coastal city of the world. We have no time to experiment with visionary energy sources: civilization is in imminent danger and has to use nuclear, the one safe, available energy source, now, or suffer the pain soon to be inflicted by our outraged planet.

His was a fair appraisal. Nevertheless, for his efforts, brickbats were hurled his way. He was breaking ranks. But worst of all, he was telling them what they did not want to hear.

A second crack came from an unexpected source and created hysteria in the ranks: Steward Brand, who was invoked in Chapter 4 as suggesting to his acolytes, to their downright chagrin, that over the next 10 years they would need to reverse their opinion and activism in four major areas—population growth, urbanization, genetically engineered organisms and nuclear power—and, as I also noted, the man has no shame. For him, science and scientists can be ignored. But he can speak for himself: "There are a great many more environmental romantics than there are scientists. That's fortunate, since their inspiration means that most people see themselves as environmentalists. But it also means that scientific perceptions are always a minority view, easily ignored, suppressed, or demonized, if they don't fit the consensus story line." That's it. No holds barred. Scientists and their studies can be ignored, dismissed, demonized. The environmentalist view will prevail. Of course, he's right. The environmentalist world view is wholly unrelated to studies of radiation and new nuclear plant design. Recall that they were quick to denounce the Chernobyl Forum's recent report as a fraud and whitewash because its findings and conclusions were totally at odds with their preferred beliefs. Yet they have won the hearts and minds of enough people to have their views prevail—and Brand has the chutzpah to say it, and put it in writing. They'll change their attitude, their position, when they are good and ready to, because at this moment they own the cat bird seat. Lovelock had it right when he said, "we have no time to experiment with visionary energy sources." Environmentalists, with their penchant for sustainable energy sources to provide the country's and world's need for electrical energy, are in effect fiddling while the climate heats up. Sustainable energy, while able to supply a small percentage of the country's energy needs, is a mouse that roars.

Brand does lay it to his followers, and his honesty is breathtaking: "The environmental movement has a quasi-religious aversion to nuclear energy." Even so, for him the way ahead is clear: "Everything must be done to increase energy efficiency and decarbonize energy production. Kyoto accords, radical conservation in energy transmission and use, wind energy, solar energy, passive solar, hydroelectric energy, biomass, the whole gamut. But add them all up and its still a fraction of enough. The only technology ready to fill the gap and stop the carbon dioxide loading of the atmosphere is nuclear power." But he wasn't finished—hear this: "Nuclear power plants are very high yield, with low cost fuel. Finally, they offer the best avenue to a hydrogen economy, combining high energy and high heat in one place for optional hydrogen generation." He also noted that "the storage of radioactive waste was a surmountable problem" [46]. Straight talk! But 30 years in arriving. So, it is painfully clear that the country will get nuclear power when environmentalists are ready to endorse it. For the moment, they control the agenda. Brand said something else, which must be remarked on, as he was right again: "Americans are so wasteful of energy that their conservation efforts can have an enormous effect."

In the spring of 2005, my wife and I headed off to Spain. During our 2 weeks in Barcelona, Seville, and Madrid, we saw again that the Spanish, as with other common market citizens, are serious about energy conservation. Hotel rooms, for example, can be opened only with a key card that also controls room lights and air conditioning. When room doors are opened, rooms remain dark until the key card is inserted into a wall-mounted sensor, which turns on the electricity. As long as you remain in your room, you control the energy use. Leave your room, the key card goes with you, or you're locked out. Removing the key card, shuts off the electricity. Step into your hallway, lights go on. Continue walking to stairs or elevator, and lights go off behind you. Some hotels have removed the option of guests turning off lights and air conditioning. As they have reduced use of coal and oil, so, too, have all buses, and many trucks, which now run on natural gas. No longer does gray/black smoke belch from tailpipes. The air in cities is clean and fresh even in the unimaginable vehicular traffic in Madrid and Barcelona. In the United States, we have yet to arrive at such a clean-air condition.

Great demands are made for ever-increasing amounts of electrical energy by users of small appliances that use fully 25% of all energy. Who are the profligate users of electrical energy? Students and teenagers are surely among them; many who consider themselves environmentalists, and who often march to save the earth—the theory—but in practice demand more and more energy to run their radios, TV, desktop computers, printers, laptops, scanners, clocks, hotplates, toasters, DVDs players, popcorn poppers, microwave ovens, video-games, lamps, hair dryers, curling irons, can openers, refrigerators, knife sharpeners, coffeemakers, and more. But that is not the end of it. In addition to this horde of energy-gobbling devices, few users are aware that many of the latest devices are on standby mode; TVs, VCRs, anything that works via a remote, as well as adapters for MP3 players and cell phones, along with chargers for battery-operated devices, and cable modems left on 24/7, suck current all day and all night. It has been estimated that in a typical household 1000 kW per year are required as they wait for the "click" to instantly come alive. Those 1000 kW per household cost us $1 billion a year to power our myriad devices in SLEEP or even in OFF mode. Until electronic appliance manufacturers develop units that do not suck up energy in SLEEP or OFF mode, the responsibility for energy conservation is in our hands. Among the worst offenders are big-screen TVs (e.g., plasma TVs) because of satellite and cable boxes that can draw up to 320 watts in OFF mode [47]. The fact is, OFF is meaningless. Most of these devices are in neutral or idling mode, with electricity running constantly. With new electronic appliances appearing on the market almost daily, and with ever more people using them, demand for electricity has skyrocketed, but few users see themselves as contributing to the demand. Indeed, conservation does begin at home. By the way, do you know the number of lightbulbs currently used in your home? I counted ours, and was amazed to find 63. How many do you have?

When finally geared up with adequate nuclear power, the United States will be free of foreign oil and the pressures it exerts on our relationships with other countries. Furthermore, the worldwide competition for oil by newly industrialized China and India continues to propel oil prices to record levels, which does nothing but disrupt our economy. If we are serious about getting off fossil fuels, and preventing the disruption of the polar regimes in search for new supplies of oil and gas, now that the permafrost is disappearing, nuclear power is the way to go. And with global warming at hand, it is predicted that hurricanes in the Gulf of Mexico will become more frequent and violent, creating and leaving untold economic damage in their wake, as well as disrupting oil production there—oil needed until nuclear power becomes widely available. To be sure, there are good and sufficient reasons to support nuclear power. Your future depends on it.

REFERENCES

1. Lufburrow, R. F., Antoine Henri Becquerel, *Encyclopedia Americana*, International Edition, Scholastic Library Publishing, Danbury, CT, Vol. 3, p. 429; see also (http://en.wikipedia.org/wiki/Henri_Becquerel#Rise_on_natural_sciences2c_discoveries_and_major_works).

2. *Radiation, People and the Environment*, International Atomic Energy Agency, Vienna, Austria, IAEA/PI/A 75/04-00391, Feb. 2004.

3. Goldstein, R., Charles Sweeney, 84, pilot in bombing of Nagasaki, dies, *New York Times* (Obits) (July 19, 2004).

4. Shigematsu, I., and Mendelsohn, M. L., The Radiation Effects Research Foundation of Hiroshima and Nagasaki: Past, present and future, *JAMA* **274**(5):425–426 (1995).

5. Shull, W. J., *Effects of Atomic Radiation: A Half-Century of Studies from Hiroshima and Nagasaki*, Wiley-Liss, New York, 1995.

6. Preston D. L., Shimizu, Y., Pierce, D. A. et al., Studies of mortality of atom bomb survivors: Report 13: Solid cancer and noncancer disease mortality: 1950–1997, *Radiat. Res.* **160**:381–407 (2003).

7. Lukacs, J. D., Nagasaki, 1946: Football amid the ruins, *New York Times* (Dec. 25, 2005).

8. Three Mile Island: The China syndrome, *Wikipedia*, The Free Encyclopedia (http://en.wikipedia.org/wiki/Three_Mile_Island#TheChinaSyndrome).

9. Walter, J. S., *Three Mile Island: A Nuclear Crisis in Historical Perspective*, Univ. Calif. Press, Berkeley, 2004.

10. Hatch, M. C., Beyea, J., Nieves, J. W., and Susser, M., Cancer near the Three Mile Island nuclear plant: Radiation emissions, *Am. J. Epidemiol.* **132**(3):397–412 (1990).

11. Muirhead, C. R., Invited commentary: Cancer near nuclear installations, *Am. J. Epidemiol.* **132**(3):413–415 (1990).

12. Talbot, E. O., Youk, A. O., McHugh, K. P. et al., Mortality among the residents of the Three Mile Island accident area: 1979–1992, *Environ. Health Perspect.* **108**(6):545–552 (2000).

13. Talbot, E. O., Youk, A. O., and McHugh-Pemu, K. P., Long-term follow-up of the residents of the Three Mile Island accident area: 1979–1998, *Environ. Health Perspect.* **111**(3):341–348 (2003).

14. *Chernobyl: Assessment of Radiological and Health Impact, Update of Chernobyl: Ten Years On*, Nuclear Energy Agency, OECD, 2002, Chap. I, The site and accident sequence. (http://www.mea.fr/html/vp/chernobyl/eo/.html.)

15. The Chernobyl Forum, *Chernobyl's Legacy: Health, Environmental and Socio-economic Impacts and Recommendations to the Governments of Belarus, The Russian Federation and Ukraine*, IAEA, Vienna, Austria, IAEA/PI/A.87/05-28601, Oct. 2005.

16. Myers, S. E., Belarus resumes farming in Chernobyl radiation zone, *New York Times* (Oct. 22, 2005).

17. Chivers, C. J., New sight in Chernobyl's dead zone: Tourists, *New York Times* A1, A4 (International) (June 15, 2005).

18. Chesser, P. K., and Baker, R. J., Growing up in Chernobyl, *American Scient.* **94**:542–547 (2006).

19. Centers for Disease and Prevention, *The Hanford Thyroid Disease Study: Final Report*, National Center for Environmental Health, Atlanta, June 21, 2002.

20. Matonoski, G. M., *Health Effects of Low-Level Radiation on Shipyard Workers: Final Report*, U.S. DOE. DE-AC02-79EV/10095, Washington, DC, June 1991.

21. Zablotska, L. B., Ashmore, J. P., and Howe, G. R. Analysis of mortality among Canadian nuclear power industry workers after chronic low-dose exposure to ionizing radiation, *Radiat. Res.* **161**:633–641 (2004).

22. Cardis, E., Gilbert, E. S., Carpenter, L. et al., Effects of low doses and low dose rates of external ionizing radiation: Cancer mortality among nuclear industry workers in three countries, *Radiat. Res.* **142**:117–132 (1995).

23. Howe, G., Zablotska, L. B., and Fix, J. J. et al., Analysis of the mortality experience amongst U.S. nuclear power industry workers after chronic low-dose exposure to ionizing radiation, *Radiat. Res.* **162**:517–526 (2004).

24. Jacobi, J., *Paracelsus: Selected Writings Bollinger Series XXVIII*, Princeton Univ. Press, Princeton, NJ, 1951.

25. Calabrese, E. J., Historical blunders: How toxicology got the dose-response relationship half right, *Cell Molec. Biol.* **51**:643–654 (2005).

26. Nair, K. M., Nambi, K. S. V., Amma, N. S. et al., Population study in the high natural background radiation area in Kerala, India, *Radiat. Res.* **152**:S145–S148 (1999).

27. Jaikrishan, G., Andrews, V. J., Thampi, M. V. et al., Genetic monitoring of the human population from high level natural radiation areas of Kerala on the southwest coast of India. I. Prevalence of congenital malformations of newborns, *Radiat. Res.* **152**:S149–S153 (1999).

28. Cheriyan, V. D., Kurien, C. J., Das, B. et al., Genetic monitoring of the human population, etc. II. Incidence of numerical and structural chromosomal aberrations in the lymphocytes of newborns, *Radiat. Res.* **152**:S154–S158 (1999).

29. Rajendran, R., Raju, G. K., Nair, S. M., and Balasubramanian, G., Prevalence of oral submucous fibrosis in the high natural radiation belt of Kerala, South India, *Bull. WHO* **70**(6):783–789 (1992).

30. Blot, W. J., Xu, Z., Boice, J. D. et al., Indoor radon and lung cancer in China, *J. Natl. Cancer Inst.* **82**(12):1025–1030 (1990).

31. Wang, Z., Boice, J. D., Wei, L. et al., Thyroid nodularity and chromosome aberrations among women in areas of high background radiation in China, *J. Natl. Cancer Inst.* **82**(6):478–485 (1990).

32. *Nuclear Regulatory Commission (NRC) Set to Award License for PFS Used Fuel Storage Facility*, U.S. NRC, Washington, DC, Sept. 12, 2005 [see also Johnson, K., A tribe nimble and determined, move ahead with nuclear storage plan, *New York Times* (Feb. 28, 2005)].

33. U. S. Government Accountability Office, *Low-Level Radioactive Waste. Future Waste Volumes and Disposal Options Are Uncertain*, GAO-04-1097T, Washington, DC, Sept. 30, 2004.

34. U. S. Government Accountability Office, *Yucca Mountain. Persistent Quality Assurance Problems Could Delay Repository Licensing and Operation*, GAO-04-460, Washington, DC, April, 2004.

35. Office of Civilian Radioactive Waste Management: Office of National Transportation, *Transportation of Spent Nuclear Fuel. Fact Sheet*, Yucca Mountain Project, Las Vegas, NV, Feb. 2005 (http://www.ocrwm.doe.gov/factsheets/doeymp0500.shtml.)

36. Wald, M. L., Nuclear power hopes to find a welcome mat again, *New York Times* (Jan 27, 2005).

37. Bruschi, H. J., The Westinghouse AP1000—final design approved, *Nuclear News*, 30–34 (Nov. 2004).

38. Kray, M., NuStart selects Grand Gulf, Bellefonte for advanced nuclear plant licenses, *NuStart Energy News*, Exelon Generation, Philadelphia, Sept. 22, 2005.

39. Wald, M. L., Baltimore energy company seeking right to build reactor, *New York Times* (Oct. 28, 2005).

40. Reiss, S., Let a thousand reactors bloom, *Wired* 160–163 (Sept. 2004).

41. General Atomics, *GTMHR, Inherently Safe Nuclear Power for the 21st Century*, General Atomics, 3550 General Atomics Court, San Diego, CA 92121-1122.

42. Webb, J., Daring to be different. India has bold plans for a nuclear future, *New Sci.* 48–50 (Feb. 19, 2005).

43. Wald, M. L., Alaska town seeks reactor to cut costs of electricity, *New York Times* (Feb. 3, 2005).

44. U.S. Government General Accountability Office, *Nuclear Regulation. NRC's Liability Insurance Requirements for Nuclear Power Plants Owned by Limited Liability Companies*, GAO-04-654, Washington, DC, May 2004.

45. Bisconti, A. S., Perceptions of energy needs drive public opinion on the USA's nuclear future, *IAEA Bull.* **46**(1):27–28 (2004); see also Nuclear Energy Institute News Release, *Nuclear Power Plant Neighbors Accept Potential for New Reactor Near Them by Margin of 3 to 1*. Washington, DC, Oct. 12, 2005.

46. Brand, S., Environmental heresies, *Technol. Rev.* 60–63 (May 2005).

47. Wald, M. L., I vant to drink your vatts, *New York Times* F1, F9 (Nov. 17, 2005).

7

POWERING THE WORLD: ALTERNATIVE AND RENEWABLE ENERGY SOURCES

Strongly held opinions often determine what kind of facts people are
able or willing to perceive.
—*Robert Waelder*

Electrical energy and the American way of life are inextricably linked. Abundant and inexpensive fossil fuel has been the cornerstone of our prosperity. But our insatiable demand for ever-increasing amounts of energy is exceeding our ability to meet that demand, which triggers the twin forces of supply and demand, sending oil prices soaring.

Since the end of World War II, our energy use has tripled. Although our 290 million people, soon to be 300 million, represent only 5% of the total world population, we use 25% of all energy used worldwide. Currently the fossil fuels—coal, oil, and natural gas—provide 85% of our *energy* consumption. The remainder is made up of noncarbon sources such as nuclear power, 8%, (representing 20% of total electricity production), with the combination of hydroelectric and wind power contributing another 6% [1]. Even though over 50% of our needs can be supplied domestically, we are forced to import large amounts of oil from OPEC (Organization of Petroleum Exporting Countries), which for the most part includes the highly volatile and politically unstable Persian Gulf States, as well as unpredictable Venezuela and Nigeria. Since the mid-1970s, imports of oil and natural gas have doubled to about >30% of our total energy use. Current global market conditions indicate that competition

Our Precarious Habitat . . . It's In Your Hands, Fourth Edition. By Melvin A. Benarde
Copyright © 2007 John Wiley & Sons, Inc.

from developing countries such as China and India, with their prodigious populations, for the now-dwindling reservoirs of oil, will force the price of oil even higher as supplies shrink, and market forces will send prices spiraling. The price of Arabian light crude oil rose from $1.85 a barrel in 1973, to $40 in 1981, to $60 in 2004, and topping $70 in 2005. At the moment there is no indication how much higher it will go, but serious predictions include $100 a barrel.

In 2003, oil accounted for about 40% of total US energy use—some 7.3 billion barrels of crude oil, meaning 20 million barrels a day [1]. If current demand continues at this pace, and there is no evidence suggesting changes in behavior, we are headed for 30 million barrels per day by 2025. In fact, the demand for petroleum products is expected to rise by 1.6% annually for the next 25 years. If such use occurs, global warming will be irreversible. Consequently, it is incumbent upon us to replace oil and coal, the mainstays of our carbon economy, with other nonpolluting, greenhouse-gas-emitting fuels, of which there are many. We need only galvanize our political will.

Non-fossil-fuel, renewable energy sources can and must replace coal and oil, if we are to become energy-independent. How much of a contribution they can make is arguable; not because of any bias toward renewables, but, as we shall see, their inherent limitations, and our voracious appetite for energy. However, a number of alternative sources, renewables, can be considered as bridge technologies that can power our cities, cars, and industries until nuclear power and hydrogen become readily available. Unfortunately, alternative energy sources have suffered from official and political neglect, without a national policy to fund research and improve competitiveness. Congress has allowed new energy development to languish. On the other hand, private interests and vision have continued to improve the technologies, bringing down costs, making them more competitive with fossil fuels.

Because of the rush of global warming, the instability in the Middle East, and the current and future prices of fossil fuel, alternative fuels and/or processes will have a place in our future, if we are to have secure, affordable, reliable, and environmentally sound options [1].

PRELUDE TO RENEWABLE ENERGY SOURCES

Electricity has become an integral and essential part of our lives, which we cannot live without. This became dreadfully apparent when Hurricanes Katrina and Rita struck the Gulf Coast states of Florida, Alabama, Mississippi, Louisiana, and Texas in 2005. Not only did the lights go out—on streets, in homes, and in offices—but refrigeration went along with it. Perishable foods perished quickly, as did frozen foods along with them. Electrically driven pumps could not deliver gasoline at gas stations, nor could water be pumped. Without generators, hospitals, schools, government offices, and libraries could not function.

Without electricity, computers, TV, and radio were useless unless they were battery-operated. Manifestly, our lives are tightly linked to electricity. All alternative, renewable energy sources ultimately produce electricity. That's the name of the game. The issue then becomes, how much electricity can the various sources produce? Bear in mind that electricity is not an alternative energy source. Electricity is produced by various types of energy sources, as, for example, when a spinning turbine drives a generator that produces electricity by converting mechanical energy into an electric current. A spinning magnet inside a coil of copper wire creates an electric current that either changes direction, creating an alternating current (remember Faraday?), or remains in one direction, producing a direct current of electricity, which is then sent over wires to homes, office buildings, and industry, providing the current needed to light our way, as well as plug into, to run our scores of appliances. So, again, given our quenchless thirst for electricity, the open question remains: Can renewable, sustainable sources satisfy that demand? We shall see.

ENERGY SOURCES

The Wind at Our Backs

Wind is energy in motion, and wind power, the extraction of kinetic energy from wind and its conversion to useful energy, most often electrical, is also a form of solar energy as wind is created by the sun's heating the earth unevenly. As the sun heats the earth, hot air rises and cooler air rushes in, creating a pressure differential that drives air from one point or location to another, causing wind to blow. The magnitude of the blowing wind is the consequence of the pressure gradient between two locations. When the wind blows, its energy can be captured. The operational term is "when" —as the wind doesn't always blow. Consequently, wind energy must be viewed as an intermittent source. Nevertheless, wind and wind power have an ancient and honorable history.

Windmills were used in the Middle East by the eleventh century; in Europe, by the thirteenth; and extensively by the Dutch, in the fourteenth. But it was in Denmark by the 1890s that literally thousands of windmills were being used for generating electricity [2].

In the American West, from the 1880s onward, farmers and ranchers used small water-pumping windmills, and in rural areas in the 1920s, small windmills generated electricity for home appliances. However, with President Franklin D. Roosevelt's Rural Electrification Administration, in the 1930s, which electrified the countryside, windmills were laid to rest. The actual demise of windmills was due to an accident to a windmill operated by a public utility in Rutland, Vermont. In 1945, this windmill lost a blade while spinning, and the blade was hurled almost a thousand feet. That was the end of windmills until

Figure 7.1. An offshore farm of hundreds of wind turbines in the North Sea off Denmark's northwest coast. (Figure courtesy of Vestas Wind Systems—Vestas Americas.)

the early 1970s, when the worldwide oil embargo sent oil prices soaring, and with it, worldwide interest in wind power revived [2].

A turbine, no longer a typical mill, now uses the mechanical energy provided by the wind to produce electricity. The turbine works the opposite of an electric fan. Instead of using electricity to move air, producing a cooling breeze, the wind turbine uses wind to make electricity. It's the wind that turns the long blades (not referred to as "propellers") that are connected to a drivetrain or shaft that drives a generator within the turbine housing, producing electric current. Figure 7.1 shows a field of turbines or mills, and Figure 7.2 displays inner workings of the turbines.

All turbines operate within a range of windspeeds, from about 8–16 to 65 miles per hour (mph). Under 8 mph, the rotors (blades and hub together) cannot turn the blades fast enough to generate electricity. The controller, (see Fig. 7.2) starts up the turbine at windspeeds of about 8–16 mph, and shuts off at about 65 mph, as turbines cannot operate at windspeeds of 65 and higher, as the generators overheat [3,4].

The energy sent to the generator by the spinning blades increases the voltage from 480 to 65,000 Volts. This current is sent through cables down the turbine tower to an underground transformer that boosts the voltage higher still—up to 400,000 volts (V). Higher voltages are far more efficient in trans-

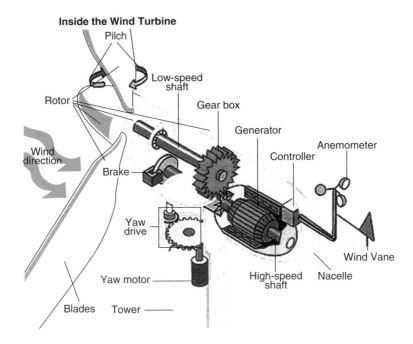

Inside the Wind Turbine

Pilch · Low-speed shaft · Rotor · Gear box · Generator · Anemometer · Controller · Wind direction · Brake · Yaw drive · Wind Vane · Yaw motor · High-speed shaft · Nacelle · Blades · Tower

Anemometer: Measures the windspeed and transmits windspeed data to the controller.

Blades: Most turbines have either two or three blades. Wind blowing over the blades causes the blades to "lift" and rotate.

Brake: A disc brake, which can be applied mechanically, electrically, or hydraulically to stop the rotor in emergencies.

Controller: The controller starts up the machine at windspeeds of ~8–16 miles per hour (mph) and shuts off the machine at ~65 mph. Turbines do not operate at windspeeds above ~65 mph because they might be damaged by the high winds.

Gear box: Gears connect the low-speed shaft to the high-speed shaft and increase the rotational speeds from ~30 to 60 rotations per minute (rpm) to about 1200–1500 rpm, the rotational speed required by most generators to produce electricity. The gear box is a costly (and heavy) part of the wind turbine, and engineers are exploring "direct-drive" generators that operate at lower rotational speeds and don't need gear boxes.

Generator: Usually an off-the-shelf induction generator that produces 60-cycle AC electricity.

High-speed shaft: Drives the generator.

Low-speed shaft: The rotor turns the low-speed shaft at about 30–60 rotations per minute.

Nacelle: The nacelle sits atop the tower and contains the gear box, low- and high-speed shafts, generator, controller, and brake. Some nacelles are large enough for a helicopter to land on.

Pitch: Blades are turned, or pitched, out of the wind to control the rotor speed and keep the rotor from turning in winds that are too high or too low to produce electricity.

Rotor: The blades and the hub together are called the rotor.

Tower: Towers are made from tubular steel (shown here), concrete, or steel lattice. Because wind speed increases with height, taller towers enable turbines to capture more energy and generate more electricity.

Wind directions: This is an "upwind" turbine, so-called because it operates facing into the wind. Other turbines are designed to run "downwind," facing away from the wind.

Wing vane: Measures wind direction and communicates with the yaw drive to orient the turbines properly with respect to the wind.

Yaw drive: Upwind turbines face into the wind; the yaw drive is used to keep the rotor facing into the wind as the wind direction changes. Downwind turbines don't require a yaw drive, the wind blows the rotor downwind.

Yaw motor: Powers the yaw drive.

Figure 7.2. An exploded view of a wind turbine's inner workings. Note the controller toward the rear. (*Source*: Department of Energy.)

porting over long-distance powerlines. In the underground cables this high voltage is moved to substations where the current is reduced. It then moves to above-ground powerlines and on for use in homes, where the voltages are 110 V for lights, TV, and small appliances and 220 V for clothes washers and dryers.

A small computer housed in a metal container attached to a tower connects to a central office and tracks the current running through the cables, recording the amount of electricity produced. It also records the speed of the spinning rotor, the temperature of the generator, wind direction, and windspeed [2]. To be effectively productive, 50–100 or more turbine towers must be situated together on what are now referred to as *wind farms*, requiring great parcels of real estate as each tower requires about 2 acres of land—in the windiest areas. As windspeed increases with altitude, the taller the tower (turbine), the more wind it will capture. Consequently turbine towers now rise a hundred feet or more. And to generate greater amounts of electricity, today's largest, commercial wind turbines have blade spans of 104 meters—343 feet from tip to tip, and can produce 5.6 megawatts (MW) of electricity—enough to power 1000 average-size homes. Recently, an experimental turbine in Germany used blades spanning 416 feet, and the General Electric Company is developing blade designs topping 462 feet. Such massive units are anticipated to produce as much as 7 MW, but these are some years away [5].

Wind turbines are generally divided into two major categories: *horizontal-axis turbines*, which resemble a windmill, and the *vertical-axis*, or *Darrieus, turbine*, which looks for all the world like a huge eggbeater. Figure 7.3 shows realistic representations of each. Currently, the horizontal-axis turbine is the most common. To generate electricity this type of turbine captures wind energy with two or three propeller like blades mounted on a rotor sitting on top of a tower often well over 100 feet in the air. The blades are made of a resin-coated fiberglass to withstand years of battering by heavy winds. These turbines have variable-speed generators, as turbines need to run at constant speed to produce a constant flow of electricity. The generator regulates this flow. Should windspeed fall below 8 mph, as it often does, the generator will shut down as it cannot maintain a constant flow. All variable-speed generators can increase the range of windspeeds under which a turbine can operate, by switching gears when windspeed changes. This type of turbine has a *yaw drive*, used to keep the rotor facing into the wind, as wind direction changes. These are also referred to as "upwind turbines" with their blades facing into the wind [6].

Vertical-axis turbines can accept wind from any direction. Furthermore, the eggbeater has its mechanical components at ground level, making them much easier to inspect and repair. One of the only eggbeater wind forms in the United States is at Altamount Pass, near Livermore, California. Because the vertical-axis wind turbine has its gearbox and generator at ground level, maintenance is simplified, but at ground level it cannot take advantage of the greater windspeeds and lower turbulence at the higher altitudes.

Horizontal and Vertical Axis Wind Turbines

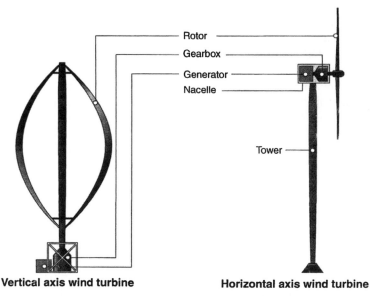

Vertical axis wind turbine   **Horizontal axis wind turbine**

Figure 7.3. Horizontal- and vertical-axis wind turbines in profile. (Figures courtesy of GAO-04-766, *Renewable Energy*.) (*Source*: Izaak Walton League of America.)

TABLE 7.1. Units of Measurement of Electrical Power

A (amperes) × V (volts) = W[a] (watts)
1000 watts = 1 kW = kilowatt
1 kilowatt = (kW) = 1000 watts
1 megawatt = (MW) = 1000 kW or 1 million watts
1 billion kW = 1 terrawatt = TW
1 kW × 1 hour = 1 kwh (kilowatt-hour)

[a] The watt is the basic unit used to measure electric power.

Utility-scale turbines, those that contribute (sell) power to a regional grid, range in size from 50 kilowatts (kW) to as large as several megawatts. Table 7.1 indicates how electrical power is measured.

Electricity production and consumption are measured in kilowatt-hours (kWh), while generating capacity is measured in kilowatts or megawatts. If a power plant that has 1 MW of capacity operates nonstop for the 8760 hours in a year, it will produce 8,760,000 kWh. An average US household consumes about 10,000 kWh a year. On average, however, wind turbines operate at 40% of their peak total hours per year because of the intermittent nature of wind and time of year. This is the motivation for higher towers, longer blades, and more units per farm, or larger farms. Wattage production is their reason for being. General Electric's Wind Energy 3.6-MW wind turbine is one of the largest ever erected. The larger the turbine, the more efficient and cost-effective. But they are also expensive.

In 2003, American wind farms generated 13 billion kWh—enough to light and serve 1.3 million households—but this was 0.3 of one percent of all the electricity generated in the United States, which represents a quadrupling of generating capacity between 1990 and 2003, to 6400 MW, and the Department of Energy projects continued growth through 2025. On a percentage basis, wind power capacity has been growing at a higher rate than other electricity generating sources. Additionally, according to the DOE, the US Midwest theoretically has enough wind power potential to meet a significant portion of the nation's electricity needs—but remains largely untapped. Two additional projections are worth contemplating. Wind energy potential is estimated at over 10,000 billion kilowatt-hours annually—that's 1×10^{13} or 10 trillion kWh—more than twice the total generated from all sources in the United States today. The potency of such numbers does inspire mulling [6]. So, for example, it is also worthy of note that annual nuclear energy production runs at about 765–770 billion kWh, and if wind power were just to equal that, it would require an area equal to the state of Minnesota. How many states would it take to achieve 10,000 billion? It does seem a bit of a stretch.

It has also been suggested that wind energy could easily generate 6% of the nation's electricity by 2020, as much as hydroelectric power does currently. Is that realistic? After all, if wind energy contributed about 0.3 of one percent of the country's total electricity in 2003, it would be necessary to increase its production 20-fold in the coming 15 years. That, too, seems like pushing the envelope, even though it has been shown that with regular maintenance, turbines should work at least 90% of the time. But it is also well known that wind doesn't always blow, and doesn't always blow at optimum speed. From Table 7.2, we see that of the seven wind power classes, only four are suitable. North and South Dakota, with the greatest wind power potential, have done little to utilize their wind's kinetic energy. The winds intermittancy and fickleness was vividly brought home to me with the account of the four-day delay in the America's Cup Race because of the lack of wind off the coast of Valencia, Spain, in April, 2007. Another major limiting factor is space. With even higher towers and larger blades, towers require greater space between one another or they will steal each other's wind. If turbines are placed in long, straight rows on flat farmland, those in front will grab the most wind, while those in back will be deprived.

Is windspeed the be all and end all? What about the overall environment? Tornadoes cut a swath through southern Indiana and northern Kentucky, flattening everything with a 20-mile-long, 0.75-mile wide path. And hurricanes are still fresh in our minds. Ice storms are no strangers to the US Northeast, as well as states bordering Canada. Adverse environmental conditions can damage turbines and increase downtime. Furthermore, although turbines can generate electricity, are they located near population centers where electricity

TABLE 7.2. Wind Energy Classification[a]

Wind Power Class	Potential	Windspeed at 50 m (mph)
1	Unsuitable	>12.5
2	Marginal	12.5–14.3
3	Fair	14.3–15.7
4	Good	15.7–16.8
5	Excellent	16.8–17.9
6	Outstanding	17.9–19.7
7	Superb	19.7–24.8

[a] Estimates of wind resources are expressed in wind power classes, at 50 m above ground level. Class 1, with 11 Eastern states is wholly unsuitable as a resource. Class 2, with 14 states from the Atlantic seaboard to the Midwest, is marginal, and 5 states are fair, leaving 20 states with good to supberb sites scattered west of the Mississippi River where population centers are few and far in between. The only superb areas are the Aleutian Islands off Alaska.

is needed? Long-distance powerlines can be more expensive than the towers. Also, how expensive are they to maintain? Is there sufficient experience with wind turbines to determine their price competitiveness per kilowatt-hour with other renewables, but especially the nonrenewables, oil and coal? Answers need to be pinned down before decisions are finalized.

Wind energy appears to be the fastest-growing energy source in the world because of its many advantages, including the fact that it's a clean source with no polluting byproducts. It is, of course, home-grown, so to speak; it does not need to be imported. No one owns it, and it is abundant. As long as the earth is heated unevenly, and there is no reason to think that that will change, there will be wind. That being the case, it can be considered inexhaustible. It also appears to be price-competitive at 0.4–0.6 cents per kilowatt-hour. For farmers and ranchers, on whose vast acreages wind farms are being located, this means additional rental income, while crops and herds go undisturbed, as both are well beneath the spinning blades.

With all that, there is a downside. Wind power must compete with traditional energy sources on a cost basis. Depending on the wind site, and its energy, the farm may not be competitive. Although the cost of wind power has decreased substantially over the past decade, the technology requires higher initial investments than do fossil-fueled plants. Of course, the major downside is the intermittency of the wind, which may not blow when electricity is needed, and wind energy cannot be stored, nor can it be harvested to meet the timing of electricity demands. Moreover, some of the best wind sites are in remote areas, far from population centers, where the demand is. In addition, as noted earlier, long-distance lines are expensive. Expense must also be considered when wind farms must compete for available land that can offer other lucrative uses, which may be more valuable than the generation of electricity.

Although wind farms appear to have little adverse impact on the environment, there is growing concern about the noise produced by the whirling blades, the aesthetic impact of hundreds of turbines, and the birds killed flying into the rotors. In fact, at the moment, a number of events typify the pros and cons of wind power.

Nimby

The International Wildlife Coalition, the Ocean Conservancy, The Humane Society, and The Alliance to Protect Nantucket Sound, are unhappy. The folks on Martha's Vineyard and Nantucket Island are not happy. Nimby, not in my backyard, has created an unlikely coalition of environmentalists and bent-out-of-shape residents from Cape Cod's Hyannis Port, all the way to Buzzards Bay. For years the Sierra Club and Greenpeace promoted wind power as a way to reduce or control the use of fossil fuels. But when there was a proposal to locate a wind farm offshore in Nantucket Sound, between ferry lanes to Martha's Vineyard and Nantucket, that was just too much, even for environmentalists dead set against coal and oil and global warming [7].

War has been declared between Cape Wind Associates of Boston, who want to install a parcel of 130 towering turbines, soaring to 420 feet about the sea, 6 miles out in the middle of Nantucket Sound, covering a 24-square-mile area, while yet another company has proposed another wind farm on Nantucket's backside. Formed up in full battle dress against these interlopers are residents and environmentalists who see looming threats to migrating birds, as well as to the habitat of dolphins and seals, as the turbine's 161-foot blades, churning at 16 revolutions per minute (rpm), would rattle the seabed. Additionally, air traffic controllers are upset, believing that small planes, of which there are many shuttling to and from the islands, would tangle with the rotors. Yachtowners were distraught that the many towers would thwart their annual Hyannis–Nantucket regatta. Adding to the displeasure was the fact that untrammeled views of the sea would be lost to the thousands of tourists who vacation there. For Cape Coders, a wind farm would be a disaster. As the *Wall Street Journal* editorialized, "The great pleasure here is watching affluent environmentalists condemn one of their pet causes just because it happens to obstruct the view from their wrap around porch." Further, as Walter Cronkite, owner of a summer home on Martha's Vineyard, intoned, "Our national treasure should be off limits to industrialization" [8]. Indeed, it's a battle between aesthetics, climate change, particulate matter, and celebrities.

It didn't take long. In April 2006, a Senate–House Conference Committee approved an amendment to a Coast Guard budget bill that effectively killed the proposal for a wind farm in Nantucket Sound. The amendment gave Governor Romney of Massachusetts veto power over any wind farm in the Sound. Score one for politicians, environmentalists, and local residents (celebrities).

Residents of the New Jersey Shore have a similar concern. With the pros and cons each pleading their special interests, the then Acting Governor Richard J. Codey, wisely seeking time for a cooling-off period, declared a

15-month moratorium until a panel could study the issue. Winergy, based in Shirley Long Island, submitted a proposal to place over a thousand towers on five sites off New Jersey's beaches: 98 off Asbury Park, 122 off Atlantic City, 298 at Avalon, and 531 along the coast from Cape May to Avalon. The price of oil and the cost of generating a kilowatt of electricity may well determine how fast these towers arise. It will also be an interesting test for Jon Corzine, the newly elected governor [9].

What has been a fixture of California's hilly midlands has crossed the Mississippi into West Virginia. The wind farm in Thomas, West Virginia, developed by Florida Power and Light, is the largest east of the Mississippi, where local zoning laws do not yet exist, and is attracting developers. Some 400 turbines could be sprouting across 40 square miles of West Virginia's scenic landscape. According to some residents, "They look like alien monsters coming out of the ground."

For wind farms to be financially viable, they must be huge, which to many residents means a blight on the countryside, and a source for chopping up birds. One farm in California was dubbed the "Condor Cuisinart." The fact that they produce such small amounts of energy and the unreliability of wind, along with the inability to store energy, are dominant complaints. Here, too, divisions within environmentalist ranks has caused shock and dismay given the belief that "there are appropriate places for everything." After all, "You wouldn't want one in Yosemite Park, or in Central Park" [10]. Manifestly, the reason wind farms are springing up, is the huge financial incentives provided by the recent government subsidies such as the federal tax credit. This credit allows companies to deduct 1.8 cents from their tax liability for every kilowatt-hour they produce for 10 years. The savings are huge—and a great motivator.

Discord is not limited to the United States. The Battle at Trafalgar (October 1805) remains sacred to many people, especially the British. Cape Trafalgar, Spain, south of Cadiz, where Admiral Nelson won an epic naval victory against the French two centuries ago, has become the site of another battle at Trafalgar, where privately owned wind parks have sprouted on the hills behind the Cape. The current rush to develop wind farms is off shore, especially near the spot of the famous battle, where winds rushing through the Straits of Gibraltar provide substantial wind every day. Incentive also comes from the European Union, which has been pushing for alternative energy sources to supplant fossil fuels as part of its commitment to the Kyoto Protocols. But the tuna fishermen are challenging the proposed 400 towers, which, they argue, would make their work far more difficult by perturbing the migration of the tuna, and by forcing their small boats to venture further out into the treacherous waters near Gibraltar. Given the fact that "there is good money in wind," the fishermen's grievances will more than likely fall on deaf ears [11].

Carlisle, Keswick, and Shap form a triangle of towns at the fringes of England's compellingly scenic Lake District, a stone's throw from Scotland. Shap is also the center of Britain's giant wind power dispute. Here again, the campaign is between Britain's desire to affirm its reduction in CO_2 emissions inline with its commitment to Kyoto, while deriving 10% of its electrical power from renewable sources by 2010, and its desire to maintain its wilderness heritage. Here in Shap, a 1500-foot-high hill known locally as Whinash is at the center of the rumpus, and where anti-wind-power sentiment runs high, objecting to the 27 turbines (each 370 feet high) planned to be mounted on its summit. The $100 million proposed Whinash wind farm has again divided the environmental movement, turning former allies against one another. The essential argument questions whether wind power is nothing more than a costly experiment that enriches the people who build the farms, without seriously reducing greenhouse gases. Here, as in Cape Cod, Whinash is considered the wrong place for turbines, in addition to the demonstrated fact that the wind does not blow steadily enough to generate constant power, and government subsidies cost consumers far more than gas, coal, or nuclear power plants [12]. As with many battles, the outcome remains uncertain. Nevertheless, wind power is not without its backers and advocates.

China, with its vast, unobstructed terrain, its colossal need for energy, and its obligation to reduce its gargantuan CO_2 emissions, second only to those of the United States, is investing heavily in wind farms. Its Huitengxile wind farm in Inner Mongolia, northwest of Beijing, has 96 turbines, producing 68 MW, with plans calling for 400 MW by 2008. Harvesting power from the wind is a high priority for the Chinese government, and readily understandable, with the stiff breezes that are a permanent physical presence in this area of Asia. By 2020, China expects to supply 10% of its energy needs via renewable sources, with wind leading the way. Wind farms have been installed in six heavily populated provinces around the country, and the demand for turbines and towers is so great that manufacturers cannot keep pace. The government's targets call for 4,000 MW by 2010, and 20,000 MW, [20 gigawatts (GW)] by 2020 [13]. China has set out on an exceptional national policy that it intends to meet.

Turbines and towers stretch even skyward to catch more wind. The higher, the better. New York City is reversing that trend. They're placing turbines underwater—tidal turbines, to capture the kinetic energy of the 6-mph current flowing in Manhattan's East River. Six miles an hour may not sound like much, but actually it is one of the fastest flows of any waterway on the East Coast. Six turbines were to have been submerged in the East River alongside Roosevelt Island in August 2004, to produce 150 kW, and by 2006, 200–300 turbines (each 15 feet tall), turbines would have been deployed below the surface to produce 10 MW of electricity for 8000 homes locally. Verdant Power, of Arlington, Virginia, the company producing the turbines, has finally managed to cut through the bureaucratic inertia and obtain the necessary permits that will allow them to place the turbines in the river in March 2006. After that, it would take 9–18 months of data gathering before several hundred are placed in.

A 10-MW field would save New York City the equivalent of 65,000 barrels of oil annually, and reduce CO_2 emissions by 33,000 tons [14]. The residents love the idea because it will clear the air that they refer to as "asthma alley," and because the energy production is local—no foreign or long-distance providers to rely on. Although the kilowatt-hours will not be price-competitive at startup, the price is expected to drop as more families sign on. Unfortunately New York City does not have China's determination. Perhaps now that the elections are over, that will change.

We will not have heard the last of underwater turbines without considering Alexander M. Gorlov. Professor Gorlov, emeritus professor of mechanical engineering at Boston's Northeastern University, arrived from Russia with commendable experience credentials, and including work on the design and construction of Egypt's Aswan High Dam. That dam convinced him that large-scale dams were not the answer for the generation of electricity via falling water, which led him to develop his triple-helix turbine. This novel turbine won him the prestigious American Society of Mechanical Engineers 2001 Thomas A. Edison Patent Award, which the ASME declared had the potential "to alleviate the world-wide crisis in energy."

Gorlov's helical turbine (see Fig. 7.4), often described as resembling a DNA spiral, was designed to harness the kinetic energy of flowing water. Electricity, as we know, can be generated anywhere that water flows; oceans and rivers are sources of enormous amounts of potential energy, and are far more dependable than wind or solar energy. Furthermore, Gorlov's turbine does not depend on the direction of a current, which means that it needs no reorientation with changing tides.

Figure 7.4. A Gorlov helical turbine before application of an anticorrosion epoxy. Alexander M. Gorlov stands behind his turbine. (Photo courtesy of Richard S. Greely, St. Davids, PA.)

Hanging well below the surface, or sitting on the bottom of a channel, the whirling blades of these flowthrough turbines, coupled to generators via cental shafts, set them spinning, producing electricity without the need for fuel, and with no noise, no air or water pollution, no obstruction to surface vessels, and no harm to marine life. Measuring 1 meter in diameter by 2.5 m high, (36 × 40 inches) and sitting in a barrel-shaped cage, its helix-shaped blades, unlike a propeller, present a rectangular cross section, to current flow. The total electric power that could be generated from tidal estuaries and rivers is simply (and theoretically) tremendous.

If the speed of the turbine blades attains 3.6 knots, the average power produced would reach 4000 watts or 4 kilowatts (kW). Over a 20-hour day (assuming 4 hours of slack water, depending on location), 80 kilowatt-hours (kWh) of electrical energy could be generated.

The future of the Gorlov Turbine (GHT), now being field-tested in South Korea—a country in the throes of an energy crisis, as it possesses no fossil fuels and must import all its oil and gas—may well depend on its success there, where the fast-moving tides on its west coast could generate huge amounts of much-needed electrical enegy. If it succeeds, the South Korean government plans to install thousands of underwater turbines generating thousands of megawatts of power. Furthermore, this helical turbine can produce hydrogen and oxygen by electrolysis. An array of 40 GHTs running 20 hours a day at 3.6 knots could produce enough hydrogen to drive a fuel cell car 300 miles. Is it too good to be true?

The bottom line with both wind- and tide-water-generated electricity is reliability. The question most often avoided is: When does the wind blow, and when does the tide reverse? When the wind will blow is unpredictable. "West Texas, for example, is notoriously windy, but mostly at night and in the winter, when the electric market is glutted with cheap power from coal and nuclear plants. Peak electric load occurs in summer, during the day." In New York City, there supply shortages are expected during the 6 hours of slack tide, when the tide causes the water to change direction, making the current too slow to turn the turbine's rotors. The same problem afflicts wind farms around the country and world—when slack winds are unable to spin turbine rotors beyond 15 rpm when 20 is optimum. Furthermore, wind energy plants produce only 30–40% of their capacity [15]. A panacea they are not. Nevertheless, farming the wind has come of age and can provide clean energy, and together with other renewables should make a fair contribution to the country's energy needs.

Solar Power

The sun in the morning, and . . . solar power . . . can the sun provide?

For atmospheric scientists, the sun may be a globe of gas, but that globe, that star, around which our planet wanders, is our source of light, heat, and life. Undeniably, the light and heat from that fiery furnace whose surface temperature reaches 9.981°F (55.27°C) has made life on earth sublimely comfortable, as only 1 kW of energy, a minuscule amount of its radiant heat, falls on a square yard of earth each day. Over the course of a year, this 1 kW contributes a thousand times more energy than the combined sources of energy produced by coal, oil, gas, nuclear, wind, hydroelectricity, and other sources of heat/electricity-producing power plants.

Most of that 1 kW of solar energy is absorbed by the oceans—which cover 72% of the planet's surface—and the belt of lands circling the equator. Nonetheless, sunlight is everywhere, is inherently clean, and is infinitely renewable; yet only within the past 50 years has it began to be seriously exploited.

With the availability of lenses, it was quickly learned that the sun's rays could be focused and concentrated, producing sufficient heat at the focal point for paper to burst into flame.

A solar-powered water heater was invented by Clarence M. Kemp, and was offered to the public in 1891. Many are in use in the US West and Southwest, but the heat generated is insufficient to boil water, which means that if solar energy is to be used to produce electricity, it must develop temperatures of at least 212°F (100°C). To do so, solar energy must be concentrated—focused.

Parabolic trough systems can concentrate the sun's energy via long, rectangular U-shaped mirrors. The mirrors are tilted toward the sun, focusing sunlight on a pipe running along the center of the trough. This heats the medium—oil or molten salt (liquid sodium)—flowing through the pipe. The heated medium transfers its heat to water in a conventional steam generator, which spins, producing an electric current [16].

A second type of concentrator is the dish/engine system, which uses a mirrored dish, similar to a large satellite dish. This highly polished and reflecting surface—the concentrator—collects and focuses the sun's heat onto a receiver that absorbs the heat and transfers it to a fluid within the engine. The heat causes the fluid to expand against a piston or turbine that spins, producing the mechanical energy needed to run a generator or alternator.

Yet a third type of concentrator, the solar power tower system (see Fig. 7.5), employs a huge field of highly reflecting mirrors (1800–2000 total) heliostats, each measuring 7 × 7 meters (22 × 22 feet), that concentrate sunlight onto the top of a tower situated in the center of the field. As the sun rises and crosses the sky, the mirrors tilt and swivel to continuously harvest the rays and focus them onto the tower, or receiver, where molten salt, a mixture of sodium and potassium nitrate, reaches a temperature of 1000°F. Unlike water, molten salt retains heat efficiently so that it can be stored for days before being converted to electricity, which means electricity can be produced on cloudy days or hours after sunset. Ergo, storage is a key element in the alternative energy equation. Unfortunately thousands of heliostats laid out in circles require 120–130 acres of land, which can be found only in barren deserts or on the great plains (see Ref. 18 or 19).

Concentrating collectors are good as far as they go, but are impractical for generating the levels of electricity demanded currently and projected for the

Figure 7.5. A solar power tower showing the hundreds of mobile, highly polished helio-stats—concentrating tracking mirrors—arrayed around the Tower, and the white-hot central power receiver at the top of the tower. Heat energy directed to the receiver is absorbed by molten salt, which generates steam that drives the generator. (Figure courtesy of Sandia National Labs.)

future. A device that can convert or change sunlight directly into electricity is needed.

With no moving parts, no mirrors, no heat transfer fluid or generators, no polluting chemicals or particles, and no land-use constraints, photovoltaic cells may be too good to be true. To understand photovoltaics, it may be helpful to go back a few years, to 1839, when Edmund Becquerel, son of Henri Bec-querel, discovered the process of using sunlight to produce an electric current in a solid material. Without knowing it, Edmund had discovered the photoelec-tric effect, but which is attributed to Heinrich Hertz in 1897, and for which Albert Einstein won the Nobel Prize in Physics in 1905—for explaining the theory behind it. With these great minds at our service, comprehension may not be difficult. So, for example, Einstein showed that the kinetic energy of an ejected electron was equal to the energy of the incident photon minus the energy required to remove the electron from the surface. Thus, a photon of light hits a surface, transfers almost all its energy to an electron, and the elec-tron is ejected with that energy less whatever energy is required to get it out of the atom and away from the surface. The photoelectric effect has many practical applications including the photocell and solar cells.

Solar cells, usually made from specially prepared silicon, act like a battery when exposed to light. Individual solar cells produce voltages of abut 0.6–

0.7 V, but higher voltages can be obtained by connecting many cells together. The photovoltaic effect is the conversion of radiation, sunlight, into electric power via absorption of a semiconducting material. How do they work?

When sunlight shines on a photovoltaic (PV) cell, it may be reflected, be absorbed, or pass through. Only the absorbed light generates electricity. The energy of the absorbed light is transferred to electrons in the atoms of the PV cell. With this new energy the electrons escape from their normal positions in the atoms of the semiconductor–PV material and become part of the electric flow or current.

A "built-in" electric field provides the force, a voltage needed to drive the current. To induce the built-in electric field within a PV cell, two layers of differing semiconductor materials are placed in contact with one another.

Time Out for Crystal Chemistry

Silicon has four valence electrons in its outermost orbit that bond with each other to form a crystal. In a crystalline solid, each silicon atom normally shares one of its four valence electrons in a covalent bond with each of four neighboring silicon atoms. In this way, the solid consists of units of five silicon atoms: the original atom plus the four other atoms with which it shares valence electrons.

In a crystalline silicon cell, two thin layers must be sandwiched together to create an electric field. To create the field a p-type (positive) silicon layer is placed against an n-type (negative) layer. To do this, a process of doping is required in which atoms of other elements are placed in each layer. Phosphorus, with five valence electrons is placed in the n layer, providing a free electron; a phosphorus atom occupies the same place in the crystal lattice formerly occupied by the silicon atom that it replaced. When phosphorus atoms are substituted for silicon in a crystal, many free atoms become available. Substituting, doping, the p layer with boron—which has three valence electrons— leaves a "hole" (a bond missing an electron) that is now free to move around the crystal. So, there are free electrons in the n layer and free "holes" in the p layer, both capable of movement. Neither layer can form an electric field by itself. Each is a semiconductor that must be sandwiched with the other; p-type semiconductors have an abundance of positively charged holes, and the n-type have an abundance of charged electrons. When n- and p-type silicons come together, excess electrons move from the n-type side to the p-type side, resulting in a buildup of a positive charge along the n-type side of the interface and a buildup of negative charge on the p-type side. With the flow of electrons and holes, the two semiconductor layers act like a battery, creating an electric field at the surface, where they meet—this is the p/n junction. The electric field causes the electrons to move from the semiconductor toward the negative surface, making them available for the electric circuit. At the same time, the holes move in the opposite direction, toward the positive surface, where they combine with incoming electrons [17]. Photovoltaic cells exploit the fact that certain metals emit electrons and generate electricity without moving parts, or the need to store or convert heat energy, as well as the fact that most are made of silicon, which is almost inexhaustible, is another plus. But are PVs cost-effective, and what level of energy can they produce?

A PV or solar cell is the basic building block of a solar electric system. As an individual PV cell is small, typically producing no more than a watt of power, it is necessary to connect them into modules and larger groups or arrays. The size of an array depends on the amount of sunlight available in a specific location, and the needs of users. Solar power has meant photovoltaic panels, and is "retail" power; that is, it has not been transmitted to homes or businesses via a central grid. Photovoltaic panels on an individual roof supplies power directly, and only to that roof.

Not only has the PV market been growing briskly; a team of researchers at the University of Toronto has created PV cells containing lead sulfide, which can absorb infrared light, enabling the system to harvest a wider portion of the electromagnetic spectrum. This will surely help reduce costs. Of additional excitement is the fact that Japanese investigators were able to induce solar cells to split water molecules and collect hydrogen, while a British company took that development several steps further, announcing a 10-fold improvement in efficiency of water splitting. If this can be brought to commercial applications, it would not be necessary to fill a fuel cell car at a hydrogen fuel station. It would only be necessary to tap into the solar panel on the roof of your house [18]. Wouldn't *that* explode the market potential!

Photovoltaic panels currently convert sunlight into 3 gigawatts (GW) of electricity. Although use is growing at 40% a year, it remains, literally, a drop in the total bucket of need, and solar panels are only 10% efficient at turning sunlight into electricity. To satisfy its current electricity demand using current technology, the United States would need 10 billion square meters of PV panels at a cost of $5 trillion dollars—an outlandish number considering it is half the country's gross domestic product. Furthermore, storing sun-derived energy is a major obstacle. Because electricity cannot be stored directly, it must be converted to another form such as the electrochemical form of batteries. But considering the total energy required, that, too, would necessitate massive numbers and drive the cost skyward. It is estimated that for solar energy to compete worldwide, it must be 50 times lower than fossil-fuel-based electricity, and more efficient—seemingly contradictory ends. Until then, continuing research to bring costs down, and efficiency up, is the way ahead. But that is the global problem. Locally, the concern is of a different dimension. Currently the United States is in a building boom with homeowners ready to spend $10,000– $20,000 for solar power systems of 2000–5000 watts. But with all the customers lined up, there is a severe shortage of PV cells, created by a shortage of raw materials that could last up to a year. Simultaneous with this, a bill in the California legislature aims to put solar power in 50% of new homes within the next 13 years. The Million Solar Roofs Legislation will subsidize the installation of solar equipment with a goal of putting 3000 MW of solar energy to

work by 2018. New York, New Jersey, and Connecticut already provide subsidies to solar power users. But the legislation in California will dwarf all the
others. In addition to state efforts, in August 2005, the US Congress passed an
energy bill that offers a tax credit of up to $2000 for homeowners who install
solar equipment. Consequently the future looks exceedingly bright for solar
power in the United States if the raw material shortage can get worked out.
The shortage of PV materials has been exacerbated because photovoltaic
electricity is used to power businesses, boats, recreational vehicles, highway
signs, and cell phone towers, in addition to homes, which means that prices are
rising sharply as supply and demand continue out-of-sync [19]. Three years
ago, you couldn't give solar away. Things, they are a'changing.

Falling Water

The Greeks had a word for it: *kinetikos*, which means "of motion," and *kinetic*
means "relating to motion." Kinetic energy is the energy associated with
motion. Flowing water contains energy that can be captured and used to do
work. Waterwheels are considered the first rotor mechanism in which a force,
water flowing over a large wheel, spins a shaft driving mills to grind grain, or
moving water to irrigate fields. Water power is in fact the oldest and most
highly developed of all renewable energy sources. Waterwheels are the ancestors of today's massive hydroelectric installations, which brings flowing water
up-to-date by providing the power to create electricity. Flowing water that can
be harvested and turned into electric energy is called *hydroelectric power*
or *hydropower* and is the country's leading renewable energy technology.
Of all the renewable power sources, it is the most reliable, efficient, and
economical.

The most common type of hydroelectric power plant employs a dam on a
river to store water in a reservoir. Water released from the resevoir behind
the dam flows down with tremendous force through a turbine, spinning its
blades, activating a generator, which produces electric current. Transmission
lines from the generator conducts electricity to homes and commercial buildings. After passing through the turbine, the water reenters the river on the
downside of the dam. Damming a river controls the amount of water entering
the turbines so that supply and demand can be approximated, increasing efficiency and producing electricity at minimum cost.

Hydroelectric plants can generate a few hundred watts to more than
10,000 MW, depending on the volume of water flowing through the turbines
and the "head"—the vertical distance from the impounded lake or river to the
turbines below. At Hoover Dam, 30 miles southeast of Las Vegas, Nevada, the
water of Lake Mead impounded behind the dam drops 590 atmospheres (atm;
of pressure) as it rushes to the turbines. Although it generates over 4 billion
kWh of electricity annually, serving 1.3 million people, Hoover Dam is no
longer the world's largest dam, having been overshadowed by Brazil's Itaipu,

on the Panama River, where Brazil, Argentina, and Paraguay meet. Now the world's largest, generating more power than 10 nuclear power plants, Itaipu supplies 26% of all electric power used in Brazil and 78% of Paraguay's electricity. This massive dam generates 75 terrawatts (TW) of electricity annually. But both Itaipu and Hoover Dams will be dwarfed by the Three Gorges Dam—on China's Yangtse River, east of Shanghai in Hubei Province—which is scheduled for completion around 2010. It is designed to deliver 20,000 MW of electricity, contributing 10% of China's total electricity supply [20].

A pumped storage plant is another type of hydropower plant. It uses two reservoirs and can store power. Power is sent from a power grid into the generators that spin the turbines backward, which causes the turbines to pump water from the lower of the two reservoirs to the upper reservoir, where the power is stored. To use the power, the water is released back down into the lower reservoir—lake or river. This forceful water then spins the turbines forward, activating the generators to produce electricity. It is during the periods of low electricity demand, such as nights and weekends, that energy is stored by the reversing turbines, pumping water from the lower to the upper reservoir.

Hydropower, while essentially emission-free, can have undesirable effects, such as fish injury and death from passage through turbines. Fish passing through turbine blades are subjected to potentially lethal conditions including shear, turbulence, pressure, and cavitation—bubble formation generated by whirling blades. Although many turbines are reaching their operational life expectancies, and are due to be replaced with turbines that incorporate fish-friendly designs that should increase fish survival, that will take years [21]. Nevertheless, increased use of hydropower will reduce use of fossil fuels with their release of CO_2 into an already overburdened atmosphere.

The Chinese Yangste River project offers a cautionary tale. When completed, it will substantially reduce the burning of coal; however, it has already forced over a million people to relocate as the dam's reservoirs have submerged some 100 towns. Also, dams can spread disease, as was demonstrated by the Aswan High Dam on the Nile River in Egypt, which caused the snail-borne worm *Schistosoma mansoni*, to spread from the dam along the Nile to the Mediterranean, causing many new cases of the debilitating disease, schistosomiosis. In addition, the Aswan Dam prevented the normal flow of the nutrients, nitrates, and phosphates, from reaching the Mediterranean, disrupting the sardine fishery industry. Engineers failed to consider these environmentally unfriendly possibilities when designing the dam; and now, there appears to be some evidence that dams generate both carbon dioxide and methane as a consequence of microbial activity in the reservoirs. But this remains to be firmly established. These faults notwithstanding, dams and hydropower can contribute mightily to a decarbonized world. The current world capacity of 600–700 gigawatts (GW) is a fraction of the estimated terawatts potentially possible if all available water resources were brought into use. While hydropower is increasing worldwide at a modest rate, less than 2%, resistance to constructing gigantic dams stems from the negative experiences of the Aswan and Three Gorges Dams.

An idea that is gaining currency is the microhydroelectric power system that can produce sufficient electricity, up to 5 MW for homes, farms, or ranches, and which would have little or no adverse impact on the land or its people. An idea to look forward to.

Beneath the Earth's Crust

There's a lot of heat beneath our feet. Albeit at some depth. Our earth appears to have layers; the crust, the topmost layer varies in thickness; deeper than 40 miles in mountainous areas, and less than 3 miles deep under the oceans. Beneath the crust, the upper and lower mantles stretch 2100 miles toward the center, while the outer core plunges another 1300 miles, and the inner core extends yet another 700 miles to the ultimate center, for a total of 4000+ miles from the top of Mount Everest to the white-hot solid nickel/iron core. And the heat is directly proportional to the depth. The deeper the hotter. At the surface we are insulated from the heat by the fairly thick crust. But the mantle is hot; about 650–1250°C (1200–2280°F). Further down, at the outer core's magma, temperatures climb precipitously to 5400–9000°F (3000–5000°C). At the inner core, the center the earth's temperatures remain unknown.

This heat is apparently the consequence of the earth's formation and consolidation some 4–5 billion years ago, and the isotopic decay of the unstable elements uranium, thorium, and potassium, which suggests a natural nuclear power plant with an ongoing chain reaction, which also suggests the potential for limitless energy. Perhaps. However, as we know from normal home heating, heat moves from warmer to cooler areas, and heat also flows upward, from the core to the cooler surface.

Well below the surface, in the hot-rock magma region, and in such areas of the world as Iceland, east Africa, western Arabia, Central Asia, the US West Coast, and Hawaii, there are deep cracks and faults in the crust that allow rain and snow to seep underground where the water is heated by the hot rocks and then circulates back up as hot springs, geysers, and fumaroles. In some areas, however, the hot water is blocked from reaching the surface by an impermeable layer of rock. That hot water, at 700–800°F, hotter than the hot springs and geysers, forms an underground reservoir that can be tapped for its available energy. Production of geothermal energy begins here, as exploitation of geothermal heat is currently limited to a depth of 3 miles. At this depth the temperature increases by 30–35°C per kilometer. By now, it is apparent that all power plants use hot water/steam to spin turbine blades, which in turn drive generators that produce electric current, including geothermal power plants.

The first geothermal energy plant in the United States was Pacific Gas and Electric's 11-MW plant that began operating in September 1960, in the area known as the Geysers, some 70 miles north of San Francisco. Since then, over 600 wells have been drilled up to depths of 10,000 feet.

Three basic types of geothermal plants are currently in use: dry steam, flash steam, and binary cycle. Dry steam is produced when trapped water is super-heated by the magma, and high pressure forces it upward and is released at the surface as a hot gas with little to no liquid content, which is piped directly to a turbine/generator unit. The concept is simple, as is the process. Currently dry steam plants draw from underground reservoirs using steam over 300°F, which is piped directly from underground wells to the power plant, where it is directed into a turbine/generator unit. The concept is simple, and so is the process. Currently there are only two underground resources of steam in the United States: the Geysers in northern California and Wyoming's Yellowstone National Park, where the Old Faithful geyser's, timely spoutings have thrilled tourists for years. Since Yellowstone is protected from development, the only dry steam plants in the country are at the Geysers [22].

Flash steam plants are the most common, using reservoirs of hot water with temperatures above 360°F (182°C) under pressure. This hot water flows up through wells, and as it flows upward it is "flashed," quickly depressurized, to produce steam, which then powers a turbine. The water is then injected back into the reservoir, making this a sustainable source.

Binary cycle plants, a bit more sophisticated, use water at temperatures of 225–360°F (107–182°C) to boil a heat exchanger fluid, the secondary, or binary fluid, usually one of the hydrocarbons, isobutane or isopentane. The secondary fluid vaporizes, causing the turbines to spin, then it, too, is recycled through the heat exchanger, the water is returned to the reservoir, and the cycle con-tinues. Energy produced by binary plants, which are becoming more prevalent, costs about 0.5–0.8 cents/kWh. Hybrid plants, combining flash and binary, provide 25% of the electricity used on the island of Hawaii.

In 2003, a total 2800 MW of power was generated; not quite 4 times as much as wind and solar combined. A productive geothermal field produces about 20 tons of steam or several hundred tons of hot water per hour.

The chief problems in producing geothermal power involve mineral deposi-tion—silica and calcium carbonate scale, on the piping and turbine blades, as well as corrosion. In addition, extensive withdrawal of hot water from rocks can cause ground subsidence of up to several meters. Another disturbing factor for local residents are the heavy releases of hydrogen sulfide, methane and ammonia, that subterranean steam carries to the surface. Hydrogen sulfide, in addition to conveying objectionable odors, induces rapid degradation of con-crete, plastics, and paint of local buildings. Methane, of course, is a greenhouse gas whose contribution is unwanted.

Currently, the share of total world electric capacity contributed by geother-mal plants has yet to attain one percent, and even when operating at full capacity, its benefits will accrue to only 30–40 countries. But geothermal energy can also be used for heating buildings and entire towns—including Oakland and San Francisco. It is being used to heat greenhouses where crops are being raised and for drying crops, heating water in fish farms, and pasteurizing milk. Indeed, there is a place for geothermal energy, but it will always be local and on a small scale. Yet a new report produced by the Western Governors Association projects a goal of 8 GW of electricity for the western states by 2015 [23]. This may be reality-based, or wishful thinking. Again, time holds the answer.

Biomass and "Green" Energy

The question as to what constitutes *biomass* has at least two responses; the short answer is, all living things are biomass. All animals, plants, and we humans are the organic matter that characterizes biomass. The more specific, industrial response defines biomass as the organic matter produced by crops, roots, stems, seeds, and stalks, along with animal metabolic wastes, and refers to materials that do not go into food products, but do have alternative commercial uses. Again, the concern is for energy, bioenergy, the energy inherent in the carbohydrates, fats, and proteins that constitute all organic matter.

Biomass has been used for energy since fire was discovered and people used wood to cook food and keep warm. Although wood remains the largest biomass energy source today, other sources include food crops; grassy and woody plants; residues from agriculture such as cornstalks, straw, sugarcane, bagasse, nutshells, and dung (manure from cattle, poultry, and hogs); yard clippings; and the organic components of municipal and industrial wastes. Even the methane emanating from landfills can be used as a bioenergy source. The fact is, the entire surface of the earth is covered with biomass—and because it is renewable, its potential for energy is nothing less than tremendous—at least 10 times the total amount of energy consumed worldwide annually from all sources. Think about that!

Biomass can be used for fuel, power production, and commercial and consumer products that would otherwise be made from fossil fuels, providing an array of benefits. For example, bioenergy has the real potential to reduce greenhouse gas emissions. Burning biomass releases about the same amount of carbon dioxide as burning coal or oil. But—and this cannot be taken lightly—fossil fuels released CO_2 captured by photosynthesis millions of years ago—an essentially "new" greenhouse gas. Burning biomass, on the other hand, releases the CO_2 balanced by the CO_2 used in the photosynthetic process that more recently created it. Furthermore, use of biofuels, as we shall see, can reduce our dependence on imported oil. Since the US economy is so closely linked with petroleum (importing over 60% of its oil), small fluctuations in oil prices or disruptions in supplies can have an enormous impact on our economy. Consequently biomass offers an alternative to foreign oil, providing national energy security, economic growth, and environmental benefits. Yet another benefit is its renewability. Utilizing sunlight energy in photosynthesis, plants metabolize atmospheric CO_2 to create new biomass, producing an estimated 140 billion metric tons annually. Combustion of biomass forms CO_2, which plants use to form more biomass. The overall process is referred to as the *carbon cycle.* From an energy viewpoint the net effect of the carbon cycle is to convert solar energy into thermal energy that can be converted to more useful forms such as electricity and fuels. Recall, too, that coal and oil were originally created from biomass deep in the earth over millions of years ago, and, because of the long periods needed for their formation, they are not considered renewable. Even though coal and oil replaced biomass fuels because of their higher energy content and ease of handling, with oil production peaking and for the reasons noted above, biomass is once again in ascendance.

Biomass is a complex mixture of carbohydrates, fats, and proteins, along with small amounts of sodium, iron, calcium, and phosphorus. However, the main components of biomass are the carbohydrates—some 75% dry weight, and lignin, the remaining 25%, but these do vary by plant type. The carbohydrates are primarily the long-chain cellulose and hemicellulose fibers that impart structural strength, and lignin, which holds the fibers together. Plants also store starch, another long-chain polymer, and fats as sources of energy. Starch and cellulose are polysaccharides, long, repeating chains of a single sugar, glucose, used by living systems as their source of biochemical energy. Hemicellulose, in the cell walls of plant tissue, is composed of the sugars, xylose, arabinose, mannose, and glucose. Lignan, another cell-wall polymer, combines with hemicellulose to bind cells together and direct water flow [24]. Fats are triglycerides, three long-chain fatty acids attached to a molecule of glycerol. The major part of the fat molecule are the fatty acids which have a chemical structure similar to that of hydrocarbons—petroleum. Hydrocarbons, composed solely of carbon and hydrogen, release water, CO_2, and heat when burned in air. The more oxygen involved, the more energy released.

The composition of biomass leads directly to its use as a biofuel, but given the renewability of biomass, a brief diversion is in order to reacquaint ourselves with photosynthesis. In a word, *photosynthesis* is the biological conversion of light energy into chemical energy. Sunlight is absorbed by chlorophyll in the chloroplasts of green plant cells, which produce carbohydrates from water and CO_2 taken from the atmosphere. In its most simplified form, the process can be represented by the formula

$$6CO_2 + 6H_2O \text{ chlorophyll} \rightarrow C_6H_{12}O_6 + 6O_2$$
$$\text{sunlight}$$

in which six molecules of carbon dioxide and six of water combine to produce one molecule of glucose and six molecules of oxygen. Globally, this process produces about 220 billion dry tons of biomass annually. Of course, it is the photosynthetic process that provides the oxygen that allows all living things, including us humans, to breathe.

Curiously enough, how plants obtain the nutrients needed for growth is as much a mystery today—for too many people—as it was before Johannes

Baptiste Van Helmont (1579–1644), a Dutch alchemist performed his enlightening experiment. Van Helmont placed a willow tree weighing 5 pounds (lb) in a clay pot in which he also placed 200 lb of soil. Five years later, after watering the willow as needed, it weighed about 169 lb, even though the soil in the pot lost only 2 ounces (oz). Clearly, the tree had gained weight from the water, not the soil—although the role of sunlight and carbon dioxide were not yet known [25].

In the photosynthetic process, sunlight stimulates the pigment chlorophyll found in the chloroplsasts, where sunlight also reacts with the CO_2 the plant breathes in via the stomata—microscopic holes in the leaves—and with water the plant absorbs from its roots. During a series of light and dark reactions the water molecules are split into hydrogen and oxygen; the hydrogen combines with carbon dioxide to produce glucose, which will become the building blocks for starch and other long-chain, complex carbohydrates. The excess oxygen is released to the atmosphere during the process of respiration.

It is the carbohydrates that are of interest today, as they can be used directly as fuel, burning wood, or converted into liquids and gases—ethanol and/or methane.

Two types of alcohol can be produced from biomass: (1) ethanol, C_2H_5OH, grain alcohol, which can be made by the microbial fermentation of sugarcane, sweet sorghum, cassava, sweet potatoes; and, of course, (2) maize or corn, which provide the hexose sugars that are metabolized by yeast cells to alcohol. Methanol, methylalcohol, CH_3OH, wood alcohol, originally made by the distillation of wood chips, is now made from natural gas. Most current interest however is in fuel ethanol production involving the use of cornstarch—in the United States (and sugarcane in Brazil, where, as we shall see, straight gasoline is no longer used to power motor vehicles), which is a two-step process in which starch is converted to glucose and the glucose is then microbioally metabolized to alcohol. In 2003, corn-produced ethanol reached 2.81 billion gallons—not all that much given our enormous usage.

In addition to ethanol, biodiesel fuel is a domestically produced, renewable fuel for diesel engines derived from soybean oils. The typical soybean oil methylester contains a number of fatty acids as shown in Table 7.3, and contains no petroleum.

TABLE 7.3. Typical Soybean Oil Methylester Profile

Fatty Acid	Weight Percent	Molecular Weight	Formula
Palmitic	12.0	270.46	$C_{15}H_{31}CO_2CH_3$
Stearic	5.0	298.52	$C_{17}H_{35}CO_2CH_3$
Oleic	25.0	296.50	$C_{17}H_{33}CO_2CH_3$
Linoleic	52.0	294.48	$CH_3 (CH_2)_4 CH=CHCH_2CH=CH(CH_2)_7CO_2CH_3$
Linolenic	6.0	292.46	$CH_3 (CH_2CH=CH)_3 (CH_2)_7CO_2CH_3$

Biodiesel is made by the process of transesterification, in which glycerine is separated from the fat or vegetable oil. The process byproducts are methylesters, the chemical designation of biodiesel. (Esters, by the way, are a class of organic compounds formed by the reaction of organic acids and alcohols.) It is worth recalling that when, the German engineer Rudolph Diesel unveiled his prototype engine that bears his name, he fueled his engine with peanut oil, an idea that never took off because gasoline became so plentiful and so cheap. The smell of biodiesel fuel is another selling point, smelling as it does like a kitchen, not a garage.

As for the kitchen, northern Biodiesel, in Ontario, NY, is opening a plant to turn cooking oils and agricultural waste into biodiesel fuel. Brent Baker's 1989 International Blue Bird School Bus smells like a barbecue now that he uses only leftover cooking grease from french fries, fried chicken, and fish. Baker has logged thousands of miles driving his highly visible bus around the country, highlighting the relative ease of shifting away from gasoline. The ride is smooth and quiet, and when the gas gauge so indicates, he simply pulls into a roadside diner and empties out the grease in their dumpsters—with permission, of course. Baker is not alone in believing that vegetable oil could be a valuable source of fuel. Cars and trucks of more than 40 federal and state agencies across the country, including the U.S. Postal Service and the EPA, run on a blend of vegetable oil and diesel fuel [26].

More importantly, perhaps, Willie Nelson, my rock idol, drives a Mercedes, but the exhaust of his diesel-powered vehicle smells like french fries or peanuts, depending on the alternative fuel that happens to be in his tank—no longer a "gas" tank, just a tank; a fuel tank, filled with "biowillie," which, as Nelson sees it, is a domestic product that can profit farmers and help the environment. Biodiesel is picking up around the country: Cincinnati began using B30, 70% conventional diesel and 30% vegetable oil. The state of Minnesota now requires all diesel fuel sold in the state to be 2% biodiesel. In the Seattle area, B20, 20% vegetable and 80% diesel, is being used. Willie Nelson's company, Willie Nelson Biodiesel-B20-at Carl's Corner, Hillsboro, Texas, is aimed at truckers. His biodiesel is currently being sold in four states, and is fueling buses and trucks for his tours. Willie doesn't expect to get rich on his biodiesel as there just isn't enough used grease or vegetable oil to go around [27].

Since fatty acids from plant fats and petroleum hydrocarbons have similar chemical structures, plant-based fats can be coverted into a liquid fuel similar to diesel fuel. Fatty acids obtained from soybeans or other oil-rich vegetation are reacted with methanol to form fatty methylesters, or monoalkylesters of long-chain fatty acids. As these methylesters have nearly equivalent energy content and chemical structures similar to those of petrochemical diesel fuels, they have been dubbed "biodiesel fuels." They not only work well in diesel engines but also have lower particulate emissions than does gasoline.

Ethanol and biodiesel can be blended with or directly substituted for gasoline or diesel fuel. Use of these biofuels reduces toxic air emissions, greenhouse gas buildup, and dependence on imported oil, while supporting agriculture and rural economies. Unlike gasoline and diesel, biofuels contain oxygen—ergo, adding biofuels to petroleum products allows the fuel to combust more completely, reducing air pollution.

Because the vast bulk of biomass consists of cellulose, hemicellulose, and lignin, compared to starch and sugar, the U.S. Department of Energy (DOE) is spear-heading a national effort to develop methods to break down cellulose and hemicellulose into their component sugars. Doing so will allow biorefineries to biologically process the monosaccharide sugars to ethanol. The ability to use cellulosic materials in addition to sugar will open the way for far greater production of ethanol. Lack of available agricultural acreage to grown corn is a current bottleneck for expanded use of ethanol. So, in October 2005, the DOE awarded $92 million for six genomic projects that seek solutions to more efficient microbial metabolism of the long-chain polysaccharides that currently resist breakdown. To develop solutions, it is essential to know the details of microbial biochemical pathways under normal and environmentally modified conditions. It is anticipated, and rightly so, that these newly funded projects will develop this much needed information [28].

In his State of the Union address to the nation in February 2006, President Bush touted switchgrass, *Panicum virgatum*, a cereal grain and member of the millet family, currently used as fodder for cattle, horses, sheep, and goats, as another crop useful for ethanol production. This perennial grows in great abundance on the prairies of the Great Plains. Considering that there is so much of it, and that it is easy to grow, it can be a valuable addition to corn.

The substitution of ethanol for gasoline in passenger cars and light vehicles is only in its infancy in the Unite States. South of the border, ethanol is well into its adulthood. In Brazil, the world's largest commercial biomass-to-energy program is in full bloom. Would anyone have imagined that engines that run only on gasoline would be banned in Brazil, or anywhere else, for that matter? Brazil has decreed that automobiles and light trucks must use either all ethanol, or gasohol, a blend of 78% gasoline and 22% ethanol. Although ethanol from sugarcane has been used as engine fuel in Brazil since 1903, it was the institution of the "Pro-alcool" program in 1973, that Brazil's addiction to imported oil was abruptly halted. For years the government feared disruption of supplies from foreign countries. When the world oil crisis struck in 1987, Brazil was financially wrecked, as they had been importing 80% of their oil. Faced with an economic crisis, not unlike what is happening in the United States today, the decision was taken to substitute alcohol for oil. The "Pro-alcool" program went into effect and the outcome was a new sugarcane—alcohol industry. Over the ensuing decades, the "Pro-alcool" initiative has served as a model for other countries; led to the creation of high quality jobs, cleared the air, became an effective tool for managed energy costs, and the country learned to integrate new fuels into existing commercial end-use systems [29]. The United States has much to learn from Brazil. We need only shore up our political will. Of course, the oil companies are not going to be happy, but the United States does not exist to please oil companies. Certainly not now with oil prices bouncing between $60 and $70 per barrel. In fact, governments and sugar producers around the world have beaten a path to Brazil looking for help in developing their own ethanol industries. India, the world's number 2 sugar producer after Brazil, is rushing to spread the use of ethanol as it also seeks to reduce dependence on imported oil, as automobile ownership rises sharply.

E85 is the new kid on the fuel "block" in the United States. As extensive as our communication system is, most people are unaware that over four million american cars are in fact flex-fuel cars, designed to run on E85, a blend of 85% corn-based ethanol and 15% gasoline. The next time you fill up with gas, take a hard look at the gas flap; there may just be a sticker the size of a business card announcing the fact that this is a flex-fuel car that can run on E85. Unfortunately E85 is unavailable at most of the 180,000 gas stations in the United States. Only about 500 carry it, and they've mostly where the corn grows [30]. When, not if, ethanol becomes cellulose-based, there could easily be a hefty shift from gasoline to ethanol. Glucose-based ethanol simply cannot provide the multibillion gallons required for the many millions of American vehicles, and even when cellulose bares its biochemical secrets, biofuels will be unable to replace oil/gasoline completely because sufficient agricultural area is simply unavailable. But together with other renewables, we will become far less dependent on the vagaries of foreign oil. However, it does take two-thirds of a gallon of oil to make a gallon equivalent of ethanol from corn. Therefore, one gallon of ethanol used in gasahol displaces about one-third of a gallon of oil or less. Not that much of a savings. If one were to bet on a technology that could displace foreign oil/gasoline, the difficult-to-digest celluloses and hemicelluloses could do it, once new enzymes become available.

Corn is a many splended thing—beyond ethanol. Polylactides are yet another of its gems. Polylactides, or polylactic acid, is a corn-based transparent plastic without a whiff of petrochemicals. Refining corn can yield a plastic that is readily maleable into cups, containers, candy wrappers, cutlery, and medical sutures.

Polylactide, an alternative plastic, is Cargill/Dow's latest attempt to wean us off oil. At their new plant in Blair, Nebraska, cornstarch is hydrolyzed to glucose, which is then microbially fermented to lactic acid. Squeezing out water from lactic acid yields lactide molecules that form into long chains that can be spun into fiber for clothing and bedding, or molded and shaped into clear, glossy, rigid cups, bottles, knives, forks, and spoons. Wal-Mart, Del Monte, Wild Oats, and Newman's Own Organics have already opted for these "green" plastics. The next time you're in a supermarket, check the molded vegetable containers for the Nature Works trademark, which is beginning to make a welcome appearance. Sales at Nature Works, the Cargill subsidiary that produces the plastic, grew 200% in the first half of 2005, over the same period in 2004 [31]. These environmentally friendly plastics take less than 2 months to degrade in a landfill, compared to petroleum-based plastics, which can take ages to decompose. These yellow nibblets on a cob surely look like corn, but actually they're all-chemical—waiting for creative minds to unleash their bounty—in addition to tickling our palates.

Creativity is what the DuPont Company of Wilmington, Delaware, is using in its quest to switch us from a hydrocarbon to a carbohydrate economy, which has much to recommend it, as carbohydrate is infinately sustainable, and of course, hydrocarbon is not. DuPont believes that biology can solve problems that chemistry can't.. So, for example, they will be marketing Sorona (propane diol) made from glucose, a product used in carpet fibers that offers greater dye absorbtion and stain resistance than does the petrochemical version that they've been selling. They are also developing plant-based hair dyes and nail polishes that will not adhere to skin, surgical glues that can stop bleeding, and a textile fiber made from sugar that will act and feel like cotton—cotton candy? Industrial biotechnology is DuPont's new direction. Perhaps it will take the country along with it [32]. Another way is via BioButanol, a crop-based fuel that may just jump to the head of the biofuels line. Developed by DuPont in conjunction with BP (British Petroleum), biobutanol is inherently better than ethanol because it has as much energy per gallon as gasoline. Although called biobutanol because it's produced from biomass, it is no less butanol, $C_4H_{10}O$, a four carbon alcohol. The fermentation of biomass via the bacterium *Clostridium acetobutylicum*, is relatively simple and inexpensive. *Cl. aceto-butylicum* is the same microbe that Prof. Chaim Weizmann used to obtain acetone from horse chestnuts for explosives for the British military during the first World War. Given appropriate nutrients microbes can be induced to produce almost any substance.

Fuel Cells

Is the venerable internal-combustion engine on the way out? It's entirely possible, and it's long overdue. But replace it with what? Is there such a thing as a truly "green" car? (Consider, for example, the GM model shown in Fig. 7.6) Fuel cells are widely believed to hold the key to pollution-free motor vehicles and longer-running electric devices. Fuel cells can provide energy for devices as small as a laptop computer, or as large as utility power stations, and just about anything in between, inducing every type of motor vehicle. Just what is this sweeping development that holds promise of repowering our world?

Fuel cells use the chemical energy in hydrogen to generate electricity. That's the short answer. A tad more technical response informs us that a fuel cell is an electrochemical energy conversion device that converts hydrogen gas and oxygen from air into water and heat, and in the process, produces electricity. So elegantly simple. Is it too good to be true? If hydrogen were readily available, and inexhaustible, as it may well be, the generation of electricity would no longer require fossil fuels, turbines, or generators, and we'd live in a cleaner, quieter world. Hydrogen could revolutionize the world. The idea borders on fantasy—or does it?

GM HydroGen3

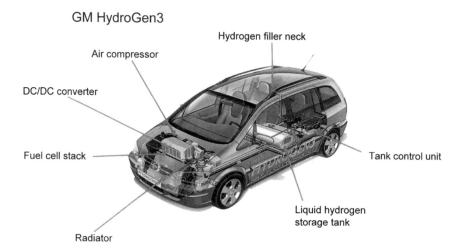

<u>Figure 7.6.</u> The GM HydroGen3 automobile appears to be the prototype of all hydrogen fuel-cell-powered vehicles. (Figure courtesy of the General Motors Corporation.)

During 1874/75, Jules Verne, the eminent and prolific French novelist, wrote his three-volume masterpiece *The Mysterious Island*, based on the true story of Alexander Selkirk's survival (remember Robinson Crusoe?) on an uninhabital island for 5 years. *The Mysterious Island* is an enthralling tale of five men and a dog, who escape from Richmond, Virginia, during the Civil War, in a hot-air balloon, landing on a faraway and uncharted island, where they learn to survive. One day, as winter was approaching and the men were sitting near a glowing fireplace after dinner, smoking their pipes, one of them asks Cyrus Smith, an engineer, what might happen if one day the earth's supply of coal will be exhausted, with no more machines, no more railways, no more steamships, no more factories, no more of anything that the progress of modern life requires. "What would take its place?" asked Pencroff. "Do you have any idea, Mr. Cyrus?" "What will people burn in place of coal?"

"Water," answered Cyrus Smith. "Water broken down into its component elements. And broken down by electricity. Yes, I believe that water will one day be used as fuel, that the hydrogen and oxygen of which it is constituted will be used, simultaneously or in isolation, to furnish an inexhaustible source of heat and light, more powerful then coal can ever be" [33]. Imagine. Water, the coal of the future. Here we are 132 years after Jules Verne predicted a future in which all our needs would be supplied by an inexhaustible source of hydrogen. The world is ready for a hydrogen economy to interdict the dislocations of global warming, to strengthen national energy security, and to improve energy efficiency. Just what is a hydrogen economy, and what do we know about hydrogen and its availability?

The hydrogen economy is a world in which hydrogen, a component of many organic chemicals, and water, is readily available to everyone, everywhere, and the United States is no longer dependent on a single source of fuel, as hydrogen is produced in abundance domestically, cleanly, and cost-effectively, from a range of sources such as biomass, nuclear power, and water. In this economy, hydrogen-powered fuel cells will be as common as gasoline was in the twentieth century—powering our homes, cars, trains, planes, ships, offices, and factories. Developing countries have hydrogen-based systems to fuel their energy needs.

With fuel cells as the means whereby hydrogen will be used to produce electricity, the question for us is: How do fuel cells work, and how do they differ from batteries, if they do? We know that batteries are power storage devices, with all their chemicals are inside, and are converted to electricity when switched on. This means that eventually the battery will "go dead," and must be recharged or replaced. Recharging is limited, as eventually they are no longer rechargeable. A fuel cell combines its hydrogen fuel, and oxygen from air to produce electricity steadily for as long as fuel is provided, which means that it can never "go dead" and needs no recharging. As long as hydrogen and oxygen flow into the cell, electricity will always flow out. The fuel cell provides DC (direct current), voltage that can be used to power motors, lights, or any appliance, large or small. Most fuel cells systems consist of four basic components:

- Fuel cell stack
- Current converter
- Fuel processor
- Heat recovery system

Additional components control humidity, temperature, gas pressure, and wastewater. Our concern is with the basic four. The fuel cell stack generates electric current from chemical reactions within the cell. A typical fuel cell stack can consist of hundreds of cells. The amount of power produced depends on fuel cell type, cell size, temperature at which it operates, and the pressure at which the gases are supplied to the cell.

The fuel processor converts fuel into a form usable by the cell. If hydrogen is fed into the system, a processor may be needed only to filter out impurities in the gas. Current inverters and conditioners adapt the electrical current from the cell to suit the electricial needs of the application: a simple electric motor, an automobile engine, or a complex utility grid. Since fuel cells produce direct current, and our homes and businesses use alternating current, the direct current must be converted to alternating current. Both AC and DC must be conditioned, which means controlling current flow in terms of amperes, voltage, and frequency to meet the needs of the end user.

As fuel cells generate heat as a byproduct, a heat recovery system may be used to produce steam or hot water that can be converted to electricity via a traditional turbine/generator system. Because heat is captured, the overall efficiency of the system increases. But this depends on the type of fuel cell system. Of primary importance is the fact that fuel cells are significantly more energy-efficient than are combustion-based power generation processes. A conventional combustion-based power plant typically generates electricity at efficiencies of 30–35%; fuel cell plants can generate electricity at efficiencies of up to 60%. When used to generate electricity and heat (cogeneration), they can attain efficiencies of 85%. Internal-combustion engines in today's automobiles convert less than 30% of the energy in gasoline. Vehicles using electric motors powered by hydrogen fuel cells attain efficiencies of 40–60% [34].

Before describing the various types of fuel cell systems, we can take a deeper look at the system. The heart or core of the fuel cell is its stack, which is made of many thin, flat cells layered together. While the term *fuel cell* often refers to the entire stack, in fact, it refers only to the individual cells. A single cell produces a small amount of electricity, but many cells stacked together provide sufficient power to drive a car. Increasing the number of cells in a stack increases the voltage; increasing the area of the cells increases the current.

When hydrogen is fed to a fuel cell, it encounters the first catalyst-coated electrode, the anode, where hydrogen molecules release electrons and protons. The protons migrate through the electrolyte membrane to the second catalyst-coated electrode, the cathode, where they react with oxygen to form water. The electrons can't pass through the electrolyte membrane to the cathode. Unable to do so, they travel around it. This traveling around of electrons produces the electric current.

The electrodes (anode and cathode), catalyst, and polymer electrolyte membrane together constitute the membrane electrode assembly of a polymer electrolyte membrane (PEM)—the fuel cell type of current concern. Each membrane assembly consists of two electrodes, anode and cathode, and an exceedingly thin layer of catalyst.

The anode, the negative side of the fuel cell, conducts the electrons freed from the hydrogen molecules so that they can be used in an external circuit. Channels etched into the anode disperse the hydrogen gas equally over the surface of the catalyst. The cathode, the positive side of the fuel cell, contains channels that distribute oxygen to the surface of the catalyst. It also conducts the electrons back from the external circuit to the catalyst, where they can recombine with the hydrogen ions and oxygen to form water.

The polymer electrolyte membrane (PEM), a specially treated material that looks like saran wrap, conducts only positively charged ions, and blocks the electrons. This is the key to fuel cell technology; permitting only the necessary ions to pass between the anode and cathode. All electrochemical reactions in a fuel cell consist of two separate reactions: an oxidation half-reaction at the anode, and a reduction half-reaction at the cathode:

$$\text{Anode side:} \quad 2H_2 \rightarrow 4H + +4e^-$$

$$\text{Cathode side:} \quad O_2 + 4H + +4e^- \rightarrow 2H_2O$$

$$\text{Net reaction:} \quad 2H_2 + O_2 \rightarrow 2H_2O$$

Normally these two reactions would occur slowly with the low operating temperatures of the PEM cell. To speed things up, to facilitate the reactions, each electrode is coated on one side with a catalyst layer that increases the velocity of the reaction of hydrogen with oxygen. This catalyst is most often made of platinum powder, which is tremendously expensive—one of the most pressing current problems.

When an H_2 molecule contacts the platinum catalyst, it splits into two H^+ ions and two electrons (e^-), which are conducted through the anode, making their way through the external circuit—powering a motor, then returning to the cathode side of the cell. At the same time, on the cathode side, oxygen gas (O_2) is being forced through the catalyst, where it forms two oxygen atoms. Each has a strong negative charge that attracts the two H^+ ions through the membrane, where they combine with the oxygen atom and two of the electrons from the external circuit to form a water molecule [35]. This reaction in a single fuel cell produces about 0.7 V. To raise this meager voltage to levels required to power an engine, many cells are combined to form a fuel cell stack, as shown in Figures 7.7a–7.7d. These are the essentials. Fuel cells are simple; no moving parts, yet theoretically capable of powering the world, cleanly and efficiently and limitlessly. But not for some time. A number of problems must be overcome. Fuel cells are far more costly than fossil fuels, but the rising price of oil is narrowing the gap. The durability of fuel cells has yet to be established; they must show themselves to be at least as durable as current engines, and be able to operate efficiently in the dead of winter, and heat of summer. The size and weight of current fuel cells are still excessive for automobiles of all sizes. Today fuel cell stacks are extremely large and heavy in order to produce the required current to power cars and light-rail tracks. Their size and weight include the stack (with its membranes, bipolar plates, catalysts, and gas diffusion media) but not the ancillary components; namely, the heat exchanger, humidifiers, sensors, and compressor/expander. Pressure to reduce both size and weight comes from rising oil prices and global warming concerns. There isn't a car manufacturer that hasn't invested heavily in fuel cells. Within the coming decade the problem should be solved as there are a number of different types of fuel cells being developed. In fact, recently, chemists at the University of North Carolina developed a new, much improved polymer membrane, based on perfluoropolyester that doesn't dissolve when the water content rises, and which can operate efficiently at 120°C, well above the standard 80°C. This new polymer conducts protons 3 times better than does the standard Nafion (DuPont) polymer membrane, which will markedly improve the performance of automotive fuel cells [36].

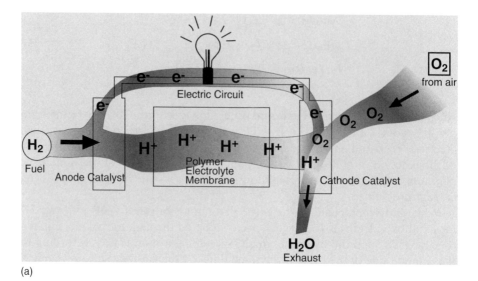

(a)

<u>Figure 7.7.</u> Parts (a) and (b) illustrate the basic principle of a fuel cell. The cell consists of two electrodes sandwiched around an electrolyte. Oxygen passes over one electrode and hydrogen over the other, generating electricty, heat, and water. Hydrogen is sent into one side of a proton exchange membrane (PEM), which splits it into a proton and an electron. The hydrogen proton travels through the membrane [Follow the arrows in part (b)], while the electron enters an electrical circuit, creating a DC electrical current. On the other side of the membrane, the proton and electron are recombined and mixed with oxygen from room air, forming water. Because there is no combustion in the process, there are no emissions other than that of water, making fuel cells an extremely clean and renewable source of electricity. [Part (a) courtesy of Fuel Cell 2000 (Breakthrough Technologies Institute), Washington, DC, part (b) courtesy of Ballard Power Systems, Burnaby, British Columbia, Canada.] Part (c) shows a 25-W fuel cell, with a stack of three cells. The voltage from a single cell is about 0.7 V, enough for a common lightbulb. When cells are stacked in a series, the operating voltage increases to 0.7 V multiplied by the number of cells stacked. (Photo courtesy of the U.S. DOE National Renewable Energy Lab., Golden, CO.) Part (d) shows a 5-kW fuel cell, with its assembly of many individual cells locked together. This large cell is manufactured by Plug Power, Latham, NY. (Photo courtesy of the U.S. DOE's National Renewable Energy Laboratory, Golden, CO. Warren Gretz, Photo credit.) Part (e) shows a Plug Power stacked fuel cell installed in a stationary electric power unit. [Photo courtesy of Fuel Cell 2000 (Breakthrough Technologies Institute), Washington, DC.]

Types of Fuel Cells

Fuel cells are categorized by the type of electrolyte they use: a chemical that conducts charged ions from one electrode to another. In fact, the type of electrolyte determines the kind of chemical reactions that occur in the cell, the type of catalysts required, the temperature range in which the cell operates,

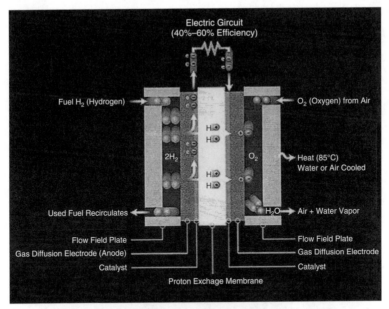

(b)

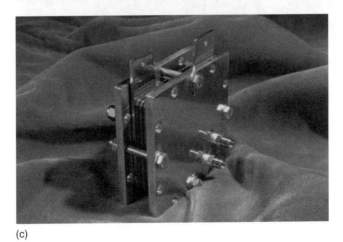

(c)

Figure 7.7. *Continued*

and the fuel required. A number of fuel cells are under development, each with its specific advantages, limitations and potential applications.

Currently six systems are being pursued. In addition to PEM, there are

Direct methanol fuel cells, powered by methanol mixed with steam and fed directly to the anode. Methanol is easier to transport and supply, but it is only now beginning to be studied.

(d)

(e)

Figure 7.7. *Continued*

Alkaline fuel cells, which use potassium hydroxide in water as the electrolyte. This was the earliest of the fuel cells and was used by NASA since the 1960s for its space program. But it is too easily degraded by carbon dioxide, and is extremely expensive, which means that it is unlikely to be used commercially.

Phosphoric acid fuel cells, which use liquid phosphoric acid as the electrolyte. This cell does not provide sufficient current or power to drive automobiles, and its longer warmup time makes it doubly unsuitable and expensive. It is currently used for small, stationary power generation.

Molten carbonate fuel cells, which are being developed for use in natural gas and coal-based power plants and for military applications. These high-temperature fuel cells operate at 650°C (1200°F) using a molten carbonate/salt mixture as their electrolyte. Although these are highly corrosive, breaking down cell components relatively rapidly, they appear to be highly efficient.

Regenerative fuel cells, which employ electricity from solar power to perform the electrolysis of water that yields the hydrogen and oxygen needed. This fuel cell is being developed by NASA for its special applications.

As Table 7.4 indicates, each of these fuel cells has its unique advantages, applications, and limitations. Of one thing we can be certain—several will become realities and take their place supplying a range of electrical applications. At the moment, PEM and hydrogen is the frontrunner.

With hydrogen projected as the fuel of the future, several questions arise: Where will it come from, will there be a plentiful supply, is it safe, and can it be stored?

Hydrogen is the most abundant element in the universe, but it is not a free element. All hydrogen is bound to chemicals in plants and animals, and is found in fossil fuels—coal, oil, and natural gas, and, of course, bound to oxygen as water. Consequently there are a number of straightforward ways to obtain it, including producing it directly from water via electrolysis and solar and nuclear thermochemical splitting of water by extremely high temperatures. Most of the hydrogen used in the United States currently is produced by steam reforming of natural gas. Thermochemical conversion processes can produce hydrogen from hydrocarbons and biomass-derived gases. Clearly, there is no foreseeable lack of hydrogen. In fact, it could be said to be limitless. The only concerns are which are the most economically sound ways of producing abundant supplies. The long-term goal is to obtain hydrogen from renewable sources such as nuclear power—of which a modified Pebble Bed reactor looks promising, and is being tested by the U.S. Department of Energy. Wind and solar power are also being pursued as additional means of procuring adequate supplies. However, and interestingly enough, producing hydrogen from biomass, for example, also produces large amounts of unwanted carbon monoxide. With researchers in hot pursuit of solutions, a team of chemists at the University of Wisconsin (Madison) recently found that using a gold catalyst reactor removes the carbon monoxide. Because the conventional procedure employs high-temperature steam to react with carbon monoxide, forming CO_2, this new gold catalyst procedure should obviate the need for steam, bringing production prices down, and increasing the safety of the process [37].

To the question as to whether hydrogen is a safe fuel, the ready response must be that when handled properly, hydrogen is a safe fuel, and like all fuels, must be treated with respect. Hydrogen can be stored in metal or chemical hydrides, or as a gas or liquid in pressurized tanks at 5000 pounds per square inch (lb/in^2; psi). With automobiles, in the event of a collision, the chance of a

TABLE 7.4. Comparison of Fuel Cell Technologies

Fuel Cell Type	Electrolyte	Operating Temperature	Applications	Advantages	Disadvantages
Polymer electrolyte membrane (PEM)	Solid organic polymer polyperfluorosulfonic acid	60–100°C (140–212°F)	Electric utility Portable power Transportation	Solid electrolyte reduces corrosion and management problems Low temperature Quick startup	Low temperature requires expensive catalysts High sensitivity to fuel impurities
Alkaline (AFC)	Aqueous solution of potassium hydroxide soaked in a matrix	90–100°C (194–212°F)	Military Space	Cathode reaction faster in alkaline electrolyte, so high performance	Expensive removal of CO_2 from fuel and airstreams required
Phosphoric acid (PAFC)	Liquid phosphoric acid soaked in a matrix	175–200°C (347–392°F)	Electric utility Transportation	Up to 85% efficiency in cogeneration of electricity and heat Can use impure H_2 as fuel	Requires platinum catalyst Low current and power Large size/weight
Molten carbonate (MCFC)	Liquid solution of lithium, sodium, and/or potassium carbonates, soaked in a matrix	600–1000°C (1112–1832°F)	Electric utility	High efficiency Fuel flexibility Can use a variety of catalysts	High temperature enhances corrosion and breakdown of cell components
Solid oxide (SOFC)	Solid zirconium oxide to which a small amount of yttria is added	600–1000°C (1112–1832°F)	Electric utility	High efficiency Fuel flexibility Can use a variety of catalysts Solid electrolyte reduces corrosion and management problems Low temperature Quick startup	High temperature enhances breakdown of cell components

fuel leak is less likely than that from a gasoline tank, due to the structure of the storage tanks. Additionally, hydrogen is lighter than air, which would allow it to quickly dissipate should a leak occur. Also, it is far safer than gasoline, which ignites rapidly with a spark. Hydrogen does not produce smoke when burning. Inhalation is the primary cause of death in gasoline fuel fires and explosions. Moreover, a hydrogen fire radiates little heat, and hydrogen is nontoxic and nonpoisonous. As for storage and distribution, as hydrogen demand grows, economies of scale will likely make centralized hydrogen production more cost-effective than will distributed production. But that will require a cost-effective means of transportation and delivery. Currently, hydrogen is transported by road via cylinders, tube trailers, cryogenic tankers, and pipeline. Increased delivery will require high-pressure compressors for gaseous hydrogen and liquefactors for cryogenic hydrogen—both have significant costs and inefficiencies associated with them, which must be reduced before hydrogen can become competitive and widely available.

Delivering bulk hydrogen to hydrogen (gas) stations is one thing; use as a fuel for fuel cells in motor vehicles is quite another. A "fuel cell automobile" is in electric vehicle that uses hydrogen as fuel, rather than a battery, to produce electricity. The fuel cell utilizes a catalyst to perform a chemical process (no moving parts) that, as we have seen, combines hydrogen and oxygen to produce electricity. Hydrogen gas can be safely stored within a car or truck in a pressurized tank, and is supplied to the fuel cell at less pressure than gasoline is to a fuel-injected engine. Fuel cell vehicles are twice as efficient as gasoline-driven vehicles and produce zero tailpipe emissions—other than water. In a direct hydrogen fuel cell vehicle, hydrogen is piped to the fuel cells, where it combines with oxygen from an air compressor. An electrode coated with a catalyst splits the hydrogen into electrons and protons. The movement of the electrons generates electricity, which is sent to a traction inverter module that converts the electric energy into mechanical energy that turns the wheels. This is the flow plan that drives the Ford Focus FCV and P2000. However, the challenge facing the marketing of mass-produced cars is to provide hydrogen to the stack module. Two alternatives appear possible: direct, onboard storage of hydrogen, and/or onboard reformation of hydrogen from liquid fuels such a methanol or gasoline. Direct hydrogen provides the greatest environmental benefit, and the simplest fuel cell system, but requires the development of an extensive hydrogen infrastructure that would go head-to-head with existing gas stations owned by the oil companies, who are not ready to pass into history. Onboard reformation of hydrogen would use a liquid fuel option with an established fueling infrastructure, but that requires additional onboard complex chemical fuel processing. Both require additional development to determine final cost estimates. Furthermore, safety is of primary concern. Several safety systems are available. The Ford Focus incorporates four hydrogen sensors: two in the trunk, one under the hood, and another in the passenger section. If a leak is detected, the car simply shuts down. In addition, eight small fans continually vent the car while driving, filling at a hydrogen gas station, and when the fuel door is opened.

Again, at this time, the major deterrent to mass marketing of a fuel cell vehicle is the lack of infrastructure. A recent study by DOE/Ford found that factory-built hydrogen refueling stations capable of supporting 100 cars could produce hydrogen that is cost-competitive with gasoline. These hydrogen refueling stations would produce hydrogen by steam reforming of natural gas, or via electrolysis of water, and would utilize existing natural gas and electric power infrastructures. An initial hydrogen infrastructure based on on-site natural gas reformers could account for 10–15% of all conventional filling stations in the country today. This small percentage would be sufficient to support mass production of direct hydrogen fuel cell cars. As is evident, nothing is simple. A great deal of planning and money and public confidence will be needed for transition to a hydrogen economy. Until these long-term solutions are sorted out, hybrid vehicles can (will) serve as a bridge to reduce emissions and ease dependence on fossil fuels.

What is a hybrid? Any vehicle is a hybrid when it combines two or more sources of power. Most hybrid cars run off of a rechargeable battery and gasoline. Essentially, it works this way. In Toyota's Prius, turning on the ignition activates the electric generator/starter, which cranks up the internal-combustion engine (ICE). When the ICE is warm, it automatically shuts off. The electric motor is now operating, and will remain in all-electric mode until 15 mph. If the car remains at low speed, it will effectively be an electric car with no gasoline being used, and no tailpipe emissions. Above 15 mph and into cruising speed, the gasoline engine powers the vehicle, and provides power to the battery for later use. All the while, the onboard computer is making decisions about when to go to gasoline and when to electric. When the car stops at a traffic light, both the gasoline engine and the electric motor shut off to save energy, but the battery continues to power auxiliary systems, such as air conditioning and dashboard panel. This technology can reduce emissions by 80%, and allows traveling for 400–500 miles on a tank of gasoline.

Evidently there are many ways to design a hybrid system; every car manufacturer has several designs in the works. In some hybrids, the batteries power electric motors, which drive all four wheels. In others, the engine may drive the rear wheels while the batteries power the electric motors running the front wheels. In yet others, the batteries or other electric source acts as auxiliary power. The Ford model "U" concept car is propelled by an ICE that's optimized to run on hydrogen fuel instead of gasoline. Because there are no carbon atoms in the fuel, combustion produces no hydrocarbons or CO_2 emissions. It is entirely possible that with further refinement, many of these hybrids will have air leaving their tailpipes that is actually cleaner than the air entering the engine.

Of course, fuel cells are not limited to automobiles. Manhattan Scientifics, Inc., has developed a fuel-cell-powered mountain bike that uses hydrogen and air as fuel and emits only water vapor as a waste product. The *Hydrocycle* has a range of 40–60 miles on a flat surface, and can achieve speeds of up to 18 mph. Because a fuel cell stack powers its electric motor, it's extremely quiet—

with no moving parts, and does not need to be recharged, only refilled. Here again, the lack of hydrogen gas stations makes this more than an inconvenience. Nevertheless, the company sees major markets for the bike in Asian cities, where bicycles and gas-powered scooters make up a major portion of vehicle traffic, and would help reduce the severe air pollution of many Asian cities.

It remains only for the future, and not too far into it at that, for computers, cars, buses, home appliances, and industry to be powered by hydrogen fuel cells.

OVER THE HORIZON

While the world awaits hydrogen from diverse sources to refuel and drive it, clever scientists around the world are quietly working to glean the secrets of flower power. Just over the horizon, perhaps no more than a decade or two, yet another source of hydrogen energy will become available. Photosynthesis will have yielded up its mysteries. Photosynthesis, the most successful solar converting mechanism on earth, is a potential source of unlimited renewable energy. Artificial photosynthesis will make it possible to produce limitless quantities of hydrogen and/or other energy rich fuels from water, clearly and cheaply. Keep your eyes on the horizon.

HAS THE FUTURE ARRIVED?

Can there be total energy self-sufficiency? In other words, can a country supply its heat and electricity needs by the renewable, domestic sources described above? The idea has all the earmarks of fantasy. It's just to good to be true. But—and however—it does appear to be happening—in one country: Iceland.

Three renewable sources are working for Iceland: geothermal, hydropower, and hydrogen fuel. Imported fossil fuel is on its way out. Currently, Iceland obtains hot water from its many geothermal wells, and falling water from hydroelectric dams, to provide heat for homes and heat for hot showers, and to light its streets and buildings through its long, dark winter. In fact, 95% of its heating and lighting needs are supplied by geothermal and hydropower. Now, the Icelanders are working with hydrogen fuel cells to drive buses, cars, and ships, as a means of completely dispensing with imported oil. When—not if—this occurs, the country will be fossil-fuel-free. Surrounded as they are by water, Icelanders see their future tied to hydrogen—to power everything. For them, the hydrogen economy is the way to go, and see it as reality by 2050—or sooner. Currently they produce hydrogen overnight by electrolysis, using electricity from Reykjavik's (their capital city) power grid to split water into hydrogen and oxygen. The hydrogen is then stored as compressed gas. City buses with rooftop hydrogen tanks pipe hydrogen to the fuel cells, which produce sufficient electricity to drive a motor behind the rear wheels, for an entire day without refueling. Not only are the buses exceptionally quiet; their water vapor emissions are so clean that they are drinkable [38]. Obviously Iceland is leading the way to a hydrogen economy, and most assuredly, Jules Verne had it right.

REFERENCES

1. *Meeting Energy Demands in the 21st Century: Many Challenges and Key Questions,* U.S. Government Accountability Office, GAO-05-414T, Washington, DC, March 2005.

2. Woelfle, G., *The Wind at Work: An Activity Guide to Windmills,* Chicago Review Press, Chicago, 1997.

3. Thomann, G.,Wind power, in*McGraw-Hill Encyclopedia of Science and Technology,* Vol. 19, New York, 2003, p. 559.

4. *How Wind Turbines Work,* U.S. Dept. Energy EERE Information Center, Wind and Hydropower Technology Program (available at http://eere.energy.gov/windandhydro/wind_how.html; accessed 2/2/05).

5. Talbot, D., Wind power upgrade: Turbine technology looms larger on the horizon, *Technol. Rev.* 21–22 (May 2005).

6. *Renewable Energy: Wind Power's Contribution to Electric Power Generation and Impact on Farms and Rural Communities,* U.S. Government Accountability Office, GAO-04-756, Washington, DC, Sept. 2004.

7. Dean, D., A seashore fight to harness the wind, *New York Times* 22 (Nov. 14, 2004).

8. Burkett, E., A mighty wind, *New York Times Mag.* 48–51 (June 15, 2003).

9. Kelley, T., On Jersey shore, residents debate whether windmill power plants will be on their horizon, *New York Times* (May 29, 2005).

10. Seelye, K. Q., Windmills sow dissent for environmentalists, *New York Times* A28 (June 5, 2003).

11. Simons, M., Where Nelson triumphed, a battle rages over windmills, *New York Times* A4 (Jan. 10, 2005).

12. Cowell, A., Menacing the land, but promising to rescue the earth, *New York Times* (July 4, 2005).

13. French, H. W., In search of new energy source, China rides the wind, *New York Times* (July 26, 2005).

14. Urbina, I., In search of new power source, city looks under water, *New York Times* (July 10, 2004); also personal communication with Trey Taylor, Nov. 8, 2005.

15. Wald, M., Wind power is becoming a better bargain, *New York Times* 27 (Feb. 13, 2003).

16. National Renewable Energy Laboratory, *Concentrating Solar Power. Learning about Renewable Energy and Energy Efficiency* (http://www.nreligor/learning/re_csp.html; accessed 1/12/05).

17. *An Atomic Description of Silicon, and, Built-in Electric Field*, US. Dept. Energy, Energy Efficiency and Renewable Energy (`http://www.eere.energy.gov/solar/photoelectriceffect.html?print`; accessed 11/03/05).

18. Musser, G., More power to solar. Photovoltaic advances make the ever lagging technology more of a competitor, *Sci. Am.* **293**(6):53–54 (2005).

19. Dixon, C., Shortages stifle a boom time for the solar industry, *New York Times* A10 (Aug. 5, 2005).

20. Three Gorges Dam, *Wikipedia*, The Free Encyclopedia (`http://en.wikipedia.org/wiki/_Three_Gorges_Dam`; accessed 4/18/06); also Itaipu Dam: Text. Joel Sampaio, J., and Silva, L. A. N. (`http://ce.eng.ust.edu/pharos/Wnders/modern/itaipu.html`).

21. *Hydropower: Direct Effects of Sheer Strain on Fish*, Idaho National Laboratory, Hydropower Program (`http://hydropower.inel.gov/turbines/direct_strain.shtml`).

22. *About Geothermal Electricity, and, Geothermal Electricity Production, Geothermal Technologies Program*, National Renewable Energy Laboratory. (`http://www.mrel.gov/learning/re_geo_elec_production.html?print`).

23. *Clean and Diversified Energy Initiative*, Geothermal Task Force Report, Western Governors Association, Jan. 2006 (`www.westgov.org/wga/initiatives/cdcac/Geothermal-f`).

24. *Understanding Biomass as a Source of Sugars and Energy*, Biomass Program, USDOE. (`http://www.eere.energy.gov/biomass/understanding_biomass.html?print`; accessed 11/3/05).

25. Ensminger, P. A., Photosynthesis, in *The Gale Encyclopedia of Science*, Thomson/Gale, Detroit, 2004, Vol. 5, p. 3065.

26. Feuer, A., On this freedom ride, fuel comes from the fryer, *New York Times* B1, B3 (Dec. 27, 2004).

27. Hakim, D., On the road again: His car smelling like french fries, Willie Nelson sells biodiesel, *New York Times* 1, 4 (Dec. 30, 2005).

28. *Energy Department Awards $92 million for Genomics Research*, U.S. DOE, Oct. 3, 2005 (`http://www.energy.gov/engine/content.do?`) (also PUBLIC_ID=18904&BT_CODE=PR_PRESSRELEASE&TT_CODE=PRESSRELEASE).

29. Goldemberg, J., Monaco, L. C., and Macedo, I. C., The Brazilian fuel-alcohol Program, in *Renewable Energy Sources for Fuels and Electricity*, Johansson, T. B., Kelly, H., Reddy, A. K. N., and Williams, R. H., eds., Island Press, Washington, DC, 1993, Chap. 20, pp. 844–863.

30. Hakim, D. The new prize; alternative fuels, *New York Times* 1, 7 (Aug. 10, 2005).

31. Deutsch, C. H. Saving the environment, one quarterly earnings report at a time, *New York Times*, C1, C3 (Nov. 22, 2005).

32. Deutsch, C., Dupont looking to displace fossil fuels as building blocks of chemicals, *New York Times* (Feb. 28, 2006).

33. Verne, J., *The Mysterious Island* (transl. J. Stump), The Modern Library, New York, 1959.

34. *Fuel Cell System. Energy Efficiency and Renewable Energy*, U.S. DOE (`http://www.eere.energy.gov/hydrogenandfuelcells/fuelcells/fc_systems.html?print`).

35. *Chemistry of a Fuel Cell. How Fuel Cells Work. How Stuff Works.* (http://science.howstuffworks.com/fuel-cell.htm/printable).

36. Service, R. F., New polymer may rev up the output of fuel cells used to power cars, *Science* **312**:35 (2006).

37. Kim, W. B., Voitl, T., Rodriguez-Rivera, G. J., and Dumesic, J. A., Powering fuel cells with CO via aqueous polyoxometabolites and gold catalysts, *Science* **305**:1280–1283 (2005).

38. Vogel, G., Will the future dawn in the north? *Science* **305**:966–967 (2004).

8

ENVIRONMENTAL ISSUES: PERCEPTIONS AND POLITICAL ASPECTS

> The problem is not just what we don't know, but
> what we don't know that ain't so.
> —*Mark Twain*

In the third edition of *Our Precarious Habitat*, almost two decades ago, I wrote that, "In his *Politics of Pollution*, J. Clarence Davies, III, raised the question: What is pollution? "The very definition of pollution," he said, "hinges on politics." That immediately subverts objectivity and places scientists on precarious ground. Davies went on to note that "pollution can not be defined with any scientific or mathematical finality as the definition hinges on the concept of human use . . . dependent on the publics decision as to what it wants to make of environment. If there was a homogenous public, the problem would take on manageable proportions." He remarks, too, that "it becomes a political decision, a voicing by the community of its concept of the public interest. And finally, underlying much popular discussion is the idea that pollution is the artificial befouling of the pure state of the environment. This is reflected in the many appeals to return our air, or water or soil—to its natural state."

Over the past two decades it has become increasingly evident that too many people either fail to understand that there are macro- and microenvironments, namely, the macro external physical world of air, water, light, soil, food, and animals, and the microenvironment of the self—ourselves—with our behaviors, our habits, our genomes, with its plethora of anomalies; which contribute singly or in combination to illness, injury, and death. For the most part, however, as shown in Chapter 1, it is the latter, the microenvironment, that contributes the vast majority of our leading causes of illness and death. It would be salutary and fruitful if this division were more widely recognized.

Evident also is the fact that environment, pollution, and politics are not only linked but also inseparable. And it is further evident that it isn't the community, as Davies believed, that voices its concept of the public interest, but actually vociferous pressure groups with their own agendas that politicians respond to, believing that these groups represent votes and voters, the lifeblood of politicians.

For 30 years, political leaders at the national level, Senators and Congressmen, grievously betrayed the country by failing to objectively consider the needs for electrical energy, dragging their feet on nuclear power plants, and placing us in the grip of the adverse effects of global warming, by following the bidding of environmentalist organizations dedicated to the rejection of nuclear-derived energy.

The uncompromising repudiation of culling deer populations by local and state officials fearing reprisals at the polls, has caused the loss of human life via motor vehicle accidents involving drivers and passengers crashing into deer, as well as the increase in Lyme disease given the spectacular increases in deer populations now that their natural predators have been effectively eliminated. Culling deer to protect both the human population, and many deer from starvation due to their density and lack of food, is not a crime against humanity. Here again community interest and intent are deflected by small, organized groups.

Although it is rare individuals who would not support healthcare benefits for veterans of our armed services exposed to harmful chemicals, dioxin, as in Agent Orange, has been a controversial issue since both clinical and epidemiologic evidence have indicated that Agent Orange is not the cause of Vietnam veterans' illnesses. Nevertheless, legislators have ignored the evidence and passed legislation compensating veterans and stating that Agent Orange was indeed the cause of their disabilities.

Women in Long Island, New York, have long believed that their breast cancer rates were inordinately high, and that some risk factor in the Long Island environment must be the cause. So certain were they, and so adamant, that they pressed their Congressman for an epidemiologic study with unrelenting pressure. The Congressman turned up the heat on the National Cancer Institute. After a 3-year investigation, and hundreds of thousands of tax dollars, no links or associations between their breast cancers and the physical (macro)environment were found. Needless to say, the women remain unconvinced, preferring to believe that external, rather than internal, factors—the microenvironment—are responsible.

Accordingly, this chapter considers a clutch of troubling issues that continue to confuse and bedevil the public. It will seek to pin down and render judgments on the basis of the best available current evidence dealing with air pollution—indoor and outdoor, high powerlines; Love Canal and hazardous wastes, mercury, lead, dioxin, and asbestos. We shall also delve again into the connection between environment and cancer.

INDOOR AIR

We begin with indoor air as we spend most of our time indoors. That alone confers importance to indoor air. And, again, it's the numbers. I suspect that 93, 5, and 2 will come as a surprise and shock. It may be difficult to believe, but most of us spend 93% of our day—a 24-hour day—indoors; 5% in transit (cars, buses, trains, planes), and, believe it or not, only 2% of our lives outdoors. You need only keep an accurate hourly log of your daily activities, from the time you arise in the morning to the time you call it a day, to confirm these numbers. Indoors includes home, school, work, entertainment, visits, the mall, and so on. I kept a log for 10 days, and I'm a believer.

Indoor environments, wherever they may be, have much to recommend them as potential risk factors. So, for example, cats, dogs, birds, rodents (a variety), cockroaches, cows, horses, and dust mites do live among us, as do such biological agents as fungi (molds), microbial toxins, houseplants, and pollen, along with infections agents such as rhinovirus, respiratory syncytial virus, *Chlamydia trachomatis*, and pneumonial and mycoplasmas. Nor can a host of chemicals be overlooked. Pesticides, volatile organics, fragrances, tobacco smoke, nitrogen and sulfur oxides, formaldehyde, and cleaning fluids of every conceivable type, are common to homes and offices. Not to be excluded is dust-carrying microbial toxins, and humidity, both high and low—indeed, more than enough potential risks to affect women, men and children.

A hacking cough, sneezing, wheezing, breathlessness, chest tightness, a choking feeling, struggling to breathe—life-threatening respiratory failure. Asthma! Asthma, which erupts at unpredictable intervals, is distinguished by overactive immune cells in the airways—trachea, bronchus, and bronchioles. When these cells are switched on by allergic triggers—cockroaches, dust mites, microbial endotoxins, animal dander, humidity, and many others—they can induce a sudden inflammation of the airways, which swell and narrow. Muscles around the bronchioles tighten, producing bronchial spasm, with the production of thick, sticky mucus; sufferers gasp for breath. These attacks recur regularly and frequently because the environmental triggers are commonly present.

Asthma is a serious and widespread condition, having attained epidemic proportions, and although much has been learned about its pathophysiological underpinnings, it has resisted therapeutic resolution. Obviously a good deal more of its immunologic basis remains to be discovered.

Why asthma? In 2002, an estimated 31 million people in the United States had been diagnosed as having had asthma in some point in their lives. That's over 10% of the total population. Of these 31 million, some 9 million are children. Worldwide, asthma is estimated to affect as many as 300 million. But that is only the tip of the iceberg; asthma is the sixth ranking chronic health condition in the United States, accounting for fully a quarter of all emergency room visits. It is the leading cause of school absences, and the fourth leading cause of work absences. Adults miss about 3 million workdays a year because of asthma, and about 500,000 are hospitalized annually, of which some 5000 die—deaths that should be preventable.

Women account for 65% of the 15–20 million yearly cases of asthma in the United States. From 1982 to 1996, asthma increased by 97% among women compared to 27% among men, and twice as many women died from their attacks as men. One of the mysteries of asthma is why it is more common and more severe in African Americans, as well as in people with low incomes living in urban areas. Yet another unanswered question is why asthma appears to be increasing, especially in young children, both in numbers affected and severity of episodes. As many as 7–10% of all children are believed to be asthmatic.

Asthma Triggers: Are They Causal?

Many studies have attempted to elucidate relationships between environmental exposures in early infancy and risk of allergic sensitization in later life. These studies are based on the theory that an individual's genetic predisposition to allergic disease is activated or enhanced by early allergen exposure.

An *allergic response* is one in which components of the immune system react strongly to a normally inoffensive foreign substance. Some of us inherently have a greater-than-normal tendency to mount an IgE response. Those who do are referred to as *atopic*. Atopy appears to be the strongest single risk factor identified so far, increasing the risk in those affected by 10–20 times. IgE antibodies call forth a type I hypersensitivity, which involves inflammatory reactions engendered by IgE antibodies bound to mast cells. This reaction results from the release of a cascade of cytokines, and is an immediate reaction, following exposure to an antigen trigger. This reaction is also known as an *allergy*, and antigens that stimulate allergies are termed *allergens*. Curiously enough, these hypersensitive allergic responses are far higher in the more affluent, Western, industrialized countries than in less developed countries, a seeming contradiction requiring explanation.

Sensitization to common indoor allergens is strongly associated with asthma, and one of the most common are pets, some 100 million of which are found in about 70% of US households, and include cats, dogs, hamsters, gerbils, birds, and a diversity of other companions such as snakes, turtles, and iguanas.

The National Institutes of Health estimates that 1 in 10 people (i.e., 10% of the population) in the United States is allergic to their pets, and those with asthma are especially vulnerable, with some 30% allergic to animal dander [the skin flakes on an animal's skin or fur (especially cats)]—actually dandruff.

Too many people assume that pet hair is responsible for their wheezing and sneezing. It isn't. Again, it's proteins! The proteins are found in the secretions of epidermal glands, which become airborne and are inhaled. Proteins in both saliva and dried urine that stick to pet hair after the animal grooms (licks) itself, can also induce allergic responses considering that of all our time indoors, 30% is spent in bedrooms, where pets often sleep; close human–pet contact is frequent and ongoing.

For years the steady mantra of physicians cautioning parents that lowering children's risk of developing allergies later in life, required "a pet-free home." Current research, however, impugns that advice. In fact, pets may be protective. Researchers from the Division of Allergy and Immunology, Department of Pediatrics, Medical College of Georgia (Augusta), conducted a prospective study (an epidemiologic study, discussed in Chapter 9) of healthy, full-term infants enrolled in a health maintenance organization (HMO) in suburban Detroit, and followed yearly to a mean age of 6.7 years. Of 835 children initially in the study at birth, 474 (57%) completed follow-up evaluations at age 6–7. The researchers found that children raised in a house with two or more dogs or cats in the first year of their lives had not only less allergic sensitization to dog and cat antigens as determined by skin-prick tests and allergen-specific IgE levels, but also less sensitization to allergens in general at ages 6–7. They also remark that children who lived in a house with two or more dogs and cats during their first year had a decreased prevalence of positive skin-prick test responses and IgE antibodies to pollens—ragweed, grasses, molds, and dust mites [1]. This study supports another, extensive study undertaken by physicians at the University of Basel, Switzerland in cooperation with researchers in Germany and Austria. They conducted their study in rural farm areas of those countries, where the environment presented extensive opportunity for exposure to microbial products. The underlying hypothesis maintains that changes in the type and degree of stimulation from the microbial environment, along with improvements in public health and hygiene, may increase the predisposition to chronic allergic conditions during childhood. Exposure to microbes can occur in the absence of infection. For example, viable and nonviable parts of microbes can be found in both indoor and outdoor environments. It is believed that these microbial substances, proteins, are recognized by the immune system and can induce a potent inflammatory response. Consequently, early exposure to microbial products may play a decisive role during the development of a child's immune response, providing a level of tolerance to animal dander and other potential allergens. Ergo, an environment rich in germs may reduce the risk of asthma. This European study collected dust from mattresses from both farming and nonfarming families and tested it for endotoxin and allergen (mite) content. Endotoxin is a lipopolysaccharide that forms the outer layer of cell membranes of all gram-negative bacteria. Recall that "gram-negative" refers to those bacteria that remain red during the gram-staining procedure. Current evidence indicates that endotoxins are also found in air, and can be inhaled and swallowed, acting as potent immunostimulatory molecules whose inhibitory effects stem from their induction of cytokines, especially interleukins 4, 5, 12, and 13, along with surfactant proteins in the lungs that enhance the developing immune response in infants, and that these endotoxin levels continue into childhood and preteen years. Endotoxin levels vary widely but tend to be highest in environments where there are farm animals such as cows, horses, and pigs, as the fecal flora of larger animals is a major source of endotoxin [2]. Complete data were available for 812 children—319 from farm families and 493 from nonfarm families. All had IgE serum measurements. Twice as much endotoxin was found in mattresses from farm families, and the investigators noted that endotoxin levels in dust samples from children's mattresses were inversely related to the occurrence of atopic asthma and hayfever (allergic rhinitis), and concluded that "a subject's environmental exposure to endotoxin may have a crucial role in the development of tolerance to ubiqutous allergens found in natural environments" [3]. These recent studies do suggest that growing up in a house with pets, or the nearby presence of large mammals, appears to substantially reduce the risk of developing positive skin test responses and IgE antibodies to allergens derived from animals. So, should we all move to farms, should the burbs become farmland again, or do all young families avail themselves of pet dogs and cats? Prudence dictates that further research in this highly active arena will shortly inform us as to the immune mechanisms involved and the extent of exposure to endotoxin necessary to ensure safety and protection against immune response resulting in a nasty asthmatic attack [3].

It is of additional interest that these researchers found, in an earlier study, a lower prevalence of asthma among inner-city Turkish children living in Germany, who lived in families less fastidious than native German families. On the other hand, researchers at the College of Public Health, University of Iowa, recently obtained and analyzed dust samples from 831 households across the country. They reported that although endotoxins were highest in kitchens, their health effects were most evident in bedrooms, where people spent more time in close contact with them. For adults, the higher concentration of endotoxins, the more likely the residents were to have asthma, and suffer from wheezing [4]. Unfortunately, this study did not ascertain the age at which asthma symptoms first appeared.

In another study, Dr. Erika Von Mutius of Munich, Germany, believed that children growing up in the poorer, dirtier, and generally less healthful cities of (Former) East Germany would suffer more from allergy and asthma than children in (Former) West Germany, with its cleaner, more modern environment. When the two Germanies were reunited in 1999, she compared their rates, and found exactly the opposite. Children in the polluted areas of East Germany had a lower allergic reactions and fewer cases of asthma than did children in the west. She found a number of lifestyle differences between the two Germanies, including family size and the more prevalent use of day care for young children. She now ascribes to the *hygiene hypothesis*, which maintains that children who are around numerous other children or animals early in life are exposed to more microbes and consequently their immune systems develop more tolerance for the triggers that cause asthma [5].

In its most basic form, the hygiene hypothesis suggests that Western, developed countries are simply too clean. This hypothesis, developed by David Strachen of the School of Hygiene and Tropical Medicine, University of London, says that as communities become progressively less rural, infants and children will no longer contact animals and their microbes that our immune systems need to develop protective antibodies to the many antigens in our environments—dirt, manure, and hair, which can elicit alarming respiratory events. The injudicious use of antibiotics and our much cleaner homes, along with obsessive avoidance of germs—bacteria, microbes generally—may have contributed to the increases in allergy, asthma, and eczema now being seen around the developed world. Have we become a nation of germophobes? It does look that way, which harkens back to Chapter 3, where it was noted that far too many people have little or no idea that germs—microbes—are both good and not so good, and that the not so good can be readily managed by our immune systems, if, as has been noted, our immune systems are sufficiently mature, having experienced microbes early in life and developed an array of protective antibodies. Trying to avoid microbes with a plethora of commercial germicides and antiseptics is an exercise in futility and self-sabotage. Germaphobia is self-deluding and nonprotective.

The importance of prior sensitization to house dust mites, cockroaches, and mold as risk factors for asthma is well established. The NIH informs us that some 4 million people are allergic to molds, but these are estimates based on self-reporting. With hard numbers unavailable, the problems actual dimensions remain questionable. What is certain is that moisture is essential for mold growth, and that these nonphotosynthetic plants will feed on—metabolize—almost anything containing cellulose, a polysacchbaride of glucose, the main component of all cell walls. Wood, cotton, paper, wallboard, all contain cellulose. In addition, our tighter, more energy-efficient homes trap moisture, if not properly ventilated. In its latest report on indoor mold, the Institute of Medicine (IOM) noted that the presence of mold in damp buildings appears to be associated (let's remember that word) with flare-ups of asthma symptoms along with coughing, wheezing, and upper respiratory tract congestion in otherwise healthy individuals. Although indoor dampness is considered inimical to health, it has been difficult to pin down. But the connection between moldy conditions and respiratory problems plus the prevalence of moisture in buildings, makes indoor dampness a public health concern. While mold spores do produce an array of chemicals, including antibiotics, they can also produce toxins: aflatoxin, ergotamine, and tricothecenes. However, the predominant indoor genera of fungi, *Alternaria*, *Fusarium*, *Cladosporium*, and *Penicillium*, are not mycotoxin producers, and whether their metabolic products can become airborne, inhaled, and swallowed is problematic. Unquestionably, mold spores do contain allergens that can provoke an allergic attack, triggering an asthmatic response. Yet here, too, the matter is arguable as much of the evidence indicates that to induce an allergic reaction the density of spores needed indoors would approach that of outdoor air in August— 10,000 spores/cm^3; wholly unrealistic indoors [6]. *Stachybotrys chartarum*, a black, slimy mold, is the current villain. It does produce chemicals toxic to lab animals, but its large and sticky spores do not hang in air. Currently, all molds, especially black mold, have caused a "mold rush," exciting contractors and lawyers as the new asbestos issue. But there is no arguing with the fact that asthma rates have soared in many US cities.

To determine whether cleaning up mold in homes can slow down or reverse the trend, federal housing officials are funding such studies. Nevertheless, mold in commercial buildings has captured the attention of the real estate establishment. The incident that focused their attention occurred at a Hilton hotel. A year after it opened in 2001, the 25-story 453-room Kalia Tower in the Hilton Hawaiian Village in Waikiki was shut down by a gray, not black, mold. The building cost $95 million to erect, with $55 million and 13 months to clean up the mold, which could readily return if the moisture problem were not appropriately addressed. Over the past several years, insured losses for mold damage has risen so sharply—from $700 million in 2000 to $3 billion in 2002, that insurance carriers no longer cover mold damage in conventional policies [7]. If there is doubt about mold and human health, there is little doubt about mold and litigation. Lawyers may not be acquainted with the fact that researchers recently modeled a maximum possible dose of mycotoxin that could be inhaled in a 24-hour period of continuous exposure to a high concentration of mold spores containing a mixture of fungal spores including *Stachybotrys chartarum*, and found that "None of the maximum doses modeled were sufficiently high to cause any adverse effect. The model also demonstrated the inefficiency of delivery of mycotoxin via inhalation of mold spores, and suggest that the lack of association between mold exposure and mycotoxicosis in indoor environments is due to a requirement for extremely high airborne spore levels and extended periods of exposure to elicit a response." They further state that "human mycotoxicoses are implausible following inhalation exposure to mycotoxins in mold-contaminated home, school or office environments" [6]. Even at 200,000 spores/m^3, more than most people could tolerate because of the odor and reduced visibility, a person couldn't inhale enough toxin in a 24-hour period to get sick. All this may not matter in a courtroom.

Indoor air is also home to dust mites and cockroach feces. Cockroaches and dust mites adore moisture and warmth. The dust mites *Dermatophagoides farinae* and *D. pteronyssinus* are almost invisible to the unaided eye, and are a well-known cause of chronic allergic rhinitis, hayfever. The allergic response is triggered by inhalation of the mites' feces. Mites, of course, are not insects, but Acarina, related to spiders, and can be found throughout homes, preferring carpets, bedding, mattresses, and blankets. Cockroaches, both the German (*Blatta germanica*) the croton bug, and *Blatta americana*, the Palmetto bug, are winged insects that prefer darkness, but are so large that they can be seen scurrying from place to place in search of food—any food.

A recent study conducted by members of 10 medical institutions across the country found that "more than 60% of children's bedrooms had cockroach allergen concentrations that exceeded 2 units per gram of dust. "Dust mite allergen, both *D. farinae* and *D. pteronyssinus*, were found in 84% of the bedrooms, and 28% of the bedrooms had a dust mite allergen concentration exceeding 2 micrograms per gram (μg/g) of dust. They also found that the majority of children who were sensitive to cockroach allergen (77%) and dust mite allergen (87%) had detectable levels of these allergens in their bedrooms [8]. Furthermore, when environmental intervention was accomplished and both dust mite and cockroach feces were removed, there was a clear reduction in asthmatic attacks and fewer days with symptoms during the 2 years of study. Interestingly enough, a recently published study on factors that affect asthma in inner-city children found that cockroach allergen worsened asthma symptoms more than did either dust mite or pet allergens. This study, mounted by the University of Texas Southwestern Medical Center, was the first large-scale study to show marked geographic differences in allergen exposure and sensitivity in inner-city children. Most homes in northeastern cities, such as Chicago, Manhattan, and Bronx, New York, had high levels of cockroach allergens, while those in the South and Northwest, such as Dallas and Seattle, had dust mite allergens at levels known to exacerbate asthma symptoms. Their data appear to confirm that cockroach allergen is the primary contribution to childhood asthma in inner-city indoor environments [9]. Where past studies focused on cockroach feces, this new study found that cockroach allergen derived from a range of sources—saliva, feces, cast skins, dead bodies, and secretions. The study also discovered that the type of housing influenced allergen levels. Cockroach levels were highest in high-rise apartment buildings, while dust mite concentrations were highest in detached homes. Curiously enough, in this study, pet dander and dust mite levels did not appear to increase asthma symptoms. The authors inform us that cockroaches are attracted by food debris and water sources, but that dust mites feed on flakes of human skin, and reside in bedding, carpets, upholstery, drapes, and other dust traps. Dust mite allergens are proteins from the mites' digestive tract and are found in mite feces. This study, funded by the National Institute of Environmental Health Sciences, to the tune of $7.5 million, had as its goal the development and implementation of a comprehensive, cost-effective intervention program aimed at reducing asthma incidence among children living in low-socioeconomic areas.

Before considering low socioeconomic status and asthma from another aspect, it may be helpful to summarize the underlyhing pathophysiological events common to all indoor allergic risk factors.

The most common allergic reaction is affected by the immunoglobulin IgE, a glycoprotein, and mast cells containing histamine; other chemicals also play a role. After an initial exposure to a specific antigen/allergen, a person who develops an allergy becomes "sensitized" to that allergen. Sensitization occurs when IgE antibody specific for that allergen attaches to the surface of the person's mast cells, making that person hypersensitive to additional exposures. This binding causes the mast cell to release histamine and a cascade of inflammatory chemicals such as cystenyl leukotrienes, prostaglandin D_2, and tryptase. Histamine stimulates mucus production, which leads to nasal and sinus congestion and causes smooth-muscle tissue to contract, which constricts the bronchioles, causing breathing difficulty. Histamine also dilates blood vessels, producing greater blood flow and resulting in redness and swelling. It also produces tearing—watery eyes and nose. To counter these symptoms, antihistamines are used. Should the hypersensitive response became excessive, as in the case of asthma, epinephrine (Adrenaline) may be necessary to prevent suffocation.

It can be thought of this way. Assume that there are three individuals, one nonsensitive, one somewhat sensitized, and a third hypersensitive via prior exposure. In the presence of an allergen the nonsensitive individual produces little to no IgE. The mildly sensitized person may produce IgE, but only enough to produce a mild or no allergic response. The third individual hypersensitized from repeated exposures will produce copious amounts of IgE on exposure, with a full-blown allergic response—which is an inappropriate physiological response in which the body mistakenly reacts to a range of allergens as though there were an invasion of microbes—ergo, an excessive immune response occurs.

So much concern and media publicity centers on childhood asthma that one could be forgiven for thinking that it was only a disease of childhood. Unfortunately, it is very much an adult disease, albeit often with different triggers, but fundamentally the same illness.

The internationally recognized rising trends in asthma and/or allergy has also documented a disproportionately adverse affect on minorities—blacks and Asians, as well as the urban poor, a phenomenon that thus far has eluded explanation. Given the rapid occurrence of these striking increases, genetic factors must be excluded. Consequently, other environmental risks must be occurring. Recently published studies suggest a direction. Researchers at Harvard Medical School and Boston Medical Center report on the effects of living in a violent environment "with a chronic pervasive atmosphere of fear and the perceived or real threat of violence on health"—especially asthma. They state that the "identification of exposure to violence as a trigger of asthma aggravation should alert health professionals caring for asthmatics in the inner city setting to inquire about a patients exposure among other known triggers" [10].

Another investigative group, from the Harvard School of Public Health and The Kaiser Foundation Research Institute, has found that socioeconomic and racial/ethnic disparities contribute disproportionately to asthma and allergy. From a population of 173,859 men and women in northern California, they found strong racial and socioeconomic differences in asthma/allergy events with African/American men and Asian men reporting far higher numbers of attacks than Caucasian men and women of comparable ages. Although they attempt to interpret these social inequalities, their findings remain unresolved [11]. However, a European 32-center, 15-nation study that enrolled 10,471 individuals aged 20–44, and assessed the association (that word again) between asthma prevalence and socioeconomic status found that community influences of living in a low-education area was directly related to asthma, independent of a person's own education level and social class. They also note the finding that increased asthma hospital admissions occur among those who are materially deprived [12]. A recent study done by the Department of Psychology, University of Washington (Seattle) may have hit upon a resolution to this conundrum. They proceeded from the knowledge that students with a history of asthma have flare-ups of airway inflammation during exam week, a period of significantly heightened stress. Exposing six volunteers to two different chemicals, methacholine (a bronchioconstrictor) and an antigen that produced an inflammatory response, each of the volunteers was given a magnetic resonance imagery scan 1 hour and 4 hours after exposure to each chemical. They found that brain activity in the early muscle contraction phase of the asthma attack differed from that of the inflammatory phase. They also learned that the presentation of words specifically associated with asthma during the inflammatory attacks also caused activity in that part of the brain dealing with emotion. Neutral or negative words did not elicit this activity. This study, although limited to six subjects, suggests a link between the brain and physiological responses in other parts of the body! That brain chemistry can have an effect on chemicals of the immune system does not appear to be so much of a stretch. But before this can be accepted and generalizable, additional and much larger studies will be needed. Nevertheless, it does open up new avenues of research, and does indicate that a range of stresses—emotions—may well be asthmatic triggers [13].

But that is not the end of it. Dr. Jonathan Spergel, a pediatrician/researcher at the University of Pennsylvania and Children's Hospital, has been studying wheezing mice. By smearing eggwhite protein, a common cause of allergies, on the skin of young mice, he induced a typical eczematous inflammation and rash. These sensitized mice were than made to sniff eggwhite protein. The control mice breathed normally, but the sensitized mice wheezed. This showed that "things went from the skin to the lungs. Through skin irritation we were inducing asthma." Other parts of the body could not induce asthma. Apparently there is something unique about skin. The theory posits that immune cells cluster around the skin, producing eczema in youngsters, and then migrate to the lymph modes, bloodstream, and lungs where they induce asthma when an allergen is inhaled. Enter Elidel (pimecrolimus). Elidel inhibits calcineurin, an early activator of the immune response, which signals IgE activity. Using an Elidel cream may short-circuit the immune system's overreaction at its origin. That's the idea behind a 3-year study currently in progress at the National Jewish Hospital in Denver. The goal of the study is to determine whether children using Elidel cream on their skin develop fewer cases of asthma [14].

Another curious development has unfolded at Yale University's School of Medicine. A surprising discovery in mice has linked a little-known class of chemicals to asthma, which appears to bolster the theory that asthma is a misplaced reaction to parasites. We humans do not produce chitin, so why the need for a chitinase enzyme? Yale's pulmonary researchers were astonished to find chitinase crystals in lung tissue of asthmatic mice. Healthy mice had none. When they gave the sick mice a chemical blocking the chitinase, the inflammation waned. Additionally, high levels of chitinase were also found in asthmatic human lung tissue, but not in healthy lung tissue. Question: Does the overproduction of chitinase bolster the theory that in asthmatics, the body wrongly senses parasites where there aren't any, and sends the immune system into overdrive [15]? If this proves out, a drug blocking chitinase formation ought to do the trick. But this is 3–4 years down the road.

In the real world mice are not allergic to cats. If that were the case, when a cat was lurking nearby, a mouse would begin to wheeze; an early warning mechanism that would give the mouse an early start for the nearest exit, which would be asking too much of Darwinian evolution. However, researchers at Virginia Commonwealth University, Richmond, bred mice to be allergic to cats and then "cured" them via a novel allergy treatment. They chemically bound a feline protein that causes cat allergies, to a human protein that blocks immune cells from releasing histamine. To test the therapy, they exposed the allergic mice to cat dander and cat salivary proteins, then injected them with the human–feline protein. A single injection was all that was needed to prevent an allergic response. If this new protein works on humans, the world will have a faster, safer, more effective treatment for human allergies. The Virginia researchers maintain that in theory this technique should work on any allergen. With that as their guide, peanut allergy is next on their list [16]. If that works, I'd expect the peanut industry folks will erect a monument celebrating these researchers—a monument made of peanuts, of course.

At the moment the asthma/allergy complex resembles nothing so much as a jigsaw puzzle. But which pieces fit together? One cannot help but wonder whether there are too many pieces—pieces that confuse, or is it confounding that lurks here.? Yet the pace is quickening, which my crystal ball views as a solution on the way; a completed puzzle, the actual picture coming into focus by 2012.

THE GREAT OUTDOORS

We now step outdoors for a portion of our 2 hours, to breathe deeply of the fresh air—fresher air. Since 1989, when last considered, the air above our cities has become remarkably cleaner. And vigilant scientists and investigators have zeroed in on the true villains corrupting our air. But let us understand that air pollution is not a twentieth–century invention:

> A young gentleman who had inked himself by accident, addressed me from the pavement and said, "I am from Kenge and Carboy's, miss, of Lincoln's Inn."
>
> "If you please, sir," said I.
>
> He was very obliging; and as he handed me into a fly, after superintending the removal of my boxes, I asked him whether there was a great fire anywhere? For the streets were so full of dense, brown smoke that scarcely anything was to be seen.
>
> "O dear no, miss," he said. "this is a London particular. A fog, miss," said the young gentleman.
>
> "O indeed," said I.

Although it is clear from this dialogue from Charles Dickens'*Bleak House* that mid-nineteenth-century London was no stranger to air pollution, it was as long ago as the year 1273 that a law was passed to control the burning of soft coal, in order to curtail local air pollution. Complaints against suffocating pollution were frequent in thirteenth–fourteenth-century England. In 1306 a proclamation was issued by Parliament, requiring the burning of wood rather than coal by the artisans and manufacturers of London during sessions of Parliament. Shortly thereafter, a man was executed for violating this early smog ordinance.

A landmark in the history of the subject was the pamphlet entitled *Fumifugium*, addressed by John Evelyn to Charles II in 1661. Evelyn described the "Evil" as "epidemically indeangering as well the Health of Your Subjects, as its sullies the Glory of this Your Imperial Seat." Evelyn suggested that factories using coal be moved farther down the Thames valley and that a green belt of trees and flowers be put around the heart of the city. "But I hear it now objected by some," Evelyn wrote, "that in publishing this Invective against the smoake of London, I hazard the engaging of a whole Faculty against me, and particularly, that the College of Physicians esteem it rather a Preservation against Infections, than otherwise any cause of the sad effects which I have enumerated." One is reminded of the French aphorism "the more things change, the more they remain the same."

The famous seventeenth-century physician Thomas Sydenham, however, had no doubt about the ill effects of London's air. "The fumes that arise," he wrote, "from the several trades managed here, but especially sulfur and fumes of sea coals with which the air is polluted, and these, being sucked into our lungs and insinuating into the blood itself, give occasion for a cough."

On December 17, 1963, some 300 years later, President Johnson signed into law the first Clean Air Act; Public Law 88-206 was an historic milestone in the control of community air pollution. It established a national program to meet the steadily growing demands for cleaner air.

Less than two years later President Johnson signed an amendment to the Act. In so doing he indicated that the federal government would assume an even greater role in guiding and planning for pollution prevention and control in the air over our cities. On signing, the President remarked: "We have now reached the point where our factories, our automobiles, our furnaces, and our municipal dumps are spewing out more than 150 million tons of pollutants annually into the air we breathe—almost one-half million tons a day."

Continuing to fire its broadsides, Congress enacted the Clean Air Act of 1970. This required all states to submit to the Environmental Protection Agency a plan to achieve the Act's ambient air quality standards. The primary intention of the act was to control air pollutants that adversely affected human health. It specified that not only should National Ambient Air Quality Standards (NAAQS) be set, but that they be set at levels sufficient to protect human health with an adequate margin of safely.

In 1989, I followed that, stating, "That was 25 years ago, and we still do not know what the dangers are; consequently we have little idea of what constitutes a safe level of any specific pollutant." It would take another 4 years for the culprits to finally be identified, then another 8 years for new standards to be promulgated, and as 2005 passed into history, yet a newer set of standards were being debated. But we are getting ahead of the story.

President Johnson informed us that in the 1960s our daily activities were responsible for spewing out over 150 million tons of pollutants annually. By the year 2000, human enterprise released 2.5 billion tons of particulate matter (aerosols and flyash), 180 million tons of sulfur dioxide, and 40 million tons of oxides of nitrogen. The operational term is *particulate matter*—soot. High levels of particulate air pollutants have been associated with excess morbidity and mortality for centuries. More recently three severe pollution events strengthened the connection between particulate matter and illness and death.

The Meuse Valley

The Meuse Valley of Belgium, scene of some of the bloodiest battles of World War I, is a heavily industrialized area. Blast furnaces, glass factories, lime furnaces, and sulfuric acid and artificial fertilizer plants spew a variety of contaminant chemicals into the atmosphere. During the first week of December 1930, a thick fog blanketed most of Belgium. The air was especially stagnant in the river valleys, particularly along a 15-mile stretch of the Meuse. Three days after this abnormal weather condition began, residents began to report shortness of breath, coughing, and nausea. Thousands became ill; the exact number was never ascertained. About 60 people died. Again, deaths were primarily among the elderly and those with chronic illnesses of the heart and lungs. Once the fog lifted, no new cases occurred.

Because this was the first notable acute air pollution episode of the twentieth century, public health scientists were unprepared. A study of the area after the incident suggested that the effects on health had been caused by a mixture of the sulfur oxides, sulfur dioxide gas, and an aerosol of sulfur trioxide. This has never been fully substantiated, and such an episode has not recurred in the Meuse Valley over the ensuing 75 years.

Donora, Pennsylvania

The episode in Donora occurred during the last week of October 1948. On the morning of October 27, the air over Donora, 30 miles south of Pittsburgh in a highly industrialized valley along the Monongahela River, became very still and fog enveloped the city.

The air was trapped, and it remained so for 4 days. In addition to sulfur dioxide, nitrogen dioxide, and hydrocarbons from the burning of coal for heating and electricity, the air contained the effluents from a large steel mill and a large zinc reduction plant, where ores of high sulfur content were roasted. During the period of the inversion these pollutants piled up. As they did, severe respiratory distress occurred in the older members of the population. Eye, nose, and throat irritations were common. Twenty people died in a period in which only two deaths were expected. Autopsies of many of those who died showed chronic cardiovascular disease. This confirmed the fact that preexisting heart disease increased the chances of serious illness during an acute air pollution episode. Before the weather changed and broke the inversion, 5910 of the 12,000 inhabitants of Donora were sickened.

London—Again

The highly inefficient burning of soft coal in open grates by the citizens of London was primarily responsible for the fog that blotted out their capital on December 5, 1952. The city, located on a gently sloping plain, is not hemmed in by hills, as are Donora and the Meuse Valley communities. The flow of air over London is not impeded by topographic barriers. But for 5 days a strong inversion and fog enveloped the city to such a degree that the "ceiling" was no more than 150 feet high. Within 12 hours after the fog had settled over London, residents began complaining of respiratory ailments. By the time the inversion lifted on December 9, 4000 deaths in excess of the normal rate for a 4-day period had been recorded in the greater London area. Some striking differences between this episode and earlier ones were noted. The increase in mortality was not confined to the very old. Although the highest increment was among those over 45, deaths occurred in all age groups. Another difference was in the rapid onset of illness: 12 hours as compared to 48 and

72 hours in the earlier episodes. A third difference was the increased death rate in London compared with Donora and Meuse Valley.

Yet another difference was the fact that London experienced an acute episode a second time and a third time. In 1956, 1000 people died, and in 1962, 700 deaths occurred. The 30% reduction is believed due to the strenuous efforts made to reduce sulfur oxides.

The Six Cities Study: The Culprit Identified

In 1974, researchers from Harvard University's Schools of Public Health and Medicine mounted a sixteen-year prospective epiemiologic study to try to resolve a range of insufficiencies and lack of comparability in existing studies of the effects of air pollution on human health. The Six Cities Study, as it came to be called, consisted of two major elements: the level of six primary pollutants—suspended particulate matter, SO_2, CO, NO_2, photochemical oxidants, and hydrocarbons, and, to prospectively determine effects of these pollutants on the health of adults and children [17].

Using several criteria, six cities were chosen and paired, "to obtain comparable data from a variety of communities with differing levels of pollution." Portage, Wisconsin and Topeka, Kansas were the "clean air" pair; Watertown, Massachusetts and Kingston-Harriman, Tennessee, the "slightly less clean air" pair; and Steubenville, Ohio and Carandolet, at the southern tip of St. Louis, the "dirty air" pair.

Once the cities were selected, a cohort of well-characterized adults age 25–74 was gathered. For this, a unique random procedure was used in each city to obtain a sample of the population. A total of 18,079 individuals were drawn from the six cities; 11,040 satisfied the eligibility criteria for participation, and 5480 were interviewed. The numbers from each city were fairly comparable; from 2174 in Watertown to 3583 in Kingston-Harriman. The final sample consisted of 8111 white adults 25–74 age. In addition to the adults, children were also involved. All first- and second-graders in each community, including those in parochial schools, were selected, and all were surveyed annually; pulmonary function tests were also done annually. Older children in fourth grade and higher were questioned about smoking habits.

Air sampling devices were set up in centrally located stations in each community. Size-selective aerosol samplers collected fine particles ($2.5\,\mu m$) and the larger ($15\,\mu m$) before 1984, and $10\,\mu m$ beginning in 1984. In addition to health-related questionnaires, interviews were conducted by local interviewers, and all adults received yearly pulmonary function tests, usually at their homes using portable meters run by trained nonprofessionals. The study went forward prospectively for 16 years.

What did the study find? First, mortality was "most strongly associated with cigarette smoking," and "air pollution was positively associated with death from both lung cancer and heart disease, but not with death from other causes considered together" [18]. These findings were wholly unexpected. Although lung function was diminishing, the mortality from cardiac events were dys-rhythmias and heart attacks. It was the fine particles from power plants and other sources of burning fossil fuels—automobile exhausts and home heating, which showed the strongest associations with these deaths. With the appear-ance of the published article in the *New England Journal of Medicine*, The American Petroleum Industry went ballistic, and battle lines were drawn. A second 7-year war was set in motion—which shall be described, but first bear in mind that we are not defenseless, passive recipients of all manner of parti-cles. We do have a number of defenses. The nose is both an air conditioner and a filtering agent. As inhaled air moves upward into the nostrils, the high-velocity airstream makes a sudden change of direction as it flows toward the throat. As it moves over and impacts on the moist, mucus-covered surfaces of the turbinates, it deposits large particles, which cleanse the airstream. Hair in the nostrils also filters out many of the larger particles. But the smaller aero-dynamic particles, 2.5–10 μm in size, can be carried along in the airstream to the pharynx. Here again, the airstream makes a sharp turn downward as it moves into the trachea, the windpipe. With this change of direction the air-stream continues to impact on the moist, mucus-covered surfaces, which trap additional particles, further cleansing the airstream. As the airstream flows down the trachea, it impinges on a carpet of cilia, short, hairlike fibers. Embed-ded within this carpet are mucus-secreting goblet cells, which keep the cilia moist. The cilia have a characteristic biophasic beat; a fast-forward flick, and a slow recovery phase. This beating, always forward, 1000 times a minute, about 16 flicks a second, produces a wavelike motion, referred to as the "mucociliary elevator," which carries mucus-entrained particles, those larger than 2.5 μm back up to the throat, where they can be either swallowed or spit out. This system works well in those who don't smoke. Cigarette smoke paralyzes the cilia and reduces the output of mucus. Loss of the mucociliary elevator permits particulates to move deeper into the respiratory tract.

The trachea splits into the left and right bronchi, leading to the left and right lungs. The bronchus now divides into finer and finer treelike branches, the bronchioles, ending at the alveoli, the airsacs where gas exchange occurs; oxygen from the red blood cells pass into the alveoli and carbon dioxide passes out and onto the redblood cells to be carried up and exhaled out. There are some 800 million alveoli in both lungs (~400 in each lung). Their combined surface area is about 84 square meters, equal in area to a tennis court. The alveoli themselves resemble nothing so much as clusters of grapes around the closed inner endings of the bronchioles. Each alveolus is about $\frac{1}{1000}$ inch in diameter (25 μm); placing 40 side-by-side makes a "bunch" $\frac{1}{25}$ th of an inch—1 μm across, about the size of an *E. coli* bacterium. Alveoli walls are thinner than the thinnest tissue paper, measuring about 4 μm—$\frac{1}{6000}$ th of an inch—half the diameter of a red blood cell. This extreme thinness permits rapid and effi-cient transpiration of gases.

As alveoli cannot have mucus around them to trap particles, as it would impede gas exchange, specialized white blood cells, macrophages, become available in response to chemical signals indicating the presence of foreign matter. They move rapidly toword the bronchioles and alveoli, seeking to engulf and remove particulates, should they be present. What are the chances that particulates are there? Well, how much air do we breathe in a day? Normal, healthy individuals breathe about 8–9 liters of air per minute. At 60 minutes per hour, that would be about 480 liters per hour, and in a 24-hour day, forgetting for the moment that we breathe less while sleeping, we would take in some 11,000—12,000 liters. Let's round this off to 10,000 even though all of us are not the same size. Of course, we can continue multiplying by week, month, and year. The salient question is, How many particles are in each liter, and how many finally arrive at an alveoli? Even with one particle per liter per day, slipping by our defenses we'd end up with 10,000 reaching the many alveoli. So, depending on where we live, a clean city or a dirty one, the fine particulate level could be substantial.

Just what are we breathing as particulate matter? The air we inhale can contain a conglomeration of combustion products from forest fires, wood-burning power plant emissions, motor vehicle emissions, home heating, wind-blown soil, seaspray, volcanic emissions, ground-up concrete, rubber tire particles, and metal dust. A cubic centimeter, the size of a sugar cube, can contain thousands or hundreds of thousands of invisible particles. Particulate matter (PM), includes dust, soot (carbon) smoke, and liquid droplets. Many of the particles are so small that light microscopes cannot resolve them; electron microscopes are needed.

Particles less than $10\,\mu m$ (0.1–$2.5\,\mu m$) in diameter ($PM_{0.1-2.5}$) pose a health concern because they can be inhaled and reach and accumulate in the alveoli. Particles less than $2.5\,\mu m$ are referred to as *fine particles* and pose the greatest danger, because they can slip past the mucociliary elevator, then pass along the bronchioles and into the alveoli, where they initiate problems. The coarser particles, 2.5–$10\,\mu m$, derive from crushing and grinding operations, and dust from paved and unpaved roads. Although referred to as "coarse," they are still small enough to wreak havoc in the lungs. At the opposite end of the size spectrum are the ultrafine particles (UFPs) measuring less than $0.1\,\mu m$ in diameter—smaller than bacteria, and about the size of an average virus parti-cle. These UFPs are a major source of particulates in the alveoli, and as they have a high surface area–mass ratio, which can lead to enhanced biological activity. Their small size may allow them to pass directly into the circulatory system, which could allow than to be distributed systemically.

Particulate matter, the key ingredient of polluted air, has been estimated to kill more than 500,000 annually around the world, and some 60,000 in the United States. To prevent this staggering loss of life, it is necessary to under-stand and characterize the hazardous particles and determine how they under-mine health. The particles derived from the burning of fossil fuels have a carbon core coated with a layer of hydrocarbons, metals, nitrates, and sulfates that may well play a role in toxicity. Although these UFPs are potentially the most dangerous, and are the largest source of air pollution in urban areas, current air quality monitoring agencies track and regulate only the 10- and 2.5-μm particulates [19].

Respiratory effects of PM include the triggering of inflammation in the smaller airways, which appears to lead to the aggravation of asthma, chronic bronchitis, and airway obstruction. PM currently stands accused as a cardio-vascular risk. For example, a research team at the Keck School of Medicine, University of Southern California (Los Angeles) found evidence of an associa-tion between atherosclerosis and ambient air pollution. For exposures to $PM_{2.5}$ of 5.2–26.9 μg/cm^3, carotid intimamedia thickness (CMIT), a measure of sub-clinical atherosclerosis, increased by 3.9–4.3%. However, among those 60 years and older, women, and nonsmokers, the CMITs were far larger; from 5.7 to 26.6, with an average of 15.7%. This published study is the first epidemiologic evidence of an association between atherosclerosis and air pollution. Given the large population exposed to fine and ultrafine PM, and given that cardio-vascular disease is the leading cause of death in the United States, it is essential that these results be confirmed by additional studies around the country [20].

Although a number of risk factors have been advanced to explain the adverse health effects of PM, airway inflammation has the greatest experimen-tal support. In both human and animal studies, inhalation of particles produces inflammatory effects and allergic responses, which may proceed via oxidative stress.

Inflammation and oxidative stress—the formation of reactive oxygen species—may explain a portion of cardiovascular disease, such as the growth of plaque formation, but much more needs to be learned before final regula-tory efforts at standard setting can be finalized.

Back to Six Cities

What did the Six Cities Study accomplish? It's easy to say, "a great deal". In fact, it was solely responsible for forcing major revisions of the EPA's Clean Air Act, and set a standard for evidence-based environmental decisionmaking. The study uncovered the fact that cardiopulmonary problems in children and adults were, and are, occurring at levels below existing standards; and that minute PM was the most dangerous of all air pollutants. But gratification that comes with publication, after 16 demanding years, was short-lived. Attack was swift. The American Petroleum Institute challenged the study as shaky science. According to the API, the link between the fine particulates and cardiovascu-lar deaths was not upheld by lab animal studies. Animals inhaling fine particles failed to show adverse effects. Animal models of heart disease were needed. Engineers and aerosol physicists at Harvard took up the challenge, creating an ambient particle concentrator—a contraption as large as a room that collects outdoor air and concentrates particles simulating moderate to heavy pollution levels. They then exposed susceptible animals—those with heart disease and asthma—to typical particulates. Responses were compelling: changes in electrocardiogram readings, blood pressure, and heart muscle tissue. With these results they conducted a study of hundreds of people in Boston who had implanted defibrillators to ascertain their responses to fine particulates. Defibrillators differ from pacemakers in that they deliver a telling shock in the event of an abnormal heart rhythm, and record the date and time. They found that even in Boston, a relatively particulate-clean city, the higher the particle level, the more often shocking occurred, even at levels below the EPA standard [21].

While these studies were in progress, the American Lung Association sued the U.S. EPA, claiming that the Clean Air Act required the EPA to set air quality standards every 5 years on the basis of the best available scientific data, to protect the public health. Particle standards hadn't been changed since 1970. A federal court ordered the EPA to review the new evidence on PM. While the EPA dragged its feet on this, the API, the Oil Industry Association, demanded that Harvard deliver all its collected raw data to them for review. As the Six Cities Study was financed by federal funds, the oil industry believed it was within its rights. Harvard refused, noting that their study predated the law, and that the confidentiality of the study's participants would be violated, but they offered to provide the data to an independent and impartial third party—the Health Effects Institute (HEI), in Cambridge, MA.

The HEI was set up and funded by the EPA and industry in 1980, to resolve just such disputes. The HEI obtained and reanalyzed the data and in 2000, presented its report. Prior to this report, the EPA "ignited a fire storm when it declared (in 1996) that tens of thousands of people were dying each year from breathing tiny particles of dust and soot-and issued tough new regulations to crack down on these pollutants" [22]. A furious oil industry went to court to prevent new standards from being promulgated. Now comes the HEI study—solidly behind the EPA and Harvard, and strongly implicates PM in excess deaths. The HEI study went well beyond the Harvard data and found that death rates in the 90 largest US cities rose on average 0.5% with each 10 $\mu g/m^3$ increase in particulates less than PM_{10}. This study further strengthened the case against particulates because the size and inclusiveness of the study dispelled any idea that the adverse health effects had been caused by pollutants other than PM_{10}. While the increase in deaths was small in each city, studies of the long-term effects of particles found that it aggregated to roughly 60,000 deaths a year, and found also that the death rates were highest in the Northeast, lowest in the Southwest, which makes good sense as the Northeast is the region with the greatest number of industrial plants, which moved the EPA to target the even finer (2.5-μm) particles. But, as noted, when the EPA moved to set a standard in 1996, industry groups objected fiercely. Nevertheless, EPA moved ahead with the standard, but sanctioned a 5-year delay. In 1999, a federal Court of Appeals supported the EPA's $PM_{2.5}$ standard, but industry lawyers took the case to the Supreme Court, where the standards were ultimately upheld. As 2005 passed into 2006, the new standards have yet to be enforced. In fact, President Bush recently proposed new air quality regulations for fine particles that pose especially serious risks to children with asthma, frail older individuals, and other high-risk groups. The newly proposed standards, one for annual exposure based on a daily average, and another for exposure over 24 hours, are currently in a 90-day public comment period, which should lead to a final rule by the EPA in September or October 2006. But I wouldn't hold my breath. After all, the Harvard Six City Study was published in 1993: 14 years ago. Nothing comes easily, least of all protecting the public's health. It appears that the proposed annual standard for the smallest particles may remain at the 1997 level ($15\,\mu g/m^3$), and the 24-hour standard could drop from 65 to 35. The EPA has also proposed eliminating the annual standard for PM_{10} and lowering the daily standard to 70 from $150\,\mu g/m^3$, which it believes represents the best available data. Several scientists believe that these are not the most protective standards, while industry spokespersons are firm in their belief that stronger standards would yield limited benefits. So, what else is new? In March 2006, a federal Appeals Court overturned a clean air regulation issued by the Bush Administration that would have allowed power plants, refineries, and factories to avoid installing new pollution controls to offset any increased emissions caused by repairs and replacement equipment. The court simply rejected any easing of rules on clean air.

Industry attitudes notwithstanding, a number of new studies support the particulate hypothesis. Among the most cogent is the Children's Health Study, a prospective monitoring of 1759 children (average age 10 years) from schools in 12 southern California communities that measured lung function annually for 8 years. The objective of the study was to ascertain whether exposure to air pollutants adversely effects the growth of lung function during the period of rapid lung development that occurs between the ages of 10 and 18. The 12 communities represented a wide range of exposures to particulates, ozone, nitrogen dioxide, and acid vapor. The researchers from both California and British Columbia found that air pollution was significantly associated with deficits in lung development. This study clearly implicates fine particulate matter and associated combustion-related pollutants as largely responsible for the adverse effects of air pollution [23].

A group of investigators from the United States and Canada determined that long-term exposure to fine particulates was strongly associated with mortality attributable to is chronic heart disease, cardiac arrest, heart failure, and dysrhythmias. For these causes of death, a $10\,\mu g/m^3$ elevation in fine particulates was associated with 8–18% increases in death rates, and with even larger risk for smokers compared to nonsmokers [24].

Furthermore, an expert panel convened by the American Heart Association concluded that "At the very least, short-term exposure to elevated PM significantly contributes to increased acute cardiovascular mortality, particularly in certain at-risk sub-sets of the population. Hospital admissions for several cardiovascular and pulmonary diseases acutely increase in response to higher ambient PM concentrations. The evidence further implicates prolonged exposure to elevated levels of PM in reducing overall life expectancy on the order of a few years" [25].

Six Cities Yet Again

A follow-up to the Six Cities Study was reported in March 2006. Dr. Francine Laden and colleagues of the Harvard School of Public Health followed 8096 residents of the six cities, aged 25–74, from 1974 to 1998, and found that for each decrease of 1 µg/m^3 of soot, less than PM$_{2.5}$, rates of cardiovascular disease, and respiratory illness decreased by 3%, which translates to extending the lives of 75,000 people a year. This clearly supports the call for reducing the clean air standards well below 15 µg/m^3 [26].

The recently published study of the Women's Health Initiative by researchers at the University of Washington (Seattle), found that long-term exposure to fine particles, less than PM$_{2.5}$, were closely linked to cardiovascular disease [27]. These particles induce inflammation that leads to stroke and heart disease. However, and unfortuately the EPA will not be reviewing its current standards until 2011; much too long to wait for those who will be inhaling these malevolent particles.

The World Trade Center

The dust and fumes from the catastrophic collapse of the World Trade Center released an estimated million tons of dust, smoke, fumes, and soot generated from pulverized concrete, metals, silica, and organic materials within the buildings, appeared to have wreaked havoc on the lungs of thousands of firefighters, police officers, and World Trade Center employees who were on site on 9/11, and days and weeks after. The unimaginable concentration of dust and particles were an ongoing concern for thousands of people, including residents living in apartment buildings nearby. In addition to dust and particulates, hazardous chemicals such as polycyclic aromatic hydrocarbons (PAHs), including benz(a)pyrene, crysene, and benz(b)fluoranthene, have produced carcinogenic, mutagenic, and genotoxic effects in lab animal studies.

Shortly after the collapse, the EPA collected air samples at or near ground zero to monitor fine PM. However, a team from the Department of Environmental Science and Engineering, School of Public Health, University of North Carolina (Chapel Hill) analyzed 243 of the collected samples. According to Dr. Stephen M. Rappaport, the team leader, the PAH levels declined rapidly, indicating the unlikelihood of long-term risks of cancer from PAH exposure. Also discovered was the fact that workers engaged in cleanup efforts could have been exposed to much higher levels of PAH than those in the collected air samples, and thus would have higher cancer risks. Also, because PAH levels were high for weeks after the collapse, the potential for adverse reproductive effects cannot be discounted among women who were pregnant during that period. The researchers caution that indoor air cannot be overlooked as a source of PAH as huge levels of dust seeped into both residential and commercial buildings, and could likely become a source of exposure [28]. A major factor limiting adverse health effects was the fact that 90% of the airborne particles were larger than 10 µm, a size that settles out of the air quickly. That is why Dr. Rapaport could estimate the risks of cancer as 0.157 cases per million people over 70 years near ground zero, and could increase to 0.167 cases per million people among cleanup workers—far less than one in a million. Will cancer clusters develop around ground zero? That's an appropriate question, as that study was done in 2001. By mid-2006, following wide concern that the dust at ground zero appeared to be responsible for firefighters (who worked without respirators) continuing to fall ill, there is still no convincing evidence that the dust is responsible for any long-term cancer threat. But there do appear to be indications that the heavy dust, smoke, and ash that the first responders inhaled is responsible for increased numbers of cases of sarcoidosis, which scars the lungs, markedly reducing lung capacity. Curiously enough, among New York City police who were also among the first responders at the World Trade Center, there does not appear to be an increased incidence of sarcoidosis, but this may be a matter of interpretatation. A more complete account of the ill effects of the WTC disaster may not be known until 2020 [29].

Holy Smoke

Then there is holy smoke, and its perils. According to researchers at the University of Maastrict, The Netherlands, public health can be adversely affected by burning candles and incense in churches, and to test their idea, they sampled the air in a Dutch church. Before the church service began, levels of PM_{10} were 3–5 times higher than normal roadside levels. After 9 hours of continuous candle-burning, and several puffs of incense, PM_{10} levels increased 13–20 times roadside levels. At its peak, the PM_{10} level in the church's chapel was over a mg/m^3 of air; more than 20 times the European Union's recommended 24-hour limit for outdoor air. In addition, they found that in both the basilica and the chapel, PAH levels were higher than outdoor levels, "and increased by a factor of 4 and 10 after burning incense and candles, respectively" [30]. Professor de Kok, the lead author of this study, informs us that "it cannot be excluded that regular exposure to candle or incense-derived particulate matter results in increased risk of lung cancer or other pulmonary diseases." However, thus far no studies have evaluated the actual risk to health of candle- and/or incense-generated particulates and PAH. Members of holy orders who spend the best part of their lives inside churches and monasteries may well be at increased risk of pulmonary disease.

China Wins

China holds the title of being the world's worst air polluter. Not exactly a much-sought-after prize. But China's headlong rush toward industrial domination has them building cheap and inefficient coal-fired power plants to obtain the electricity required to run their booming industries. The belching smokestacks are sending even more coal dust and sulfur oxides into the air—theirs and ours. Winds carry the pollutants across China, to Korea, Japan, and—yes—to California. It is feared that pollution levels in China could more than quadruple within 15 years, which translates into a greater density of particulates along the Pacific Rim, but especially in California, where, according to the U.S. EPA, almost 25% of the particulate matter permeating Los Angeles' air can be traced to China. As China continues to assault the air, as much as 33% of California's air pollution will have been made in China.

A recent study found that 400,000 people die prematurely in China every year from diseases linked to air pollution. This pollution problem could become a major embarrassment if China fails to meet its environmental targets for 2008, when the Olympic Games open there.

Industrial pollution can be difficult to control because local officials ignore emission standards to appease polluting factory owners who pay local taxes, and there is yet no political will to enforce standards as long as the economy is booming along [31].

A MÉLANGE OF ENVIRONMENTAL ISSUES

Extremely Low-Frequency (ELF) Magnetic Fields

Wherever electricity is generated, transmitted, and used, electric and magnetic fields are created. It is impossible to use electrical energy without creating these fields; they are an inevitable consequence of the world's reliance on electricity, and exist wherever any form of electricity is used. To say that electricity is an essential part of our lives cannot be overstated.

The earth itself is surrounded by a static magnetic field that varies between 25 and 65 microteslas (mT), and is hundreds of times greater than the alternating magnetic fields produced by the 110-volt (V) current in our homes, which produces about 0.01–0.05 mT.

As electromagnetic fields from high-voltage transmission lines, household appliances, and more recently mobile (cellular) phones, have received wide media attention, with concerns raised about possible adverse health effects of such exposure, we shall consider the underlying factors pertaining to electric and magnetic fields, and concern ourselves with the assessment and/or potential risks to health. After all, the human brain is an electromagnetic organ, and the most complex; shouldn't some effect be expected?

We begin by raising the questions, What is an electromagnetic field, what is the meaning of field, and how do they work? Additional questions will, of course, follow.

Electric fields exist around all copper wires and electrical appliances whenever they are plugged into a supply of electricity. On the other hand, magnetic fields are produced only when current flows and power is being used. Electric fields are produced by *voltages* and magnetic fields by *currents*. The higher the voltage—or greater the current—the stronger the field produced. These are, of course, invisible lines of force, but can become evident as static when driving an automobile near or under high-voltage transmission lines with the radio on. At much lower intensity, these fields also surround all electrical appliances. Twenty or thirty years ago it was not possible to operate an electric drill or clothes iron near a radio without static. Today that interference has disappeared because of the shielding provided.

As electric fields are produced by voltage, let us consider voltage—electric pressure, the ability to do work, measured in volts (V) or kilovolts (kV), which is equal to 1000 Volts. In the United States most domestic voltage is set at 110. Air conditioning systems pulling more energy often require 220 V. Home appliances are plugged into wall receptacles working on 110 V. When a lamp or toaster is switched on, an electric current flows that is measured in amperes (A). Think of voltage as water pressure in a hose with its nozzle in the OFF position. Turning the nozzle on allows the water to flow. In a copper wire, that's current.

With lamps, toasters, radios, TV's, coffeemakers, dishwashers plugged into wall receptacles, but switched off, an electric field exists around the cord or wire. But these fields are shielded by the covering around the wire, and field strength decreases rapidly with increasing distance from the wall receptacle. Turn these devices on, and current flows and magnetic fields develop around the cord, and exist along with the electric field. These magnetic fields are measured in gauss (G) (after the German mathematician Karl Frieidrich Gauss, 1777–1855) or teslas (T) (after Nicola Tesla, 1856–1943, a Croatian scientist/mathematician, and an American citizen), and its strength also decreases rapidly with increasing distance from the source. Magnetic field levels in either the domestic or occupational environments are less than 1 G; on the order of a milligauss, a thousandth of a Gauss. A person standing 4 feet from a copy machine may receive 1 mG of energy. At the copier, the magnetic field can be as high as 90 mG:

$$1 \text{ tesla (T)} = 10,000 \text{ G}$$

$$1 \text{ millitesla (mT)} = 10 \text{ G}$$

$$1 \text{ microtesla } (\mu\text{T}) = 10 \text{ mG}$$

To convert from microtesla to milligauss, multiply by 10. The gauss unit is most often used in the United States. The tesla is the internationally accepted unit.

Figure 8.1 depicts the magnetic field levels of common household appliances. Of particular importance is the fact that the electromagnetic spectrum spans a vast range of frequencies. Electric and magnetic fields can both be characterized by their wavelengths, frequencies, and amplitude or strength.

Electric Power and EMF

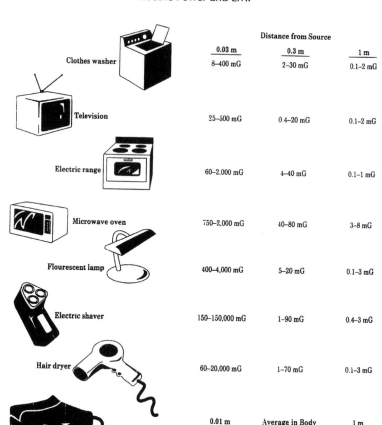

	Distance from Source		
	0.03 m	0.3 m	1 m
Clothes washer	8–400 mG	2–30 mG	0.1–2 mG
Television	25–500 mG	0.4–20 mG	0.1–2 mG
Electric range	60–2.000 mG	4–40 mG	0.1–1 mG
Microwave oven	750–2,000 mG	40–80 mG	3–8 mG
Flourescent lamp	400–4,000 mG	5–20 mG	0.1–3 mG
Electric shaver	150–150,000 mG	1–90 mG	0.4–3 mG
Hair dryer	60–20,000 mG	1–70 mG	0.1–3 mG

	0.01 m	Average in Body	1 m
Electric blanket	100 mG	15 mG	< 1 mG

Figure 8.1. Magnetic field levels near household appliances. [Figure courtesy of the Electric Power Research Institute (EPRI).]

Figure 8.2 shows the waveform of an alternating electric or magnetic field. The direction of the field alternates from one polarity to the opposite and back in a period of time of one cycle. Wavelength describes the distance between a peak on the wave, and the next peak of the same polarity. The frequency of the field measured in hertz (Hz) (after Gustave Ludwig Hertz, a German physicist, 1887–1875) describes the number of cycles occurring in one second. Electricity in the United States alternates through 60 cycles per second or 60 Hz. Recall that radiation includes X rays through ultraviolet light, visible light, infrared (heat), and microwaves, television to radiowaves along with electromagnetic waves—all forms of electromagnetic energy. A property

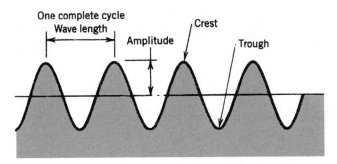

Figure 8.2. A waveform of an alternating electric or magnetic field in a period of one cycle.

distinguishing different forms of electromagnetic energy is the frequency expressed as hertz—60 Hz carries little energy, has no ionizing effects, and has little to no thermal effects. Figure 8.3 provides several representations of the electromagnetic spectrum. Note that 60 Hz is at the tail end, close to zero frequency. Figure 8.4 shows the location of cellular phones, close to that of radio and TV broadcasts. The radio and microwave fields are different from the extremely low-frequency (ELF) electromagentic fields produced by high powerlines and most domestic appliances. Furthermore, all frequencies below ultraviolet are *nonionizing*, which means that they have insufficient energy to break chemical bonds. Concerns about EMFs and health most often focus on two distinct nonionizing frequency bands: the 60-Hz band, the frequency of electric power systems, and the 1-GHz band, where mobile phones operate. Typically, mobile phone frequencies range between 900 MHz and 1.8 GHz. Their power is measured in watts per square meter (W/m^2) or more often million $\mu W/m^2$, as opposed to electric fields, which are measured in volts/meter (V/m). Again, background fields in most homes arise from low-voltage electric wiring, while higher-power fields are produced by overhead powerlines. As for mobile phones, the highest fields stem from the phones themselves.

The main effect of electric fields can cause small currents to flow in the body but are not large enough to interfere with the action of nerves in the brain or spinal cord. In fact, a recently published article in the *American Journal of Psychiatry* informs us that individuals with diagnosed bipolar disorder exposed to the oscillating magnetic fields of an MRI scanner (a device used to produce high-resolution images of internal organs and tissues) reported improved moods after the scanning [32].

As all of us have experienced, very high electric fields can cause "microshocks" when metal objects are touched, much the same as walking across a nylon carpet does. These can be annoying or playful, but are not dangerous. With mobile phones, frequencies, if high enough, can induce heating—but current phone designs limit this to body temperature: 98.6°F (37°C).

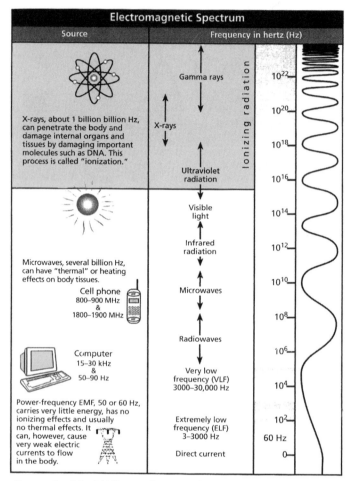

The wavy line at the right illustrates the concept that the higher the frequency, the more rapidly the field varies. The fields do not vary at 0 Hz (direct current) and vary trillions of times per second near the top of the spectrum. Note that 10^4 means 10 x 10 x 10 x 10 or 10,000 Hz. 1 kilohertz (kHz) = 1,000 Hz. 1 megahertz (MHz) = 1,000,000 Hz.

Figure 8.3. A representation of the electromagnetic spectrum. (Figure courtesy of the National Academy of Sciences, Washington, DC.)

An interesting and instructive experiment that can, and actually should, be done in everyone's home every few years is a check of both the magnetic fields around all appliances, and a check of voltages at each receptacle. Rather than a Gauss meter and voltammeter, avail yourself of a multimeter, one that can measure both. Hold the meter close to each of your appliances—the microwave oven, the toaster, dishwasher, radio, TV, electric iron—which, of course, must be switched on. Readings on the Gauss scale in milligauss should be near or at zero. Be sure to check the cords and take several readings, close to the

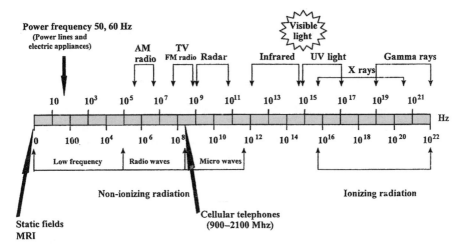

Figure 8.4. The electromagnetic spectrum showing the location of communication appliances at specific frequencies. (Figure copyright © 1996, Electric Power Research Institute, TR-106626, *Electric and Magnetic Fields: Invisible Risks?*)

appliance and 1, 2, and 3 feet away. The appliance that will surely get you a magnetic field reading will be your electric fan. Why? Because most electric fans are unshielded. Make sure that you place the meter at the rear of the fan. Then step back slowly and watch the readings drop precipitously. As for the voltages, don't be surprised to see them fluctuate between 100 and 110 V. That's normal. But higher and lower readings suggest trouble and call for professional involvement. A neighbor lost three TV sets because excessively high voltages were coming into her home and knocking out the sets.

With the above as background, our next set of questions concern the possible danger to our health from these fields or zones of energy. Do these fields around our hair dryers, electric blankets, electric heaters, power tools, and mobile phones—anything that requires the flow of electric energy— increase the risk of cancer, or neurodegenerative disease in adults and/or children?

Evaluating Potential Health Effects

To determine whether exposure to a physical or chemical agent can cause disease, scientists can obtain valuable information from animal experiments, studies of isolated cells in vitro, computer simulations, clinical studies, and human population (epidemiologic) studies. Between them, the necessary and sufficient information needed to render a judgment should be available. More often than not, a combination of studies is needed to obtain the necessary data, as no single type of study is definitive. Direct human studies carry greater weight than do animal studies, cell cultures, or computer simulations.

If some condition or illness can be produced in animals exposed to a specific agent, especially at levels closely approximating those experienced by humans, this strengthens the case against the agent. Cell cultures may provide information about a biochemical pathway that leads to an illness, but bear in mind that these are disembodied cells, far removed from a total, complex human being. Similarly with animal studies. Animals are not humans; they have different genomes, which means they and we have a plethora of different proteins and biochemical pathways. But sound data gathered from a variety of sources requires respect.

Clinical studies are the gold standard. They deal directly with people, and the best of them are controlled trials in which a test group, those exposed to an agent, are compared to an unexposed group of the same age, sex, and a host of other matched characteristics. In studies of electric and magnetic fields the test group, volunteers, are exposed to these fields at a number of levels including those normally encountered as well as levels higher than are regularly experienced. Measurements of heart rates, brain activity, hormonal levels, and nerve conduction are examples of the variables examined.

Epidemiologic studies deal with groups of people who have had an exposure and those who have not. These studies attempt to uncover risk factors that could have contributed to a specific condition. These studies seek to detect patterns and associations, and are considered valuable instruments for identifying risks to health.

Indirect measures, surrogates for human and animal studies, have also been used to determine whether EMFs are hazardous to health. Spot measurements, usually lasting only 30 seconds, obtain readings of EMF levels in rooms throughout a house. The values are averaged and used as the exposure level that a person could receive. The problem here is that spot measurements cover only a moment in time, then make a mental leap (of faith?) by assuming that this measurement is what a child or adult actually received, and are thus responsible for some untoward condition. It is reasonably certain that spot measurements underestimate actual exposures.

Yet another surrogate of residential exposure is distance to powerlines. This measurement assumes that the powerline is the major source of a person's exposure.

Wiring codes use five categories to evaluate magnetic field exposure in homes: very high, high, low, very low, and underground; each represents a decreased level of residential exposure. This scheme classifies home exposure according to the estimated current-carrying capacity of nearby powerlines, wire thickness, and the distance from the lines from a house. A newer version of the code scheme employs computer models to integrate distance from home to powerline, configuration of lines, and the power carried by the lines. All of these proxies fail to obtain actual exposure to residents, and offer nothing more than dubious estimates of human exposure.

Earlier I noted that the earth's magnetic field varies from 25 to 65 µT. It must be understood that the magnetic fields found near high powerliens and feedlines—those that go from the transmission line to homes—as well as those around household appliances, are 100 and more times less intense than the earth's magnetic fields. The current entering residential buildings is AC (alternating current), which reverses direction 60 times a second; that's the 60 Hz noted earlier that the earth's magnetic field does not do. The forces generated in the human body by 60 Hz AC when a person stands under high-tension powerlines are in fact thousands of times less intense than the forces generated by the variable electric fields of the heart, brain, and muscles. In addition, the forces generated in our bodies by alternating current are thousands of times too weak to move DNA or RNA molecules, and trillions of times to weak to sever chemical bonds or cause a mutation—the initial step in cancer development [33].

So, what have direct measurements of magnetic and electric fields revealed with respect to cancer and neurodegenerative disease from residential exposure? In 1993, Ahlbom and Feychting of Sweden's Karolinska Institute (Stockholm), who have been involved in EMF/ELF research for over two decades, reported on a study of every child under the age of 16 who had lived on a property located within 300 meters of the 220- and 400-kV powerlines in Sweden during the period 1960–1985. A total of 538 controls were also selected. Exposure assessment was done by spot measurements and calculation of the magnetic fields generated by the powerlines. They state that "For central nervous system tumor, lymphoma and all childhood cancers combined, there was no support for an association. However, based on historical calculations, that is, calculating the EMF's from historical records of power usage, and cases of childhood cancer that occurred years in the past, they report a positive association for leukemia, but when all childhood cancers are combined, the association vanishes" [34].

In 1997, the *New England Journal of Medicine* published a study done by the National Cancer Institute on EMFs and childhood leukemia. Twelve researchers from medical institutions around the country, led by Dr. Martha S. Linet of the NCI, enrolled 638 children with acute lymphoblastic leukemia (ALL), the most common childhood cancer, and a matched group of 620 healthy controls, all under age 15, in a study of residential exposure to magnetic fields generated by nearby powerlines. As the published study describes it, previous studies found associations between childhood leukemia and surrogate indicators of exposure to magnetic fields (wire coding) but not between childhood leukemia and measurements of 60-Hz residential magnetic fields. The NCI study sought to rectify that flaw. This blinded study determined that the risk of ALL "was not increased among children whose main residences were in the highest wire code category, compared with the lowest." In addition, "the risk was not significantly associated with either residential magnetic field levels or wire codes of the homes mothers resided in when pregnant with the subject" [35]. This study followed close on the heels of a review of literally thousands of studies by the National Research Council, which concluded that "the current body of evidence does not show that exposure to these fields presents a human health hazard." Accompanying the NCI study was an editorial giving the *NEJM*'s view of the EMF–health relationship. It stated that "The 18 years of research have produced considerable paranoia, but little insight and no prevention. It is time to stop wasting our research resources. We should re-direct them to research that will be able to discover the true biologic causes of leukemia that threaten our children" [36]. The deputy editor was appealing to researchers to end studies of EMFs and health as being nonproductive and of little public health value.

In fact, studies had been declining until Paul Brodeur's screeds, *Annals of Radiation*, appeared in the *New Yorker* magazine, in 1989, followed by his misbegotten books, *Currents of Death* and *The Great Power Line Cover-up; How Utilities and the Government are Trying to Hide the Cancer Hazard Posed by Electromagnetic Fields*, which did nothing but prey on public fears [37]. Unfortunately, Brodeur writes well, but has the remarkable talent of excluding all published studies with negative conclusions. Nowhere does he mention that the evidence against EMFs having a biological effect of any magnitude is strong. Most disturbing is the fact that the positive findings are marginal, and of overarching importance, and that few, if any, have been replicated. Nevertheless, Brodeur's work packs a wallop, and the news media have a field day disparaging local, state, and federal officials and agencies for failing to protect the public, while Brodeur laughs all the way to the bank. It is this type of controversy that keeps the studies flowing.

But let us continue to review additional published studies. A team of researchers from the Schools of Medicine, Stony Brook University, and University of North Carolina investigated the potential relationship between EMFs and breast cancer among women on Long Island, NY. Using a case–control study, they enrolled 1365 women (663 cases of breast cancer and 702 controls). Most of the women were in their 50s and 60s, and had lived in their homes for at least 15 years. Of particular concern was the fact that breast cancer rates have been high in northeastern states for over 40 years, and rates of breast cancer on Long Island are among the highest in New York State. As a result of efforts by local breast cancer coalitions, a congressional mandate resulted in a comprehensive study of possible environmental factors and breast cancer on Long Island. EMFs were among them. EMF exposure measurements were obtained by spot measurements in three locations: the front door, the bedroom, and the room where they spent most of their time other than in the bedroom and kitchen, as well as proximity to powerlines, and ground–current test load. The study's authors inform us that they could find "no association between EMF levels (wire coding or measurements) and breast cancer risk. This study thus provides no empirical evidence suggesting that residential EMF exposures contribute to the risk of breast cancer on Long Island" [38].

Incidentally, it is widely believed that Long Island, NY has an extraordinarily high breast cancer rate. Fortunately, that is not the case. According to the

New York State Department of Health, between 1994 and 1995, there were 117 cases of breast cancer per 100,000 women on Long Island. Several areas of the Northeast have higher rates.

In yet another published study, Ahlbom reported on a pooled analysis of nine studies of magnetic fields and childhood leukemia by an American–Canadian consortium. They found that 99.2% of children residing in homes with exposure levels less than $0.4\,\mu T$ had no increased risk, but that 0.8% of children exposed to levels above $0.4\,\mu T$ had an elevated risk. The explanation, they say, for the elevated risk is unknown, "but selection bias may have accounted for some of the increase" [39].

There is general agreement among researchers that there is no known mechanism by which EMFs might influence or affect cancer development. This conundrum has eluded all attempts at elucidation for over 25 years. With the recent study by Chinese researchers from the ZheJiang University School of Medicine and Electro Magnetic Laboratory, Hangzhou, China, new light may have been cast on the problem. They maintain that as it is the protein complement that ultimately determines cellular phenotype; they exposed a human breast cancer cell line to a 50-Hz, 0.4-mT ELF magnetic field for 24 hours to ascertain changes in the protein profile. They found that there was a change in protein expression, but that it appears to have been the first such result; many more such studies are needed to verify their findings [40].

In their recent study of cell phone use and risk of acoustic neuroma—a tumor of the eighth cranial nerve, the vestibulocochlear, Christensen and coworkers [39] note that electromagnetic radiation from cell phones can penetrate the skull, depositing energy 4–6 centimeters (cm) into the brain, which can potentially result in heating of the tissue up to 0.1°C. This has raised concern that these fields may initiate or promote cancer. This Danish study, part of the 14-country, interphone study, included 107 patients with this neuroma, comparing them to 212 matched controls. The results of this prospective, nationwide study "do not support an association between cell phone use and risk of acoustic neuroma" [41]. There is no evidence for it. Additionally they state that "There are no differences in socioeconomic characteristics between participants and non-participants among either patients or controls. This is reassuring because long term use of cell phones might have been related to higher income or higher education, thus introducing selections bias" [41].

The Stewart Report, produced by the Independent Expert Group on Mobile Phones, concluded that "There is one substantial established risk to health from mobile phone technology, namely through increased incidence of motor vehicle accidents when drivers use mobile phones. Since the chance of an accident appears to be equally elevated for hands-free and hands-held use, this effect is almost certainly due to the distracting effect of the conversation, rather than to interference with steering the vehicle or to a direct influence of RF radiation on the brain" [42]. They also maintain that "the epidemiologic evidence currently available does not suggest that RF (radio frequency) causes cancer. This conclusion is compatible with the balance of biological evidence which suggests that RF fields below guidelines do not cause mutation, or initiate or promote tumor formation."

With regard to biological evidence, researchers at the School of Medicine, Suleyman Demirel University, Isparta, Turkey, studied the effects of 900- and 1800-MHz EMFs on serum nocturnal melatonin levels in adult Sprague–Dawley rats. Of 30 rats, divided into three groups of 10 each, one group that received 900 MHz, another 1800 MHz, and a third, a control group were "sham-exposed." Exposures of 30 minutes per day continued for 5 days/week for 4 weeks. Rat serum was measured by radioimmunoassay. The investigators reported that their results "indicate that mobile phones emitting 900 and 1800 MHz EMF's have no effect on nocturnal serum melatonin level in rats" [43].

Rarely mentioned in public discussions of EMF is the tangible fact, noted in Chapter 1, that cancer incidence has been declining for at least two decades, while electrification of the country has been increasing. It has been estimated that during 1900–1992, the country experienced a 300-fold per capita increase in electrical usage. Thus, it is reasonable to infer, and suggest, that if the rapidly increasing electrification of the country, with high powerlines everywhere, were in fact related to childhood or adult cancers, the country should by now have witnessed a readily observable epidemic of leukemia, brain, and other cancers, which, of course, it has not. This widespread electrification has occurred in all western European countries as well. Why this piece of good news has been given a very silent treatment by the mass media is yet another conundrum.

Finally, one of the first principles of science maintains that it is simply impossible to prove the nonexistence of something. So, if EMFs have no detrimental effects, the absence of an effect will, of course, be impossible to prove. The best that can be said is that the existence of the phenomenon in question is highly unlikely and that the evidence in its favor is poor, which is the current condition of EMFs and their relationship to human health [33].

Hazardous Waste and the Love Canal Follow-up Study

In *Healthy People*, his report to President Carter in 1979, Surgeon General Julius Richmond stated unequivocally that "the Health of the American people had never been better." Shortly thereafter, *Time* magazine published the first of its ghoulish covers announcing and publicizing "The poisoning of America." Malevolent chemicals were everywhere, the accompanying featured piece claimed, and there was no safe harbor. The first cover, on September 22, 1980, simply wasn't a strong enough dose. On October 14, 1985, *Time* struck again with an even more ghoulish cover. These strikingly dramatic covers, using deadly expressive scare tactics, dramatized "Those toxic chemical wastes," with front-page covers flagrantly displaying a person immersed up to his nose in toxic waste. Below the surface, only a skeleton remained. Although the 1980 cover had birds wheeling in a blue sky and greenery on the far shore, by 1985, the skeleton was up to its ears in scouring waste, and what remained of its face and hair was now sullied. The trees had yellowed and a solitary bird flew overhead. Death and dying was the dreadful news. The takeaway message of those egregious covers was that we were drowning in toxic waste and reaping its foul effects. *Time*'s editors knew that between 1980 and 1985, the ill effects had increased outrageously. They knew it, and they wanted the country to know it. Here are the publisher's words: "If the cover of this weeks issue looks familiar it is not because your imagination is playing tricks on you. Five years ago, Time used an almost identical image to illustrate a cover story on the environmental dangers of toxic wastes. The decision to reprise the earlier cover, an unprecedented step for the magazine, was prompted by the heightened sense of urgency about the problem." The ghoulish scene depicted in 1980 "has been slightly altered by the artist to account for the passage of time. The tide of pollution has edged higher and now threatens the victim's eyes. Skyscrapers have risen on the background, two trees have yellowed, and cattails near the shore have died," and "we wanted to show," the artist explained, "that above ground things may look O.K, but beneath it's death."

Time's writers who visited waste sites were struck by the eerie nature of the newly created wastelands. Associate Editor Kurt Anderson was most impressed by Times Beach, Missouri, which was evacuated in 1982/83, after being contaminated by dioxin. "Times Beach," said Anderson, "looks as if a neutron bomb hit it. Houses are standing, windows are broken, some toys are still scattered around, but nobody walks the streets." Senior correspondent Peter Stoler and Chicago correspondent J. Madeline Nash experienced a strong sense of déjà vu as they updated their 1980 reporting. Stoler was reminded of John Brunner's 1972 science fiction novel, *The Sheep Look Up*, which described a world that was poisoning its air. "I thought it far-fetched," recalls Stoler. "Now I wonder if Brunner shouldn't have been more hysterical." Nash, who traveled from the Stringfellow Acid Pits in California to burned Fly Bog in New Jersey, was particularly disheartened by the Lone Pine landfill in New Jersey: "I had to wear a pair of protective boots," she says. "Around me were sticky rivulets of toxic ooze. There was an unmistakable scent of solvent in the air." Smelly, sticky ooze. It must be toxic. How could it not be? Oozy, sticky, and smelly; you can feel it right through your boots. Isn't that how Nash knew? Did the truth matter? *Time* had surely cast its message; the country was awash in toxic chemicals and they were exacting a prodigious toll of illness and death. *Time* knew this in the early 1980s. But did anyone else? The U.S. Surgeon General surely didn't. As he stated it, "The health of the American people has never been better."

But the year 1991 etches *Time*'s covers and its faux messages in stark relief. Six years after the second cover, the National Academy of Sciences (NAS) published volume I, in the series Environmental Epidemiology, which bears the title, *Public Health and Hazardous Waste*— "hazardous," not "toxic." This is not a subtle difference. These terms are neither synonymous nor interchangeable. "Hazardous" refers to the possibility of harm, risk; "toxic" means poisonous, producing a deadly or injurious effect: two very different concepts that are too often glossed over. Well, what did NAS tell us? They say [44]:

> Part of our modern heritage, "is the increasing volume of waste created by all industrial societies." Today, there is also unprecedented concern over the potential consequences for public health and the environment of the exposure to wastes that are deemed hazardous under a variety of regulatory regimes. According to recent opinion polls, the American public believes that hazardous wastes constitute a serious threat to public health. In contrast, many scientists and administrators in the field do not share this belief. On the basis of its best efforts to evalblic health concerns remain and critical information on the distribution of expouate the published literature relevant to this subject, the Committee cannot confirm or refute either view. A decade after implementation of Superfund (Comprehensive Environmental Response, Compensation and Liability Act, CERCLA, 1980), and despite Congressional efforts to redirect the program, substantial pusures and health effects associated with hazardous-waste sites is still lacking.

That last line bears repeating: "Substantial public health concerns remain and critical information on the distribution of exposures and health effects associated with hazardous-waste site is still lacking." Perhaps for the NAS, but certainly not the editors of *Time*, who were so certain that that they socked it to the public a second time. But let us not interrupt NAS' conclusion. Listen to them [44]:

> Whether Superfund and other hazardous-waste program actually protect human health is a critical question with respect to federal and state efforts to clean up hazardous wastes. To answer this question requires information on the scope of potential and actual human exposure to hazardous wastes and with the health effects that could be associated with these exposures. Based on its review of the published scientific/medical literature, the Committee finds that the question cannot be answered. Although billions of dollars have been spent during the past decade to study and manage hazardous waste sites in the U.S., an insignificant portion has been devoted to evaluate the attendant health risks. This has resulted in an inadequate amount of information about the connection between exposure and effect.

Can that be right? Ten years of searching and billions of dollars later and they still didn't know! How could that be? *Time* had the answer in 1980. The rivulets of ooze were toxic—a word never used by the NAS. The mother of all bitter pills is that *Time*'s deplorable message remains impressed in the public's consciousness, as does the legacy of Love Canal: both darlings of the mass media. Of course, *Time* has never concerned itself with NAS' findings, nor has it deemed necessary to bring its contents to public attention; and who reads NAS reports, anyway? Is it any wonder, then, that the public continues to hold tightly their erroneous beliefs about risks to health from chemical wastes? After all, *Time* has spoken. But, as we shall see, these erroneous beliefs have consequences; they have taken their toll, especially among residents living near waste sites. The Environmental Protection Agency (EPA) estimates that one in four Americans lives within 4 miles of a hazardous-waste site. To protect the public's health, Congress passed the Comprehensive Environmental Response, Compensation, and Liability Act (CERCLA) of 1980, which established the Superfund program to clean up the most seriously contaminated of these sites. In addition, in 1984, Congress amended the Resource Conservation and Recovery Act (RCRA) to add a corrective-action program to clean up contamination at facilities that treat, store, and dispose of hazardous waste. Since the inception of these two programs, EPA has overseen the cleanup of over 5000 hazardous-waste sites across the country. At many of these sites, however, EPA has selected cleanup remedies that leave at least some waste in place because the Agency believes that it is impossible, impractical, or too costly to clean up the contaminated property so that it can be used without restriction.

This cleanup warrants a nod of approval for its positive environmental effects. But it is incumbent on us to inquire into the human health effects of living hard—by waste sites as the disposal of increasing quantities of hazardous an/or toxic wastes has raised serious concerns, and fears, about possible ill effects for anyone in their vicinity.

Do dump sites constitute a threat to health? How would we know? If they do, what type of threat(s) do they pose?

Whereas the NAS overview had serious concerns about waste sites, there was little hard data to permit interpretation, let alone conclusions. Over the ensuing 16 years, new studies have appeared that allow discussion and critical appraisal. Accordingly, a number of studies will be presented, but we begin with an overview of all studies published since 1980. This key review is the work of Professor Martin Vrijheld of the Environmental Epidemiology Unit, Department of Public Health and Policy, London School of Hygiene and Tropical Medicine, and covers 50 studies done in North America and Europe [45].

A widespread problem in epidemiologic studies of waste sites is the lack of, or insufficient information about, individual exposures. Although there are many waste sites, few have been evaluated for their chemical content and the extent to which they may be, or are, releasing chemicals. Moreover, where it is known that chemicals have leaked from the site, there is little information available about the extent to which nearby residents have been exposed. The few studies that have attempted to measure specific chemicals in blood or urine samples of nearby residents have not found increased levels of volatile organic chemicals, mercury, or poly(chlorinated biphenyl)s (PCBs). This cannot be passed over lightly, because in too many instances surrogate measures such as proximity to site, or living in an area with a waste site, have been used as measures of exposure, and that make the mental leap to harm, which often leads to misclassification of exposures of diseased and nondiseased individuals.

Added complications occur because populations may be exposed to low doses of chemicals over long periods of time, making it exceedingly difficult to establish an association, let alone cause and effect. An additional problem is the difficulty establishing routes of exposure to site chemicals. As the drinking-water supply of residents near waste sites rarely originates in the local area, chemical migration into groundwater is not a concern, which requires other routes be considered. Few seem to recall, or are concerned about, the salient fact that the drinking water at Love Canal was not contaminated—not polluted. Failure to mention or consider this does affect decisionmaking.

Currently little is known about aerosol exposures of dust and gases from waste sites. Moreover, because of long latency periods as in the case of cancer, many people—families and individuals—may well have migrated in or out of the exposed areas between time of exposure and time of diagnosis, which will also lead to misclassification of exposures. Also, while it is easy to obtain birth weight information from birth certificates, many risk factors have been documented as associated with low birth weight—parental smoking, socioeconomic status, nutritional factors, parental height—which can readily serve as confounding factors (to be discussed in Chapter 9), giving biased estimates of association with residence close to a waste site.

Birth defect determination requires great caution as the defects are so varied in type, and are known to arise from many different risk factors; consequently, there is little basis for assigning a specific birth defect to a specific risk factor or exposure.

Professor Vrijheld comments on several studies conducted at Love Canal. One study reported a threefold increased risk of low birth weight for children from 1965 to 1978 who were exposed during gestational life to the Love Canal area, compared to children born elsewhere Data were analyzed separately for homeowners and for renters so that groups of socioeconomic status were compared. The risk of low birth weight was significantly increased for homeowners alone. This finding, he notes, is difficult to interpret as there are no strong reasons to believe that homeowners would be more susceptible than renters to the effects of toxic chemicals. Furthermore, both homeowners and renters reported increases in birth defects among their infants—and, as he points out, information on birth defects relied primarily on reports by parents.

In another study, increased prevalence of seizures, learning problems, hyperactivity, eye irritation, skin rashes, abdominal pain, and incontinence in children were higher among those living close to Love Canal compared to controls from other areas. Again, these conditions were reported by parents. This study was conducted 2 years after residents of Love Canal had become aware of the hazardous-waste problem, when media and public interest were high and people were being evacuated. "This," he writes, "makes it likely that the results were biased by differential reporting of health problems."

Yet another study compared cancer incidence (new cases) for the Love Canal area with data for the entire state from 1955 to 1977, and found no increase in cancer rates at Love Canal for any organ, including leukemia, lymphoma, and liver cancers, which were believed to be the cancers most likely to result from chemical exposures at the site. He maintains that the study is limited in that no information was available on the confounding factors such as smoking and socioeconomic status.

The complicated relation between worry, odor perception, and subsequent symptom reporting related to waste sites is a vexing problem requiring explanation given the far higher reported symptom rates among neighbors residing near waste sites, especially as those studies are so weakly positive.

Epidemiologists at the State of California Department of Health Services focused their concerns directly on this issue, by conducting a follow-up survey of 193 residents living near the McColl waste site, and comparing it to an area 5 miles away. Results from this survey were compared with results from a similar survey carried out 7 years earlier. They found that off-odors were detected less than half of the time in the 1988 survey than in the 1981 survey, but that reported symptoms of illness were greater in 1988 than 1981. They found that living near a waste site and being worried about the environment, rather than toxic effects of chemicals from the site, explained the excess symptoms reported [46]. In a second study, they collected data by door-to-door interview from over 2000 adult residents living near three different waste sites, analyzing them for self-reported "environmental worry" and frequency of perceiving petrochemical odors. Positive findings were observed for headache, nausea, eye and throat irritation, and degree of worry and odor perception. Odors appear to serve as a sensory cue, and the collected data clearly show a role for both environmental odors and environmental worry in the genesis of symptom complaints near hazardous-waste sites [47].

How, then, can the higher symptom rates observed among people residing near waste sites be explained, especially as these studies are so weakly positive? In their third report, we learn of the five studies of exposure-related illness conducted around hazardous-waste sites across the state. Their findings were unusually striking. They tell us that although there was no evidence of excess cancer or birth defects, there were far greater numbers of subjective symptoms among residents near waste sites compared to control neighborhoods. They concluded that stress alone from environmental anxiety produced these excess symptoms. Furthermore, they found that reported symptoms, including headache, toothache, and eye irritations, are odor-triggered, and can be reversed only by dealing with the sources of odor. In addition, lawyers, they say, pursuing tort litigation cases on the part of plaintiffs, have alluded to a "hazardous-waste syndrome," during court proceedings, suggesting that odors are responsible for their self-reported symptoms. The researchers maintain that such a hypothesis is biologically implausible [48].

These studies are a major contribution to knowledge of hazardous-waste sites and their purported adverse health effects, as they separate chemical effects from psychological effects, which is of primary importance in decision-making and policy setting.

Professor Vrijheld's conclusions bear attention. He reminds us that, "For several reasons evidence is limited for a causal role of landfill exposures in the health outcomes examined despite the large number of studies." Existing epidemiologic studies are affected by a range of methodologic problems, potential biases, and compounding factors, making the interpretation of both positive and negative findings difficult. Lack of direct exposure measurements and resulting misclassification of exposure affect most studies and can limit their powers to detect health risks. Moreover, evidence for a causal relationship between waste site exposure and cancers is weak, and most studies concern waste sites where no local drinking-water wells were present and potential exposure was either airborne or through other routes such as direct contact and consumption of home-grown vegetables.

The hazardous-waste site/health connection has suffered from anecdotal, and activist, promotion of what appears to be a nonexistent connection. Here we are in 2007, 16 years after the NAS exposition, and 25 years beyond *Time* magazine's initial cynical cover depicting the poisoning of America—a nonevent. What have we learned since then? The best available data indicate that if there are adverse effects to residents near waste sites, the effects are either nonexistent or too few to be detected by epidemiologic studies. We have also learned that much of the reported effects, *self-reported* at that, derives from worry and psychological stress, provoked more than likely by uninformed, anecdotal fanciful accounts in the mass media, made to appear substantive. The time is long overdue for *Time* magazine's publisher to present its readers an evidence-based, authentic accounting of the current state of the hazardous-waste–health relationship. Such a detailing would have a salutary effect on the mental health of the worried well.

Love Canal Revisited

Love Canal is the mother of all dump sites. Its reputation as a notorious cancer and birth-defects-producing area has circled the globe. Is its reputation deserved? Was it necessary for the media and environmentalists to create a cancer epidemic and public panic? Was it really necessary for then Congressman Al Gore to refer to Love Canal as a "very large cancer cesspool," or for Michael Brown, reporter for the *Niagara Gazette* and author of the infamous book *The Poisoning of America by Toxic Chemicals*, to be prosecutor, judge, and jury pronouncing Love Canal guilty as charged (as he charged); or for Lois Gibbs, the formidable community activist, who made a career of Love Canal, to declare that Love Canal was the immediate cause of irreparable harm to its residents? Both Brown and Gibbs accepted the anecdotal, self-reported accounts by residents as revealed truth and broadcast them to a stunned public. Were these people and their pronouncements more hazardous than the waste?

Enter the Love Canal follow-up health study. Yogi (Berra) had it right. "It ain't over till its over" —and as the other fellow said, "it ain't over till the fat lady sings." In this instance the fat lady has arrived in the guise of the Love Canal follow-up health study—an attempt to provide reliable answers to the many unsubstantiated claims surrounding Love Canal's supposedly adverse effects on its residents' health. It won't be over until the follow-up study makes its final report public in 2007. But what is the follow-up study, and what has it learned thus far?

In 1997, the Agency for Toxic Substances and the Disease Registry, a branch of the NCDC, both in Atlanta, awarded a grant to the New York State Department of Health Division of Environmental and Occupational Epidemiology, to gather data about exposure and routes of exposure to Canal residents and attempt to locate all former residents. That was the first part of the study. The second part was to determine the actual number and type of health outcomes of Canal residents and compare them to other New York residents. They would also measure, now that highly sophisticated analytic tools have become available, levels of chemicals in frozen blood samples, and determine whether the levels are related to the results of routine blood chemistry tests done in the late 1970s.

To their everlasting good fortune, they found a cache of 961 frozen blood sera samples taken in 1978 that have been linked to names and addresses of Love Canal residents of the Emergency Declaration area, and which have been tested for liver enzymes, chemicals from the Canal such as di-, tri-, and tetrachlorobenzenes, and $\alpha,\beta,\delta,\gamma$-hexachlorohexanes, and correlating them with 13,000 soil and sediment samples also obtained in 1978.

In addition, they have extensive files of personal interviews dealing with smoking habits, alcohol use, occupation, and other personal information. The Department is also checking the New York State Cancer Registry for unbiased determinations of cancer in those individuals still living in New York State. They have also sent names to cancer registries in other states to ascertain if cancer has occurred in those who left New York State. Adverse reproductive effects, birth defects in children and grandchildren, another potential response to polluting chemicals, are being determined.

The investigators have already gathered a cohort of 6000 residents, 96% of whom have been interviewed. Because schoolchildren were playing in soil at the 93rd Street school site, concern about soil as a source of chemical contamination was raised early on. Soil sampling conducted in 1978 by the State Department of Health showed low-level surface contamination. Soil sampling done years later indicated that soil contamination was primarily subsurface and that children had no contact with that, nor was the groundwater contaminated.

Thus far it has been determined that the overall cancer rate of former Love Canal residents was no greater than that of upstate New Yorkers, but lower than those of Niagara County residents. A total of 281 cancer cases were reported among Canal residents, where 313 would have been expected in upper New York State, and 321 in Niagara County residents. Furthermore, when splintering cancer rates by organ sites, some are higher among Love Canal residents, others lower. So, for example, lip, mouth, and throat constituted 1% of all cancers among Love Canal residents, but 3% for upstate New Yorkers. Similarly, bone, connective tissue, skin, and breast totaled 16% at Love Canal, and 19% upstate. Conversely, lung and respiratory organ cancers were responsible for 22% of all cancers at Love Canal and 18% among upstate New Yorkers. Stomach, intestine, and other digestive organs represented 22% of all cancers at Love Canal, and 21% among upstaters.

Among Niagara County residents, 27% had cervix, uterus, ovary, prostate, and bladder cancers, while upstaters had 25% and Love Canal, 29%. On the basis of as yet incomplete data, reproductive results indicate that the average birth weight of Love Canal babies was the same as the upstate New York and Niagara County averages, and the rate of prematurity for Love Canal women was the same as upstate New York and Niagara County. The rate of birth defects for Love Canal was 3%, and for upstate New York and Niagara County, 2%.

Data have now been obtained from out-of-state cancer registries. Texas, Florida, Arizona, Pennsylvania, Ohio, California, Virginia, and North Carolina show lower cancer incidence among residents who moved to those states, compared to those who remained. But it was also lower *before* they left New York.

Mortality data are also less than spectacular. By September 2000, they showed that the death rates (adjusted for age and sex) of Love Canal residents was the same as that for upstate New Yorkers, but slightly lower than Niagara County residents. When causes of death are grouped and compared to the County and upstate rates, some causes are higher and some lower, which by chance alone is to be expected. So, although the final tabulations remain to be broadcast, startling changes are not anticipated. While few are aware of this work in progress, its exceptional, but totally unexpected, revelations may be overwhelming for people who, over the past decades have been taught or made to believe that Love Canal was (is) synonymous with mischief and misery: human-made misery. These new numbers will not only be difficult to swallow, but will also be suspect, seen as a whitewash by a cabal of government officials and academic hacks. Three decades of fanning the flames of hazardous and toxic wastes as fated sources of illness and death may be too difficult to surmount. I'd like to be wrong on this one. It will also be illuminating to see how this follow-up study is treated by *Time* magazine and the media generally.

A further consideration is warranted. Love Canal was also the ultimate example of the fear of dioxin. Soil samples were taken from properties directly across from the Canal. In soil sampling and/or animal feeding studies, especially where low levels are involved, scientists use milligrams per kilogram (mg/kg) as the unit of quantity. In scientific notation, this ratio is equivalent to parts per million (ppm). Smaller still is the microgram (μg), a thousandth of a milligram. The ratio μg/kg is referred to as *parts per billion* (ppb), while *nanograms*, a thousandth of a microgram, are parts per trillion (ppt), and parts per quadrillion; a minuscule amount is a femtogram. Understanding these microscopic levels is essential if we are to make sense of any discussion of trace substances. It is necessary to understand that:

$$ppb = 1000 \times less\ than\ ppm$$

$$ppt = 1000 \times less\ than\ ppb$$

$$ppt = 1,000,000 \times less\ than\ ppm$$

$$ppq = 1,000,000,000 \times less\ than\ a\ ppm$$

These are not childish gibberish, but actual measurable amounts. Of course, it is confusing. It's just the reverse of dollar amounts, where trillions are more than billions and billions more than millions; and parts per billion, trillion, and quadrillion are staggeringly microscopic amounts of anything. Zero may no longer exist given the current ability to detect a half-dozen or fewer molecules.

Indeed, it is understandable that residents of Love Canal were mortified to learn that they had dioxin on their properties at levels of parts per billion. Billions! It sounds horrendous. Did anyone even attempt to explain the evident confusion? Certainly not the media.

Soil samples from homes across from Love Canal contained levels of dioxin from 1 to 20 ppb. Recall, too, that it was the levels of dioxin obtained from feeding studies of guinea pigs that tagged dioxin as the most poisonous of chemicals. But the amount that felled guinea pigs was a thousand times greater than that found at Love Canal.

Dioxin, 2,3,7,8-tetrachlorodibenzo-*p*-dioxin (TCDD) is a specific member of a family of structurally and chemically related poly(chlorinated dibenzo-*p*-dioxin)s (PCDDs), poly(chlorinated dibenzofuran)s (PCDFs), and certain poly(chlorinated biphenyl)s (PCBs). Some 419 types of dioxin-related compounds have been identified, but only 30 or so are considered to have significant toxicity, with TCDD the most toxic. At room temperature TCDD is a colorless, crystalline solid.

Early on it was determined that the lethal dose required to kill 50% (LD_{50}) of guinea pigs was 1 ppm: 22 ppm for male rats, 45 ppm for female rats, 70 ppm for monkeys, 114 ppm for mice, and 5000 ppm for hamsters. The monkey, a primate, is less sensitive than the guinea pig, a rodent, and the hamster, another rodent, is remarkably more resistant than the guinea pig. But mice and rats are also rodents, which shows great diversity of response to dioxin just within the order Rodentia. Obviously there are substantial differences in biochemical systems and metabolic pathways between species. The tantalizing question is Which species of lab animal is predictive of human effects? Or, are human effects unpredictable from animal responses?

The Incident at Meda

Saturday morning, July 10, 1976, was not unlike other Saturday mornings in Meda, Italy. Meda, 15 miles north of Milan, with a population of less than 20,000, is a quiet, rural community. In 1976, 160 of its citizens worked for Industrie Chemiche Meda Societa Anonima (ICMESA), a subsidiary of Givaudin, whose parent company, Hoffmann-LaRoche, is a Swiss pharmaceutical firm headquartered in Basel.

The plant at Meda produced trichlorophenol (TCP) for use in producing hexachlorophene. Plant workers knew that the safety valve of the 10,000-liter (L) reactor vessel had not been provided with protective devices, but few gave it a second thought; production had proceeded smoothly for years. Sometime between noon and 2 P.M. a runaway reaction began in the reactor; although the plant was shut down for the weekend, the reactor had been charged with ingredients necessary for production of a new batch of TCP. An unusual amount of pressure developed within the reactor. Without the protective units, the rupture disk burst and hot gases, vapor, and particulates exploded into the air above the plant. Dense white clouds formed, and the prevailing winds carried the chemical mist to the southwest. Although the incident occurred in Meda, Seveso became the "epicenter" of the toxic rain. By Monday, children began to complain of red, burning skin lesions. Rabbits, cats, birds, poultry, and frogs were dying. But it was not until the following Monday, July 19, that ICMESA officials acknowledged that 2,3,7,8-tetrachlorodibenzo-p-dioxin (TCDD) had been expelled along with TCP and sodium hydroxide.

Between July 10 and 20, no one was alerted, nor was medical attention provided anyone to control for possible TCDD toxicity. Given an almost complete lack of public health facilities and organization, the inhabitants were exposed for approximately 3 weeks before the first group of children and pregnant women were evacuated [49].

For at least 30 days after the release, the only clinically observable signs were the florid skin lesions, showing second- and third-degree burns. Chloracne began to appear 2 months later. New cases occurred well into spring 1977. For some children chloracne persisted into 1979. One of the most pronounced effects of several of the chlorinated dioxins, first described in 1899, is its induction of blackheads (comedones), characteristically arrayed on the cheeks—particularly on the prominences beneath the eyes, spreading to the chin, behind the ears and back of the neck.

Because data from laboratory animals indicated that birth defects were a prominent sequelae of TCDD, obstetricians and pediatricians in Lombardy were alerted. Under Italian law, reporting of birth defects is mandatory. Nevertheless, between 1972 and 1978, one to two defects were reported in a population of approximately 1800 births per year within the four most exposed towns; this was an unusually low rate. However, Italy is well known for under-reporting birth defects. During 1977, a total of 38 birth defects were reported in 11 cities: 7 in Seveso and 16 in Meda. Of the three zones of contamination established (high, medium, and low), Seveso was in the high zone and Meda in the medium. Obviously, with so large a surge in numbers of defects, some-thing untoward appeared to have occurred. But the data are curious. Meda, with less than half the exposure, had more than twice the level of defects. In addition, doublings of defects were reported from Bergamo, Como, Brescia, and Cremona in the northeastern reaches of Lombardy, in the opposite direc-tion of the path of the drifting chemical cloud.

Unfortunately, although of inestimable value, because of the fear of birth defects that the publicity generated, 30 medically induced abortions were performed, 4 among women of Seveso. None of the embryos manifested any type of malformation. Is TCDD fetotoxic? Animal data say that it is, and that tumors (cancers) were to be expected. By 1984, 8 years after the accident, examinations showed that birth defects, spontaneous abortions, deaths, and tumors were not induced by the heavy "rain" of TCDD and TCP. It is especially pertinent to consider that levels of TCDD averaged 20 ppm in Seveso. That is equivalent to 20,000 ppb, an astronomical concentration. As 8 years may be too brief a period for effects to express themselves, the accident became an ongoing study for Italian scientists. In 1998, 22 years later, researchers at the University of Milan reviewed the long-term effects of heavy dioxin exposure. They noted that "the amount of TCDD released has been the subject of con-flicting estimates," and "the results reported . . . do not provide conclusive evi-dence of long-term effects on the exposed subjects of the Seveso accident. For this reason, the follow-up period for the mortality and cancer incidence studies has been extended . . . and molecular epidemiolgy studies have been initiated." However, they did find that TCDD blood plasma levels obtained from both men and women in the most contaminated zone were at the ppt and ppq (femtogram/kilogram) levels [50].

But the study continued. In 2003, the researchers reported on maternal serum dioxin levels and birth outcomes in the cohort of women exposed to dioxin in Seveso. However, associations of TCDD and lowered birth weight are somewhat stronger, albeit nonsignificant. It remains possible that the effects of TCDD on birth outcomes are yet to be observed, because the most heavily exposed women were the youngest at follow-up and therefore less likely to have yet had a post explosion pregnancy" [51]. Bear in mind that young teenage girls, the 13–18-year-olds in 1976, are now 43–48 years old, well past their childbearing years. Consequently birth outcomes—defects, low birth weight, small stature—are unlikely to appear. Moreover, whether animal data are translatable to women appears to be decidedly shaky. But what about men? Does paternal exposure to TCDD raise the risk of adverse birth out-comes? A group of researchers at the National Institute for Occupational Safety and Health (NIOSH), together with the Department of Health and

Human Services, Hanoi, Vietnam, delved into this concern recently because there is evidence that the most profound effects on the fetus may be paternally rather than maternally influenced. The men participating in the study had worked with Agent Orange, a phenoxyherbicide widely used as a defoliant in Vietnam, and which was contaminated with dioxin as a consequence of the herbicide manufacturing process. This study looked at pregnancy outcomes among wives of male chemical workers who were highly exposed to chemicals contaminated with TCDD, comparing them to results from nonexposed men living in the same neighborhood. Data on lab animals suggest that developing tissues are highly sensitive to TCDD, so association between TCDD exposure and adverse reproductive outcomes are biologically plausible. Again, however, the numbers are meaningful. So, for example, when the Air Force began using 2,4,5-TCP in Vietnam in 1962, the defoliant was "coded" as Agent Green. The drums containing the herbicide had green bands painted around them. Three years later, "Orange" took its place, following agents Pink, Purple, Blue, and White. Estimates indicated that Green contained dioxin at a level of 66 ppm. Agent Purple contained 33 ppm, and from 1965 to 1970, Agent Orange, with 2 ppm, was used. At the time airforce crews and nonflying civilian personnel were employed to prepare and broadcast the defoliant. Some of the groups received daily exposures of 2 ppm. Because both airforce and civilian workers were mostly men, there has been increased concern about male reproductive outcomes associated with exposure to TCDD.

In the fourth biennial update of exposure to herbicides, the NAS concluded that there was insufficient or inadequate evidence to determine whether there is an association between paternal herbicide exposure and low birth weight or preterm delivery [52]. They also concluded that there was limited evidence of an association between paternal exposure to herbicides and spina bifida, and evidence was inadequate for other birth defects [50]. The results of the NIOSH study "does [do] not support a causal relationship between low birth weight and high paternal TCDD exposure. Analysis showed a somewhat protective effect of paternal TCDD concentration with respect to pre-term birth." They also inform us that the men in their occupational study group were exposed to TCDD at substantially higher concentrations at the time of conception, ranging up to 16,340 picograms per gram (pg/g). The results indicate that TCDD is unlikely to increase the risk of low birth weight or preterm delivery through a paternal mechanism [53]. Next, there's a further consideration.

Viktor A. Yuschenko, now President of the Ukraine, was poisoned by political opponents. Tests conducted at the Rudolfinerhouse Hospital in Vienna, Austria, confirmed that he had been poisoned by dioxin, most likely laced in his food, which provided a reasonable explanation for the array of pain and disturbing conditions that plagued him during the last 3 months of his presidential campaign.

The hospital chief, Dr. Michael Zimpfer, said that Mr. Yuschenko's blood dioxin level was more than 1000 times the upper limits of normal, and that his severe abdominal pain suggested that he had ingested the chemical. Clearly this had been an attempt on his life. Trace amounts of dioxin are common contaminants in soil and food and can remain in the body for years sequestered away in fatty tissue. When a sample of Yuschenko's blood was sent to a lab at the Free University in Amsterdam for further analysis, it was discovered that the dioxin was a single type, not a mixture of dioxins, which immediately informs us that his condition was in no way accidental, as the dioxin was so pure. Further, the analysis at the Free University revealed that the blood level of the sample sent from Vienna was 6000 times higher than the highest normal levels. In fact, this was the second highest level ever recorded in a human being. Interestingly enough, physicians in Amsterdam also believed that there would be no lasting change to Yuschenko's internal organs, but his skin condition could take years before returning to normal. Recent photographs of President Yuschenko are proof-positive that he has indeed returned to his previous handsome self. However, with such astronomic levels, how is it possible that he is he still with us [54]?

The World Health Organization's published statement on the health effects of dioxin warrents repeating. They say that on the basis of epidemiologic data, dioxin is categorized as a known human carcinogen, but that TCDD does not effect genetic material and there is a level of exposure below which cancer risk would be negligible [55]. Nevertheless, the U.S. EPA maintains that there is no safe level of dioxin. This discrepancy, along with the many published studies, has been the subject of ongoing debate for the past 20 years. Clearly there is a dilemma in the numbers—and numbers do matter. Chemists can devise analytic methods sensitive enough to measure parts per quadrillion and below, which raises the question, When is nothing present? If ill effects are not found at parts per million, than parts per billion and trillion must be meaningless. That should have been the message at Love Canal, and the message of the numbers should also go out to those concerned about dioxin in incinerator stack gas effluents, where levels of dioxin have been measured in parts per quadrillion.

As the twentieth century faded into the twenty-first, new occupants moved into refurbished homes around Love Canal. With the publication of the follow-up health study's reassuring new data, more will follow, and the fat lady's pear-shaped tones will cease; silence will pervade a stunned public. We shall also keep an eye on Mr. Yuschenko.

Chemical Agents

Hydragyrum

Mercury has a bad reputation. Is it deserved? An answer lies ahead. Mercury has a venerable paternity, as it is one of the eight ancient metals along with gold, silver, iron, tin, copper, lead, and zinc. The Romans called it *hydragyrum*, liquid silver, from which its chemical symbol, Hg, derives. Mercury is found naturally in the earth's crust as mercuric sulfide, in the mineral cinnabar; it has been used for over 3000 years as the red-brown pigment vermilion (also called Chinese Red), which was used to make the red-brown ink of the Dead Sea scrolls. The Romans called the pigment *minium,* and because this red-brown color (depending on how it was ground) was the dominant color in small paintings, they were referred to as *miniatures*, and manuscript titles in red were called *rubrics*, from the Latin, *ruber*, for red.

The early use of mercury was likely as an alloy metal with gold and silver to form amalgams. The Romans burned their used garments that contained gold thread and extracted the gold from the ash with mercury.

Mercury exists in three forms: elemental mercury, inorganic mercury salts, and organic mercury. Elemental mercury is a silver-gray liquid, the only metal liquid at room temperature, and that vaporizes readily when heated. Commonly referred to as "quicksilver," it is used in thermometers, thermostats, switches, barometers, and batteries. Elemental mercury accounts for most occupational exposures. It forms a number of mercuric and mercurous salts, most of which readily dissociate into ions in the body. Mercury can also bind to carbon to form organomercury compounds; methyl mercury is the most important in human exposure.

The major sources of atmospheric mercury are the burning of coal, smelting, and mining, but it is dispersed naturally through weathering of rocks, erosion, volcanic emissions, and offgassing of the earth's crust. Incineration of garbage, including medical, agricultural and municipal wastes also release mercury into the environment. Figure 8.5 diagrams mercury cycling in the marine environment. Once Hg^{2+} enters the aquatic environment, it settles into sediments where methanogenic bacteria convert it via biomethylation into organic methyl mercury, which fish absorb via their gills while feeding on aquatic plants.

While mercury's reputation was tarnishing almost as it was worked, archeological data clearly indicate human exposure, where it was never worked. Hair mercury levels from tenth-century arctic area mummies are as high as 4 ppm. As for those who worked closely with mercury, the danger was evident. Bernardino Ramazzini, professor and physician at the University of Padua, and the author of *Diseases of Workmen (DeMorbis Artificum Diatriba*, 1700), described the muscle degeneration and skin inflammation of the mirrormakers on Murano, an island off the coast of Venice, as well as the illness in surgeons who treated syphilis with mercury. Also, results for the madhatters of the eighteenth and nineteenth centuries, who worked with mercurous nitrate used in turning fur into felt for women's hats, are extensively documented. Inhaling the fumes, due to poor ventilation, these victims developed severe and uncontrolled muscle tremors and twitching limbs, called the "hatters' shakes."

However, it wasn't until the mid-twentieth century that robust evidence from large epidemiologic studies and two major human disasters pinned down mercury's affinity for the central nervous system. Any mercury compound released into the environment becomes available for potential methylation to CH_3Hg by soil and marine microbes. The mercury concentration in fish at the

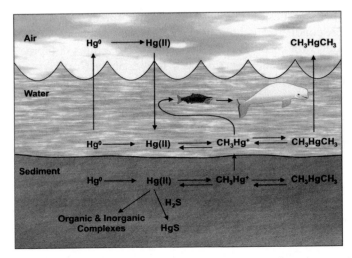

<u>Figure 8.5.</u> The cycling of mercury in its several forms in an aquatic environment. CH_3Hg+, methylmercury ion; CH_3HgCH_3, dimethylmercury; $Hg(II)$, mercuric mercury; HgO, elemental mercury; H_2S, hydrogen sulfide; HgS, cinnabar. (Figure adapted from *Risks in Perspective—Mercury*, The National Research Council, 2000.)

top of the food chain is typically biomagnified up to 100,000 times the concentration in surrounding waters. Thus it was in May 1956, that Minimata disease became a household name worldwide. People living along the shores of Minimata Bay, Kyushu, Japan, were consuming substantial amounts of mercury-contaminated fish and shellfish. The source of the methyl mercury was the effluent from the Chisso Chemical Company using mercury as a catalyst in the production of acetaldehyde. Although the level of mercury in the bay was not all that high, it was concentrated as it ascended the food chain from shellfish to finfish, the staple diet of the villagers [56]. Abnormal gait, impaired speech, loss of muscular coordination, deafness, and blindness were the primary symptoms. Infants were born with cerebral palsy–like symptoms. Cats eating fish lost muscle function; some ran about madly, jumping into the bay and drowning. It was later realized that the floating fish and shellfish seen earlier were also mercury-related fatalities.

In October 1971, Iraq imported more than 95,000 tons of methylmercury-treated grain seed: 73,201 tons of wheat seed and 22,262 tons of treated barley, shipped from Mexico to Basra, Iraq, and distributed free of charge to all 16 provinces, between September and November 1971—after the growing season. Rather then holding the seed for later planting, much of it was ground into floor and used to make home-baked bread.. The first case of CH_3Hg poisoning appeared in late December. During the following 2 months 6530 people were hospitalized and 459 died. Many other victims were turned away from the hospitals [57].

In the final decades of the twentieth century, three vigorous epidemiologic studies were conducted to determine whether subtle neurological effects could be associated with chronic, low-level exposure to a fetus in utero.

One study was conducted in the Faroe Islands, in the North Atlantic between Iceland and Scotland; another, in the Seychelle Islands, in the Indian Ocean between Tanzania and India's southwest coast; and a third, in New Zealand.

In the Faroe Island study [58], neither neurologic nor clinical MeHg-related abnormalities were found in 917 children evaluated at 7 years of age. The median maternal hair mercury level was 4.5 ppm. Methylmercury levels in this cohort occurred via consumption pilot whale meat. Yet neurodevelopment test scores of highly exposed children were normal. Levels of CH_3Hg in the whale meat had averaged 1.6 ppm.

In the Seychelles study, conducted jointly with pediatricians from the University of Rochester, no adverse neurodevelopmental effects were detected in 643 children, even though maternal hair mercury levels reached 66 ppm, far higher than the Faroe Island levels [59]. Mothers in the Seychelles reported eating 10–14 fish meals per week, yet fish mercury levels averaged 0.3 ppm. The New Zealand study found associations with MeHg similar to those of the Faroes [60]. However, in developing its reference dose (RFD) to be used for populations in the United States, the EPA chose the Faroe Island study as a baseline. In 2001, the EPA announced its RFD—the safe dose that can be consumed every day over a 70-year lifetime without ill effects. This daily dose is 0.1 mg/kg. This decision over rode any known health benefit of fish consumption such as omega-3 fatty acids, believed to be protective of heart and stroke events. But bear in mind that the mercury warnings are meant primarily for women of childbearing ages, as diet is the primary route of exposure for the population. It is also worth considering that in Alaska, where fish consumption is both frequent and high, during the period 2002–2004 biomonitoring of hair of pregnant women revealed a 79% incidence of levels below 1 ppm; 92% of the women had levels below 2 ppm, and 83% had levels below 1.2 ppm—the level corresponding to EPA's reference dose. On the basis of these levels, the state of Alaska advised all Alaskans, including pregnant women, women who were breastfeeding, and children, to continue unrestricted consumption of fish from Alaskan waters.

In 2005, the Harvard Center for Risk Analysis convened an expert panel to evaluate whether altering fish consumption by women of childbearing age and other adults was or was not beneficial, given the concern for organic mercury. The panelists were concerned with balancing risk of mercury consumption and potential for harm to a developing fetus, versus the benefits of fish as a source of omega-3 polyunsaturated fatty acids that may confer health benefits. They concluded that men and women of childbearing age could easily switch to fish of low to no mercury content, as well as reducing the frequency of fish meals, but not totally abandon fish, given its proven benefits [61, 62]. Clear enough, but recently a female high school student, who we shall call

Erica Spiritos, in her junior year, told me that a teacher informed the class that if a woman planned to become pregnant, she should stop eating fish 3 years prior to the pregnancy to ensure that mercury would not adversely affect the fetus. Clearly, this is first-order nonsense. Nevertheless, if a high school teacher believes this, and feels so strongly about it that she passes it along to her students, there is ample reason to wonder where such nonsense was obtained, and whether it can be rectified. What do you think?

As for non-fish-eaters, the major source of mercury exposure is elemental mercury from mercury-containing dental amalgams. But this is minuscule, and as a recent study revealed, does not increase the risk of low-birth-weight infants [63]. Nevertheless, mercury's reputation is well deserved. The less of it in air, soil, and water, the better for all of us.

Lead

Lead, and its well-documented adverse health effects, is also not a twentieth-century invention. Lead is also one of the ancient metals, and has been an important metal for 6000–7000 years. The ease with which it can be worked, its remarkable malleability, and its durability account for its early use as a construction material. Also, when lead ores are heated, they produce silver, which accounted for a major portion of lead mining. Many of the so-called silver mines of antiquity were in fact lead mines. It has been estimated that the Romans mined some 60,000 tons of lead ore annually for 800 years—approximately 50 million tons over the period from 600 BCE to 200 CE. The major Roman mines were in southern Spain, and their mining operations released some 5% of the mined lead into the atmosphere causing a veritable global pollution.

The Romans had a word for their lead, *plumbum*, which goes beyond *lead pipe*, to mean waterworks, and the lead pipes and lead-lined aqueducts that brought water from its sources to their towns and homes. Originally, a plumber was a person who made and repaired lead piping. The use of lead-piping for their water supplies was widespread during the height of the Roman Empire some 1900 years ago. Further, although the use of lead-pipe water systems was a major human advance, it was also a serious threat to health. If the water was acidic, as it usually was, and is, and not constantly flowing, significant leaching of lead into the drinking water readily occurred. But both the Romans and the Greeks exposed themselves to yet a greater risk, coating their copper cooking pots and wine drinking vessels with lead to prevent the leaching of copper, which had a bitter taste. They boiled and concentrated fruit juice and preserves in these lead-lined containers, which contained acetic acid among the organic fruit acids, and which readily combines with lead to form lead acetate (sugar of lead), a sweet-tasting chemical that adds a pleasant taste to food and drink, but also poisons. As was written at the time, "From the excessive use of such wines arises paralytic hands"; today this is known as "wrist drop." It also induced gut-wrenching pains, fatigue, impotence, depression, and death.

Lead poisoning didn't end with Rome. Colonial America was plagued by lead poisoning, which was referred to as "the dry grypes," a severe bellyache without diarrhea. "Dangles," a term coined by Ben Franklin, referred to the wrist drop seen in older men who worked in the printing trade and inhaled lead fumes as the lead type was melted. In fact, Franklin was the first to associate lead type with the symptoms. Additionally, the production and consumption of rum, a favorite liquor in the Colonies was a great source of colic and paralysis of arms and legs. When physicians in New England examined distilleries, they found that the still heads and spiral tubing—the "worms"—were made of lead, which the acidic rum was leaching in large amounts. But no matter its lamentable reputation, it continued to be used. In 1887, 64 cases of lead poisoning with 8 deaths, via adulteration of food, were reported in Philadelphia. Lead chromate ($PbCrO_4$) was used extensively as a coloring agent by bakers for their delicious, highly sought-after yellow buns. For the most part, however, lead poisoning was believed to be an adult-only malady; until 1892, when Dr. J. Lockhart Gibson, a physician in Brisbane (Queensland), Australia, in Holmsian-like sleuthing, connected residential paint with childhood lead poisoning. Unfortunately, his report was published in an obscure, Australian medical journal and was not rediscovered until the 1930s. At the time, however, no one believed that lead toxicity, plumbism, could afflict children, let alone infants.

Many of the homes in Brisbane were set on pilings and had porches with long handrailing painted with white lead paint: (litharge, $PbCO_3$, lead carbonate, also known as "dead white," leading to the ground, which became warn, flaky, and powdery in the hot Brisbane sun. According to Dr. Gibson, the great danger was to be found "in adhesion of the paint either by nature of its stickiness or by nature of its powdery character, and finger and nails by which it is carried into the mouths of children, especially in the case of those who bite their nails, suck their fingers, or eat with unwashed hands" [64].

But change doesn't come easily. It had to await the publications of Dr. Herbert L. Needleman, a pediatrician at the School of Medicine, University of Pittsburgh, to sound the tocsin—that exposure to lead in childhood is associated with deficits in central nervous system functioning, deficits in psychological and school performance, and that these deficits persisted into adulthood [65]. Again, the data came as a shock; they were unwanted, and he was vilified.

Lead was used as a pigment and drying agent in "alkyd" oil-based paint. About two-thirds of the homes built before 1940, and one-half of the homes built from 1940 to 1980, contained heavily leaded paint, with walls, windows, and doors having as many as a dozen layers of leaded paint. In 1978, the Consumer Products Safety Commission lowered the legal maximum lead content in most paint to 0.06%—a trace amount.

ENVIRONMENTAL LEAD. Although lead occurs naturally via weathering of lead-containing ores, especially Galina, most environmental levels derive from human activity, with the greatest increases occurring between the 1940s and 1990s as a consequence of the increased number of motor vehicles and their use of leaded gasoline.

Lead was added to gasoline as an antiknocking additive, in the form of tetraethyl lead, a complex mixture of 4–12 aliphatic and aromatic hydrocarbons, whose molecular structure contained an atom of lead to which four ethyl groups (C_2H_5) were covalently bonded, so that is far more organic than inorganic lead—both of which are central nervous system poisons. Tetraethyl lead, a product of the Ethyl Gasoline Corporation, breaks down in the internal-combustion engine, liberating free lead. As an antiknocking additive, it increased environmental lead enormously, releasing literally millions of tons of lead from automobile and truck exhausts. In 1979, motor vehicles released 95 million kilograms of lead into the air. In 1989, when the use of lead was limited but not yet banned, 2.2 million kilograms were released, but this impressive reduction was also accompanied by a stunning reduction in children's blood lead levels (BLLs) as shown in Figure 8.6. The correlation between sales of leaded gasoline and BLLs is nothing short of phenomenal. As demand for leaded gasoline crashed, energy companies quit the leaded gas market. This dramatic reduction in Bll's must stand as one of the great public health successes of recent memory. The mean BLL of children, 15 µg/dL (micrograms per deciliter) in the 1970s, now averages less than 2 µg/dL.

Lead enters the environment through releases from mining lead and other metals, and from factories that make or use lead, lead alloys, or lead com-

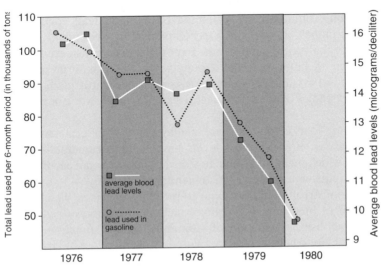

Figure 8.6. Striking correlation between sales of leaded gasoline and blood lead levels in children. (Figure adapted from the *Harvard School of Public Health Magazine*.)

pounds. It is also released during burning of coal and oil. Once lead falls onto soil, it adheres strongly to soil particles and remains in the upper layer of soil. Migration of lead particles from soil to groundwater occurs when acid rain releases the lead.

Indeed, lead has become an integral part of our environment. Its ubiquity derives from the wide use of ammunition—bullets and shot, storage batteries, solder, pigments, covering for underground telephone cables, and as shielding for X-ray and nuclear radiation. All motor vehicles, cars, buses, trucks, and boats—the hundreds of millions of them—use lead storage batteries. The positive plates, made of lead oxide, and the negative plates of spongy lead, are alloys; 91% lead and 9% antimony. Fortunately, over 80% of these are recovered and recycled as a source of lead metal, but a portion does escape into the air.

Although inhalation of lead is a major route of exposure for both adults and children, ingestion is the most common route for infants and youngsters. In preschool children, lead is frequently ingested via hands, toys, dust, soil, and other objects that lead directly for the mouth, posing serious health threats. Estimates of amounts of soil ingested ranges from 5 to 20 mg per day. Children with pica (Latin for magpie), the habitual eating of nonfood objects, are at even greater risk as they can consume as much as an ounce of soil a day. Of greater concern is the peeling and flaking of leaded paint in much of the country's older housing that kids eat—because it tastes good. It's sweet. But lead in infants and young children has an affinity for their central nervous systems (CNSs). With the publication of Dr. Needleman's initial report, a new idea was born; very low lead levels could damage the CNS in ways never before imagined. Later, he found that very low BLLs at age 2 were directly associated with intelligence and well-being in early adulthood. Prior to 1970, lead poisoning, plumbism, was defined by a blood lead concentration of 60 μg/dL or higher, a level often accompanied by abdominal pain (colic), anemia, encephalopathy, and death. Exposures once considered safe are no longer so. The definition of lead toxicity was reduced to 40 μg in 1971, 30 in 1978, and 25 in 1985, and currently stands precariously at 10 μg/dL, as numbers of published studies continue to suggest that 5 μg and lower can induce cognitive deficits.

Lead in any of its compounds are small molecules that can easily cross the placental barrier. Elevated lead levels have been documented in umbilical cord blood, which means that lead ingested by mothers can accumulate in the developing fetus with consequent congenital afflictions.

In the bloodstream, lead can bind to proteins, enzymes, which are critical to a number of biochemical functions such as formation of red blood cells, energy metabolism, production of vitamin D, kidney function, and others. Losses of red blood cells is directly related to anemia; loss of vitamin D leads to reduced weight and height. Circulating lead takes the place of calcium, thereby preventing normal functioning. Most distressing is the fact that many children with elevated BLLs appear asymptomatic. The effect of lead is largely silent. Yet in spite of the silence, there is functional impairment of both the central and peripheral nervous systems—brain and spinal cord are adversely affected, showing themselves through loss of hearing, mood swings, headaches, sleep disturbances, lethargy, and reduced attention span, which are also associated with IQ scores. The fact is, lead serves no useful purpose in the human body. Although "getting the lead out" has been quite successful and beneficial, there remains a way to go.

A National Health and Nutrition Examination Survey (NHANES) for the period 1999–2002, obtained BLLs of 16,825 individuals, and compared them with BLLs of 13,472 individuals obtained in the 1991–1994 survey. Elevated BLLs were defined as greater than 10 µg/dL. For 1999–2002, the overall elevated BLLs for the US population was 0.7%: a decrease of 65% from the 22% in the 1991–1994 survey. Nevertheless, during the 1999–2002 survey period, children aged 1–5 years had the highest prevalence of elevated BLLs—1.6%, indicating that over 300,000 of them remain at risk of potentially harmful lead levels. However, overall, indications are that the BLLs continue their decline, and it is evident that this is directly related to the decline in numbers of homes with lead-based paint. The findings of the latest NHANES report show decided progress toward achieving the national health objective of eliminating blood lead in children by 2010 [66].

Ben Franklin had it right: "The mischievous effect from lead," he wrote, "is at least above 60 years old; and you will observe with concern how long a useful truth may be known, and exist, before it is generally received and practiced on." How long, indeed.

Asbestos

On October 20, 2005, a jury in California found Gurlock Sealing Technologies, of Palymyra, New York, and Kelly-Moore Paint Company, of San Carlos, California, jointly liable for Robert Treggett's malignant lung cancer. The jury recommended that Tregget receive a total of $36.6 million dollars. Mr. Tregget, 60, allegedly contracted his lung cancer through asbestos exposure on a U.S. Navy submarine and while remodeling a home. Tregget had chemotherapy and surgery to remove his right lung after being diagnosed with lung cancer in 2003.

An eight-member Middlesex County, New Jersey, jury awarded William Rhodes, 68, and his wife Doreen, $10 million dollars on August 11, 2005, as compensation for their pain and suffering caused by the mesothelioma, and as punitive damages against Universal Engineering of Manville, New Jersey. Mr. Rhodes had worked as a boiler repair technician for 30 years, during which time he used asbestos—containing products produced by Universal Engineering.

It began with Clarance Borel. *Borel v. Fiberboard Paper Products Corporation* was the groundbreaking case in all asbestos litigation. Mr. Borel was an asbestos insulation worker from Texas, where he was first exposed to asbestos in 1936. In 1969, he became disabled by asbestosis. A year later mesotheliona was diagnosed. Because compensation was strictly limited under Texas law, he sued several asbestos manufacturing companies, seeking full recovery of compensatory damages. The dam was about to break.

In court, Borel maintained that he and his coworkers had not been informed of the hazards of asbestos, and that respirators were not available, and when, years later they were available, they were not required to wear them as they were awkward and uncomfortable. His lawyers noted that none of the asbestos manufacturers had tested the effects of asbestos dust on workers, nor had they determined whether the workers' exposure exceeded threshold limits. Mr. Borel died before the trial ended. Nevertheless, the jury returned a verdict favorable to his widow and family. The dam burst. Tens of thousands of asbestos personal injury cases, settlements, and trials followed, and thousands remain in progress.

What is this material that can induce such fatal conditions and for which juries feel constrained to award unprecedented compensation?

Asbestos is not a single mineral; rather, it is a generic term describing six naturally occurring fibrous silicates, for the most part silicates of magnesium, calcium, iron, and/or sodium, with an unmatched resistance to fire. The six include chrysotile, crocidolite, amosite, tremolite, anthophyllite, and actinolite. But the common commercial products are chrysotile $MgSiO_5$ (OH_4), which currently accounts for 99% of world production, while crocidolite Na_2 $(Fe_2^+,$ $Mg)$ $Fe_3^+Si_8O_{22}$ $(OH)_2$, and amosite (Fe_2^+, Mg) Si_8O_{22} $(OH)_2$, account for the remaining 1%, if that much. Chrystotile, white asbestos, is the product of choice because of its soft, silky fibers, chemical and thermal stability, high tensile strength, flexibility, and low electrical conductivity. In addition, it does not evaporate, dissolve, burn, or react with other chemicals, nor is it your common garden variety of plant or animal fiber, such as cotton, wool, flax, hemp, or hair, which are carbon-containing compounds. Asbestos contains no carbon and is thus inorganic; the others are all organic molecules that the human body can readily metabolize. The qualities that made it so desirable also make it difficult for the body to remove—difficult, but not impossible. However, these fibers are also tasteless, odorless, invisible, and nonirritating to the throat. You simply do not know that they are present. Consequently, asbestos can be a dangerous material. But again, it's the dose that is consequential.

Asbestos was not, is not, used in its raw, fibrous state. It is added to vinyl, plastic, asphalt, cotton, and cement. It can be spun into yarn, woven into fabric, braided into rope (its fireproof threads are stronger than steel), or fabricated into corrugated or flat sheets for use as building materials as well as piping. Its heat- and corrosion-resistant qualities have been so beneficial and so desirable that between 1900 and 1980, over 36 million tons were used worldwide in over 3000 products.

It's fire-resistant qualities, which have never been surpassed, were known to the Egyptians, Greeks, and Romans, who dubbed it the miracle material. The Greeks, we are told, used asbestos for wicks in their temple lamps and candles and were amazed that the flames did not consume the wicks. Consequently they called the substance "sasbestos," meaning inextinguishable, or unquenchable. In the ninth century, Charles the Great Charlemagne was known to clean his asbestos tablecloth by throwing it into a fire [67].

In the early 1930s widespread use of asbestos in homes, schools, and offices began. Fire protection was the compelling reason. But it was during World War II (1939–1945), with its demand for a vast fleet of naval and cargo vessels, when thousands of workers were employed in shipyards across the country as insulators and pipefitters, working in cramped, confining holds of ships applying the fire-retardant asbestos-containing material (ACM), often without benefit of respirators and adequate ventilation, and smoking as they worked, that many of them became the cases of asbestosis, lung cancer, and mesothelioma that began to appear 20, 30, and 40 years later.

But it was with Public Law 99-519, AHERA, The Asbestos Hazard Emergency Response Act, signed into law by President Ronald Reagan on October 22, 1986, that asbestos became a household word and a highly emotional issue, and the asbestos abatement industry was born, as was the pandemic of mediagenic disease.

Because of the public concern for what was perceived as ill effects to their school-age children by 1 or 2 asbestos fibers in the air, AHERA required that the countries 100,000 plus schools conduct inspections by appropriately trained and certified building inspectors to identify both friable (crumbly) and nonfriable ACM. With this information, the schools were required to prepare a management plan which would contain their schools strategy for managing asbestos safely. That was EPA's original intent. Unfortunately, by this action EPA fostered the notion that any level, any amount of asbestos was a threat to health. Political expedience impeded impartial scientific judgment. That exposure did not necessarily mean illness was totally lost in the onslaught to remove asbestos from schools, homes, and office buildings, even though 2 years after AHERA, the EPA stated that "although asbestos is hazardous [that word again], the risk of asbestos-related disease depends upon exposure to airborne asbestos fibers," and furthermore, on the basis of the available data, "The average airborne asbestos levels in buildings seem to be very low. Accordingly, the health risk to most building occupants also appears to be very low." In fact, the EPA itself found asbestos air levels in federal buildings to be essentially the same as levels outside these buildings. They stated emphatically that "Removal is often not a building owner's best course of action to reduce asbestos exposure. In fact, an improper removal can create a dangerous situation where none previously existed" [68]. But who was listening? Certainly not the media, who might have been counted on to get EPA's message out, but were more concerned with fanning the flames of fear of disease, which abetted lawyers in their steady stream of lawsuits.

Yes, asbestos fibers are known to have adverse health effects. The mechanism by which fine particulate matter gains deep penetration into the lungs was described earlier. Because of their long, filamentous, silky nature, asbestos fibers can readily penetrate down to the alveoli. Asbestos fibers, with their fishhooklike ends, fasten onto alveoli walls. In response to this irritation, the body, attempting to protect itself, begins a process of walling off, which is a double-edged sword in that wall thickening also reduces the passage of oxygen and CO_2. This walling off, referred to as *fibrosis*, in the case of asbestos, is *asbestosis*. Shortness of breath occurs on walking, climbing stairs, and as it progresses, on lying down; a form of suffocation. Asbestosis is most common in men over 40 who have worked in asbestos-related occupations for long years. Cigarette smokers are at greatest risk of developing asbestosis. It is worth recalling that each lung consists of some 400 million alveoli (800 million in both lungs). If one, two, or three fibers enter an alveolus, causing irritation, it will be of no consequence. But again, this noteworthy message fails to get to the public, for whom a single fiber becomes a fearsome thought.

Mesothelioma is something else again. The lungs and abdominal organs lie in hollow cavities. For the lungs, it's the pleural cavity; for the abdominal organs, it's the peritoneum. Both cavities are lined and covered by a clear, saranlike tissue. It is in this tissue, the mesothelial tissue, that the rare malignancy mesothelioma arises. In the 1960s mesothelioma was reported to be due to exposure to crocidolite by South African asbestos miners. It is a nasty tumor characterized by shortness of breath, coughing, weight loss, and incapacitating pain, invariably resulting in death a year or two after diagnosis. Exposures had usually occurred 20–40 years earlier and may have been relatively brief, but heavy. Smoking does not appear to be connected to mesothelioma. Mesothelioma cases increased steeply from the 1970s through the mid-1990s then leveled off. This mesothelioma epidemic was the result of high-level occupational exposure from the 1930s through the 1960s in workplaces populated almost exclusively by men: "It is unlikely that low environmental exposure to asbestos, or any other type of low-level asbestos exposure, is associated with more than a negligible risk of mesothelioma" [69].

Epidemiologic studies have conclusively documented the association of bronchogenic carcinoma, lung cancer, with asbestos. Here again, it is the asbestos workers who smoked who have the highest risk and incidence of lung cancer compared to smokers not exposed to asbestos, or asbestos-exposed nonsmokers.

Unfortunately, asbestos, especially asbestos litigation, has inspired a clutch of unsavory characters. No subject has raised more emotion in the field of occupational health in the past 35 years than asbestos, which has provided the media, the legal profession, radiologists, and removal contractors with a bountiful harvest. It is well known and amply documented that reading X rays, radiographs, is a tricky business. Also, because the federal government needed to keep track of coal miners with black lung disease, the National Institute of Occupational Safety and Health (NIOSH) was concerned about the competence of physicians to read chest X rays, it instituted a week-long training program of X-ray reading, with final exams, including reading of X rays. Those who passed the rigorous test were certified as "B" readers. Lists of "B" readers in all 50 states are available from NIOSH.

In diagnosing asbestos-related diseases the most important medical tool is the chest X ray. In 1990, physicians from Johns Hopkins Medical School, West Virginia Medical Center, and the University of Southern California collaborated on a radiologic reevaluation of cases of alleged asbestos-related disease. But what prompted that? Why reevaluate X-ray films? During 1986, the National Tire Litigation Project sent physicians around the country in mobile X-ray vans to screen rubberworkers for asbestosis; a scheme calculated to provide cases for a group of lawyers. In one location, 700–750 workers submitted to a screening. Approximately 440 of the workers examined filed legal claims for an asbestos-related injury. Such numbers of positive X-ray findings represents a prevalence of 60%. If this is correct, it would constitute an epidemic of massive proportions. Consequently the physician/investigators sought to confirm or refute the radiologic findings. For this reevaluation they obtained the X-ray films of the tire workers and presented them to three independent board-certified "B" readers, who were not told the background of these films, but were required to interpret them using guidelines developed by the International Labor Organization. In as much as the technical quality of an x-ray film can influence a diagnosis, an assessment of film quality was made. One of the readers found 35, or 8% of the films to be unreadable, but most importantly they found the original prevalence to be mistakenly high. How high? The prevalence of likely asbestos-related conditions according to each of the readers was 3.7%, 3.0%, and 2.7%. "A large proportion of the re-evaluated cases had completely normal chests for their ages," and "the vast majority of abnormalities found were non-occupational in origin and consisted of conditions to be expected in an aged population." They concluded that the best estimate is that possibly 16, but more realistically, 11 of the 439 cases, some 2.5%, may have a condition consistent with asbestos exposure [70].

Echoing Reger, in 2004, researchers at John Hopkins and West Virginia published their findings of a study comparing "B" readers interpretations of chest X-ray films with those of "B" readers employed by plaintiff's lawyers. They obtained 492 chest radiographs that had been read as 95.9% positive for lung changes. All were presented to each of six independent "B" readers who had no knowledge of the history of these radiographs. The six independent readers found 4.5% positive [71]. In an accompanying editorial, the question was raised; "Is something rotten in the courtroom?" as the differences between the groups is far beyond chance variability. Listen to their concern: "the radiologic community itself has an obligation to conduct further investigations to determine whether the integrity of 'B' reader radiologists has indeed been breached, and if so, to repair the breach,implement measures to prevent it from happening again, and restore integrity to our nobel and proud profession" [72].

Fast-forward to 2006. Mounting evidence indicates that many asbestos injury claims, and there are thousands of them, are not genuine; a lawsuit machine appears to be at work. In July 2005, Judge Janis Graham Jack of Federal District Court, Corpus Christi, Texas, in a withering decision threw out 10,000 plaintiffs' claims in a giant silicosis lawsuit because she found, as noted in the case *Re: Silicia Products Liability Litigation. Order No. 29. Addressing Subject Matter Jurisdiction*, some expert testimony and sanctions, from 1994. A physician, Dr. Ray A. Harron, had found "opacities and scars in all zones of Mr. Kimble's lungs, consistent with asbestosis." In 2002, the patient, Mr. Kimble, was X-rayed again, this time in connection with the current silicosis litigation. Dr. Harron again read Mr. Kimble's radiograph and determined that Mr. Kimble's lungs had opacities and scars consistent with silicosis. Just as the defendants prepared to introduce a packet of eight more identical asbestos/silicon reversals by Dr. Harron, Dr. Harron stated to the defendant's attorney, "if you're accusing me of fabricating these things, I think that's a serious charge." When the court responded that the defendants seemed to be making that accusation—and the defense council agreed, Dr. Harron asked for representation. The court ended its testimony at that point in order to allow Dr. Harron to hire an attorney. As Judge Jack stated [73]:

> The court finds that filing and then persisting in the prosecution of silicosis claims while recklessly disregarding the fact that there is no reliable basis for believing that every plaintiff has silicosis constitutes and unreasonable multiplication of the proceedings. When factoring in the obvious motivation—overwhelming the system to prevent examination of each individual claim and to extract mass settlements—the behavior becomes vexatious as well. Therefore, the court finds that O'Quinn (Plaintiff's Council) has multiplied the proceedings . . . unreasonably and vexatiously, and his firm will be required to satisfy personally the excess costs, expenses and attorney's fees reasonably incurred because of such conduct.

The *New York Times* noted that Dr. Ray A. Harron reviewed as many as 150 X-ray films a day, one every few minutes at $125 each—some $20,000 a day, earning millions of dollars over the years. Some of his reports supported claims by more than 75,000 people seeking compensation for asbestos-related lung injury. Over the last 30 years, more than 700,000 claims have been filed involving inhalation of asbestos, and more than $70 billion has been spent on asbestos litigation, and $49 billion as compensation. Dr. Harron gave up his medical practice for reading X-ray films full time and averaged 6400 X-ray readings yearly over the past 10 years [74].

One reason for the wave of claims was the realization on the part of claimants, their attorneys, and trade unions that many asbestos manufacturers had already been driven into bankruptcy, and settlement funds were drying up. As many as 90% of current claimants have no signs of serious illness, but are filing claims while there is still hope for some compensation. Clearly, asbestos-related illness is not as widespread as corruption among lawyers, physicians, and their plaintiffs.

Environmental Cancer and Cancer Clusters

The relationship leading causes of death and the environment was broached in Chapter 1, questioning whether any of the top dozen leading causes of death could be identified with an environmental risk factor. We now revisit this question, focusing on cancer, the country's second leading cause of death, but eliciting the greatest fear.

The fact that cancer incidence has actually been declining for the past 20 years is lost on most people. For the public, cancer is seen as the leading cause of death, and elicits nothing so much as fear and apprehension: a condition to be avoided at all costs. But how does one avoid an environmentally induced illness? Perhaps that depends on our understanding of that slippery locution, "environment." Toward that end, we shall pursue the cancer–environment link as it may recommend ways of reducing the community burden of cancer—a consummation to be devoutly wished for. Unfortunately, advances in cancer therapy, the war on cancer, has not been as rewarding as that for heart disease and other chronic ailments. Cancer is usually a disease of old age; it often takes years to develop, and has a multitude of causes. Factors inside and outside the body contribute to its development. In the National Cancer Institute's recently published brochure, *Cancer and the Environment*, they inform us that "exposure to a wide variety of natural and man-made substances in the environment accounts for two-thirds of all cases of cancer in the U.S." They further state that "these environmental factors include life style choices like cigarette smoking, excessive alcohol consumption. poor diet, lack of exercise, excessive sunlight exposure and sexual behavior that increases exposure to certain viruses." Would most people agree that this is the environment they have long believed is the source of their fear and apprehension? It surely isn't the environment that environmental activists rail against. But let us not drop this thread. The NCI further maintains that

> The importance of environment can be seen in the differences in cancer rates throughout the world and the change in cancer rates when groups of people move from one country to another. When Asians, for example, who have low rates of prostate and breast cancer and high rates of stomach cancer in their native countries, immigrate to the U.S., their prostate and breast cancer rates increase over time until they are nearly equal to or greater than the higher levels of these cancers in the U.S. Similarly, their rates of stomach cancer fall, becoming nearly equal to the lower U.S. rates.

The scientists at the NCI believe that, "Life style factors, such as diet, exercise, and being overweight play a major role in the trends for breast and prostate cancer, and infection with the Helicobactor plyori bacterium is an important risk factor for stomach cancer. And recently, they note too, the rapid rise in rates of colorectal cancer in Japan and China suggests an environmental cause such as lifestyle factors."

Smoking, they say, is linked to cancers of the lung, bladder, mouth, colon, kidney, throat, voicebox, esophagus, lip, stomach, cervix, liver, and pancreas, and that certain factors within the body render some people more likely to develop cancer than others. Some people either inherit or acquire altered genes, abnormal hormone levels, or a weakened immune system. "Each of these factors may make an individual more susceptible to cancer." Have we been looking in the wrong places for answers? There are more than 100 types of cancer, and one of the most challenging areas of current research is trying to identify the unique combination of environmental exposure and inherent susceptibility or resistance, that protects one person but allows another to develop cancer.

Tobacco is a major success story. The proportion of the US population that smoke has been declining for decades, and over the last 10 years, lung cancer rates have also started to decline. As many people stop smoking, cancer rates of many of the sites noted above will also drop.

Perhaps a most significant finding was a recent report by members of Harvard University's School of Public Health in collaboration with the universities of Auckland, New Zealand, and Queensland, Brisbane, Australia. They analyzed the 7 million deaths from cancer worldwide in 2001, and determined that 2.43 million (35%) could be attributable to nine potentially modifiable risk factors: overweight and obesity; physical inactivity; low fruit and vegetable intake; smoking; alcohol use; unsafe sex; indoor smoke from household use of solid fuel; contaminated injections; and finally, urban air pollution, specifically related to fine sooty particles less than 10 μm in size. Eight of the nine clearly relate to the environment of the self, while one deals with the outdoor environment. But they do say that more than one in every three of the 7 million cancer deaths worldwide are due to modifiable risk factors, with smoking and alcohol use having particularly important roles in all countries. They also indicate that such important cancers as prostate, kidney, melanoma and lymphomas were not attributable to any of the (mine) risks assessed, nor did they assess some fairly well known occupationally related exposures, which are responsible for 102,000 cancer deaths worldwide—1.4% of the total. Nor can their results be compared with the Doll and Peto estimates of 80–90% of all cancers due to personal environmental factors (see Chapter 1, Ref. 62, and discussion in text), as Doll and Peto's estimates apply only to the United States, while theirs are worldwide. Obviously, however, both the NCI and the Harvard collaboration study point overwhelmingly in a single direction—the importance of modifying injurious personal behavioral factors if the burden of cancer is to be significantly reduced [75]. The message is clear; we must look to ourselves if we are to humble cancer. This message requires wide distribution, but I suspect it will not be joyfully received. That the fault dear Brutus, lies with us, not some vague physical, military/industrial collusion, does not fit with the decades-long preferred dogma.

This discussion would be remiss if childhood cancers were not mentioned. The rarity of childhood cancers, and the fact that data on both incidence and deaths are too often classified using a system more appropriate for adults, makes available statistics difficult to interpret. Be that as it may, little is actually known about the etiology of childhood cancer, although numerous possible risk factors have been advanced. With two exceptions, there is no good evidence for any large increase in incidence for any form of childhood cancer over the past several decades other than Kaposi's sarcoma in Kampala, Uganda, one of the regions severely affected by the AIDS epidemic. The other exception is thyroid carcinoma in Belarus, The Ukraine, where the currently reported incidence indicates a 50-fold increase over previous decades. This may be partly explained by the explosion at Chernobyl, and improved screening procedures. Otherwise no important risk factors appear to have involved children [76].

Cancer Clusters

Any community believing that it has an unusual number of cancer cases naturally looks for an environmental cause—the soil, the water, the air. Since the beginning of human societies, disease—any disease in a community—always strikes more than one person. Consequently, clusters are to be expected, as the many AIDS cases in Kampala, and thyroid cancers in Belarus, Legionnaires' disease in Philadelphia, influenza in season, and the many outbreaks of bacterial food poisoning at picnics, all of which are readily explained as having a commonality of exposure in place and time.

Clusters do stir the imagination, and distinguishing true clusters from chance or perceived ones can be difficult, but in the case of cancer, the types of cancer in a cluster helps disentangle the episode. However, from the theory of random sampling we learn that clusters can be expected to occur simply by chance alone. Using probability theory, Dr. Raymond R. Neutra, of the California Department of Health, calculated that 17% of the 29,000 towns and census tracts in the United States, will have at least 1 of the 80 recognized types of cancer elevated in any given decade producing 4930 chance clusters. Furthermore, increasing the difficulty for epidemiologists seeking to discover associations and/or causes is the fact that the great majority of clusters involve fewer than 10 cases. Also all too often, cases rounded up by community residents include cases diagnosed prior to the individuals moving into the neighborhood, or cases found by activists seeking them out. Geographic boundaries are then drawn around them, as was the case of the Texas sharp shooter who fired at the side of a barn, then painted a ball's eye around the holes. Such manipulation does produce seemingly related cases, but in fact is nothing more than anthropogenic (synthetic) artifacts. A revealing example recently occurred in

Long Island, NY, where the incidence of breast cancer (see Chapter 8) found no commonality of risks, or risks specifically related to breast cancer, yet community activists were absolutely certain that risk factors in the physical environment were responsible for their breast cancers. A somewhat similar incident, in Woburn, Massachusetts, became a national cause celibre (and later a movie). Residents of Woburn, 12 miles northeast of Boston, were agitated over what they perceived to be a large number of childhood leukemia cases, and tagged the tapwater along with hazardous-waste sites as the culprits.

In their pursuit of the investigation that ensued, epidemiologists from the Massachusetts State Department of Health and the NCDC in Atlanta interviewed parents of the children with leukemia, and two matched controls for each case, inquiring about medical history, mother's pregnancy history, school history, and environmental exposures. No significant differences were found between cases and controls. In fact, no child with leukemia had had contact with a hazardous-waste site, and the contaminants of the wells supplying the drinking water were not leukemogenic [77].

Dover Township, New Jersey, residents remain upset and unconvinced that their water supply was not found to be the cause of the large number and diversity of childhood cancers. A 6-year, $10-million-dollar study could not explain the different types of cancers diagnosed among their children, nor could the study relate the cancers to either water or air pollution. The Agency for Toxic Substances and Disease Registry developed a computer model of the water supply that is the most detailed of its kind ever to be developed for monitoring a water system. The model was so exacting that it was able to track, on an hourly basis, the source and amount of tapwater each family had been exposed to for over 34 years. Even though the study appears to have vindicated the water company supplying the tapwater, a $8-million-dollar settlement was negotiated out of court for the plaintiffs, because there was such strong belief by residents that there must have been exposure with resultant illness [78].

Lawyers and distressed parents notwithstanding, each type of cancer has certain known and/or unknown suspected risk factors associated with it, so that a suspected cancer cluster is more likely to be a true cluster, rather than a chance occurrence if it involves a number of cases of a specific type of cancer, rather than several different types, or it involves an increased number of cases of a specific type of cancer in an age group that is seldom affected by that type of cancer. We humans have a deep-seated tendency to see meaning in the common variations that always appear in small numbers—small samples. True clusters are themselves rare events. But—and this is a sizable but—perception, lawyers, politics, and the mass media do muddy the waters. From the many locally perceived and reported cancer clusters, it has been learned that misclassification, gerrymandering, and the Texas Sharp Shooter do become unwitting allies. Most unfortunate, however, is the media's undergirding local cancer clusters, promoting superficial and emotional anecdotes contributed by residents that are attention grabbers, and sell newspapers, but that only obfuscate fact from fiction. Time passes, the issue fades, but the idea, the memory of toxins remains fixed in the public consciousness. Scientists become the goats for failing to obtain positive results agreeing with residents' perceptions, and doing so rapidly, while the media become the heroes for getting the cluster local and national attention. The country's escalating health consciousness may well motivate people to report perceived clusters more often where in fact there are none—and with the media's cooperation, more melodramatic cluster coverage can be expected.

The takeaway message is clear; "environment" requires reconsideration and reinterpretation. Pogo had it right, but we weren't listening. So, here it is again: "We have met the enemy, and he is us."

REFERENCES

1. Ownby, D. R., Johnson, C. C., and Peterson, E. L., Exposure to dogs and cats in the first year of life and risk of allergic sensitization at 6 to 7 years of age, *JAMA* **288**(8):963–972 (2002).

2. Braun-Fahrlander, C., Riedler, J., Herz, U. et al., Environmental exposure to endotoxin and its relation to asthma in school-age children, *NEJM* **347**(12):869–877 (2002).

3. Bach, J. F., The effect of infections on susceptibility to autoimmune and allergic diseases, *NEJM* **347**(12):911–918 (2002).

4. Thorne, P. S., KulhanKova, K., Yin, M. et al., Endotoxin exposure is a risk factor for asthma, *Am. J. Resp. Crit. Care Med.* **172**:1371–1377 (2005).

5. Von Mutius, E., Martinez, F. D., Fritzch, C. et al., Prevalence of asthma and atopy in two areas of west and east Germany, *Am. J. Resp. Crit. Care Med.* **149**:358–384 (1994).

6. Kelman, B. J., Robbins, C. A., Swenson, J., and Hardin, B. D., Risk from inhaled mycotoxins in indoor office and residential environments, *Intl. J. Toxicol.* **23**:3–10 (2004).

7. Pristine, T., High rise alert: There's a fungus among us, *New York Times* C8 (June 2, 2004).

8. Morgan, W. J., Crain, E. F., Gruchella, R. S. et al., Results of a home-based environmental intervention among urban children with asthma, *NEJM* **351**(11):1068–1080 (2004).

9. Grucella, R. S., Pongracic, J., Plant, M. et al., Inner city asthma study: Relationships among sensitivity, allergen exposure, and asthma morbidity, *J. Allergy Clin. Immunol.* **115**(3):478–485 (2005).

10. Wright, R. J., and Steinbach, S. F., Violence: An unrecognized environmental exposure that may contribute to greater asthma morbidity in high risk inner city populations, *Environ. Health Perspect.* **109**(10):1085–1089 (2001).

11. Chen, J. T., Krieger, N., Van den Eeden, S. K., and Quesenberry, C. P., Different slopes for different folks: Socioeconomic and racial/ethnic disparities in asthma and hay fever among 173,859 U.S. men and women, *Environ. Health Perspect.* **110**(Suppl. 2): 211–216 (2002).

12. Basagana, X., Sunyer, J., Kogevinas, M. et al., Socioeconomic status and asthma. Prevalence in young adults. The European Community Respiratory Health Survey, *Am. J. Epidemiol.* **160**(2):178–188 (2004).

13. Rosenkranz, M. A., Busse, W. W., Johnstone, T. et al., Neural circuitry underlying the interaction between emotion and asthma symptom exacerbation, *Proc. Natl. Acad. Sci. USA* **102**(37):13319–13321 (2005).

14. Chase, M., Starting early: A new treatment to prevent asthma is only skin deep, *Wall Street J.* A1 (June 24, 2004).

15. Zhu, Z., Zheng, T., Homer, R. J. et al., Acidic mammalian chitinase in asthmatic Th2 inflammation and IL-13 pathway activation, *Science* **304**:1678–1681 (2004).

16. Zhu, D., Kepley, C. L., Zhang, K. et al., A chimeric human-cat fusion protein blocks cat-induced allergy, *Nature Med.* **11**(4):446–449 (2005).

17. Ferris, B. G., Speizer, F. E., Spengler, J. D. et al., Effects of sulfur oxides and respirable particles on human health, *Am. Rev. Resp. Dis.* **120**:767–779 (1979).

18. Dockery, D. W., Pope, C. A., Zu, X., Spengler, J. D. et al., An association between air pollution and mortality in six U.S. cities, *NEJM* **329**(24):1753–1759 (1993).

19. Nel, A., Air pollution-related illness; effects of particles, *Science* **308**:804–806 (2005).

20. Kunzli, N., Jerrett, M., Mack, W. J. et al., Ambient air pollution and atherosclerosis in Los Angeles, *Environ. Health Perspect.* **113**(2):201–206 (2005).

21. Shaw, J., Clearing the air, *Harvard Mag.* 29–35 (May/June 2005).

22. Kaiser, J., Evidence mounts that tiny particles can kill, *Science* **289**:22–23 (2000).

23. Gauderman, W. J., Avol, E., Gilliand, F. et al., The effect of air pollution on lung development from 10 to 18 years of age, *NEJM* **351**(11):1057–1067 (2004).

24. Pope, C. A., Burnett, R. T., Thurston, G. D. et al., Cardiovascular mortality and long-term exposure to particulate air pollution, *Circulation* **109**:71–77 (2004).

25. Brook, R. D., Franklin, B., Cascio, W. et al., Air pollution and cardiovascular disease. A statement for health care professionals from the Expert Panel on Population and Preventive Science of the American Heart Association, *Circulation* **109**:2655–2671 (2004).

26. Laden, F., Schwartz, J., Speizer, F. E., and Dockery, D. W., Reduction in fine particulate air pollution and mortality: Extended follow-up of the Harvard Six Cities Study, *Am. J. Resp. Crit. Care Med.* **173**:667–672 (2006).

27. Miller, K. A., Siscovick, D. S., Sheppard, L., et al., Long-term exposure to air pollution and incidence of cardiovascular events in women, *NEJM.* **356**(5):447–458 (2007).

28. Pleil, J., Vette, A. F., Johnson, B. A., and Rappaport, S. M., Air levels of carcinogenic polycyclic aromatic hydrocarbons after the World Trade Center disaster, *Proc. Natl. Acad. Sci. USA* **101**(32): 11685–11688 (2004).

29. DePalma, A., Tracing lung ailments that arose with 9/11 dust, *New York Times* (May 13, 2006).

30. DeKok, T. M. C. M., Hogervorst, J. G. F., Kleinjans, J. C. S., and Briede, J. J., Radicals in the church, *Eur. Resp. J.* **24**:1069–1073 (2004).

31. Yardley, J., China's next big boom could be foul air, *New York Times* 3 (Oct. 30, 2005).

32. Rohan, M., Parow, A., Stoll, A. L. et al., Low-field magnetic stimulation in bipolar depression using MRI-based stimulator, *Am. J. Psychiatr.* **161**:93–98 (2004).

33. Taubes, G., Fields of fear, *Atlantic Monthly* 94–108 (Nov. 1994).

34. Feychting, M., and Ahlbom, A., Magnetic fields and cancer in children residing near Swedish high voltage powerlines, *Am. J. Epidemiol.* **138**:467–481 (1993).

35. Linet, M. S., Hatch, E. E., Kleinerman, R. A. et al., Residential exposure to magentic fields and acute lymphoblastic leukemia in children, *NEJM* **337**(1):1–7 (1997).

36. Campion, E. W., Powerlines, cancer and fear, *NEJM* **337**(1):44–46 (1997).

37. Brodeur, P., *The Great Power Line Cover-up*, Little, Brown, Boston, 1993.

38. Schoenfeld, E. R., O'Leary, E. S., Henderson, K. et al., Electromagnetic fields and breast cancer on Long Island: A case-control study, *Am. J. Epidemiol.* **158**:47–58 (2003).

39. Ahlbom, A., Day, N., Fechting, M. et al., A pooled analysis of magnetic fields and childhood leukemia, *Br. J. Cancer* **83**(5):692–698 (2000).

40. Han, L., Quinli, Z., Yu, W. et al., Effects of ELF magnetic fields on protein expression profile of human breast cancer cell MCF7, *Sci. China Ser. C. Life Sci.* **48**(5):506–514 (2005).

41. Christensen, H. C., Schuz, J., Kosteljanetz, M. et al., The cellular telephone use and risk of acoustic neuroma, *Am. J. Epidemiol.* **159**(3):277–283 (2004).

42. Stewart, W., *The Stewart Report: Mobile Phones and Health.* Sect. 5, Scientific evidence. Independent expert group on mobile phones (accessed at `www.iegmp.org. uk`).

43. Koyu, A., Ozguner, F., Cesur, G. et al., No effects of 900 MHz and 1800 MHz electromagnetic field emitted from cellular phone on nocturnal serum melatonin levels in rats, *Toxicol. Indust. Health* **21**:27–31 (2005).

44. National Research Council, Environmental Epidemiology. Vol. **1**, *Public Health and Hazardous Wastes*, National Academy Press, Washington, DC, 1991.

45. Vrijheld, M., Health effects of residence near hazardous waste landfill sites: A review of epidemiologic literature, *Environ. Health Perspect.* **108**(Suppl. S1):101–112 (2000).

46. Lipscomb, J. A., Goldman, L. R., Satin, K. P. et al., A follow-up study of the community near the McColl Waste Disposal Site, *Environ. Health Perspect.* **94**:15–24 (1991).

47. Shusterman, D., Lipscomb, J., Neutra, R., and Satin, K., Symptom prevalence and odor-worry interaction near hazardous waste sites, *Environ. Health Perspect.* **94**:25–30 (1991).

48. Neutra, R., Lipscomb, J., Satin, K., and Shusterman, D., Hypotheses to explain the higher symptom rates observed around hazardous waste sites, *Environ. Health Perspect.* **94**:31–38 (1991).

49. Reggiani, G., Anatomy of a TCDD spill: The Seveso Accident, in *Hazard Assessment of Chemicals: Current Developments*, Vol. 2, Tendra Sexena, J., ed., Academic Press, New York, 1983, pp. 269–342.

50. Landi, M. T., Consonni, D. et al., 2,3,7,8-Tetrachlorodibenzo-*p*-dioxin plasma levels in Seveso 20 years after the accident, *Environ. Health Perspect.* **106**(Suppl. 2):625–632 (1998).

51. Eskenazi, B., Mocarelli, P., Warner, M. et al., Maternal serum dioxin levels and birth outcomes in women of Seveso, Italy, *Environ. Health Perspect.* **111**:947–953 (2003).

52. National Academy of Sciences, 2003, *Veterans and Agent Orange: Health Effects of Herbicides Used in Vietnam: Update 2002*, Institute of Medicine, Washington, DC (available at www.nap.edu/books/0309086167/htm; accessed 4/15/05).

53. Lawson, C. C., Schnorr, T. M., Whelan, E. A. et al., Paternal occupational exposure to 2,3,7,8-tetrachlorodibenzo-p-dioxin and birth outcomes of offsprings. Birth weight, pre-term delivery and birth defects, *Environ. Health Perspect.* **112**:1403–1408 (2004).

54. Rosenthal, E., Liberal leader from Ukraine was poisoned, *New York Times* (Dec. 12, 2004); see also *BBC News*, Deadly dioxin used on Yuschenko (http://news.bbc.co.uk/go/pr/fr/1/hi/world/europe/4105035.stm).

55. World Health Organization, *Dioxins and Their Effects on Human Health*, Fact Sheet 225, June 1999 (www.who.int/mediacentre/factsheets/fs225/en/index.html).

56. Workshop on Risk Assessment Methodology for Neurobehavioral Toxicity, Evolution of our understanding of methyl mercury as a health threat, *Environ. Health Perspect.* **104**(Suppl. 2):367–379 (1996).

57. Bakir, F., Damluji, S. F., Amin-Zaki, L. et al., Methyl mercury poisoning in Iraq, *Science* **181**:230–234 (1973).

58. Grandjean, P., Weihe, P., White, R. I. et al., Cognitive deficit in 7-year-old children with prenatal exposure to methyl mercury, *Neurotoxicol Teratol.* **19**:417–428 (1997).

59. Marsh, D. O., Clarkson, T. W., Myers, G. J. et al., The Seychelles study of fetal methyl mercury exposure and child development: Introduction, *Neurotoxicology* **16**(4):583–596, 629–638 (1995).

60. Kjellstrom, T., Kennedy, P., Wallis, S., and Mantell, C., *Physical and Mental Development of Children with Prenatal Exposure to Mercury from Fish. Stage 2*, National Swedish Environmental Protection Board Dept. 3642, Solna, Sweden, 1989.

61. Arnold, S. M., and Middaugh, J. P., *Use of Traditional Foods in a Healthy Diet in Alaska. Risks in Perspective*, 2nd ed., Vol. 2, Mercury, State of Alaska; reprinted in Epidemiology **8**(11):1–47 (2004) (www.epi.Alaska.gov).

62. Willet, W., Fish: Balancing health risks and benefits, *Am. J. Prevent. Med.* **29**(4):320–321 (2005).

63. Hujoel, P. P., Lydon-Rochelle, M., Bollen, A. M. et al., Mercury exposure from dental filling placement during pregnancy and low birth weight risk, *Am. J. Epidemiol.* **161**(8):734–740 (2005).

64. Gibson, J. L., Notes on lead-poisoning as observed among children in Brisbane, *Proc. Intercolonial Med. Congress Australia* **3**:76–83 (1892).

65. Needleman, H. L., Gunnoe, C., Leviton, A. et al., Deficits in psychologic and classroom performance of children with elevated dentine lead levels, *NEJM* **300**:689–695 (1979); see also *Annu. Rev. Publ. Health* 55:209–222 (2004).

66. Blood lead levels—United States 1992–2002, *Morbidity/Mortality Weekly Report* **54**(20):513–516 (May 27, 2005).

67. De Boot, A. B., *Gemmarumet Lapidum. Lugduni Bataborum*, 1647; see Benarde, M. A., *Asbestos: The Hazardous Fiber*, CRC Press, Boca Raton, FL, 1990.

68. U.S. EPA, *Managing Asbestos In Place*, "*The Green Brook*," 20T-2003, *Pesticides and Toxic Substances* (TS-799), Washington, DC, 1990.

69. Price, B., and Ware, A., Mesothelioma trends in the United States: An update based on surveillance, epidemiology, and end results program data for 1973–2003, *Am. J. Epidemiol.* **159**(27):107–112 (2004).

70. Reger, R. B., Cole, W. S., Sargent, E. N., and Wheeler, P. S., Cases of alleged asbestos-related disease: A radiologic re-evaluation, *J. Occup. Med.* **32**:1088–1090 (1990).

71. Gitlin, J. N., Cook, L. L., Linton, O. W., and Garret Mayers, E., Comparison of "B" readers interpretations of chest radiographs for asbestos-related changes, *Acad. Radiol.* **11**:843–856 (2004).

72. Janower, M. L., and Berlin, L., "B" readers radiographic interpretations in asbestos litigation: Is something rotten in the courtroom? *Acad. Radiol.* **11**:841–842 (2004).

73. Jack, J. G., in re: *Silica Products Liability Litigation*, in the U.S. District Court for the Southern District of Texas, Corpus Christi Division, 28 USC, Paragraph 1927, July 2005, MDL Docket 1553.

74. Glater, J. D., Reading X-rays in asbestos suits enriched doctor. *New York Times* (Nov. 29, 2005).

75. Danaei, G., Vander Hoom, S., Lopez, A. D. et al., Causes of cancer in the world: Comparative risk assessment of nine behavioral and environmental risk factors, *Lancet* **366**(9499):1784–1793 (2005).

76. Draper, G. D., Kroll, M. E., and Stiller, C. A., Childhood cancer, in *Trends in Cancer Incidence and Mortality*, Doll, R., Fraumeni, J. F., Jr., and Mrir, C. S., Cold Spring Harbor Press, Cold Spring Harbor, New York, 1994.

77. Cutler, J. J., Parker, G. S., Rosen, S. et al., Childhood leukemia in Woburn, MA, *Public Health Reports* **101**(2):201–205 (1986).

78. Peterson, I., Tom's River still asking a question: "Why us?" *New York Times* F4 (Dec. 24, 2001).

9

SCIENCE FOR THE PEOPLE; THE PEOPLE FOR SCIENCE

Loyalty to petrified opinion never broke a chain or freed a human soul.
—*Mark Twain*

We are poised on the threshold of great change. The twenty-first century will see the unlocking of impressive new knowledge. But how will it be translated and transmitted to a distrustful, doubting public? Will it be accepted? Will the current lack of trust of scientists prevail, or be surmounted? These are not trivial questions, as they can have a serious impact on our lives.

The public is inundated with a plethora of opinions, editorials, commentaries, talking heads, news articles, magazines, TV, radio, private mailings, and—oh, yes—advertisements promising all manner of personal improvements. How are they to distinguish fact from fiction? How, in fact, can a muddled public even consider individuals offering sound information at odds with the conventional wisdom, afflicted as they are by mental pollution? Change is sorely needed. But where to begin? For openers, with the published scientific literature, and the mass media's rush to be the first to announce a "breakthrough," they are ill-equipped to separate wheat from chaff, which there is a great deal of—and with the Internet, and its world without walls; no rules, no controls, no editors as gatekeepers. How can the public meet this new information age head on, able to ask the necessary probing questions to avoid being deluded, and fleeced, by the many fleecers who, like death and taxes, are forever with us, because there are always sheep to be shorn. Shearing needs slackening. A

Our Precarious Habitat . . . It's In Your Hands, Fourth Edition. By Melvin A. Benarde
Copyright © 2007 John Wiley & Sons, Inc.

dollop of critical thinking would work wonders. Consider, for example, what is being thrown our way in the form of simple numbers, which in print take on a reality all their own. Daniel Okrent, Public Editor of *The New York Times*, notes that "in Mexico, drug trafficking is a $250 billion dollar industry," or that the "economic impact of a new stadium proposed by New York's Mayor would generate about $400 million a year," or that New Yorkers purchase more than $23 billion in counterfeit goods each year. Are these numbers credible? Do the math. For counterfeit goods, this works out to be $8000 per household, a number ludicrous on the face of it. As are the others. The numbers all too often just don't add up.

Similarly, too many of us are ready to believe anything a celebrity tells us. So, for example, Prince Charles of England has no unease in promoting such cancer cures as coffee enemas or carrot juice, and many follow his princely bogus advice [1].

Poor information leads to bad decisions that take their toll in little or no health benefits, wasted tax dollars, and little to no environmental benefits. Reporting of science and medicine by the media needs overhauling, and scientific/medical journals are the place to start. Medical journals are the primary source of information about advances and modifications that can change the way physicians practice their craft. Scientific journals are a means of communication among researchers, keeping them current, and showing those in Germany, China, Israel, Australia, Spain, and the United States, the world over, the various ways investigators have dealt with the same or similar problem. The most important facet of any published research paper is its methodology. This is the potential pitfall of all publications. Ergo, few, if any media writers and commentators can penetrate this. For them the conclusion is all. In fact, journal articles were not, are not, meant to be reading material for the public. Dr. Judah Folkman, of Boston's Children's Hospital, was eminently correct when he said, "Mouse experiments do not belong on the front pages of the New York Times." He could have added, any newspaper's front pages. Premature hope is the last thing the public needs or wants. A good deal of the mistrust of science and scientists stems from the media's reporting of published reports that are only preliminary findings, but which have been hyped as serious breakthroughs, only to be contradicted a year later. A fed-up, disgruntled public blames the scientists, when the media should be targeted. Foul!

Dr. Drummond Rennie, Deputy Editor of the *Journal of the American Medical Association*, is clear about it: "There is still a massive amount of rubbish" in the journals. He was referring to the fact of peer review, the attempt by many, not all, journal editors to weed out shoddy work, methodological errors, biases, claims unsupported by the data, and outright fraud. The peer review system trends to set a barrier for authors to publish truly novel findings, but too often authors fail to discuss the limitations of their data. Premature announcements in the media by scientifically immature journalists (writers?) provide false reassurances that what is published in scientifically sound [2]. If that weren't enough, environmentalists have won the hearts and

minds of the media so that coverage of cancer causes is skewed toward manufactured environmental risks that cause few, if any, cases of cancer. In their
zeal to inform us, they have created anxiety, alarm, and mistrust, year after
year. Thus, over the past decade we have been treated to a diversity of potential causes of cancer in our communities such as dioxin, asbestos, coffee,
hotdogs, liquor, magnetic fields, and cellular phones, which, as we shall see,
have not withstood the test of epidemiologic studies, but there have been no
retractions of any kind, or discussions of failure to establish any of these as
causes of cancer.

Tentative reports become facts once the press gets a hold of an article—and
the many caveats laced into a publication are lost on their way to the press-
room or podium. As a matter of fact, the decades-long search for links between
environment and illness has yielded little. Where, then, should we turn?

Jean Piaget, the Swiss psychologist, offers guidance: "The great danger
today," he wrote, "is of slogans, collective opinions, ready-made trends of
thought." "We have to be able," he continued, "to resist, individually, to criti-
cize, to distinguish between works in progress, and completed pieces." That's
what journal articles are, works in progress, and that's the takeaway message.
No, it isn't the scientists who deserve our ire; although there is exasperation
aplenty to go around, blame for the current pandemic of mediagenic disease
must be heaped on a mindless communications media who foist works in
progress on an unsuspecting, gullible public.

Just how gullible are we? Fourteen-year-old Nathan Zohner, at the time a
freshman at Eagle Rock Junior High, won top prize at the Greater Idaho Falls
Science fair, where he elucidated the dangers of dihydrogen monoxide:

- It can cause excessive sweating and vomiting.
- It is a major component of acid rain.
- It causes severe burns when in gaseous form.
- Accidental inhalation can kill you.
- It has been found in tumors of terminal cancer patients.

At the Fair, Nathan asked 50 people whether they would support a ban
on this substance. The result: 43 favored a ban, 6 were unsure; 1 opposed a
ban, and why his opposition? Because dihydrogen monoxide is H_2O—water.
Zohaer's project was titled "How gullible are we?" His conclusion: "I'd say
they're extremely gullible. They need to pay more attention" [3].

I was smitten with this simple, illuminating project. But is gullibility the
essential problem, or is it shallowness, lack of scientific literacy? Lack of scientific literacy can be fixed. How can it be approached? Consider this. Over
the past several decades, declines in Europe's stork population have been
documented. During the same period, the European birth rate also declined.
For those who believe that storks deliver babies, this can be unsettling. For
nonbelievers, the decline of storks and babies requires further analysis, as do

the recent Volvo ads in newspapres and TV across the country carrying the headline:

Swedish Automaker Linked to Nations's Overpopulation.

There was an accompanying bar chart showing an increase in population from 1950 to 2000. The ad informs us, in a hard-news-like column, that "with people living longer than ever and the U.S. population continuing to climb at alarming rates, some experts are pointing to Swedish automaker, Volvo, and their obsession with safety as a root cause of this trend." They also state that Volvo officials are unapologetic about this trend and relationship. A wonderfully clever ad. Separating cause and effect, from spurious associations, is the domain of epidemiology. As we shall see, using a range of investigative instruments, epidemiologists attempt to relate cause with effect, but this can be a daunting business. Nevertheless, much can be gleaned, as we pursue this.

Circumstantial evidence—and it is circumstantial, indirect—indicates that cigarette smoke causes lung cancer, but not everyone who smokes gets lung cancer. Nevertheless, the risk increases with the number of cigarettes smoked: the dose—and, as the Swiss alchemist/physician Paracelsus informed us, the dose makes the poison. But the link between smoking and lung cancer is not supported by a direct, cause–effect relationship. How, then, is the smoking–lung cancer relationship buttressed? Retrospectively, for the most part, which means that epidemiologists gathered groups of people, those with lung cancer, the cases, and those without, the controls, matching the groups for age, sex, race, socioeconomic status, and educational levels, to ensure a high degree of comparability. Once the groups are established, they are interviewed and questioned to determine potential risk factors that could have caused the lung cancer. A major issue is smoking habits: numbers of cigarettes smoked per day, and years smoked. The preferred outcome would be for the case group to smoke a great deal, and the control group not to have smoked, or to have smoked little. The more cigarettes smoked, the greater the frequency of lung cancer cases to be expected.

Literally hundreds of these retrospective, case–control studies have been conducted around the world, and published. Other studies have examined lung tissue from cancer patients, comparing them to tissue sections from individuals free of lung cancer. The object is to determine differences, and if these differences can be related to smoking. Yet other studies subjected animal, human, and microbial cells to cigarette smoke in vitro to ascertain untoward effects. Dogs, trained to smoke, showed changes in lung tissue similar to those seen in human tissue. So, from thousands of studies conducted in dozens of countries, the conclusion was drawn, and a reasonable one it was, that smoking was an exceptionally strong risk factor for lung cancer, and that the risk increased with the number of cigarettes smoked. Additional strength for this association came from results of women smokers. Indeed, the evidence is strong, but the association remains circumstantial, and consequently offers an opportunity for anyone to question the relationship. Tobacco companies have made good use of this window of opportunity in newspaper and magazine ads implying that the smoking–lung cancer relationship is nonsense and certainly not established as a direct causal relationship.

These people are manufacturers of uncertainty—planting doubt, which plays well with a public lacking the ability to respond to protestations of doubt. Many smokers, to their chagrin and affliction, have bought into this specious argument. Was there, is there, no way to establish a direct cause–effect relationship? Yes and no. Mostly no. How would a study seeking to establish a direct relationship be conducted? I posed this question regularly to medical, nursing, and graduate students in my epidemiology classes. It may have seemed simple, but it wasn't. A high level of imagination and creativity were called for.

As they raised questions, and became more creative, they generally came up with a solution, but agreed that establishing a cause–effect relatioinship was impossible. Why impossible? If we are to demonstrate unequivocally that smoking is a cause of lung cancer (I say *a*, rather than *the*, because lung cancer can be induced by more than a single risk factor), we must be certain that our study participants inhale smoke, and only cigarette smoke; otherwise we are back to square one, trying to disentangle smoke from other risks. How do we assure such a circumstance? Again, comparing groups of people is essential. No surrogates, no animals, no retrospective interviews and faulty memories to contend with, and no inherent bias.

One of the groups must be smokers, another nonsmokers. Of course, we must begin our study with people who have never smoked, and they must be made to smoke. Each group must have sufficient numbers to assure reliable conclusions. How many? Do we know the number of people expected to get lung cancer annually? Data from past years suggest about 50 per 100,000 people. That can translate to 5 per 10,000. For us, 5 are too few; 50 is a reasonable number, especially if we develop a strong, tight study design. Undoubtedly, there would be objections from those claiming that 50 are too few, that 100 would be better. It's something to consider. At a minimum, our study will require enrolling 100,000 people who smoke. For the sake of cost and efficiency, I would be willing to go with 50,000 controls, the nonsmokers. So, we are now looking at a study with 150,000 participants. Actually, I would prefer another 50,000 in the smoking group because it would be nice to establish groups of smokers by number of cigarettes smoked per day: half a pack, one, two, and three packs. This would yield a dose–response curve that would add immense strength to the cause–effect relationship. Is 200,000 any more burdensome that 150,000? That, too, is worth pondering.

How will we choose our participants? This is vital. Volunteers? Researchers have learned to be wary of volunteers. They simply do not reflect the population as a whole. We need an appropriate cross section of the general population. We could assign numbers randomly, but would those selected participate? Now, consider this. Smoking-related lung cancer has a latency period of 15–40 years. This means that from the initial insult, the initiation of smoking, an interval of 15–40 years may elapse before a single case of lung cancer appears. Now, that is of considerable concern.

We must also have both male and female participants, roughly of equal numbers, and none of our enrollees can yet have had their first cigarette. What does that imply? Knowing as we do that smoking often begins at age 12, possibly 11, our "subjects" should be no older than 10, and guaranteed not to have smoked. Are the dimensions and difficulties of this study, as well as its complexity, beginning to take shape? Let's go on. We've only just begun.

We need at least 100,000 boys and girls ages 9–11 who have never smoked. We must assign them to a smoking or nonsmoking group. In this they have no choice. We must also be certain that the smokers actually smoke, and the nonsmokers do not. Now, how many cigarettes must the smokers smoke every day, and for how many years? Evidence gathered over the years suggests a minimum of 20 per day—one pack—for 5–10 years.

Before the study gets under way, we must establish our endpoint. What will our clinical criteria be for lung cancer? Criteria of a positive finding must be agreed on in advance by all physicians who will read chest X rays and diagnose clinical symptoms. This prior agreement will preclude bias and confounding. If this study is to yield unequivocal results, we must be certain that smokers smoke the cigarettes assigned, and that the nonsmokers do not inhale a single cigarette. Can we be certain of this? Another question: Where is this study to be undertaken? That's a key! Can each participant live at home, attend school, carry on a normal social life, and promise to smoke or not to smoke as is required by the test group? Certainly not. This study is too important to trust to anyone's word; too many temptations lurk about. What if someone forgot that he or she had a cigarette, or didn't feel like having one today? We must have assurances, but how can they be obtained? The only possible way to do so would be to lock up all the participants, and the only practical way to do this would be in small groups around the country, with assigned monitor-supervisors. This incarceration would have to continue well into each participant's adulthood; none could leave before age 50. By then, our 50 lung cancer cases should have appeared.

By now you've concluded, as my students did, that such a study is both morally and ethically reprehensible, and logistically and feasably impossible. It would never fly in a democratic society and, more than likely, not even be possible in any dictatorship in today's world. And rightly so. It's mad. Why? Because the information to be obtained, a direct cause–effect relationship that cigarette smoke causes lung cancer, is worth neither the effort nor the enormous expense to achieve it, given the exceptionally strong circumstantial evidence already in place.

But—and this is the mother of all "but"s—without this type of study, the questions related to radon in homes, leukemia and electromagnetic fields, and diet and heart disease and cancer, as well as dietary supplements and many other such issues, can never be finally and directly settled. A window of opportunity will be used, is used, frequently to advantage; it is the cigarette manufacturer's dream come true. It also keeps the pot boiling on a number of other illnesses said to be environmentally related.

Given the current prohibitions involved in human studies, ethics committees must approve a study's plans before one can begin. Many are denied. Consequently the absolute certainty that many desire cannot be obtained. It wasn't always this way. In the early years of the twentieth century, pellagra, with its four "D"s—dermatitis, diarrhea, dementia, and death—took 10,000 lives per year, primarily in our southern states, and in Europe. When Dr. Joseph Goldberger was dispatched to South Carolina and Georgia by the U.S. Public Health Service (the forerunner of our current Department of Health and Human Services) to determine the cause of pellagra, which all experts believed was microbial in origin, Goldberger set up a series of human experiments, including one at the Rankin Farm, a prison in South Carolina, in which healthy male convicts were asked to participate. As Goldberger believed pellagra to be due to a dietary deficiency, the volunteers at the farm were put on a diet that he believed to be pellagrous. He was right, lack of niacin, a B vitamin, produced dermatitis and diarrhea. That study could not possibly be done today. For its trenchancy, it is worth recounting another human experiment that Goldberger, and the world, profited from. In this study, the three princes of serendip intervened.

At one orphanage in Georgia, where pellagra was rampant, Goldberger ordered a change in diet, one that he hoped would cure pellagra. It worked. Over 7 months, the cases cleared and no new cases developed. Enter the three princes. With the study over, the orphanage administrator decided it was time to return the children to their traditional diets. Over the ensuing 9 months, 40% of the children developed at least two of the pellagra "D"s. When Goldberger learned of this, he ordered the diet changed back to the one he had initially prescribed. Fourteen months later the orphanage was again pellagra-free. Epidemiologists today would give their eyeteeth for a study showing *on/off/on switching*, which drives an association directly into causal territory. It was the occurrence of this on/off/on switching at the orphanage that gave Goldberger the idea of reversing the experiment and attempting to produce in healthy men, which, of course, he did at Rankin Farm [4].

Dr. John Snow, whom we met earlier, was another who was smiled on by good fortune, and knew what to do with it. In 1849, when cholera struck London, Snow realized that two companies supplied the city with potable water. One company drew its water well above London's filthy Thames River, which received the city's raw sewage. The other company took its water just beyond the city. By interviewing the residents and learning which company supplied their water, he was able to show, by calculating death rates, that the downstream company's water supply accounted for 5 times the cholera deaths than the company drawing its water upstream. In one area of London where both companies supplied water, the death rates were midway between those in which either company was the sole provider. Snow's data clearly showed that whatever it was that was causing cholera (well before bacteria were identified) was waterborne. This led to public health efforts to clean up the Thames. Here, the association between water supply and cholera was strong, but not directly causal. Cleaning up the Thames further strengthened the association.

In August 1854, when cholera struck again, Snow was ready. As cases developed, he spotted the location of each household on a map. As the days passed into weeks, the spots grew thicker. He then located the sources of water and marked the location of each community pump on his map. Spots were heavy around one pump, and extremely light around another. Snow found that a brewery near that pump had a deep well on its site where its workers, who lived in the area, got their water. But the brewery workers also slaked their thirst with beer. He realized that these men and their families were, in effect, protected from the community's cholera-laden water. Conversely, the heavily spotted area on the map fairly shouted out— "it's the Broad Street pump, in Golden Square!" As Snow tells it, "I had an interview with the Board of Guardians of St. James's parish, on the evening of Thursday, 7th September, and represented the above circumstances to them. In consequence of what I said, the handle of the pump was removed on the following day." This, of course, prevented anyone from fetching more water from this pump. Too many microbiology textbooks, and the media, continue to maintain that Snow removed the handle and stopped the epidemic in its tracks. This, as the Reverand Henry Whitehead, a contemporary of Snow's, informs us, was not the case. Listen to him: "It is commonly supposed" he says, "and sometimes asserted even at meetings of medical societies, that the Broad Street outbreak of cholera in 1854 was arrested in mid-career by the closing of the pump in that street. That this is a mistake is sufficiently shown by the following table [see Table 9.1], which, though incomplete proves that the outbreak had already reached its climax, and had been steadily on the decline for several days before the pump handle was removed" [5].

Here again, strong evidence of an association indicts the water supply, but it is not directly causal. Yet it would be difficult to deny the relationship, especially as 1849 and 1854 add their separate strengths to establish that water was the vehicle for transmitting the infecting agent. Over the years this association has gone from strength to strength as the bacterium was finally identified.

Earlier I noted that low-fat diets were believed to be protective against heart disease and cancer. Of course, this belief is everywhere. Our supermarkets are heavily laden with low-fat foods. An industry has grown up around it. Restaurants offer menus with low-fat foods to satisfy the demand, and the media daily promote the notion that foods can prevent chronic illness. Low-fat foods have become an icon of this decade. But does evidence support the association? Not a lot. Most rest on flimsy anecdotes. In February 2006, the *Journal of the American Medical Association* published a study that on the face of it puts the association to rest. This was not just another study. It was initiated and conducted by the Women's Health Initiative of the National

TABLE 9.1. The Broad Street Outbreak

Date	Number of Fatal Attacks	Deaths
August 31	34	4
September 1	142	72
2	128	127
3	62	76
4	55	71
5	26	45
6	28	40
7	22	34
8 (pump closed)	14	30
9	6	24
10	2	18
11	3	15
12	1	7
13	3	13
14	0	6
15	1	8
16	4	6
17	2	5
18	3	2
19	0	3

Source: Ref. 5.

Institutes of Health, who funded it to the tune of $415 million, and enrolled 48,835 postmenopausal women ages 50–79 who were followed for over 8 years. At the conclusion of this randomized, clinical trial, the gold standard of epidemiologic studies, it was determined that those women on a low-fat diet (19,541) had the same rates of breast cancer, colon cancer, heart disease, and stroke as did those 29,294 women who ate whatever they preferred. This study tackled a dicey issue, as the public had come to believe that if they ate healthy, they would stave off the current crop of chronic diseases. However, over the past decade, study after study failed to show benefits of eating fiber, vitamins, and even veggies. The fact that diet alone would not of itself be protective, never made a dent in the public consciousness. Interestingly enough, in this latest study, the women were not trying to lose weight, and their weights remained fairly constant (weight loss did not occur). Furthermore, the widely held belief that a high-carbohydrate, low-fat diet leads to weight gain, higher insulin, and blood glucose levels, and an increase in diabetes, even as calories are equal to those of higher-fat diets, was not borne out.

Although this study enrolled only women, the colon cancer and heart disease results appear to apply equally to men, as men and women respond similarly to dietary fat. While there was a small but statistically nonsignificant reduction in breast cancer among those on the low-fat diet, the women will continue to be followed to see if more time is needed for a more positive effect to emerge [6–8]. In today's climate, had major differences between the diet groups been observed in the course of the study, the study would have been stopped, and the women on the higher-fat diet would have been switched to a low-fat regimen. Also, it would, by the way, be prudent for proponents of the Mediterranean diet to recall that it has never been tested in a large-scale, randomized fashion. Again, we are asking and expecting more than can be delivered by food alone to ward off chronic illnesses.

With the publication of the startling—and, to many—shocking revelation that a low-fat diet failed to prevent the top three leading causes of death, the media folks took to the streets of Manhattan asking passersby for their reaction to that news. The consensus was confusion. "Every year or two we get conflicting information." I can sympathize with that, but there is a real problem here. In their confusion, the people blame scientists and physicians for not getting it right the first time, not realizing the complexity of the problem, and the difficulty in obtaining direct proof. The villains are not the scientists; the villains are the media who rush premature studies to the public before the ink is dry; on studies never meant for public dissemination—and the media prefer not to report on negative studies. We shall pick up on this shortly.

As previously noted, all published studies are not of elevated or equal quality. To help determine whether an association is no more than an increased numeric relationship between some condition and a risk factor, or as noted, obtrudes on causation, a set of criteria have been widely accepted in an effort to make inferences regarding causality more objective and less vulnerable to individual interpretation. These criteria were assembled by Sir Austin Bradford Hill, a British physician with a bent for analysis. In his quintessential publication, "The environment and disease: Association or causation," Hill laid out seven guidelines that ought to be considered in establishing a relationship between environmental exposure and effect—accepting an association as causal; Hill didn't see this as an all-or-none requirement. He remarked that "the whole chain may have to be unraveled or a few links may suffice. It will depend on the circumstance" [9]. Although flexibility underpins the criteria, as we shall see, they are nevertheless rigorous.

As epidemiologic studies seek to determine whether an association exists between a factor (exposure) and a condition (risk factor), and if so, the strength of that association, the strength rests on a number, a value, termed the *relative risk*, a measure of the incidence rates among those with and without the risk factor. This relative risk (RR) is a ratio of the incidence rates of those exposed to the incidence rates of the unexposed:

$$RR = \frac{IR - exposed}{IR - unexposed}$$

So, for example, the annual death rates per 100,000 people for heavy smokers is 166; for nonsmokers, 7 (RR = 166/7 = 23.7).

This is an exceptionally high RR, considering that an RR of 1 means that there is no difference between exposed and unexposed. So high a number, which is rarely obtained, all but demands an acknowledgment of cause. Most epidemiologic studies obtain RRs between 1 and 3; far too many between 1 and 2, become headlines. An RR of 1.38, for example, is heralded as 38 percent higher among the exposed, and therefore must surely indicate a risk factor at work. That 38% is judged to be a huge difference between the two groups. But is it real. It may be; but it may not be. As we shall see, more often than not, the difference is due to internal errors. Consequently many journals will no longer publish studies that have not attained an RR of at least 3, for reasons noted below. So, we fast-forward to Hill's criteria:

1. *Strength of Association.* A large RR increases the likelihood that an association is causal. If bias is minimal, and the RR is above 10 or 15, a direct cause–effect relationship virtually certain.

2. *Consistency.* If the association has been confirmed by other studies using different methods and different populations, the chance of causal association increases.

3. *Specificity.* If the association is specific to one risk factor and one disease, the chance of a causal association increases.

4. *Temporality.* Of primary importance is the fact that the exposure must have occurred before the disease. As most epidemiologic studies are retrospective, enrolling already exposed individuals who have a condition, causality can rarely be concluded unless the strength of association is 20 or more.

5. *Biological Gradient.* This may be the most difficult to satisfy. Establishment of a dose–response relationship would more than likely clinch it. The converse maintains that with the cessation of exposure, or reduction in dose, the condition abates. This would strengthen the association á la Goldberger.

6. *Plausibility.* The association should not fly in the face of current biologic knowledge. If, for example, heavy smokers had a lower lung cancer rate than did light smokers, cigarette smoke could no longer be a risk factor for lung cancer.

7. *Coherence.* The association fits with what is known about the natural history of disease. Lung cancer in men began occurring about 30 years after they started smoking. A similar pattern exists for women who also began smoking about 30 years after men, and haven't yet given it up [9].

The extent to which these criteria are met determines the degree of confidence in accepting an association as causal. These criteria seek to provide support where human studies are impossible, as well as prevent specious associations from gaining unwarranted attention as, for example, storks and low birth rates, and dietary supplements and positive health benefits. Journalists would do well to insinuate these criteria into their laptops, referring to them before going to press, or before remarking on a newly published study on the nightly news. Perhaps even more important, a scientifically literate public could prevent or reduce nonsense from becoming entrenched, by using these criteria to raise impolite questions.

Because bias can mask or exaggerate an association, and because it is here that many commentators fail their listeners, bias requires elaboration. My *Collins English Dictionary* defines *bias* as an irrational preference or prejudice. Synonyms include partiality, favoritism, unfairness, intolerance, and discrimination. My *Dictionary of Epidemiology* defines *bias* as any effect at any stage of investigation tending to produce results that depart systematically from the true values. The term *bias* does not carry an imputation of the experimenters desire for a particular outcome. Of course, our concern is with the latter, and given the fact that there are a variety of biases, which creep into epidemiologic studies, affecting their outcome, and more importantly perhaps, their affect on policy- and decisionmakers.

So bias is an internal error resulting in an incorrect estimate, over or under, of the strength of an association. The validity (correctness of measurement or labeling) of any study depends on the accuracy with which enrolled subjects are assigned to one of four possible categories: diseased, absent or present, and exposed, absent or present. This may seem simple enough, but many conditions do not show overt symptoms or signs so that under- or overdiagnosis can put the wrong people in wrong categories. While cancer of the lung is well defined, pelvic inflammatory disease is less so, and misclassification can be substantial. Additionally, if we wanted to know what percentage of the population has heart disease at any given time and we conducted a field survey by asking passersby if they had heart disease, we would obtain a number, but it would be seriously flawed because of the number of people with heart disease too ill to get up and walk the streets. We have a bias problem as the "field" sample selectively eliminates a number of people with heart disease. Results will be underestimated. Similarly, opinion sampling prior to the 1948 presidential election, with its reliance on telephone surveys, was biased, resulting in the prediction that Thomas Dewey would defeat Harry Truman. Phone subscribers in 1948 were far more likely to be Republican than Democratic in their political affiliation, and the researchers failed to detect and correct for this bias. Telephone surveys also favor those with phones and those who do not have to work and can be at home to answer the survey questions. Fewer responses will come from poorer families.

Perhaps even more distorting of data and data interpretation is the question of who gets into a study. *Selection bias* refers to ways that participants are accepted into and managed in a study.

Selection bias invalidated a grand experiment undertaken in Scotland in 1930. The Lanarkshire Milk Experiment concluded that school children given supplemental milk gained more weight than did those who did not receive the milk. Seventy-seven years ago, the effects of milk consumption were not at all known. In the experiment, 0.75 pint of raw milk was given daily to 5000 children. Another 5000 received an equal amount of pasteurized milk (a new idea in the 1930s). Both groups received grade A tuberculin-tested milk. Another 10,000 children were enrolled as controls, receiving no additional milk. All 20,000 were weighed and their heights were measured at the beginning and end of the study, which lasted 4 months.

The two groups that received the milk gained more weight and height did than the controls. But the study had begun in mid-February and ended in June. Children were weighed with their clothes on. It was later learned that sympathetic teachers had selectively placed the poorer children in the milk groups. The poorer children also had fewer winter clothes, so they weighed less in February. In June, the weight difference between the rich and poor children disappeared. The poorer children looked as if they had gained more weight than actually they had [10].

The epidemiologic landscape is afflicted with yet another source of bias. Publication bias can skew our full understanding of a subject if only positive data are published. Journal and media editors prefer positive to negative findings; they make for more dynamic headlines. Also, researchers often fail to submit negative or equivocal results, believing that they are uninteresting or that they will be rejected. The lack of negative finding affects the balance and consistency of research. As conclusions about risk are often based on the balance of published results, the lack of negative data can produce false impressions of health effects where none actually exist [11].

So, for example, the Naval Shipyard Workers study, "Health Effects of Low-Level Radiation in Shipyard Workers," a well-designed and well-conducted study with appropriate numbers of participants to obtain good statistical power, was a collaboration between the Department of Energy, which funded the $10 million study, and the Department of Epidemiology School of Public Health, Johns Hopkins University. Although the DOE required a final report from the investigating team, which it obtained, the decision to publish the study's methods and results in a peer-reviewed scientific/medical journal was left to Professor Genevieve Matanoski, the study's principal investigator. Since 1991, when the report was submitted to the Department of Energy, she has had little interest in getting the information, the data, to the scientific community or the public generally: in this instance, data showing that low levels of gamma radiation are not detrimental to human health. An unfortunate decision on her part [12].

Given the intensity of the backlash against nuclear energy, this finding would have been a major contribution to public understanding and scientific knowledge. Lack of information, positive or negative, can be a significant omission in the decisionmaking process. And taxpayers may well ask what the point is of giving away their tax dollars for studies from which they reap no benefit. It is also unfortunate that NIOSH did not pursue, as it had planned, a follow-up study to determine the continuing state of health of the shipyard workers, who were young men at the time of the study and found to be in good health. A follow-up study would have added even greater confidence about low levels of radiation if the results had continued to show ongoing good health. There is still time.

Yet another source of unreliability arises from information bias, the difference in the way data on exposure or outcome are obtained from cases and controls. So, again, bias results when members of the exposed and nonexposed groups report their exposures with a far different degree of completeness or accuracy. Mothers whose pregnancy results in an infant with a birth defect are more likely to recall medications and injuries, which can readily lead to false associations between a potential risk factor and a congenital anomaly.

Measurement and observer bias occur all too frequently, as in measuring height with shoes on, or weight with clothes on, or blood pressure taken with blood pressure cuffs that are too large or to too small, or when two different observers read the same X-ray film differently. Subjective perceptions can result in low reliability. Interview bias occurs when an interviewer asks questions differently of members of the case and control groups, eliciting responses of a different kind and degree. Self-reporting also creates additional problems, for example, people who underreport the number of alcoholic drinks consumed daily or weekly, or the participants who overreport the number of sexual encounters during a week or month. Self-reporting of weight, food consumption, and blood pressure can be badly underreported.

As if that were not trouble enough, studies are further undermined by confounding. *Confounding* is the confusing (mixing) of the effect of particular exposure on the studied disease with a third factor, the confounder. When a confounder is at work, what appears to be a causal relationship between an exposure and an illness is something else altogether—the confounder. The confounder is a variable associated with the exposure and independent of that exposure and is itself a risk factor for the disease. A study showing that families owning cappuccinomakers have healthier babies may be flawlessly executed and produce "statistically significant" results. No competent epidemiologist, however, would accept such a study as proof that the presence of a cappuccinomaker in the home improves infant health. Obviously, it is the more affluent families that can afford cappuccinomakers; affluence and health are directly related. In this instance, the socioeconomic status of the study participants has acted as a confounder. Confusing two supposedly causal variable contributes mightily to study error such that it deserves yet another example. Gray hair has been associated with the risk of myocardial infarction, or heart disease. Age, however, increases both the amount of gray hair and the risk of heart disease. Group differences in age can and do confuse or confound comparisons of two populations. Confounding can lead to either apparent differences between groups when they do not exist, or no differences when they do exist. As we have seen, studies can determine that consumption of vegetables rich in β-carotene and vitamin C, for example, protect against cancer. More veggies, less cancer. But is it due to the carotene, fiber, or something else in vegetables?

Or are the test subjects who eat more veggies younger, less likely to smoke, and more likely to exercise, and do they eat less fat? Eating vegetables and being cancer-free may readily be confounded by a number of other possible risk factors. The sources of bias and error are many and frequent. We shall also see why the media's reporting of only published conclusions, which are often given to overstatement and more certainty than the data permit or internal bias warrants, can be so misleading to a public looking to the media for assurance.

All published studies require confirmation, and until these occur, all results must be considered preliminary. That's another takeaway message. Indeed, bias is yet another required subject that the mass media must assimilate if they are to take responsibility for keeping the public reliably informed. Also, an alert public will be aware that bias can make an already weakly positive study even weaker. They will be wary of accepting anything placed before them; a dollop of skepticism is a healthy attribute.

So, we have criteria for distinguishing *associations* from *causal* phenomena, and we must be watchful of bias, confounding, and misclassification, as we gathered from the Texas sharp shooter. What, then, does this mean about the tens of thousands of articles in medical and science journals, and the thousands in magazines all over the place? Caution. Caution is the watchword.

As medical journals are the predominant source of information about new therapies and treatments, editors attempt to ensure the quality of articles their journals publish by requesting scientific peers to review manuscripts to help weed out shoddy work and methodological errors, to blunt possible biases and unsupported claims. Despite this system, errors and outright fraud do slip through, as was recently the case with a series of published papers in the journal *Science*, by Dr. Hwang Woo-Suk of the Seoul National University, Seoul, South Korea, who had fabricated evidence that he had cloned human cells. Fortunately, fraud is reasonably quickly caught and excised. Publication of biased or fraudulent articles provides false assurances that these articles are scientifically sound. There is wide agreement that far too many "weak" papers survive initial rejection only to find acceptance somewhere among the thousands of non-peer-reviewed journals. In fact, this would not be a problem if the mass media would reduce their addiction to journal articles. If the media must use journal articles, it is incumbent upon them to have science writers with a strong educational background in science, which is currently not the case. Without that, they will continue to be ill-equipped to make sense of, let alone interpret, published articles. Also bear in mind that all publications are not created equal, even though the media treats them as though they are.

But that does not alter the fact that there are such questionable claims as lemon juice cures AIDS, mobile phones can reduce male fertility, magnetic bandages make wounds heal faster, measles vaccine is responsible for autism, and genetically modified crops will create superweeds. Such claims do not emanate from peer-reviewed journals. Nor do claims of "wonder cures," or "new dangers." But they do appear in booklets and brochures like the one many of us received by mail that exhorted us to "Own the one medical book that may save your life," brought to us by Life Extension. Unpublished studies and viewpoints are of little benefit to anyone. Decisions about our health and safety, or our families' health, cannot be based on flawed work. Ergo, no matter how wonderful the work sounds, the other questions to ask is, whether the study has been peer-reviewed, and if not, why not?

If, as has been suspected, local TV news is the numero uno source of information for most people, and reporting of health news has increased significantly over the past decade, one could reasonably assume that most people would be health-literate. According to a recent report by a team of physicians from the Department of Emergency Medicine, University of Michigan, such an assumption would be premature. These physicians reviewed the tapes of 2795 broadcasts in which 1799 health stories were presented by 122 stations in the top 50 media markets, which regularly reach 165 million viewers. Although they maintain that "local TV news has the power to provide health information to most Americans," they found that the median story airtime was 33 seconds, and that, "few gave recommendations, cited specific data sources, or discussed prevalence. Egregious errors were identified that could harm viewers who relied on the information" [13]. They also suggest that studies are needed "to understand how the public comprehends and uses health information obtained from the mass media." In fact, the public is overfed with a diet of information they don't want, and starved for news they need.

Obviously, as has been noted throughtout this volume, the mass media's efforts require significant upgrading, in scientific/medical literacy. Current science and medicine are much too sophisticated for the uninitiated, which surely offers opportunities for university-trained science majors. Most assuredly, science journalism can be a fulfilling career choice. You're needed out there.

The public is faced with a plethora of opinions, editorials, commentaries, talking-heads, magazines, newspapers, TV, radio, private mailings, neighbor-to-neighbor tidbits, and ads urging the purchase of every imaginable nostrum to cure, prevent, diagnose, and treat whatever the ailment. Add to this a nationwide epidemic of "may," "alleged," and "possibly associated with," which a susceptible public translates as fact, and which their purveyors know and expect. To further confuse an already reeling public, issues are camouflaged as questions: "Can your water supply be the source of . . .?"—which gets them off the hook of an outright declaration, which, more than likely, is not the case.

Mental pollution is so widespread that it is not surprising to hear, "I don't know what or whom to believe." But it is lamentable, that as the public is encouraged to participate in community decisionmaking, far too many can bring only untutored opinion to the table. It doesn't have to be that way, but it does highlight a reality; the public and scientists hold widely differing opinions of the risks to health. What accounts for these differences, and can there be a meeting of the minds? I believe it is fair to say, that if polled, most scien-

TABLE 9.2. Comparing Everyday Risks: Activities that Increase
Chance of Death by 0.000001[a]

Activity	Chance of
Smoking 1.4 cigarettes	Cancer, heart disease
Spending 1 hour in a coal mine	Black lung disease
Traveling 6 minutes by canoe	Accident
Traveling 300 miles by car	Accident
Traveling 1000 miles by jet	Accident
Eating 40 tablespoons of peanut butter	Liver cancer caused by aflatoxin B
Living 150 years within 20 miles of a nuclear power plant	Cancer caused by radiation
Eating 100 charcoal-broiled steaks	Cancer caused by benzopyrene
Living 2 months with a cigarette smoker	Cancer, heart disease

[a] One in a million.

Source: Wilson, R., Analyzing the daily risks of life, *Technol. Rev.* **81**:45 (Feb. 1979).

tists would agree that for the public, the world is seen as far more hazardous than it actually is. The fact that we, the people, have never been healthier, seems lost on a worried, well public.

The dominant notion, which contrasts sharply with the views of scientists, is that we face greater risks today than in the past, and that future risks will be even greater. The well-documented fact is that we have progressed, advanced, left behind a past of widespread infectious disease, infant and maternal mortality and short life expectancy, for a better life for us, our children, and our grandchildren. But this is barely acknowledged or credited.

Scientists do concern themselves with numbers; the quantity of events— injuries, illnesses, deaths, attributed to specific risk factors, in their decision-making about safety or lack of it. Bear in mind that risk is a potential for harm and that determining risk requires the calculation of the probability or odds (chance) that harm will occur, as well as its frequency and magnitude. Table 9.2 describes a number of everyday risks that increase the chance of death by one in a million, while Table 9.3, estimates the number of days of life lost by yet another set of risks. Risk can be calculated for most activities and a value assigned. These estimates do reflect a degree of control that the country has gained over uncertainty, and these numbers are published to facilitate decisionmaking. However, an event may be estimated to increase the chances of illness or death by one in a million for this or that activity, but whether it can be said to be safe, is something else again.

In addition to an estimated value, determination of safety involves a personal judgment as to the acceptability of a specific risk. If a group of people decide that diving from an airplane and falling a 1000 feet before pulling their parachute's ripcord is safe, then it is safe, no matter what number has been estimated for risk of death from such an activity. Scientists cannot make a judgment of safety for anyone. They can only provide a risk estimate.

TABLE 9.3. Estimated Loss of Life Expectancy from Health Risks

Health Risk	Average Days Lost
Unmarried male	3500 (9.6 year)
Smoking 20 cigarettes/day	2370 (6.5 year)
Unmarried female	1600 (4.4 year)
Overweight (by 20%)	985 (2.7 year)
All accidents combined	435 (1.2 year)
Auto accidents	200
Alcohol consumption (U.S. average)	130
Home accidents	95
Drowning	41
1000 mrem per year for 30 years	30
Natural background radiation	8
Medical diagnostic X rays	6
All catastrophes (earthquake, etc.)	3.5
1000 mrem occupational	1

This understanding of safety is far different from my dictionary's definition, which states that "safety is freedom from risk or danger," a totally misleading concept, as there is no such thing as freedom from risk. All things in life have risk. Zero risk is unrealistic, unattainable, and unimaginable. For the public, numbers may be meaningless, as for example, 40,000 documented deaths a year form motor vehicles (I hesitate to say "accidents," an event without apparent cause, which surely does not apply to 95% of these events), appear to be acceptable, whereas a single potential death from a nuclear energy plant is unacceptable. The public's view maintains that events over which they have no control present unacceptable risks, as though they have control of motor vehicle injuries and deaths. But that is the perception. Similarly, scientists' warnings to use seatbelts while driving also appear to fall on deaf ears, as does the admonition that women stop smoking; and that waste from nuclear energy plants is not as dangerous as environmentalists and the media would have us believe. Far too many people avoid preventive vaccines, which physicians and scientists wish they would not. Yes, scientists are concerned where the public isn't, and the public is overly worried about events of little or no concern to scientists. "Safe" may well be in "the eye of the beholder," and it may have little to do with the "facts on the ground." Even moving to a middle ground would be an improvement. But that would require trust, which is currently in short supply, which raises the question, What have scientists done to warrant public mistrust? We shall try to penetrate that.

The "accident" at Three Mile Island produced no deaths, no illnesses, no attributable cancer cases. TMI was a media event of monumental proportions. It demonstrated the overarching importance and effectiveness of the media's ability to establish nationwide perceptions. Mediagenic disease was created there. Three decades down the road, TMI remains one of the salient environmental noncatastrophes of all time. But that message has yet to gain traction and wide dissemination. Why hasn't it? How would the media explain their earlier excess? It would make a nifty story, especially in color, with a lot of red faces. Who was listening to the scientists?

Furthermore, the amount of print journalism devoted to causes of death from illicit drug use, airplane crashes, motor vehicle "accidents," "toxic" wastes, and homicide, for example, is staggeringly disproportionate to the actual number of deaths attributable to these causes. The information provided may be true, but only partially so. Consequently, the public has a skewed understanding of which risks are actually most harmful and which, the least. Ergo, the media can be credited with sending the public off in a direction far afield of the prevailing scientific consensus. But scientists, too, must be held accountable. Scientists need to drop the drawbridge, come out of the citadel, and communicate with the people in their communities, meeting with them, explaining why scientists see things differently; explaining why there are no absolutes or final answers; why yes or no, black or white are alien to science, and why shades of gray, and *perhaps* and *maybe* are preferable—and they must do this in jargon-free, accessible language.

It will be difficult for the public to substitute uncertainty for the comfort of certainty, but if trust is to be resuscitated, they and the scientists must reason together. Education and communication, two sides of the same coin, can be a means of building trust. Without trust there will be no progress and no resolution. Where or how trust was lost remains elusive. Be that as it may, both must realize that to receive appropriate attention, and affect public policy, consistency between them is needed.

Interestingly enough, a recent book by two academics, published by Yale University Press, maintains that environmentalists have won the hearts and minds of journalists and their media. Coverage of cancer causes is skewed sharply toward environmental risk factors that cause few, if any, cases of human cancer. They compared two studies in which oncologists were asked to rate a diversity of environmental factors and chemicals as possible causes of cancer. In both surveys the oncologists ranked tobacco, diet, asbestos, and sunlight at the top of the risks. Aflatoxin rose from eleventh place to sixth, while pollution fell from sixth to tenth. By contrast, cancer causes in the media placed chemicals in first place, and food additives, pollution, anthropogenic radiation, and pesticides were among the top 10—all ranked low by scientists. Over and over, media rankings, which become the public's rankings, have been shown to be flawed and misleading, while the scientists lists have stood the test of time. Who, then, is misleading the public, and why does the public distrust scientists [14]?

Among scientists, self-described liberals outnumbered conservatives by 3 to 1, and Democrats outnumbered Republicans by 4 to 1, which suggests that as a group scientists' political leanings would not be expected to harbor biases that would downplay environmental cancer risks. An objective observer could be forgiven for expecting the public to place their trust in such individuals rather than environmentalists and their media handmaidens. That they haven't is surprising, and further deepens the lack-of-trust conundrum, because it is difficult to discern where and when scientists have misled the public. Scientists can be wrong, but can anyone offer an example in which public opinion was better than expert opinion? It would appear that the public does not trust expert opinion, being more comfortable with opinions of activist environmentalists, who have misled them for decades, which brings us to yet another concern.

Three problems afflict us: (1) lack of accurate, trustworthy information; (2) conflicting information; and (3) continuous misinformation. Not knowing whom or what to believe is, therefore, understandable. Let us pursue this, as there is a critical issue abroad in the land that requires comprehension and informed decisionmaking.

Whatever their discipline or area of interest, scientists seek verifiable truths of the natural world, and their objectives, methods, and results are open and public, subject to inspection, evaluation, and criticism by others. Their collected data are obtained by observation, testing, and experimentation. It is the rare or odd subject that is ever formally closed, ever completed. New knowledge is expected to replace "old." Knowledge begets knowledge. Researchers expect and anticipate that their work will be scrutinized by other scientists, and no scientist, however great his or her celebrity, is above questioning. Whatever their field of study, from aardvark to zyzomys, it is certain to be researched by a plethora of investigators from Copenhagen to Calcutta to Chicago, employing procedures of their own devising and with unique populations, all of them seeking verifiable data. Faulty work eventually gets demolished by the work of others, while work well done gains adherents and strength, as new evidence supports it. To all this, there is but a single caveat—investigations are limited to objects and events of the natural world. That's fundamental and essential—scientists are searchers for truth about the natural world, as opposed to the spiritual world. However, there are other truths: poetic and religious. The truths of religion are more like the truths of poetry than the truths of science. Truth is a general term, ranging in meaning from an exceptional idea to an indication of conformity with fact and avoidance of error, misrepresentation, or falsehood. Poetic truth is an expression of the emotions of all aspects of our lives: natural and spiritual. Religious truth deals with the divine or supernatural. Although there is more than one truth, they need not be conflicting, nor need one supplant or supersede the other; we live by, and with, more than a single truth. Nevertheless, juxtaposed to scientific truth, which seeks only to relieve nature of her "secrets," religious truth emanates from revelation and must be taken on faith and is unverifiable. For scientists, the world is tentative, provisional, and in shades of gray. Proof arrives via testing, experiment, maximized for objectivity by design. The work thrives on controversy, but advances by consensus. Religious truth is by nature absolute, black or white; either it is or it isn't. It is personal, emotional, preconceived, and subjective.

Faith is the only requirement for the acceptance of "proof." Controversy is intolerable, as Copernicus and Galileo painfully knew. St. Augustine captured and framed its essence 1600 years ago. "Nothing is to be accepted," he declared, "save on the authority of scripture, since that authority is greater than all the power of the human mind." An absolute, sweeping assertion.

But it didn't end with Augustine of Hippo (354–430 CE), otherwise known as Aurelius Augustinus, who would be surprised and elated if he knew that his dictum is alive and well in twenty-first-century America, to the chagrin of many in our public school system, where the authority of scripture is again threatening the teaching of science. Creationists are bent on persuading elected officials, the courts, and local schoolboards that evolution is a baseless, flawed notion, and that the true story of creation of the universe is to be found in the bible. Consequently we begin our pursuit of this public conflict in the courtroom, with *Kitzmiller v. Dover*.

CREATIONISM AND INTELLIGENT DESIGN

Tammy Kitzmiller, the named plaintiff in the case, joined by 11 other residents of Dover, PA (a pocket-sized, rural community directly south of Harrisburg, the state capital), whose children are enrolled in the Dover schools, sued the Dover School Board, who had voted to change the high-school science curriculum by requiring "equal time" for "creation science," "intelligent design," to be taught side-by-side with evolution in science classes.

The intelligent design offshoot of creationism (creation science) maintains that lifeforms are too complex to have been formed by natural processes, and must have been fashioned by a higher intelligence—a supernatural being, God. They argue that the intricate workings of our body's cells, and such organs as the eye, are so complex that they could not have developed gradually via the force of Darwinian natural selection acting on genetic variations. Thus, the Dover School Board Administrators read a statement—the teachers refused to do it—to the ninthgrade biology classes asserting that evolution was a theory, not a fact; that it had gaps for which there was no evidence, that intelligent design was a differing explanation of the origin of life, and that a book on intelligent design was available for interested students, who were, of course, encouraged to keep an open mind.

The 6-week trial ended in mid-December 2005. The court's decision is worth repeating, and following that, we shall deal with Darwin, evolution and creationism.

Federal District Judge John Jones, III, a Republican, appointed by President George W. Bush, declared the Board's curriculum changes unconstitutional and told the school district to abandon a policy of such "breathless inanity." "In making this determination," Judge Jones wrote, "we have addressed the seminal question of whether ID [intelligent design] is science. We have concluded that it is not, and moreover, that ID cannot uncouple itself from its creationist, and thus religious antecedents." He continued, "Both defendants and many of the leading proponents of ID make a bedrock assumption which is utterly false. Their presupposition is that evolutionary theory is antithetical to a belief in the existence of a supreme being and to religion in general. Repeatedly in this trial, plaintiffs' scientific experts testified that the theory of evolution represents good science, is overwhelmingly accepted by the scientific community, and that it in no way conflicts with, nor does it deny the existence of a divine creator." Furthermore, "To be sure, Darwin's theory of evolution is imperfect. However, the fact that a scientific theory cannot yet render an explanation on every point should not be used as a pretext to thrust an untestable, alternative hypothesis grounded in religion into the science classroom or to misrepresent well established scientific propositions." But Judge Jones was not done: "The citizens of the Dover area," he wrote, "were poorly served by the members of the Board who voted for the ID policy. It is ironic that several of these individuals who so staunchly and proudly touted their religious convictions in public, would time and again lie to cover their tracks and disguise the real purpose behind the ID policy." He concluded, saying [15]:

> Those who disagree with our holding will likely mark it as the product of an activist Judge. If so, they will have erred as this court is manifestly not an activist court. Rather, this case came to us as the result of the activism of an ill-informed faction on a school board aided by a national public interest law firm eager to find a constitutional test case on ID, who in combination drove the Board to adopt an imprudent and ultimately unconstitutional policy. The breathtaking inanity of the Board's decision is evident when considered against the factual backdrop which has now been fully revealed. The students, parents and teachers of the Dover area School District deserved better than to be dragged into this legal maelstrom, with its resulting utter waste of monetary and personal resources.

As it should be, the townspeople and students on both sides of this unfortunate issue are trying not to let the decision disrupt their friendships. Nevertheless, tiny Dover is only one of thousands of towns and cities that will consider inserting creationism/ID into their schools, as the idea of humans as only another step in the forward march of evolution, breaches and defies their staunch belief in a special creation. In February 2006, shortly after Dover, The State of Ohio Board of Education voted 11 to 4 to "toss out a mandate that 10th grade biology classes include critical analysis of evolution and an accompanying model lesson plan dealing with the ID movement, its second serious defeat in two months" [16]. The Ohio lesson plan did not mention ID, but critics contend that the critical analysis language is simply design in disguise. Creationism and ID are not going to go away; too much is at stake for them.

In different guises they will try to slip religion and God in science classes in public schools.

"Children should not be taught that we came from monkeys." Indeed, they should not. So, we begin with the question, did humans evolve directly from monkeys? The answer is a straightforword, no! Why not? We are not descended from worms, fish, or monkeys, but we shared common ancestors back in time. Way back in time. We shared a common ancestor more recently with primates other than with dogs or other mammals. To dispel the notion of direct descent from apes, we turn to the major issues presented by Charles Robert Darwin in his *On the Origin of Species by Means of Natural Selection; or the Preservation of Favored Races in the Struggle for Life* (1859). Bear in mind that before Darwin, we humans were not considered part of the natural world. We, it was believed, were a special creation.

With the publication of the *Origin of Species*, theologians were aghast at two consequences of Darwin's work. The first was that man and apes must have had a common ancestor, which dislodged man from their privileged position as created by God in his own image. The second, if natural selection were true, was that much of the argument for the existence of God based on the presence of design in nature was destroyed. As H. L. Mencken so aptly put it, "The theory of evolution aroused concern because it was disgorged in one stupendous and appalling dose," and as with the snake that swallowed the elephant, it has taken a long time to digest. For many people, digestion has not been easy or completed, and many totally reject evolution as ungodly. This is unfortunate as understanding evolution, as we shall see, does not require such rejection.

Does the fact that we humans have a passing resemblance to chimpanzees, orangutans, and gorillas disturb and unsettle people? Perhaps an even greater cause for agita is the now-revealed fact that our genomes, ours and apes', are 98.4% similar. For too many people the tenets of evolution are seen as a direct challenge to their faith. Many fundamentalist Christians and ultraorthodox Jews are vexed by the thought that human descent from earlier primates contradicts the *Book of Genesis*. Islamist creationists consider the 6-day creation story as literal truth and believe that the theory of evolution is nothing but deception. According to a Gallup Poll done by telephone interviews in February 2001, 45% of responding adults agreed that "God created human beings pretty much in their present form within the last 10,000 years or so." Evolution, they believe, played no role in shaping us. Nevertheless, most mainline Protestant denominations and the Catholic church accept the teaching of evolution. For most Christians and Jews evolution is not inconsistent with their religious beliefs.

Darwin never said that human beings were descended from apes, but that humans and apes had common ancestors long eons ago. That major distinction seems always to get lost. What Darwin actually said was that the great diversity of creatures—species—on earth was the result of repeated branching from common ancestors—plural, *ancestors*; that's fundamental and essential. The process was called *evolution*, and functioned by means of natural selection.

Most scientists agree that evolution is a statement about the history of all species and their cumulative change through time [17]. Evolution is not about progress. It is about adaptation. Herbert Spencer's coining of the notion of "survival of the fittest" (although inappropriate, it stuck) stresses survival, when the important point is the greater fecundity of the better adapted. Those better adapted leave more offspring. And, the characteristics of living organisms change over thousands of generations as traits are passed from one generation to the next.

There were two questions that Darwin could not answer in 1859, as the state of biological and genetic knowledge was scanty, but which modern biology has answered: What is the nature of hereditary transmission between parents and offspring, and what is the nature of the origin of heritable variation? Without heritable variation, natural selection could achieve nothing. The Czech monk, Johannes Gregor Mendel and his extensive and detailed studies of pea plants and their genes, held the key, and could have been known by Darwin had he read Mendel's manuscript, which lay on his desk unopened.

A central component of evolutionary theory is that all living organisms, from bacteria to flowering plants, to insects, birds, and mammals, share a common ancestor. The animal most closely related to humans is the chimpanzee. The common ancestor of both, the chimp and we humans, is believed to have lived some 6–7 million years ago. A common ancestor to humans and reptiles is believed to have lived some 300 million years ago. One of Darwin's most revolutionary ideas was that all living things are related, connected to one another as branches on a giant "tree of life." At the root of the tree is a single-celled organism, and all living things are descendants of that ancestor—dinosaurs, cats, apples, horses, alligators, and we humans. Close relatives don't always look alike, but will share more similar traits with one another, than with remotely related creatures. So, for example, rodents—rats, mice, chipmunks, gerbils, lemmings, squirrels, porcupines, and beavers—share a number of similarities that they do not share with cats or dogs. But dogs and cats, rats, elephants, tigers, *E. coli*, and the mold penicillium, all share DNA and RNA in common, which is the primary attribute of all lifeforms. In this regard, we are all linked, and this alone attests to the fact that we share a common ancestor. However, this does not "prove" evolution, because absolute proof is unattainable, but after two centuries of searching and studying the many different lifeforms, the patterns of similarity appear to be immutable, which provides the confidence that life has evolved.

Furthermore, simpler forms of life preceded more complex forms. Bacteria, then, had to be the first development some 3.5 billion years ago, followed by more complex cells about 2 billion years ago. Then a great emergence of complex lifeforms appeared—the sponges, corals, and others, followed by arthropods and mollusks, the invertebrates. Then the vertebrates appeared—

fish, followed by amphibians, followed by reptiles (saurians), then birds and mammals, and finally humans.

If evolution occurred, we would expect to find that species that lived in the remote past would be different from species alive today. Indeed, that is what is seen. In addition, only the simplest of animals would be found in the oldest geologic strata, and more complex organisms in more recent strata. Again, that is what is found. If evolution occurred, there should be connecting forms between major groups. In geologic strata this is true for vertebrates, with bony skeletons, and those earlier forms with hard shells, which left impressions in rock and shale; but not the soft marine forms that do not preserve well. If evolution occurred, we would not expect to find branched-off lineages out of place. No one has ever found mammal bones among the fish of the Devonian period. The consistency of the fossil record is one of the strongest pieces of evidence supporting evolution. Also, new features appearing in a lineage would be passed along in the same or modified form to all its descendents. But, and this is a preeminent "but," this new feature would not be present in any of the lineages that had diverged before the appearance of the new trait.

At the recent Darwin exhibit at the American Museum of Natural History in New York, the visual impact of a display of skeletal bones of five different species—a human arm, a chimp's arm, a bat's wing bones; a Goliath frog's foreleg bones, and a Harpy Eagle's leg bones was startling, so strikingly similar that it jumped up at you. The bones may be longer, shorter, thicker, or thinner, but are of the same pattern and number, for the needs of each of their unique environments. The commonality could not be mistaken. Seeing them this way sans skin, feathers, or fur tells the evident story that written words or sketches just do not convey. The continuity and further continuity are remarkable. How could there not be a common ancestor? It is essential to visit a museum with a wide collection of skeletons. When seen this way, explanations are unnecessary.

Then, of course, there are the embryos. Here again, they are strikingly similar; human, horse, mouse, and bat can barely be differentiated during their embryonic development, but are so obviously different once born. Coincidence can play no part here. Different, yes, but some snakes continue to have rudimentary leg bones; flightless birds are still with us, and a number of beetles have wings that never open. Additional evidence of evolution.

Natural selection is the way evolution proceeded, and we see it around us daily. For the skeptic who asks, can we see evolution in action, the answer is a resounding "Yes." Natural selection is a simple, straightforward concept. We see it in the great variety of dogs, cats, horses, and birds. So, why not include us humans who vary by height, strength, swiftness, visual acuity, shape of eyes, type of hair, intelligence, susceptibility to disease, and length of days?

A fact of life that was clear to Darwin was that most animal species produce more offspring than can survive. Only a few live long enough to mate. The competition for food, reproductive opportunities, and space predicts that there will be a struggle for survival. Unfortunately, Herbert Spencer (1820–1903), a British engineer and supporter of evolution, gave evolution a bad name by coining, as noted earlier, the notion of "survival of the fittest," which morphed into social Darwinism. In fact, the "struggle for survival" referred to reproductive opportunities, in which those that survive and reproduce will have their DNA passed on to future generations, which will then have an adaptive edge in the survival game. Isn't that what cloning seeks to achieve?

Common and current examples of natural selection include antibiotic resistance in bacteria, pesticide resistance in insects, and the attempt to rid Australia of its plague of rabbits.

We are all aware of the serious problem of widespread antibiotic resistance. Hospitalized patients and their physicians know that penicillin, methicillin, and vancomycin have lost their wondrous bactericidal properties as bacteria once rapidly wiped out by these antibiotics, no longer succumb to them. What's happened here? The initial bacterial susceptibility has given way to a new generation of resistant organisms. Although penicillin initially destroyed 99.9%, the small percentage that remained developed a gene for penicillinase, an enzyme that prevents the destructive action of penicillin. These new variants increased over time until penicillin had little effect, and a modified penicillin, methicillin, was brought to bear. But in time methicillin became less effective and vancomycin became the drug of choice until the bacteria developed resistance to it. There are few better examples of evolution in action than this.

In 1950 the *Myxoma* virus was introduced into Australia from England, to kill off the pesky rabbit population. For some 3 years it worked. The rabbit population took a precipitous dive. But by the fifth year, the rabbits began rebounding, as they developed antibodies against the virus. Today, Australia's rabbits are immune to the *Mxyoma* virus as the rabbits evolved immune variants.

Another interesting example is the newt–snake relationship. Newts contain a toxin in their skin that affords them protection from predator snakes that seek them out for breakfast or lunch. As snakes evolved ability to neutralize toxin, and newt populations waned, new newt generations evolved even more potent toxin and snakes avoided them, until a new generation of snakes could manage the stronger toxin, and around and around, it goes.

This adaptation to provide survival benefits is seen in the anteater's long, thin snout; the giraffe's long neck; the great variety of bird beaks (for those on a diet of grubs, insects, seeds, or leaves); the armor plating of the armadillo; and the shells of turtles. Any animal with a competitive edge, drought tolerance, and/or a thicker than average coat, could survive longer and leave more offspring than could other members of their species. These adaptations controlled which individuals would represent the species in the future, as they will be the ones to pass on their DNA. This, then, is natural selection at work.

We now confront the complexity problem that so perturbs creationists, and from which intelligent design springs.

Many arguments for the existence of God have been advanced since ancient times. The most cogent, however, was the argument from design. In his book, *Natural Theology, or Evidence of the Existence and Attributes of the Diety Collected from the Appearances of Nature* (1802), "is the best known exposition of the argument from design," the Reverend William Paley asked his readers to consider walking across a heath and accidentally striking a watch with their feet. How did the watch get there, and who could have made so complex a device? Only a watchmaker. Similarly, Paley considers the many complex organisms in nature, and wonders who could have made all these complex living and nonliving things, and his answer is a supernatural being capable of designing infinite complexity—an intelligent designer. Paley makes his point ever so clear using the human eye as an organ of such complexity, made for vision, just as a telescope was made for assisting vision, and clearly had a designer, so, too, the eye had to have a designer.

Appealingly enough, when Darwin was a young student at Cambridge University, studying for the clergy, he read Paley's book and was taken with the idea of God the designer. However, after his 8-year voyage around the world on the *Beagle* (1831–1839) collecting specimens and observing the remarkable diversity of animals and plants, he could no longer believe in a supernatural designer. For him, a designer was no longer needed; natural selection could have produced the human eye. Further, as Richard Dawkins so aptly titled his book, *The Blind Watchmaker*, for him the blind watchmaker was in fact natural selection: "Blind because it doesn't need to see ahead, does not plan consequences, has no purpose in view" [18].

For scientists with appropriate backgrounds in comparative and evolutionary biology, it is not so difficult to see how so complex an organ as the eye could have evolved. Currently the pattern is reasonably clear, as a diversity of scientists have observed the many stages in the evolution of complex eyes in different kinds of animals. Some single-celled animals, for example, have tiny light-sensitive spots on their bodies, which provide the ability to move toward or away from sources of light. Flatworms and shellfish have light-sensitive cells arranged in a cuplike eye spot. In the *Nautilus*, a genus of cephalopod mollusks, the light sensitive cup is deeper, and the opening allowing light into it is narrower, forming a primitive kind of lensless "eye," akin to that of a pinhole camera.

Once this kind of "eye" has evolved, the evolutionary steps that lead to eyes like our own can be discerned. The light-sensitive spot is covered with a transparent or translucent layer of skin. The eye cavity then fills with cellular fluid rather than air or water. Part of the region containing the fluid evolves into a lens. Finally, full, complex eyes evolve. Each of these stages as well as intermediate ones are found in mollusks, while the octopus and squid have highly developed eyes similar to those of humans [19]. Richard Morris, in his worthy and revealing book, *The Evolutionists*, informs us also that it has been estimated that the complete evolution of a complex adaptation like a fully functioning eye requires about 2000 steps over something of the order of 4000 generations. In an evolutionary line with a generation time of one year, a complex eye could evolve in less than half-a-million years. In this case, the eye could have evolved more than once. The eyes of the octopus and squid evolved independently of our own [19]. In yet another adaptation, the eyes of a flounder sit on each side of its head during larval development, but as the fish grows and flattens, the eyes wander until both become fixed on the same side of the head. Now, that's an adaptation! Take a good and close look at a flounder in the fish department, the next time you're in your favorite supermarket, and ask yourself, "Why?" The Reverend Paley had an elegant idea, but a blind watchmaker is at least equally elegant. However, lacking the requisite background in comparative and evolutionary biology, it would be inordinately difficult, nay impossible, for nonprofessionals not to consider the intervention of an intelligent designer, which is why it is immensely important that this type of information become accessible and widely disseminated; an enterprise the mass media could well perform. With the light of comprehension, there is little need to invoke supernatural intervention.

Intelligent design and creation "science," as currently advanced, need to be understood for what they are. Succinctly stated, creationism is the belief that the Christian–Judaic *Bible* is literally true. It requires that the earth and all its inhabitants, living and non-living, were created some 6000–10,000 years ago, in six 24-hour days, and that humans and dinosaurs walked the earth together, among other beastly creatures. However, creationists cannot muster any valid evidence supporting or substantiating these beliefs. They must be taken on faith alone.

For creationists, evolution is synonymous with godliness; consequently evolution must be denounced as scientific atheism, and consequently geologic, archeologic, and paleontologic evidence have no place at their table. Echoing St. Augustine, Henry Morris, recently retired Director of the Institute for Creation Research, a division of the *Bible*-based Christian Heritage College (now called San Diego Christian College), in Santee, California, made it abundantly clear that "If man wishes to know anything about creation (the time of creation, the duration of creation, the order of creation, the methods of creation, or anything else), his sole source of true information is that of divine revelation. God was there when it happened, we were not. Therefore, we are completely limited to what God has seen fit to tell us and his information is in this written word. This is our text book on the science of creation" [20]. Greater clarity hath no person. Yet in the same breath creationists have demanded that school boards revise their curricula to provide "equal time" for "creation science," to be taught alongside evolutionary biology.

Consider that without a nickel's worth of testing, experimentation, evidence, or proof of any kind, they would like students to accept the *Genesis* story as fact, and that God created the universe essentially as we see it today; that little has changed since the creation event. Creationist arguments are not only bereft of proof but are also negative, seeking only to discredit scientific evidence. Unfortunately for creationism, Henry M, Morris, age 87, its founder and mentor, died of a stroke late in February 2006. Dr. Morris, a hydrolic engineer, graduated from the University of Minnesota. It wasn't until 1961 that he began his criticism of evolution with the publication of his first book, *The Genesis Flood*, which has sold over 250,000 copies, but which is considered of no scientific merit [21].

Intelligent design is the newest conception of creationism, and follows from Reverend Paley's argument from design. For them, natural selection is theologically offensive as it either removes God from creation, or makes God an unnecessary bystander. They also contend "that intelligence is required to produce irreducibly complex cellular structures—structures that couldn't function if one part were removed, because such structures could not be produced by the incremental additions of natural selection" [17].

For this contention, the example of a mouse trap consisting of five parts—platform, spring, catch, hammer, and hold bar—is promoted as an example of irreducible complexity, and suggests that if any one of these five parts is removed, the trap can no longer catch mice. It has, of course, been shown that mouse traps of one, two, three, or four parts can trap mice. Intelligent designers make the mistake of assuming that natural selection works by adding components singly, one by one, to finally make a complex structure. Also, because there are gaps in the theory of evolution, it is not necessary to invoke God to fill in gaps. As scientists show time and again, bit by bit their efforts fill in the gaps, and will continue to do so. In April 2006, a team of scientists led by Dr. Neil H. Shubin of the University of Chicago published their findings of well-preserved skeletons of fossil fish in former streambeds in the Canadian arctic, 600 miles from the North Pole. These skeletons, from the late Devonian period—385–356 million years ago—are now named *Tiktaalik roseae*, which in the Nanavut language means "large shallow water fish," are (missing) links between fish and land animals, just as the Upper Jurassic archaeopteryx bridged the gap between reptiles and today's birds.

Tiktaalik, up to 9 feet long, is a fish in transition, having fins, scales, and other attributes of giant fish, but in the fish's forward fins there is evidence of limbs. There are the beginnings of digits, protowrists, protoelbows and protoshoulders. This fish, with its long, flat head, sharp teeth, and unfishlike ability to swivel its head, resembles nothing so much as a crocodile. Still to come are links between Tiktaalik and the earliest land animals—tetrapods. Unearthing Tiktaalik suggests that may not be far off. Let us be clear, this was not a lucky find. Luck played no part here. The Canadian Arctic site was predicted well in advance, and the team knew that the Devonian period strata would hold the skeletal remains. Paleontologists know what they are looking for and where to look. It's elegant science. For creationists who have long maintained that there are no transitional forms, this discovery will not go down easily [22].

Of course, this continues to diminish the need for God. In fact, creation science is a contradiction in terms. It is faith-based, having nothing in common with science, which seeks to explain the universe in terms of observed and testable natural mechanisms. Belief obtained by revelation is unverifiable, is untestable, and must be accepted on faith. How can the existence of an "intelligent designer" (IDer) be proved or disproved? Not possible. It must be accepted on faith. And the IDers provide few if any answers to evolutionary questions. Rather, they belittle evolutionary explanations as far-fetched. But the methods of science do reduce ignorance and works so well that it is the method of choice on every continent, and in every nation and culture. The business of science is to understand the natural world, and in doing so, will unfortunately, have consequences for traditional religious beliefs. Allowing God to be invoked as an explanation most assuredly discourages the search for other more systematic causes. If Ben Franklin and his contemporaries had been content to regard static electricity only as an expression of divine whim, we would be unlikely to have the science of electromagnetism today. The belief that there are no connections between branches of any living organisms is unlikely to ever discover connections that do exist. Perhaps the most disturbing and disagreeable side of supernatural creationism is that it stifles curiosity and blunts intellect. Is there any better example of this than in parts of the Muslim world today, in which thinking is fixed in the twelfth century with inquiry strictly limited to their bible, *The Koran*, having left its glorious past of contributing to the worlds' enlightenment many hundreds of years ago.

Judge Jones was correct in striking down creationism as religion, not science, and therefore unfit for inclusion in biology classes. Can creationism have a place in our public schools? We shall consider that momentarily, but first let us consider that word "theory," which many use as synonymous for hunch or guess.

For scientists of varied disciplines, "theory" is not as creationists would have us believe, a vacuous, tentative notion. Einstein's special and general theories of relativity, for example, have been firmly grounded by experimental proof. No educated person doubts the veracity of these theories. And the truth of particle physics was amply demonstrated early in the morning of July 16, 1945, in the White Sands desert near Alamogordo, New Mexico. The mushroom cloud was proof of that. Does the earth orbit the sun? It may have been unproven in the sixteenth century, but proof of its orbit is firmly supported today. It is no longer an unsupported theory. A flat earth cannot be supported as there is far too much direct evidence supporting the theory or fact of a global earth. As the National Academy of Science has stated it, "a theory is a well substantiated explanation of some aspect of the natural world that can incorporate facts, laws, inferences and tested hypothesis." Creationists and IDers like to dismiss evolution as an untested, flawed theory. Nonsense. Galileo and DaVinci understood this well: "By denying scientific principles," Galileo noted, "one may maintain any paradox." DaVinci further wrote "that experience does not even err, it is only your judgment that errs in promising itself results which are not caused by your experiments" [23, 24].

We return to the question as to whether the creationisms–intelligent design controversy have a place in our schools. Isn't the position that an educated, intelligent citizen know the basics of evolution and what creationism involves, a reasonable one? If so, then creationism requires discussion somewhere; our schools, with knowledgeable teachers, would be an equitable venue, but creationism would be presented as an elective in comparative religion, philosophy, even sociology or political science. Warfare between science and religion is wholly unnecessary as science and religion have their own provinces and certainly do not overlap; do not intrude on one another's turf. Science is not religion, and religion is not science. One need not be rejected for the other; faith for science, or science for faith. Most people can readily accommodate both, and with both comes a fuller understanding of our world.

Currently lacking, and essential, is a wider understanding of both evolution and creationism, and as we see, sparks are flying in our communities over these issues. With wider understanding, dialog will become possible. Here again, there is need for scientists and the public to reason together. Greater openness is an avenue to achieve it, and the mass media can be a major conduit. As can Cafe Scientifique.

THE FUTURE

Encouraging the public to participate in decisionmaking is critical, but many, lacking information either refrain, or vote their inner voice, or the conventional wisdom. The hope of the mass media's potential for public education remains unrealized. Nor have our colleges and universities opened their doors to their communities, or encouraged their faculty members to offer workshops, conferences, or other activities that would aid in bridging the knowledge gap. Until the citadel drops its drawbridge, or fills in its moat, allowing town and gown to reason together, other enterprises must prevail. We are fortunate that in this time of need, someone has stepped forward with a bright, shiny new idea.

Cafe Scientifique, like Topsy, has burst upon us; and not a moment too soon. Cafe Scientifique is based on the Cafe Philosophique movement started in France in 1992, by the philosopher Marc Sautet, who wanted a place where the people, anyone, could discuss philosophical topics. Duncan Dallas, in Leeds, England, adapted the model to discussions of science, and it didn't take long before the idea leaped across the ocean, landing in Denver, where Dr. John Cohen, an immunologist at Colorado Health Sciences Center, started the first Cafe Scientifique in the United States.

The Denver Cafe meets monthly in a large, comfortable room at the Wynkoop Brewing Company. Meetings are free, and for the price of a cup of coffee, beer, or wine, anyone can came to discuss the scientific ideas and developments that are changing our lives. Questions, of course, are the leavening of those events. Guest speakers come from the university, and from industry. The Colorado Cafe Scientifique was organized by an informal group of the President's Teaching Scholars and other faculty from Colorado University, and other institutions in the Rocky Mountain states.

Cafe Scientifique in Bozeman, Montana, is sponsored by Montana State University's College of Letters and Science, and has had speakers from as far away as the University of Texas Health Science Center. In Seattle, the Cafe is called "Science on Tap," and has speakers from the University of Washington, Oregon, and California. Cafes have popped up in Palo Alto, California, Houston and Victoria, Texas, Pittsburgh, PA, Minneapolis, and Syracuse, NY.

The beauty of this new and much-needed movement is that town and gown are finally coming together, after work, in relaxed and pleasant environments to chat, discuss, network, share a beer, glass of wine, a snack or dinner–a cup of Joe, with a shot of science—exploring and debating the latest issues in science, outside traditional academic settings. If the general public is to vote on the issues of concern locally and nationally, they need opportunities to ask questions of experts. The Cafes offer a great way to promote continuing education without exams. Many run from 6:00 to 7:30 P.M.; others from 7:00 to 9:00 P.M. Getting together enhances trust and understanding [25].

Sigma Xi, the National Scientific Research Society, hosted a gathering of Cafe Scientifique leaders in February 2006, to network and organize the growing movement. If the public is to be engaged, many more Cafes are needed, and considering the number of universities and colleges around the country, that should be easy. The main ingredients needed to run a successful Cafe are:

1. A committed organizer, or organizing team.
2. A suitable venue—bar, restaurant, cafe, or bistro. Drinks and food are essential, as are space and comfortable seating.
3. A skilled chairperson—but the emphasis is on loose structure; open, friendly, and inclusive
4. Interesting speakers. No visuals. Just talk, discussion, and libation.
5. An audience. Getting started requires advertising: the Internet, flyers, word-of-mouth. Once started, the Cafe's presence will spread like kudzu (the Asian vine). Be prepared for standing room only.

For addditional advice and links to major networks, check with: www. cafescientifique.org/startacafe.htm, and the Society of the Sigma Xi (www.sigmaxi.org) to assist with organization. Go to it, and get going. Hundreds are needed.

Unborn Tomorrow

We, the people, begin to ask questions; probing, sophisticated questions. No more believing what others prefer us to believe. A greater degree of a skepticism topped with a dollop of cynacism will define us. Change is in the air, and will do wonders for both local and national discourse. Our daily dose of blather will diminish, as probing questions, insight, and background nurture the knowledge to make connections. Scientific literacy will see to that.

Will we continue to support and elevate sports at the expense of science in our public schools? Early signs suggest more of the same, but don't give up on change. Will we see a resurgence of interest in math, physics, biology, and engineering by our high school and university students? Here, too, the future seems murky, but placing surprise high on your agenda may be a good bet.

Less dubious is the fact that basic molecular and cell biology, especially the biology of cancer stem cells currently underway at our universities and national institutes, will yield the keys to understanding the initiation of cancer and its control. Expect this before 2025.

Also, as we are not all the same genotypically, it is becoming more widely accepted that we don't all respond equally to the same medications. Deciphering the human genome led to haplotypes, common genetic variations that indicate which genes among us make us susceptible to various diseases. It is emerging that the same disease in a group of people may in fact be different at the molecular level, which means that the prognosis, response to treatment, and outcome will vary markedly. Hence, in the future, 5–10 years along, physicians will be treating the individual, not the disease. This new paradigm will relegate current standard treatment protocols to the ash heap of medical treatment, because individuals are just that: individual. So, for example, Herceptin, a new drug for breast cancer, is effective only in those women whose tumors have high levels of the protein Her2. In other women it has either no effect or unpleasant side effects. Similarly the cancer drugs Erbitux and Iressa have been effective in only 10–12% of patients—those with the requisite proteins.

If the body lacks a specific enzyme that would otherwise metabolize a drug, that drug would need to be taken in smaller amounts if it is to be cleared from the system before inducing untoward side effects. In the future, then, before prescribing a medication, physicians will check an individual's genetic profile to determine appropriate medication levels [26]. Down the road, and not all that far down, pharmacogenetics will become a major player in medical treatment. Of course, this will mean that people will have opted to permit such personal information to be made generally available. Certainty in this arena remains problematic. But bear in mind, your personal chemistry plays a major role. Your neighbor's response to a chemotherapeutic treatment may bear no relation to yours. That's another takeaway message.

Far less uncertain is the fact that low-fat diets have now been shown to be ineffective in preventing heart disease and cancer; Saw Palmetto, ineffective in treating prostate cancer [27]; glucosamine and chondroitin sulfates, ineffective in treating osteoarthritis [28]; and calcium supplements, ineffective for osteoporosis [29]. Nevertheless, far too many people will continue to support the dietary supplement industry, which, after two decades has yet to show a health benefit for *any* "nutritional" supplement.

But the future does look bright. Nuclear-derived energy will blossom, sharply reducing the level of carbon dioxide in the atmosphere. The public will also respond favorably to food irradiation and genetically modified crops and foods. Also, although I believe in the good sense of the people once their minds are unshackled from the tenacious hold of environmentalists, there will be difficulty giving up oil as the oil and gasoline producers will stubbornly protect their investments until sea levels creep above their ankles. Further, as "there are no lobbyists for the future," we need but reach back for guidance, to Edmund Burke, that canny Irishman (1729–1797), who became one of England's greatest political theorists. "The public interest," Burke informs us, "requires doing today those things that men of intelligence and good will wish, five to ten years hence, had been done." Absolutely. This bit of pregnant counsel is a bridge to the future, and should become tacked to the DNA of all those seeking public office.

Scientific literacy will become the well-trodden path to Wellsville, purveying a less stressful lifestyle. Well-being may become infectious and communicable, and the endless negative media hype will have run its course. By 2020, the country will be breathing easier as the Luddites and Cassandras among us will have been run out of town.

The twenty-first century will be the century of scientific literacy for all. "Knowledge," as James Madison so aptly put it, "will forever govern ignorance: a people who mean to be their own Governors must arm themselves with the power which knowledge gives." Scientific literacy will open vistas not previously contemplated. Our habitat will be seen as less hazardous for children and other growing things, as perception gives way to comprehension, and we "know the place for the first time."

At this juncture, it may be well to recall Carl Sagan's sage observation [30].

> An extraterrestrial being newly arrived on earth, he wrote, scrutinizing what we mainly present to our children in television, radio, movies, newspapers, magazines, the comics, and many books, might easily conclude that we are intent on teaching them murder, rape, cruelty, superstition, credulity, and consumerism.

And he went on to say,

> We keep at it, and through constant repetition many of them finally get it. What kind of society could we create, if instead we drummed into them science and a sense of hope?

POST SCRIPTUM

In the third edition of *OPH*, I noted that the Norwegian sculptor Adolf Gustave Vigeland created a statue for Oslo, Norway's city park, depicting humankind's struggle with the environment. Vigeland believed that we had yet to overcome the many forces that impinge on our lives. The statue portrays a person struggling within a confining circle unable to break free. I would have preferred to see a sizable crack in the ring of encircling forces, suggesting achivement, and overcoming, which has most assuredly occurred. A second statue portraying a mind in chains, could be added, for if we are to be finally free, it is the mental chains that must be rend asunder. Until the ignoble perceptions that we are unhealthy and being poisoned by a polluted environment are cast off, we shall be forever shackled.

REFERENCES

1. Baum, M., An open letter to the Prince of Wales: With respect your highness, you've got it wrong, *Br. Med. J.* **329**:118 (2004).

2. Altman, L. K., When peer review produces unsound science, *New York Times* D6 (June 11, 2002).

3. The numbers behind the news. How gullible are we? *Palm Beach Post* 3 (Oct. 17, 1997) [from the *Idaho Falls Post Register* (April 27, 1997)].

4. Goldberger, J., Waring, C. H., and Tanner, W. F., Pellagra prevention by diet among institutional inmates, *Public Health Reports* **38**(41):2361–2368 (1923).

5. Whitehead, H., Remarks on the outbreak of cholera in Broad Street, Golden Square, London, in 1854, *Trans. Epidemiol. Soc. (London)* **3**:97–104 (1867).

6. Prentice, R. L., and 46 others, Low-fat dietary pattern and risk of invasive breast cancer: The women's health initiative randomized controlled dietary modification trial, *JAMA* **295**(6):629–642 (2006).

7. Beresford, S. A. A., and 46 others, Low-fat dietary pattern and risk of colorectal cancer, *JAMA* **295**(6):643–654 (2006).

8. Howard, B. V., and 46 others, Low fat dietary pattern and risk of cardiovascular disease, *JAMA* **295**(6):655–666 (2006).

9. Hill, A. B., The environment and disease: Association or causation, *Proc. Roy. Soc. Med.* **95**:213–222 (1965).

10. Montori, V. M., Publication bias: A brief review for clinicians, *Mayo Clin. Proc.* **75**:1284–1288 (2000).

11. Altman, D. B., Statistics and ethics in medical research: Collecting and screening data, *Br. Med. J.* **281**:1399–1401 (1980).

12. Matanoski, G. M., *Health Effects of Low Level Radiation on Shipyard Workers. Final Report*, Dept. Energy, DE-A-AC02-79EV10095, Washington, DC, June 1991.

13. Pribble, J. M., Goldstein, K. M., Fowler, E. F. et al., Medical news for the public to use? What's on local TV news, *Am. J. Manage. Care* **12**:170–176 (2006).

14. Reynolds, T., News headlines feed on fear of cancer risk, experts say, *J. Natl. Cancer Inst.* **93**(1):9–10 (2001).

15. Goodstein, L., Judge rejects teaching intelligent design, *New York Times* A1, A21, (Dec. 21, 2005).

16. Rudoren, J., Ohio Board undoes stand on evolution, *New York Times* (Feb. 14, 2006).

17. Scott, E. C., *Evolution vs. Creationism: An Introduction*, Univ. California Press., Berkeley, CA, 2004.

18. Dawkins, R., *The Blind Watchmaker. Why the Evidence of Evolution Reveals a Universe without Design*, Norton, New York, 1996.

19. Morris, R., *The Evolutionists: The Struggle for Darwin's Soul*, Freeman, New York, 2001.

20. Morris, H. M., *Studies on the Bible and Science*, Zondervan Publishing House, Grand Rapids, MI, 1974; see also Creationism in schools. The decision in McLean vs Arkansas Board of Education, *Science* **215**:934–944 (1952).

21. Rudoren, J., Henry, M. Morris, 87, dies; a theorist of creationism, *New York Times* (March 4, 2006).

22. Wilford, J. N., Fossil called missing link from sea to land animals, *New York Times* A1, A20 (April 6, 2006); see also Shubin, N. H., Daescher, E. B., and Jenkins, F. A., Jr., The pectoral fin of Tiktoolik roseae and the origin of the tetrapod limb, *Nature* **440**(6):764–770 (2006).

23. Zelik, M., *Astronomy. The Evolving Universe*, 5th ed., Wiley, New York, 1988.

24. Boorstin, D. J., *The Discoverers*, Vol. 2, Henry N. Abrams, Publisher, New York, 1991.

25. Sink, M., Science comes to the masses (You want fries with that?), *New York Times* F3 (Feb. 21, 2006).

26. Caraco, Y., Genes and the response to drugs, *NEJM* **357**(27):2867–2869 (2004).

27. Bent, S., Kane, C., Shinohara, K. et al., Saw Palmetto for benign prostate hyperplasia, *NEJM* **354**(6):557–566 (2006).

28. Clegg, D. O. (and 24 others), Glucosamine, chondroitin sulfate, and the two in combination for painful knee osteoarthritis, *NEJM* **354**(8):795–808 (2006).

29. Jackson, R. D. (and 49 others), Calcium plus vitamin D supplementation and the risk of fractures, *NEJM* **354**(7):669–683 (2006).

30. Sagan, C., *The Demon Haunted World*, Random House, New York, 1995.

INDEX